NOBLE GASES

	IIIA	IVA	VA	VIA	VIIA	
						2 4.0 He Helium
	5 10.8 B Boron	6 12.0 C Carbon	7 14.0 N Nitrogen	8 16.0 O Oxygen	9 19.0 F Fluorine	10 20.0 Ne Neon
	13 27.0 Al Aluminum	14 28.0 Si Silicon	15 31.0 P Phosphorus	16 32.0 S Sulfur	17 35.5 Cl Chlorine	18 40.0 Ar Argon

	IB	IIB							
28 58.7 Ni Nickel	29 63.5 Cu Copper	30 65.4 Zn Zinc	31 69.7 Ga Gallium	32 72.6 Ge Germanium	33 75.0 As Arsenic	34 79.0 Se Selenium	35 80.0 Br Bromine	36 84.0 Kr Krypton
46 106 Pd Palladium	47 108 Ag Silver	48 112 Cd Cadmium	49 115 In Indium	50 119 Sn Tin	51 122 Sb Antimony	52 128 Te Tellurium	53 127 I Iodine	54 131 Xe Xenon
78 195 Pt Platinum	79 197 Au Gold	80 201 Hg Mercury	81 204 Tl Thallium	82 207 Pb Lead	83 209 Bi Bismuth	84 (210) Po Polonium	85 (210) At Astatine	86 (222) Rn Radon

64 157 Gd Gadolinium	65 159 Tb Terbium	66 162 Dy Dysprosium	67 165 Ho Holmium	68 167 Er Erbium	69 169 Tm Thulium	70 173 Yb Ytterbium	71 175 Lu Lutetium
96 (247) Cm Curium	97 (247) Bk Berkelium	98 (249) Cf Californium	99 (254) Es Einsteinium	100 (253) Fm Fermium	101 (256) Md Mendelevium	102 (254) No Nobelium	103 (257) Lw Lawrencium

This book is in the ADDISON-WESLEY SERIES IN CHEMISTRY

Francis T. Bonner, *Consulting Editor*

The material in this book originally appeared as part of *Introduction to Chemistry* by Williams, Embree, and DeBey, © 1973, 1968.

Copyright © 1975, 1968 by Addison-Wesley Publishing Company, Inc. Philippines copyright 1975, 1968 by Addison-Wesley Publishing Company, Inc.

All rights reserved. No part of this publication may be reproduced, stored in a retrieval system, or transmitted, in any form or by any means, electronic, mechanical, photocopying, recording, or otherwise, without the prior written permission of the publisher. Printed in the United States of America. Published simultaneously in Canada. Library of Congress Catalog Card No. 72-9501.

ISBN 0-201-01886-1
ABCDEFGHIJ-HA-7987654

INTRODUCTION TO THE CHEMISTRY OF LIFE SECOND EDITION

HARLAND D. EMBREE
HAROLD J. DE BEY

California State University, San Jose

ADDISON-WESLEY PUBLISHING COMPANY
Reading, Massachusetts • Menlo Park, California
London • Amsterdam • Don Mills, Ontario • Sydney

CONTENTS

ORGANIC CHEMISTRY

CHAPTER 21 THE NATURE OF ORGANIC COMPOUNDS

- 21-1 Early experiments 275
- 21-2 Vital force and synthetic compounds 276
- 21-3 Isomers 277
- 21-4 Importance of structure 277
- 21-5 Assignment of structure 279
- 21-6 Carbon bond angles 282
- 21-7 Molecules: structures, properties, and models 284
- 21-8 Rotation on carbon bonds 287
- 21-9 Writing structural formulas 288
- 21-10 Ring compounds: structures and formulas 291

CHAPTER 22 SATURATED HYDROCARBONS: THE ALKANES

- 22-1 Petroleum, a source of saturated hydrocarbons 296
- 22-2 Development of names 298
- 22-3 The systematic names 300
- 22-4 General physical and chemical properties 304
- 22-5 Combustion 307
- 22-6 Substitution by chlorine 308
- 22-7 The nature of alkane reactions 311
- 22-8 Names of alkyl halides 312
- 22-9 Common organic halides 313
- 22-10 Relation of alkanes to other organic compounds 315
- 22-11 Classes of organic compounds 316

CHAPTER 23 UNSATURATED HYDROCARBONS: ALKENES AND ALKYNES

- 23-1 Structure and occurrence of alkenes 323
- 23-2 Source and uses of alkenes 324
- 23-3 Alkene names 324
- 23-4 *Cis-trans* isomers 326

	23-5	Alkene reactions	328
	23-6	Alkynes	333
	23-7	Acetylene (ethyne)	334

CHAPTER 24 AROMATIC HYDROCARBONS: BENZENE AND RELATED COMPOUNDS

	24-1	Type of structure	338
	24-2	Names	340
	24-3	General chemical properties	343
	24-4	Reaction with chlorine or bromine	343
	24-5	Reaction with sulfuric acid	344
	24-6	Reaction with nitric acid	345
	24-7	Oxidation of side chains	346
	24-8	Heterocyclic compounds	347
	24-9	Sources of aromatic compounds	348

CHAPTER 25 ALCOHOLS, PHENOLS, AND ETHERS

	25-1	Introduction	353
	25-2	Alcohol names	354
	25-3	Common alcohols	356
	25-4	Physical properties of alcohols, phenols, and ethers	357
	25-5	Chemical properties of alcohols	360
	25-6	Phenols	363
	25-7	Ethers	365

CHAPTER 26 ORGANIC ACIDS

	26-1	Introduction	368
	26-2	Acidic properties; salt formation	369
	26-3	Water solubility of carboxylic acids and salts	372
	26-4	Reaction with alcohols: esterification	373
	26-5	Common esters, natural and synthetic	374
	26-6	Hydrolysis and saponification of esters	376
	26-7	Formation and reactions of acid chlorides	377
	26-8	Sources of acids	378
	26-9	Other acids, esters, and salts of interest	379

CHAPTER 27 THE CHEMISTRY OF FATS

	27-1	Structure and hydrolysis of fats	384
	27-2	Fatty acid constituents	385
	27-3	Fats and oils	387
	27-4	Hydrogenation of oils	388
	27-5	Saponification	388
	27-6	The solubility and cleasing action of soaps	391
	27-7	Synthetic detergents	393
	27-8	Hard water	395
	27-9	Drying oils	396

CHAPTER 28 AMINES

- 28-1 The relatives of ammonia 398
- 28-2 Basic properties 399
- 28-3 Preparation of amines 400
- 28-4 Conversion of amines to amides 402
- 28-5 The structure of proteins and amino acids 404
- 28-6 Reactions of amino acids 408
- 28-7 More nitrogen compounds of interest 410

CHAPTER 29 CARBONYL COMPOUNDS

- 29-1 Names and structure 417
- 29-2 Synthesis from alcohols 417
- 29-3 Compounds of interest 420
- 29-4 General chemical properties 420
- 29-5 Hydrogenation 420
- 29-6 Oxidation 421
- 29-7 Addition of hydrogen cyanide 422
- 29-8 Addition of an alcohol 426
- 29-9 Addition of ammonia 428
- 29-10 Addition of ammonia derivatives 430
- 29-11 Carbohydrates 431

CHAPTER 30 POLYMERS — *omit this chapter*

- 30-1 Polymers, natural and synthetic 448
- 30-2 Addition polymers 449
- 30-3 Diene polymers 452
- 30-4 Copolymers 455
- 30-5 Condensation polymers 457
- 30-6 Thermoplastic versus thermoset polymers 459
- 30-7 Cross-linked polymers 459
- 30-8 The social importance of synthetic polymers 461
- 30-9 Conservation of organic raw materials 462

CHAPTER 31 STEREOISOMERS

- 31-1 Introduction 464
- 31-2 *Cis-trans* isomers: additional examples 464
- 31-3 Stereoisomers: differences in configuration 465
- 31-4 The strange isomers of lactic acid 465
- 31-5 Polarized light and optical activity 466
- 31-6 Tetrahedral carbon to the rescue 468
- 31-7 Formulas for optical isomers 471
- 31-8 Racemates 472
- 31-9 Optical isomers in living organisms 474
- 31-10 More complex optical isomers 475
- 31-11 Summary of types of isomers 477
- 31-12 Conclusion: the importance of structure and configuration ...478

vi   Contents

BIOCHEMISTRY

CHAPTER 32 INTRODUCTION TO BIOCHEMISTRY
- 32–1   Development of biochemistry 483
- 32–2   Biochemical composition..................................487
- 32–3   Biochemical reactions 490
- 32–4   The cell and its parts 490
- 32–5   Summary..493

CHAPTER 33 CHEMISTRY OF THE CARBOHYDRATES
- 33–1   Introduction 496
- 33–2   Monosaccharides 497
- 33–3   Compounds related to monosaccharides 502
- 33–4   Oligosaccharides 503
- 33–5   Polysaccharides 505
- 33–6   Heteropolysaccharides 508
- 33–7   Summary..509

CHAPTER 34 CHEMISTRY OF THE PROTEINS
- 34–1   Introduction 511
- 34–2   Amino acids 512
- 34–3   Proteins...522

CHAPTER 35 CHEMISTRY OF THE LIPIDS
- 35–1   Introduction 534
- 35–2   Classification of lipids 534
- 35–3   Triglycerides..537
- 35–4   Phospholipids 538
- 35–5   Waxes 539
- 35–6   Steroids...539
- 35–7   Terpenes 541
- 35–8   Summary..542

CHAPTER 36 CHEMISTRY OF THE NUCLEIC ACIDS
- 36–1   Introduction 544
- 36–2   Composition and structure of the nucleic acids 544
- 36–3   Occurrence of nucleic acids 549
- 36–4   Synthesis of nucleic acids 551
- 36–5   Physical properties of the nucleic acids 553
- 36–6   Chemical properties of the nucleic acids 553
- 36–7   The importance of nucleic acids 553

CHAPTER 37 BIOCHEMICAL REACTIONS
- 37–1   Introduction 558
- 37–2   Digestion 558
- 37–3   Further metabolism 562

37-4	Energetics	562
37-5	Electron transport	565
37-6	Enzymes	568
37-7	Failure of metabolic reactions	572
37-8	Hormones	572
37-9	Summary of metabolism	573

CHAPTER 38 METABOLISM OF CARBOHYDRATES

38-1	Introduction	577
38-2	Digestion of carbohydrates	577
38-3	Absorption and further metabolism of carbohydrates	579
38-4	Glycogen synthesis and breakdown	581
38-5	Degradation of glucose	582
38-6	Glycolysis	583
38-7	The phosphogluconic acid shunt	588
38-8	Metabolism of pyruvic acid	589
38-9	Aerobic metabolism — the citric acid cycle	591
38-10	Summary of metabolism of carbohydrates	594
38-11	Abnormalities of carbohydrate metabolism	595

CHAPTER 39 METABOLISM OF PROTEINS

39-1	Introduction	603
39-2	Protein-digesting enzymes	603
39-3	Protein digestion	604
39-4	Metabolism of amino acids	605
39-5	Excretion of nitrogenous waste products	611
39-6	Specific metabolic pathways for amino acids	616
39-7	Disorders of protein metabolism	616
39-8	Summary	617

CHAPTER 40 METABOLISM OF LIPIDS

40-1	Introduction	620
40-2	Digestion of lipids	620
40-3	Absorption of lipids	621
40-4	Further metabolism of lipids	621
40-5	Synthesis of fatty acids	624
40-6	The synthesis of cholesterol	626
40-7	Disorders of lipid metabolism	626
40-8	Summary of lipid metabolism	627
40-9	Interrelationship of the metabolism of carbohydrates, proteins and lipids	629

CHAPTER 41 NUTRITION

41-1	Introduction	631
41-2	Energy requirements	631
41-3	The essential nutrients	632
41-4	Carbohydrates in nutrition	633

41-5	Proteins in nutrition	633
41-6	Lipids in nutrition	633
41-7	Mineral elements in nutrition	634
41-8	Importance of specific mineral elements	635
41-9	Vitamins	637
41-10	On being well fed	647
41-11	Worldwide nutritional problems	648

CHAPTER 42 THE CHEMISTRY OF HEREDITY

42-1	Introduction	651
42-2	Genetics from a cellular level of observation	653
42-3	Genetics from a biochemical viewpoint	655
42-4	Inborn errors of metabolism	657
42-5	The one-gene–one-enzyme theory	663

CHAPTER 43 BIOCHEMISTRY OF PLANTS

43-1	Introduction	666
43-2	Photosynthesis	667
43-3	Nitrogen fixation in plants	671
43-4	Other plant products	672
43-5	Summary	672

CHAPTER 44 THE BIOCHEMISTRY OF DISEASE AND THERAPY

44-1	Introduction	674
44-2	The infectious disease	675
44-3	Defenses against infection	675
44-4	Chemotherapeutic agents	676
44-5	Antibiotics	677
44-6	The noninfectious diseases	679
44-7	Aging	687

CHAPTER 45 CHEMISTRY OF THE ENVIRONMENT

45-1	Introduction	691
45-2	Our global environment	692
45-3	Effects of man on local environments	695
45-4	The air we breather	696
45-5	Water pollution	701
45-6	Poisonous elements and compounds in the environment	702
45-7	Resources in short supply	705
45-8	Energy sources	707
45-9	Synthetic organic compounds in the environment	708
45-10	Solution to environmental problems	712

GLOSSARY 719

ANSWERS TO SELECTED WORK EXERCISES 735

INDEX 749

PREFACE
TO THE FIRST EDITION

The growth and development of a child and of man as a thinking animal show interesting parallels. As a child becomes aware of himself as an individual, that things are happening to him and within him, he searches for explanations. Early man explained his actions, the diseases that afflicted him, and even his birth and death as being due to external influences. The stars, the gods, or even the prevalence of winds from a certain direction were given as causes for or as the basis of his behavior, of his happiness or woe.

As a child or a civilization grows, there is an increasing awareness of self. Modern, educated man tends to explain his behavior in terms of things that happen within him. He cannot deny the influence of his environment, but he realizes that it is the interaction of the chemicals of which he is composed with stimuli both external and internal that explains life.

Accumulation of scientific information about the nature of man and his behavior means we can now explain, in physical and chemical terms, significant aspects of the processes leading from a person's conception to his birth, his actions, the diseases that afflict him, and the cause of his still-inevitable death. Explanations of the unique properties of organic compounds such as DNA and proteins and the biochemical reactions involved in heredity, reproduction, diseases, and even the effects of aging are examples of the chemical interpretation of life. Other aspects of human behavior such as learning and thought or the way our behavior is affected by certain drugs cannot be satisfactorily described in chemical terms at the present time. It is the belief of scientists that further research will help explain these and, indeed, all aspects of life in physical and chemical terms.

This book presents an introduction to the chemistry of life. It discusses the kinds of compounds found in living organisms, presents their biochemical reactions, and comments on the significance of these to life. Throughout the text, sections set off with flags (◄) present details or background material. Reading them is not obligatory, but they provide a deeper understanding of textual material.

This book is intended for persons who are interested in knowing more about life. While this interest is generally pursued by taking specific courses in college, we have tried to write in such a manner that the material can be understood by anyone with a background in general chemistry and a deep interest and a determination to spend some time acquiring a knowledge of living processes. While these processes are not simple, they are understandable, and knowledge of them provides a fascinating picture of what is involved in being alive.

As a textbook, we believe this volume will be particularly useful for college courses for those who plan to teach biology or who plan to enter the health sciences. While the book may serve as an introduction to biochemistry, it is not intended for persons majoring in these fields. There are many texts that approach these subjects with the detail and rigor that are required of those who plan to make biochemistry their life work. This book represents a small but growing number of texts addressed to the person who wants to understand, but does not plan to contribute significantly to the advancement of, these sciences. This does not mean, however, that such persons will never make such contributions. Many significant discoveries have been made as the result of observations of a well-informed amateur.

This book is a shorter version of *Introduction to Chemistry*, which includes general chemistry in addition to the material presented herein. For wider adaptability of the material in this volume, the organic chemistry and biochemistry parts are also available separately as paperbacks. These texts are being published in response to requests that the material be made available in these forms. In making this presentation, we acknowledge our gratitude to Dr. Arthur L. Williams, who is our coauthor in the larger book, and without whom none of these books would have been written.

San Jose, California
November, 1968

H.D.E.
H.J.D.

TRADE NAMES

Reference is made within the text to the following trade names:

Trade Name	Registered Trademark of
ACHROMYCIN	Lederle, Inc., Division of American Cyanamid
ACRILAN	Chemstrand Corporation
AUREOMYCIN	Lederle, Inc., Division of American Cyanamid
BAKELITE	Union Carbide Corporation, Plastics Division
BENZEDRINE	Smith, Kline, and French Laboratories
CARBONA	Worthington Pump and Machinery Corporation
CHEER	Procter and Gamble Company
CLOROX	Procter and Gamble Company
COMPAZINE	Smith, Kline, and French Laboratories
CRISCO	Procter and Gamble Company
DACRON	E.I. du Pont de Nemours and Company, Inc.
DEXEDRINE	Smith, Kline, and French Laboratories
DIABINESE	United States Vitamin Company
DRANO	The Dracket Company
DYNEL	Union Carbide Corporation
EQUANIL	Wyeth Laboratories
FAB	Colgate-Palmolive Company
FORMICA	Formica Corporation (subsidiary of American Cyanamid)
FREON	E. I. du Pont de Nemours and Company, Inc.
LUCITE	E. I. du Pont de Nemours and Company, Inc.
LUMINAL	Winthrop-Stearns, Inc.
MARPLAN	Hoffmann-La Roche, Inc.
MELMAC	American Cyanamid
MICARTA	Westinghouse Electric Corporation
MILTOWN	Wallace Laboratories

Trade Name	Registered Trademark of
MONEL	International Nickel Company, Inc.
MYLAR	E. I. du Pont de Nemours and Company, Inc.
NICHROME	Driver-Harris Company
ORINASE	Upjohn Company
ORLON	E.I. du Pont de Nemours and Company, Inc.
PLEXIGLAS	Rohm and Haas Company
PUREX	Purex Corporation, Ltd.
PYREX	Corning Glass Works
SARAN	Dow Chemical Company
SERPASIL	CIBA
SPRY	Lever Brothers Company
TEFLON	E. I. du Pont de Nemours and Company, Inc.
TERRAMYCIN	Pfizer Laboratories
THORAZINE	Smith, Kline, and French Laboratories
TIDE	Procter and Gamble Company
TREND	Purex Corporation, Ltd.
VINYLITE	Union Carbide Corporation, Plastics Division
VINYON	Union Carbide Corporation, Plastics Division

INTRODUCTION TO CHEMISTRY SECOND EDITION

THE NATURE OF ORGANIC COMPOUNDS

21-1 EARLY EXPERIMENTS

By 1800, chemistry had become firmly established as a science, so that during the next half-century there was keen interest in studying the composition of substances and the manner in which one might be changed to another. Because good methods of analysis had been developed, it was possible to determine the composition of a substance with considerable accuracy. The successful analyses stimulated two important results. First, they provided the facts which led to the concepts of atomic weights and of atoms as discrete units which combined with one another to form larger particles called molecules (Chapters 5 and 6). Second, chemists were encouraged to investigate the composition of almost any sort of material they discovered in nature.

As a result of these investigations, chemists began to distinguish two groups of substances. Those derived from plant or animal sources became known as **organic** compounds. All the others, which were obtained in one way or another from the mineral constituents of the earth, were called **inorganic** substances.

The concepts of atoms and molecules, mentioned above, were developed in terms of the inorganic compounds, which had relatively simple formulas. Formulas such as NaOH or $BaCl_2$, once assigned to particular compounds, seemed to represent their constitution very well. First of all, each inorganic compound had a unique formula; only one substance, for example, had formula $BaCl_2$. Second, the ratio of 1 barium to 2 chlorines in the compound was explained by the recently proposed concept of a valence or combining power for each atom. Finally, the Swedish chemist J. J. Berzelius proposed a logical explanation of why barium combined with a substance such as chlorine, rather than with sodium or magnesium. During this period of time, there was great interest in the newly-discovered electric currents and in their effect on compounds in water solutions. Chemists had discovered that metals collected at one electrode, and nonmetals at the other electrode. Therefore, Berzelius theorized, a barium atom, because it was positive, was attracted to the negative chlorine atoms to make up the substance $BaCl_2$.

The chemists of the early 1800's were acquainted with a very large number of organic compounds. Some had been known for centuries, some had only recently been isolated, but all were fascinating. Among them were dyes, soap, vinegar, sugar, perfumes, gums, and rubber, to mention but a few. For a variety of reasons chemists were led to believe that these were a distinctly different type of compound than the inorganic substances. Not only were they produced by plants or animals, but more impressive was the knowledge that there was a tremendous number of different organic compounds and that they were made up from very few elements. <u>The elements found were carbon, hydrogen, oxygen, sometimes nitrogen, and occasionally phosphorus</u>. How could one imagine that hundreds or perhaps thousands of different organic compounds could be assembled from three or four elements? Among the inorganic compounds it was possible to find FeO and Fe_2O_3, Cu_2O and CuO. However, different combinations between oxygen and one particular metal were limited, and were explained by assuming that iron or copper had two different valences.

If one studied the formulas of even a small group of organic compounds, it seemed impossible to decide what the combining power of the atoms could be. For example, among compounds containing only carbon and hydrogen, the formulas C_2H_6, C_2H_4, C_3H_6, C_4H_{10} and C_7H_8 were all discovered, and there were many more. Did this mean that the valence of carbon in the first is 3, in the next two is 2, and in the last two, $2\frac{1}{2}$ or $1\frac{1}{8}$? Furthermore, a compound such as C_2H_6 could be transformed to C_2Cl_6 by treatment with chlorine. If a negative chlorine could be substituted for a hydrogen, apparently Berzelius' theory of positive and negative atoms did not apply to organic compounds. Anyhow, almost none of the organic substances would conduct an electric current!

21-2 VITAL FORCE AND SYNTHETIC COMPOUNDS

Because of the profusion and complexity of organic compounds, many chemists despaired of ever understanding their nature. Furthermore, no organic compound had ever been made in a laboratory from simpler compounds or from the elements, although a number of inorganic compounds had been made synthetically. Consequently, many people believed that organic compounds were formed only under the influence of a "vital force" originating in living plants or animals.

In 1828, Friedrich Wöhler made a remarkable discovery at the University of Göttingen in Germany. He attempted to prepare ammonium cyanate by means of a double decomposition reaction in a solution of ammonium chloride and silver cyanate, both of which were regarded as inorganic substances. Instead of ammonium cyanate, Wöhler obtained crystals of urea!

$$NH_4Cl + AgCNO \begin{array}{c} \nearrow NH_4CNO + AgCl \downarrow \\ \text{ammonium cyanate} \\ \searrow CH_4N_2O + AgCl \downarrow \\ \text{urea} \end{array} \qquad (21\text{-}1)$$

Previously, urea had been obtained only from the urine of animals. Now Wöhler had synthesized it without the presence of the "vital force." The synthesis was confirmed four years later when Wöhler and Justus Liebig succeeded in preparing ammonium cyanate by a different method. It proved to have properties entirely different from those of urea. However, when a solution of ammonium cyanate was boiled down to induce crystallization, the product isolated was always urea!

Within a few years, when acetic acid ($C_2H_4O_2$) and several other organic compounds were prepared by synthesis from inorganic materials, further doubt was cast on the need for a "vital force." As time passed, chemists also synthesized a number of *new* compounds; these were not merely copies of those produced by living plants or animals, but were compounds never before observed on earth. Because of their properties, however, the new compounds seemed clearly to belong in the same class as the organic compounds from natural sources. If it is not necessary for all organic compounds to be associated with living organisms or a "vital force," how then do we define "organic" compounds?

In the mid-1800's, it became apparent that the one factor common to all organic compounds was the element carbon. Therefore, we now say simply that **organic compounds are the compounds of carbon.** At the present time, more than one million organic compounds are known; that is, they have been isolated or synthesized, and their names, properties, and formulas have been described in some chemical journal.

21-3 ISOMERS

The formulas for urea and ammonium cyanate, shown in Eq. (21-1), reveal that both consist of the same amounts of identical elements, CH_4N_2O. Yet the properties of the two substances are different. So Wöhler's experiment had dispelled the need to believe in a "vital force," but had added a new complexity. Within a decade or two, many other groups of compounds related in this unusual way were discovered. It became clear that *compounds having different properties can have the same composition*. Such related compounds are called **isomers** (Gr. *iso*, equal, plus *meros*, part). This possibility contributes to the great number of different organic compounds. To consider just a few examples, there are three different compounds having formula C_3H_8O, and three compounds having formula C_5H_{12}. Corresponding to the formula $C_4H_{10}O$ there are seven compounds, each differing distinctly from the others in many of its properties.

21-4 IMPORTANCE OF STRUCTURE

We shall not be able to describe the many efforts made between 1840 and 1860 to solve the puzzle concerning the constitution of organic compounds. Suffice it to say, however, that there was much experimentation by the most skilled chemists of the era, and that many theories were proposed which later had to be discarded when it was found that they were not in harmony with all the observed facts. It turned out that the key to the whole puzzle was the idea of *structure*. We shall describe the concept as we now understand it.

If covalent bonds exist between the atoms of a molecule, the atoms are held in some particular arrangement; that is, *the molecule has a structure*. In a water molecule, for example, covalent bonds hold two hydrogen atoms in a specific way to an oxygen atom. Each hydrogen nucleus is 1.0 Ångstroms* from the oxygen nucleus, and the angle between the two bonds is 105°, as shown in Fig. 21-1. The water molecule has a definite structure.

In contrast, an electrovalent compound such as barium chloride exists as ions in either a solution or a crystal. In a solution, we do not find two particular chloride ions held

---

* One Ångstrom unit (Å) is 0.0000001 mm.

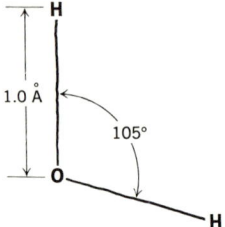

Fig. 21-1 Structure of the water molecule.

at definite distances and angles to one certain barium ion. All we know is that there are twice as many Cl⁻ ions as there are Ba²⁺ ions in the solution. It is true that a *crystal* of barium chloride has a structure in the sense that the ions are stacked together in a certain pattern, at regular distances from each other. But there is no *molecular* structure. In the crystal we cannot find distinct groups of atoms we could designate as BaCl$_2$ molecules. The two chloride ions are not definitely associated more with one certain barium ion than with another barium ion.

<u>Carbon compounds (the organic compounds) have structure because a carbon atom can form stable covalent bonds to other carbon atoms and to a number of different atoms.</u> For instance, a molecule having composition C$_3$H$_8$O can have the structure shown in Fig. 21–2(a). In that structure, the covalence of each carbon is 4, each oxygen is 2, and each hydrogen is 1. A formula such as C$_3$H$_8$O is a **molecular formula**: it states the number and kind of atoms in the molecule. A formula such as that in Fig. 21–2(a) is a **structural formula**: it shows which atom is bonded to which.

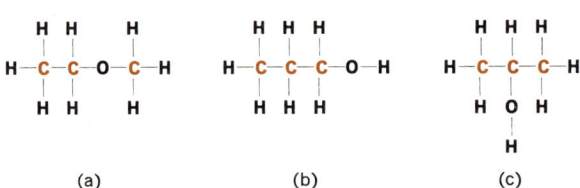

(a) (b) (c)

Fig. 21-2 Structures of the C$_3$H$_8$O isomers.

The fact that a carbon atom can form *covalent* bonds with other atoms leads us to an interesting discovery. It should be possible for three carbons, eight hydrogens, and one oxygen to be arranged into structures different from the one described in Fig. 21–2(a). The other two possibilities are shown in Figs. 21–2(b) and (c). These three different structures we have defined on paper explain the existence of three separate, real compounds having different properties.

The concept of structures based upon covalent bonds also explains why carbon can combine with other elements in so many different ratios (Section 21–1). In the molecular formulas C_2H_6 and C_4H_{10}, the ratios of the hydrogens to the carbons are different. However, when we examine possible structural formulas for these two substances, we see that all carbon atoms can have a valence of 4 in both compounds.

```
    H  H              H  H  H  H
    |  |              |  |  |  |
H — C— C— H       H — C— C— C— C— H
    |  |              |  |  |  |
    H  H              H  H  H  H
    C₂H₆               C₄H₁₀
```

Quite a few elements can form covalent bonds. However, *carbon is unique in its ability to form long chains built up of carbon atoms bonded to one another.* An example is shown in Fig. 21–3(a). The chains may be of varying lengths or they may be branched, as in Fig. 21–3(b). Furthermore, a chain of carbon atoms can twist in such a way that carbons at the ends of the chain come close to each other and can link to form a ring. Figure 21–3(c) is an example of a ring structure. For simplicity, the atoms occupying the other valences of the carbons in Fig. 21–3 are not shown. The simplified diagrams therefore represent the skeletons of the molecules.

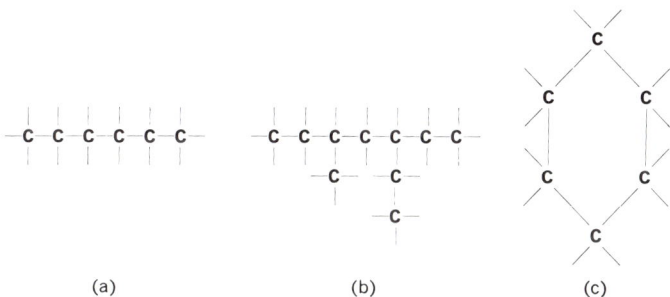

Fig. 21–3 Typical carbon skeletons in organic molecules: (a) a chain of carbon atoms; (b) a branched chain; (c) a ring.

Every organic molecule consists of some kind of carbon skeleton, to which are attached other atoms. Nearly always a large portion of the skeleton is filled out with hydrogen atoms. *The elements most often found in organic compounds, in addition to carbon and hydrogen, are nitrogen, oxygen, sulfur, and the halogens.*

21–5 ASSIGNMENT OF STRUCTURE

In the previous section, we discovered that structural formulas can correctly account for the number of different compounds that actually exist. But how do we decide which structure, of several possible, represents a real substance that we have in a bottle? We do it by observing enough different properties of the substance so that we can logically say that the facts fit one, and only one, structure. In this way we can relate, or *assign*, a structure to a real substance.

For illustration, let us use a simple case in which there are only two possible compounds having the same molecular formula. The first one we obtain from fermented fruit juice. It is a clear, colorless liquid which we will call alcohol. (We could just as well have called it something else. All that matters is that we all agree upon a name, so that each of us knows what the other is referring to. We can at least identify the substance to the extent of naming it, even though we don't know its structure.) By experimenting a little, we discover that alcohol vaporizes easily, has a distinctive odor, has a stinging taste, affects our nervous systems in a peculiar way, is very soluble in water, and burns with a hot, almost invisible flame. None of these facts tells us what the structure is, however. By observing the combustion of alcohol more carefully, we can discover that the products are carbon dioxide and water. Now suppose we burn a *weighed* amount of alcohol and then catch and weigh, separately, all the CO_2 and H_2O produced. By the appropriate calculations we can determine the amounts of carbon and hydrogen originally composing the alcohol. We find its molecular formula to be C_2H_6O.

The second compound is obtained by heating wood alcohol (a different sort of alcohol) with concentrated sulfuric acid. The reaction produces a colorless gas which we will call methyl ether. Besides being a gas rather than a liquid, this compound is decidedly different from alcohol in other ways. It is only slightly soluble in water, and its taste and odor are quite unlike those of alcohol. Methyl ether also burns readily. If we do a combustion quantitatively we find the molecular formula of methyl ether is C_2H_6O, the same as that of alcohol. In other words, the two compounds are isomers.

On paper we can write two, and only two, possible structures for formula C_2H_6O:

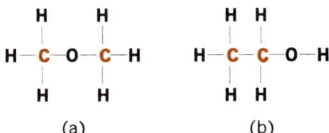

Fig. 21-4 Structures possible for C_2H_6O.

The possibility of two different structures agrees with the fact that we have isolated two real compounds, but does alcohol have structure (a) or structure (b)? Clearly, we need more information in order to make a decision. So we will treat both alcohol and methyl ether with a number of other substances; hopefully, any reactions which occur may give us some clues.

As far as we can tell, neither the liquid alcohol nor the gaseous methyl ether are changed by sodium hydroxide solution, so we try something else. Methyl ether is likewise unchanged (aside from dissolving to a certain extent) by being heated to about 90°C with concentrated hydrobromic acid solution. [The experiment is done in a sealed vessel so that the gaseous ether does not escape.] However, when alcohol is heated to 90°C with hydrobromic acid, the alcohol disappears completely, and we find a new product—an oily, colorless liquid with a sweetish, medicinal odor. Analysis shows that the new compound has formula C_2H_5Br. Now let us compare this formula with the original formula of the alcohol and attempt to write a partial equation for the reaction:

C_2H_6O + H Br → C_2H_5 Br (21-2)

alcohol   oily product

no questions on test concerning reactions.

From Eq. (21-2) it appears that one O and one H were lost from the alcohol formula and were replaced by a Br in the product. If we compare these changes with the possible structures shown in Fig. 21-4, we suspect that structure (b) may be the correct one for alcohol. The O and H which were lost *may* have been the whole OH group of structure (b), and the C_2H_5 which survived in the oily product may have been the residue of structure (b) made of the two carbons and their five hydrogens. In other words, the change may have involved something like this:

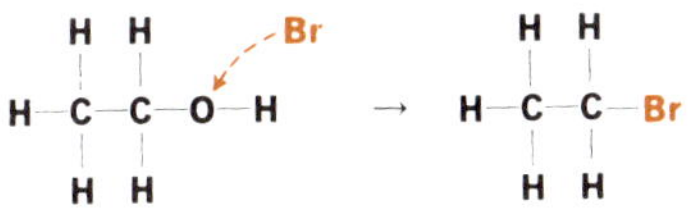

$$\begin{array}{c} H\ H \\ |\ \ | \\ H-C-C-O-H \\ |\ \ | \\ H\ H \end{array} \overset{Br}{\longrightarrow} \begin{array}{c} H\ H \\ |\ \ | \\ H-C-C-Br \\ |\ \ | \\ H\ H \end{array} \qquad (21\text{-}3)$$

Of course, we do not know, from this experiment, *how* or *why* the reaction occurred; Eq. (21-3) merely represents what we *think* is the total change. At any rate, if we *assume* structure (b) for alcohol, then Eqs. (21-2) and (21-3) seem to be a more logical interpretation of the actual experiment than anything we could write if we assumed structure (a).

However, this interpretation might be wrong. If we are to have faith in it, we must find some other supporting evidence. Sodium metal provides some revealing facts. Once again we find that methyl ether does not react, but alcohol does. When we drop a piece of sodium into liquid alcohol, there is a vigorous evolution of colorless gas. The gas is highly flammable, and thorough investigation proves that it is hydrogen. We can also isolate from the reaction mixture a white solid which is strongly basic. It has molecular formula C_2H_5ONa. Furthermore, even if we use an *excess* of sodium, the only products which result are hydrogen gas and the C_2H_5ONa compound. This must mean that *one* of the six hydrogens of alcohol is different from the other five. This makes it even more certain that the correct structure for alcohol must be Fig. 21-4(b) rather than (a).

In fact, this conclusion is especially appealing in view of the whole behavior of alcohol with sodium. It is surprisingly similar to the reaction of water with sodium. Both compounds evolve hydrogen gas and are converted to a basic product. In a water molecule, a hydrogen atom covalently bonded to the oxygen atom can be displaced by sodium, as shown in Eq. (21-4):

2HOH + 2Na → NaOH + H₂ $\qquad (21\text{-}4)$
 (a base)

The fact that structure (b) [but not (a)] also has a hydrogen covalently bonded to an oxygen apparently explains why alcohol can react with sodium in an analogous way:

2H OC₂H₅ + 2Na → 2NaOC₂H₅ + H₂ $\qquad (21\text{-}5)$
alcohol   (a base)
structure (b)

In view of all the facts observed for alcohol and methyl ether, the argument for assigning structure (b) to alcohol, and hence (a) to methyl ether, is quite convincing. We are therefore justified in using these structures to represent the compounds and to aid us in predicting their probable behavior in other situations. Of course, if the outcome of some future experiment can *not* be explained in terms of the assumed structures, the situation will have to be reevaluated. Either there must be faulty observation in the new experiment or the structures as we have been describing them are wrong.

In another case involving a more complex molecule, a greater number of experiments would have to be observed in order to accumulate sufficient facts to assign a structure. Nevertheless, the process of logically comparing the properties of the real compound with a likely structure would be the same as that described above for alcohol and methyl ether. In one way or another, every structural formula you see written in this book, or elsewhere, has been deduced from experimental investigation.

21-6 CARBON BOND ANGLES

Structural formulas of the type in Fig. 21–2 are adequate to show which atoms are bonded to each carbon atom, but do not necessarily represent exact details, such as bond angles, shown in Fig. 21–1 for the water molecule. It is important that we understand just how much our structural formulas can accurately represent, and how much we may imagine or read into these symbols when we see them.

Let us consider a compound having a composition CH_2Cl_2. Assuming a covalence of 1 for hydrogen and 1 for chlorine, the only possible arrangement is for these four atoms to be bonded to the carbon atom. Accordingly, we can write two conceivable structures, (a) and (b) of Fig. 21–5:

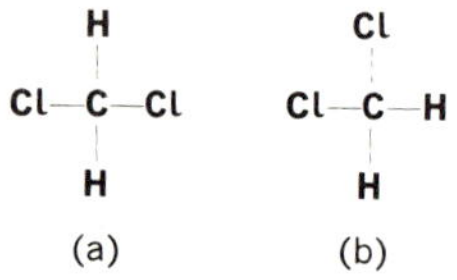

```
        H                 Cl
        |                 |
   Cl—C—Cl           Cl—C—H
        |                 |
        H                 H
       (a)               (b)
```

Fig. 21–5 Possible structural formulas for CH_2Cl_2.

A model is often useful to facilitate our thinking about a structure. One simple kind of model to represent the written structure of Fig. 21–5 consists of balls and pegs, as pictured in Fig. 21–6. The pegs represent the valence bonds (really electron pairs) holding the atoms together. The balls represent the nuclei of the atoms. (The balls, however, are much

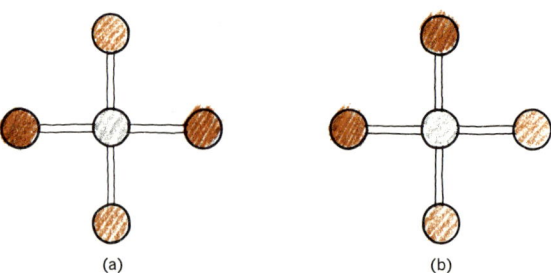

Fig. 21–6 Models of possible structures for CH_2Cl_2.

too large in proportion to the bond distance between them, and therefore too large in proportion to the size of the nuclei.) According to the models in Fig. 21–6, the angles between carbon bonds are 90°, and the CH_2Cl_2 molecules are flat; that is, the carbon atom and the four atoms attached to it are all in the same plane.

If the models are correct, there are two different structures, (a) and (b) of Fig. 21–6, having composition CH_2Cl_2. In (a) the chlorines (more accurately the chlorine nuclei) are farther apart than they are in (b). These models, however, do *not* agree with the experimental facts concerning CH_2Cl_2. No chemist, no matter how good his equipment or how carefully and skillfully he works, has ever been able to isolate more than one kind of material having composition CH_2Cl_2. Furthermore, although the structure of a molecule cannot be seen in a microscope, the same kind of information can be obtained by other modern experimental techniques, particularly X-ray diffraction studies. From such methods we find abundant evidence that the distance between the chlorine atoms is the same in all molecules of CH_2Cl_2 and that the bond angles are approximately 109° (rather than 90°). Clearly, we need to revise the models of CH_2Cl_2 proposed in Fig. 21–6.

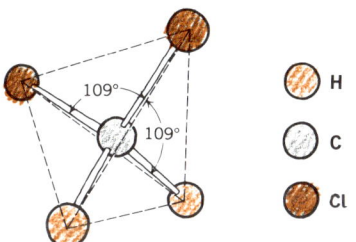

Fig. 21–7 Ball-and-peg model of CH_2Cl_2 showing the correct bond angles and therefore the correct arrangement in space of the bonded atoms.

In Fig. 21–7 is a model of CH_2Cl_2 having the correct bond angles. One striking feature of this model is the fact that the bonded atoms are spread out in space; the molecule is three-dimensional rather than planar. Another interesting aspect of this model is that regardless of which two valence bonds hold the chlorine atoms, the chlorines are always the same distance apart. The fundamental reason for all this is that the four atoms bonded to carbon spread out in space as far from each other as possible. Mathematically, it turns out that the angles between objects arranged in space in this fashion are 109° (more accurately, 109°28′).

A carbon atom is often said to have **tetrahedral** bonding. This is merely a term from mathematics which describes a proper three-dimensional model of a carbon compound such as Fig. 21–7. If we imagine lines drawn between the four atoms bonded to the carbon atom, the resultant geometric figure is a tetrahedron (Gr. *tetrahedros*, four-sided).

> The use of the term tetrahedron, or tetrahedral, does not mean that we believe a carbon atom is a sharp chunk of material with triangular sides. The dotted lines forming the tetrahedron simply represent a mapping device, like the imaginary line around the earth called the equator.

We should also point out that in the molecule CH$_4$, in which the four atoms attached to carbon are identical, all the H—C—H bond angles will be exactly equivalent and will be 109°28′. In the molecule CH$_2$Cl$_2$ the chlorine atoms are surrounded by larger electron clouds and occupy more space than do the hydrogen atoms. Therefore, the two chlorines repel each other somewhat more than do the two hydrogens, causing the Cl—C—Cl bond angle to be slightly larger than 109°28′ and the H—C—H bond angle to be slightly smaller. However, all the angles are *near* 109°, and the important three-dimensional arrangement of the CH$_2$Cl$_2$ molecule is correctly represented in Fig. 21–7.

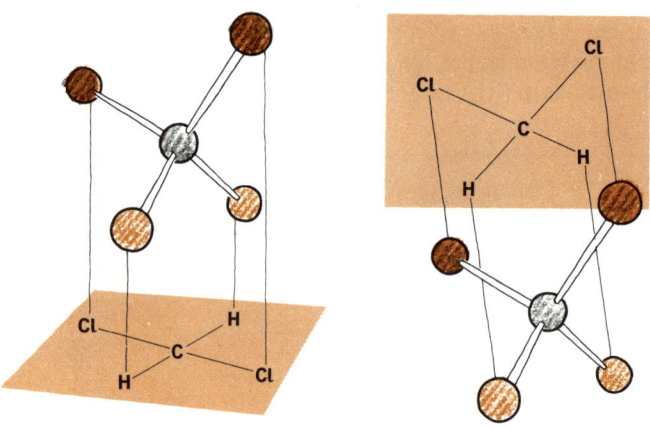

Fig. 21–8 Projections of a CH$_2$Cl$_2$ model onto flat surfaces.

Figure 21–8 shows the diagrams of a CH$_2$Cl$_2$ molecule which would result if the three-dimensional model were projected onto two-dimensional surfaces. These two diagrams are precisely the same as the structural formulas written in Fig. 21–5!

Both the professional chemist and the beginning student find it helpful to assemble models of organic compounds, so that their shapes can be visualized better. However, it is inconvenient (and for some of us perhaps nearly impossible!) to be artistic enough to draw a three-dimensional sketch of CH$_2$Cl$_2$, such as Fig. 21–7, each time we wish to represent this molecule in a chemical equation. Since we must write on a flat surface, we use for a structural formula either Fig. 21–5(a) or (b). When we write the structural formula in this way we must imagine that the actual molecule is really three-dimensional and that the formula we have written is really a projection (Fig. 21–8). We are as justified in using formula 21–5(a) as 21–5(b). We must remind ourselves that, despite the appearance of these formulas, the carbon bond angles are *not* 90° and the molecules are *not* flat. That is, the formulas do *not* represent the models of Fig. 21–6.

21–7 MOLECULES: STRUCTURES, PROPERTIES, AND MODELS

We first introduced the concept of structure solely for the purpose of explaining the great number of existing organic compounds (Section 21–4). Then we assigned structures to

21-7 Molecules: Structures, Properties, and Models

individual compounds on the basis of their properties (Section 21-5). The relation between structure and properties will be our main concern throughout the remaining chapters on organic compounds. The reason is that *structure determines properties*.

For analogy, let us consider that we are going to rent or purchase a home. The structure of a house or an apartment determines whether it is most suitable for a retired colonel, a young, single girl, or a family with four teenagers. We inspect it, if possible, to see whether the structure has the properties which will satisfy our needs. If the house cannot be seen, because it has not yet been built or is presently occupied, a model would help us visualize its properties. However, we would have to use a certain amount of imagination to understand how the real house would function. If a model were not available either, we could inspect a floor plan drawn by someone who had measured the house and observed its various properties. To look at the floor plan and visualize the properties of the real object would require an even better imagination than to look at a model.

So it is with an organic molecule, although we are limited by the fact that *no one* has ever seen the molecule itself. We can, however, observe its properties and measure it in several ways. From these observations, we can draw a structural formula (a floor plan), which symbolizes its structure and therefore suggests what its properties are. However, with the structural formula, it may not be easy to imagine certain properties, especially those depending on three-dimensional relations. Therefore, it is helpful to build a model of the molecule.

Most of us do not try to make all the observations necessary to deduce the structural formula of an organic molecule. Instead, we use the formula developed from the observations of many other chemists. In fact, even a chemist who is particularly interested in the challenging game of "structure proof" would be able to assign structures to no more than several molecules within his lifetime. For the structures of the thousands of other molecules he must use and think about, he relies on the work of others. It is highly important that we keep in mind the fact that all the models or structural formulas we use depend basically on experimental observations.

A structure that has been assigned may not correspond to the real molecule for several reasons. One of the measurements may have been in error. We may not have measured some important feature at all, because we were not aware of it. (We didn't notice the window at the rear of the house!) Or, in constructing the structural formula and model from all the data, we may have interpreted some observation incorrectly. We accept the fact that formulas and models may have limitations and could be based on downright errors, but we use them because they are so essential. We trust that sooner or later any errors will become apparent, so that we can rebuild the model in a more satisfactory way.

Now that we have a better understanding of the purpose and limitations of models, we can examine two useful types. The ball-and-peg variety was first mentioned in Section 21-6; another useful type is a space-filling model. Figure 21-9 shows a water molecule represented by the two types. The ball-and-peg model (Fig. 21-9a) is not much more than a skeleton. The locations of the nuclei are indicated by the balls, and the bonds holding them together are represented only by pegs.

The space-filling model more accurately shows the *relative size* of the atoms in a molecule. Recall that the electrons swirl about the nucleus of an atom in a very large space compared to the tiny nucleus. Also, two atoms are held together by a covalent bond when they share electrons. This can be thought of as a merging or overlapping of the electron clouds of the two atoms. In Fig. 21-9(b), a sphere of proper size portrays the space occupied by the electron cloud around an oxygen nucleus. A smaller sphere shows the volume in which the hydrogen electron may be found. If we imagine these electron clouds merging into each other, in the process of bond-formation, the result would be shown by

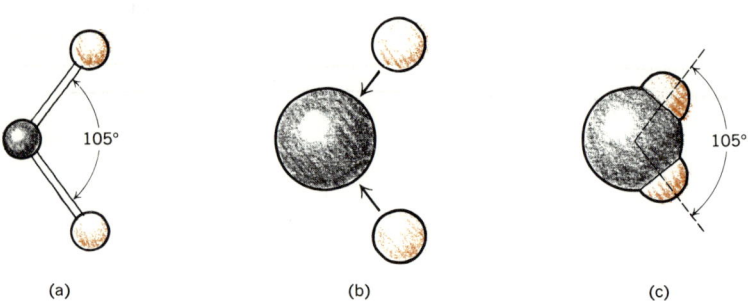

Fig. 21-9 Structure of the water molecule. (a) A ball-and-peg model. (b) Spheres representing the space occupied by electron clouds of oxygen and of hydrogen atoms. Sharing of the electron clouds results in the structure represented by (c), a space-filling model.

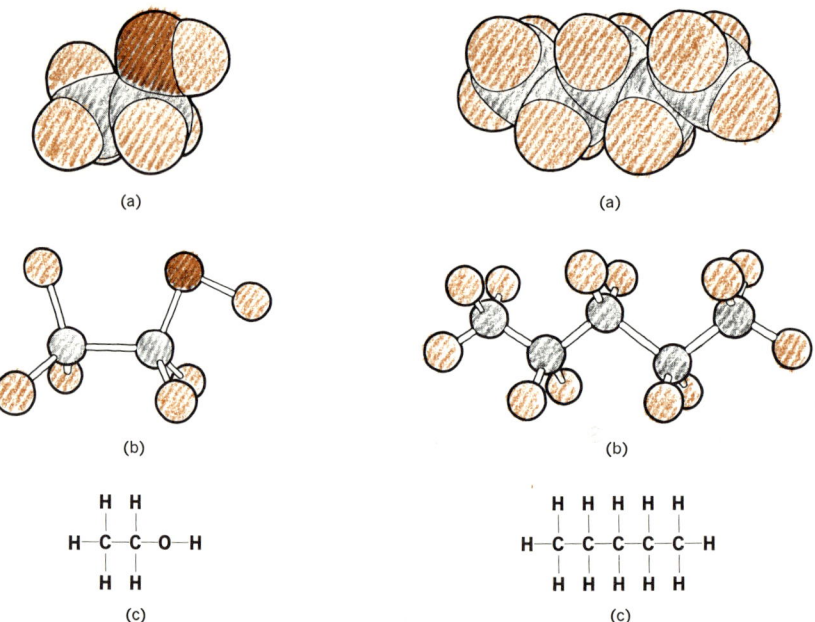

Fig. 21-10 The alcohol molecule, C_2H_6O: (a) space-filling model; (b) ball-and-peg model; (c) projection, a structural formula.

Fig. 21-11 The n-pentane molecule, C_5H_{12}: (a) space-filling model; (b) ball-and-peg model; (c) projection, a structural formula.

the model of Fig. 21-9(c). To make such a model, we can slice off sections of plastic spheres and cement the remaining portions together.

Figure 21-10 shows an alcohol molecule portrayed by models of both types. Although we can picture the shape of the molecule better from the space-filling model, the ball-and-peg model reveals the bond angles more clearly. Observe that all the carbon bond angles are 109° and, as far as we can judge without careful measurement, the oxygen bond angle is about the same. Actually, the oxygen bond angle is near 105°, as it is in water.

We can make a projection of the ball-and-peg model, a three-dimensional object, onto a flat surface. This is what an architect or engineer does when he draws a plan of an object. In Fig. 21–10 we make a projection of the alcohol molecules by imagining that we are standing above (b). By looking straight down upon (b), we would see it flattened out as in diagram (c). The resultant projection gives us a structural formula we can write on paper, that is, in two dimensions. Note that this projection of the alcohol molecule (Fig. 21–10c) is essentially the structural formula we had previously decided was satisfactory (Fig. 21–4b). Hereafter, whenever we use a structural formula, we must keep well in mind that it represents an object actually having three dimensions.

In a similar fashion, the structure of a compound called *n*-pentane can be represented by the models shown in Fig. 21–11.

21–8 ROTATION ON CARBON BONDS

One other important aspect of structure in an organic compound is that a group in the molecule (some portion of it) can rotate relative to the rest of the molecule. That is, the group can turn, on an imaginary axis through the covalent bond, without weakening the carbon–carbon bond. In a ball-and-peg model this is equivalent to twisting one carbon atom, by allowing the peg to slip in its socket, so that the carbon turns in space relative to the carbon at the other end of the peg. They do not become separated (i.e., the bond does not break) because the peg does not leave the socket.

Bond rotation is illustrated in Fig. 21–12 for the molecule called 1,2-dichloroethane.*
The molecule could be as shown in Fig. 21–12(a). A moment later it could exist as shown in (b), after one carbon (and the three other atoms it holds) has rotated relative to the second carbon. By further rotation, the molecule could exist as in condition (c). There is abundant evidence to prove that these rotations occur in a molecule very frequently. The rotation is so easy that it happens, to some degree, every time one molecule collides with another. Therefore, we cannot isolate three different kinds of molecules, (a), (b), and (c), and call them three different compounds. At any given moment a sample of 1,2-dichloroethane contains all these forms (plus many other intermediate states of rotation) and one form is rapidly and easily changing to another. However, what *is* consistent about all molecules in the sample of 1,2-dichloroethane is the *structure*, the manner in which the atoms are bonded. At all times every molecule has one chlorine bonded to the number one carbon and another chlorine bonded to the number two carbon. (On the other hand, if we had a substance in which both chlorines were bonded to the number one carbon, this *would* be a different structure and hence a different compound with a different name, 1,1-dichloroethane. The 1,2-dichloroethane could not be changed to 1,1-dichloroethane by mere rotation of groups on the carbon–carbon bond.)

The whole point of this discussion is summarized in structural formulas (d), (e), and (f) of Fig. 21–12. These formulas would result if we just projected downward the models in (a), (b), and (c), respectively. *Therefore, we can use (d), (e), or (f): each represents 1,2-dichloroethane;* one is as correct as the other. This is *because all are molecules of the same compound.* The structural formulas (d), (e), and (f) only appear to be different at first glance. When we remember that they correspond to the three-dimensional models (a), (b), and (c), and that rotation is possible, we can understand that they all represent the same type of molecule.

---

* *Dichloro* in the name indicates there are two chlorine atoms in the molecule and 1,2- means that if we call one of the carbon atoms in the molecule 1 and the other, 2, each one holds one of the chlorine atoms.

288   The Nature of Organic Compounds

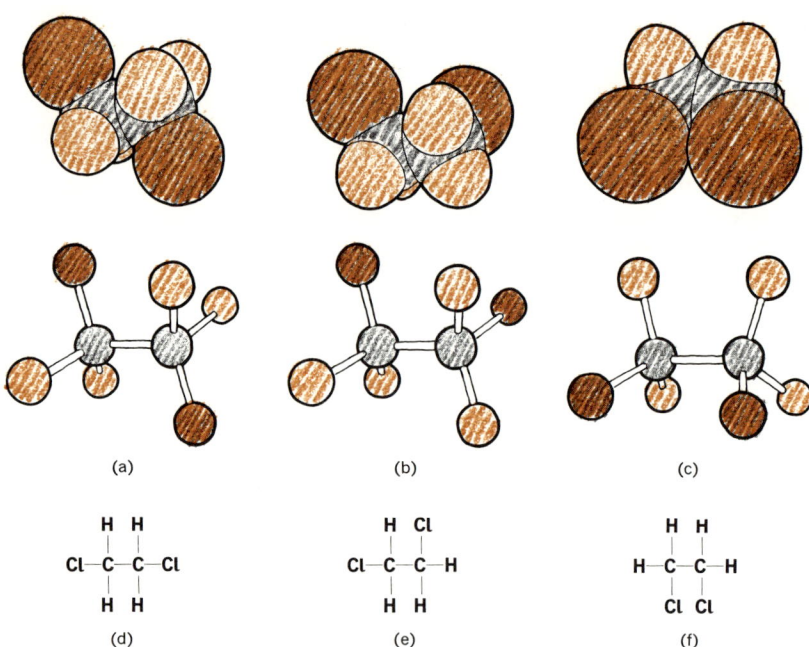

Fig. 21-12 Rotational forms of 1,2-dichloroethane. Three stages of rotation are shown in (a), (b), and (c), each represented by two types of models. Projection of these three forms yields the structural formulas (d), (e), and (f).

It is interesting to note that although various rotational forms are possible, they are not equally favorable. Form (c), Fig. 21-12, is least likely to exist because the bulky chlorine atoms are crowded when they are so close to one another. As a result, if the molecule gets flipped into form (c), it does not remain that way long. It will very soon rotate into a less crowded form. Correspondingly, form (a) is the most favorable, and (b) is intermediate.

21-9 WRITING STRUCTURAL FORMULAS

A structural formula written in the manner used in the previous sections may also be called a **valence bond structure,** because all the bonds in the molecule are shown. Once we have had a little experience with the valence bond structure and understand its relation to a model or to the real molecule, we will often find it convenient to write a more abbreviated structural formula. In general, most of the carbon valences in an organic molecule are occupied by hydrogen atoms. Since the hydrogens are so common and are also quite unreactive (as we will see in Section 22-4), a structural formula is often written in the style of Fig. 21-13(b). This is a **condensed structural formula.** In this form, all hydrogens bonded to a particular carbon are written beside the carbon, but their bonds are not shown; bonds to other atoms are shown, however.

The condensed structural formula can be written more rapidly and in less space than the valence bond structure. Also, the form of the carbon skeleton and the location of important groups (those other than hydrogens) can be more easily visualized from the con-

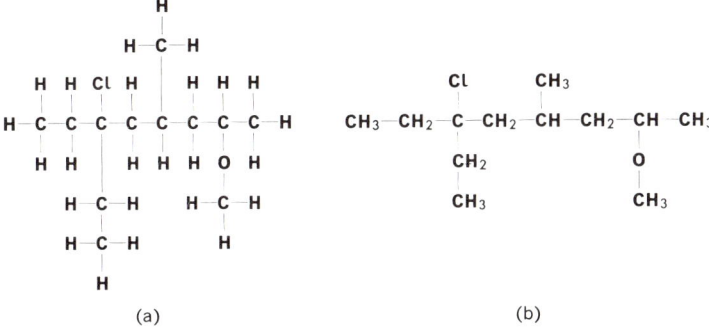

Fig. 21-13 Two types of structural formulas: (a) valence bond structure; (b) condensed structural formula.

densed structural formula. This is readily apparent if we compare the valence bond structure of the organic compound (a) with its equivalent condensed formula (b) in Fig. 21-13.

Now let us consider some compounds having the molecular formula C_5H_{12}. It turns out that more than one isomer of this composition is possible. That is, if we had five carbon atoms and twelve hydrogen atoms, each with its normal valence, we could put them together into more than one structure. Although we cannot handle individual atoms in this fashion, we can find the answers by using ball-and-peg models or by describing the results on paper, using H— to represent each hydrogen with its covalent bond, and

for each carbon and its bonds.

Figure 21-14 shows a valence bond structure and a condensed structural formula for one isomer of C_5H_{12}; the corresponding models of the same molecule are pictured in Fig. 21-11. Another structural arrangement for five carbons and twelve hydrogens is represented by Fig. 21-15; this isomer is said to have a **branched** chain carbon skeleton. In contrast, the compound in Fig. 21-14 has an **unbranched** structure, often called a

```
  H H H H H
  | | | | |
H-C-C-C-C-C-H      CH3-CH2-CH2-CH2-CH3
  | | | | |
  H H H H H
     (a)                    (b)
```

Fig. 21-14 An isomer C_5H_{12} having a straight chain (an unbranched carbon skeleton): (a) valence bond structure; (b) condensed structural formula.

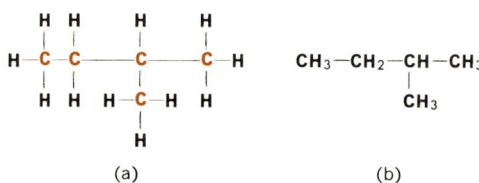

Fig. 21-15 An isomer of C_5H_{12} having a branched chain: (a) valence bond structure; (b) condensed structural formula.

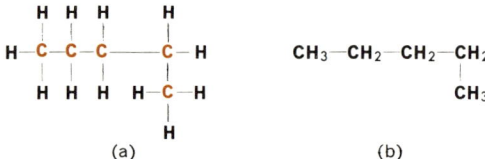

Fig. 21-16 A C_5H_{12} molecule having the same structure as that of Fig. 21-14, but in a different rotational form: (a) valence bond structure; (b) condensed structural formula.

straight chain. If we inspect a model of the latter compound (Fig. 21-11), we see that the chain is not really straight in the geometric sense, but is a zigzag. Therefore the term *straight* is used in a special way for describing organic compounds; it really means *not branched*.

One arrangement for C_5H_{12} that we might think of is diagrammed in Fig. 21-16. At first glance this seems to be different from the structures in either Fig. 21-14 or Fig. 21-15. It is indeed a different structure from that of Fig. 21-15. However, the molecule in Fig. 21-16 has the same structure as the one in Fig. 21-14. Every carbon atom in Fig. 21-16 is bonded to the same other atom as it is in Fig. 21-14. The two molecules are just different rotational forms of the same fundamental structure. (The situation is entirely analogous to the various rotational forms of 1,2-dichloroethane, described in Section 21-8). When the C_5H_{12} molecule of Fig. 21-14 collides with another object, rotation may occur at one of the bonds. The second carbon of the chain may rotate relative to the third carbon, carrying the —CH_3 group into the downward location. The molecule would then have the arrangement of Fig. 21-16. To convince yourself that this change is possible merely by rotation on one of the carbon–carbon bonds, test the theory with a ball-and-peg model and think about the relation between models and structural formulas.

The relationships between some possible isomers of C_5H_{12}, just described, can be summarized as follows: Molecules having the same molecular formula are isomers (different compounds) if their atoms are bonded differently (Figs. 21-14 and 21-15). On the other hand, if two molecules have only temporary differences in shape, due to rotation of groups on bonds, they are the same compounds (Figs. 21-14 and 21-16).

The existence of isomers creates a problem in naming the different compounds. Clearly, we need a distinctive name for each of the isomers of C_5H_{12}. This problem of nomenclature is discussed in Sections 22-2 and 22-3.

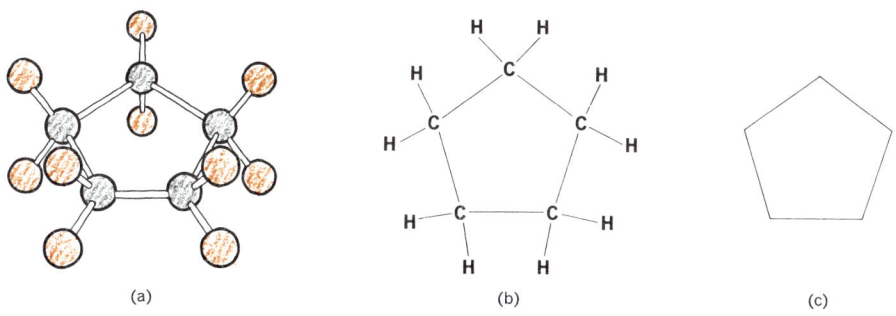

Fig. 21-17 The ring compound C_5H_{10}: (a) ball-and-peg model; (b) valence bond structure; (c) abbreviated structural formula, showing outline of carbon skeleton.

21-10 RING COMPOUNDS: STRUCTURES AND FORMULAS

We learned in Section 21-4 that two carbon atoms within the same chain could bond together, thus closing a ring. Such ring structures are of frequent occurrence in both natural and man-made compounds. Therefore, we should see how it is possible for ring compounds to exist if, as we have assumed (Section 21-6), the bonds around each carbon atom have a tetrahedral arrangement.

We will first examine the compound consisting of five carbon atoms in a ring, with ten hydrogen atoms to occupy the other carbon valences. A ball-and-peg model of the compound is shown in Fig. 21-17(a). In this arrangement all five carbons can be in one plane. If we look at the model from above, we see that the carbon skeleton forms a regular pentagon, represented in Fig. 21-17(c). Mathematically, we can calculate that the angles within a regular pentagon are each 108°. This is very close to the value of 109° for a tetrahedral bond angle. Therefore the carbon atoms can adapt themselves quite readily to a five-membered ring. Furthermore, if we look again at the ball-and-peg model we see that the H—C—H bond angles around the edge of the ring can each be 109°. These possibilities apparently explain the fact that five-membered rings are quite stable and are easily and frequently formed.

Since it is not always convenient to make an accurate model or drawing of the compound C_5H_{10}, we can use the valence bond formula (Fig. 21-17b). As usual, the written valence bond formula does not adequately portray the true three-dimensional nature of the real molecule. In this case, it does not show that the hydrogen atoms are either above or below the plane formed by the carbon atoms. However, it is easy to draw and is satisfactory once we clearly understand what it represents.

Sometimes it is convenient to use the pentagon of Fig. 21-17(c) as an abbreviated formula to represent the compound. When we do, we must imagine that there is one carbon at each corner of the pentagon and that each carbon is holding two hydrogens. In other words, in Fig. 21-17, (c) is an abbreviated version of (b).

Figures 21-18 and 21-19 imply that the bond angles of three- and four-membered rings must be distinctly different from those of a five-membered ring. Since a four-membered ring produces a square, the carbon–carbon bond angles must be distorted away

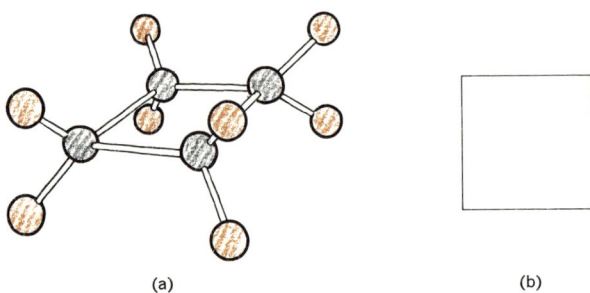

Fig. 21-18 The ring compound C_4H_8: (a) ball-and-peg model; (b) abbreviated structural formula.

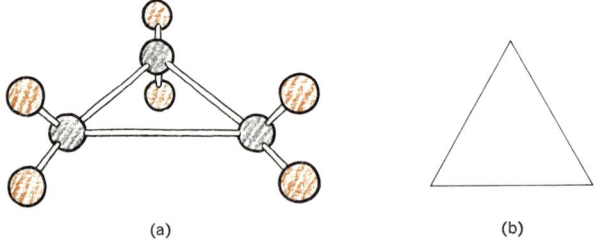

Fig. 21-19 The ring compound C_3H_6: (a) ball-and-peg model; (b) abbreviated structural formula.

---

from their ideal value of 109°, down to 90°. In the case of the regular triangle formed by a three-membered ring, the carbon–carbon bond angles must be distorted way down to 60°! Experimentally, we find that although such structures can exist, the bond distortion causes strains in the molecules, making them less stable. A three-membered ring is sufficiently unstable to cause its chemical reactions to be modified. The strain can cause the ring to break open and react in a way that is not possible in the case of a five-membered ring.

For the ring compound C_6H_{12} we can write the formulas shown in Fig. 21–20. Although we will frequently use these formulas, we find once again that they do not reveal the correct shape of the molecule very well. If it is really a regular hexagon with all the carbon atoms lying in one plane, the carbon–carbon bond angles within the ring would have to be 120°. This is far enough from 109° so that we might expect this compound to be strained also.

Actually, the six-membered ring does not have to be planar (that is, with all the carbon atoms in the same plane); it can instead be puckered, as shown by the ball-and-peg model in Fig. 21–21(a). The rotations permitted by the carbon–carbon bonds allow the carbon skeleton to assume a zigzag shape (compare with Fig. 21–11). Consequently all the carbon bond angles can remain at their optimum value of 109° when the six-membered ring is formed. The ring is quite stable, even a little more stable than a five-membered ring.

Figure 21–21(a) was drawn showing each carbon atom with two valence bonds unoccupied. In the compound C_6H_{12} each of these bonds would be holding hydrogen atoms. In other organic compounds some of the bonds may be occupied by different

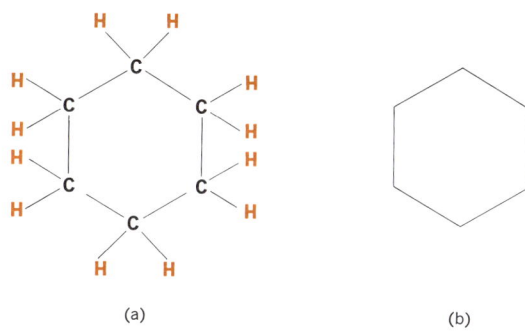

Fig. 21-20 The ring compound C_6H_{12}: (a) valence bond structure; (b) abbreviated structural formula.

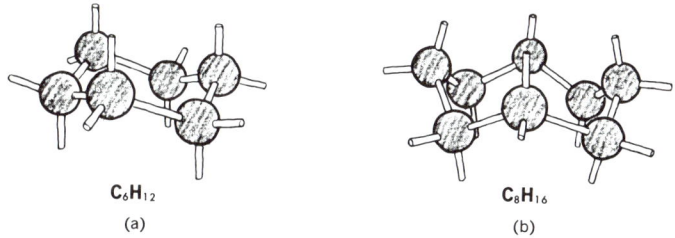

Fig. 21-21 Ball-and-peg models of puckered rings (hydrogen atoms not shown): (a) a six-membered ring; (b) an eight-membered ring.

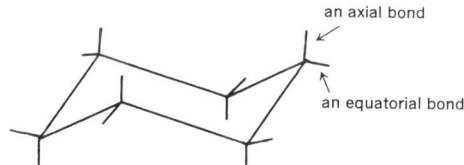

Fig. 21-22 Skeletal formula for a six-carbon ring compound. Only carbon valence bonds are shown. The unoccupied bonds could hold a variety of other atoms.

atoms or groups, such as —Cl, —OH, or —NH_2. It is apparent that one of these atoms would occupy a certain position relative to the location of the other atoms because of the puckered shape of the carbon skeleton. These subtle differences in position can sometimes have an important effect on the chemical reactivity of the ring compounds, which are common in nature and can be synthesized in the laboratory. Although a chemist or biochemist must sometimes be concerned with such effects, it is cumbersome to draw such an elaborate picture as that shown in Fig. 21–21(a). Instead, a skeletal formula of the type shown in Fig. 21–22 is often used. It shows the correct bond angles just as well.

Even larger rings can form because of the possibility of puckering. There can be some rotation at each carbon–carbon bond, permitting the skeleton to pucker into a shape which will finally allow ring closure without straining any of the bond angles. Figure 21–21(b) shows one possible shape for an eight-membered ring. Rings containing 15, 20, or even 30 atoms are known. There is no theoretical reason why rings of almost any size cannot form.

The Nature of Organic Compounds

IMPORTANT TERMS

Branched chain
Carbon chain
Carbon ring
Condensed structural formula

Covalent bond
Inorganic substances
Isomers
Molecular formula

Organic substances
Structural formula
Tetrahedral

WORK EXERCISES

1. Two isomers having composition C_5H_{12} are shown in Section 21–9. How many other isomers, if any, are possible for C_5H_{12}? Draw their structural formulas (either valence bond or condensed formulas). If you are not certain about the total number of isomers from writing structural formulas, try ball-and-peg models. Check your answer in Section 22–2.

2. From all the formulas shown below, pick out the pairs or groups of formulas which actually represent the same compound.

a) $CH_3CH_2CH_2CH_2CH_3$

b) 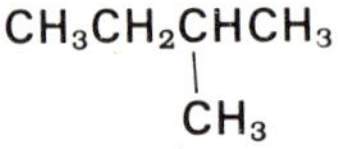 $CH_3CH_2CHCH_3$
   $\quad\quad\quad\quad\; |$
   $\quad\quad\quad\quad CH_3$

c) 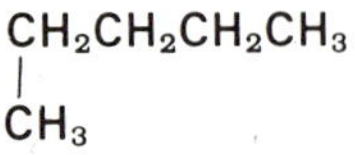 $CH_2CH_2CH_2CH_3$
   $\;\;|$
   $\;\;CH_3$

d) 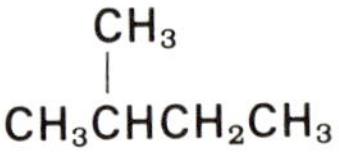 $\quad\quad\; CH_3$
   $\quad\quad\;\; |$
   $CH_3CHCH_2CH_3$

e) 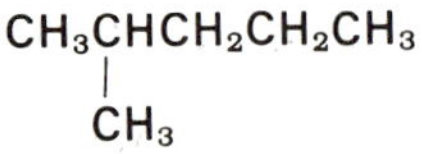 $CH_3CHCH_2CH_3$
   $\quad\; |$
   $\quad\; CH_3$

f) 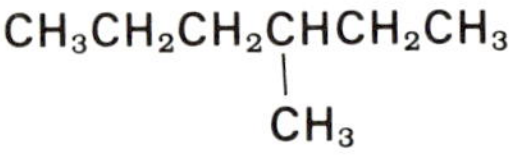 $CH_3CH_2CH_2CHCH_2CH_3$
   $\quad\quad\quad\quad\quad\;\; |$
   $\quad\quad\quad\quad\quad\; CH_3$

g) 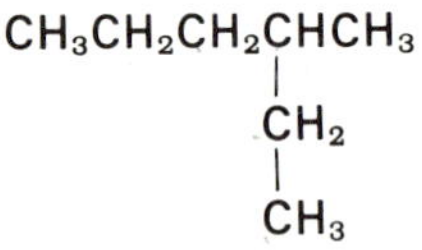 $CH_3CH_2CH_2CHCH_3$
   $\quad\quad\quad\quad\; |$
   $\quad\quad\quad\quad CH_2$
   $\quad\quad\quad\quad\; |$
   $\quad\quad\quad\quad CH_3$

h) 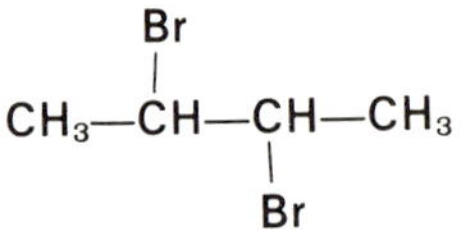 $\quad\quad\;\; Br$
   $\quad\quad\;\; |$
   $CH_3-CH-CH-CH_3$
   $\quad\quad\quad\quad\;\; |$
   $\quad\quad\quad\quad\; Br$

i) 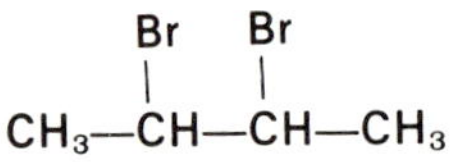 $\quad\;\; Br\;\; Br$
   $\quad\;\; |\quad\;\; |$
   $CH_3-CH-CH-CH_3$

j) 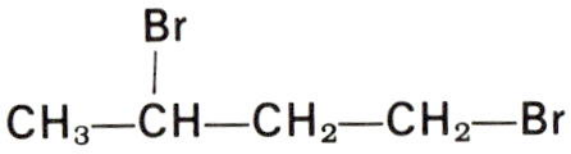 $\quad\quad\; Br$
   $\quad\quad\; |$
   $CH_3-CH-CH_2-CH_2-Br$

k) 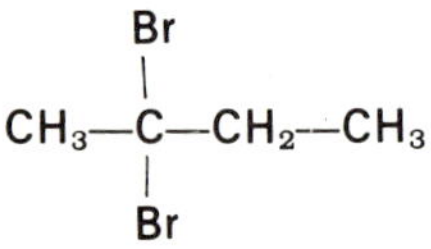 $\quad\quad\; Br$
   $\quad\quad\; |$
   $CH_3-C-CH_2-CH_3$
   $\quad\quad\; |$
   $\quad\quad\; Br$

l) 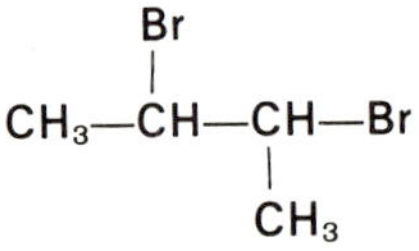 $\quad\quad\; Br$
   $\quad\quad\; |$
   $CH_3-CH-CH-Br$
   $\quad\quad\quad\quad\; |$
   $\quad\quad\quad\quad CH_3$

3. Write structural formulas to show how many different structures are possible for the molecular formulas:

   a) C_4H_9Br b) $C_3H_6Cl_2$ c) $C_4H_{10}O$

SUGGESTED READING

Asimov I. *A Short History of Chemistry*, Anchor Books, Doubleday, New York, 1965, pp. 93–96, 100–116. Describes the concepts of vitalism, isomers, valence, and structural formulas in the development of organic chemistry.

Asimov, I., *The New Intelligent Man's Guide to Science*, Basic Books, New York, 1965, pp. 433–438. An excellent brief summary of early organic chemistry, from vitalism until the development of structural formulas.

Asimov, I., *The Wellsprings of Life*, Abelard-Schuman, London–New York–Toronto, 1960, pp. 153–156. Summarizes the early controversy over inorganic versus organic substances.

Kurzer, F., and P. M. Sanderson, "Urea in the History of Organic Chemistry," *J. Chem. Educ.*, **33,** 452 (1956). The authors state "The early history of urea provides a vivid impression of the slow and difficult beginnings of chemical science . . ."

Lambert, J. B., "The Shapes of Organic Molecules," *Scientific American*, Jan. 1970.

Lipman, T. O., "Wöhler's Preparation of Urea and the Fate of Vitalism," *J. Chem. Educ.* **41,** 452 (1964).

Noller, C. R., *Textbook of Organic Chemistry*, 3rd ed., Chapters 1 and 4, W. B. Saunders, Philadelphia, 1966. Chapter 1 describes the nature and historical development of organic chemistry. Chapter 4 discusses isolation and analysis of organic compounds by both traditional and modern techniques.

Sanderson, R. T., "Principles of Chemical Bonding," *J. Chem. Educ.* **38,** 383 (1961). An extremely detailed outline for the student interested in exploring the concept more deeply.

Wasserman, E., "Chemical Topology," *Scientific American*, Nov. 1962 (Offprint #286). Chemists have succeeded in linking one ring compound through another ring. This article discusses the accomplishments and some of the intriguing possibilities.

SATURATED HYDROCARBONS: The Alkanes

22–1 PETROLEUM, A SOURCE OF SATURATED HYDROCARBONS

Ancient civilizations in many areas of the world made use of the oily materials which they discovered oozing from cracks in the earth. In some localities the crude oil was quite fluid. In others, the seepage was chiefly asphalt (also called pitch or tar). In still other places natural gas escaped and frequently became ignited. These "eternal flames" gave rise to numerous myths and religious cults. Whether the materials were gas, oil, or pitch, they were all related and all arose from deposits of **petroleum** (L. *petra*, rock plus *oleum*, oil).

By 5000 B.C., the Egyptians were already using asphalt as one ingredient for embalming. Asphalt was also well known in the region of the Euphrates river before 4000 B.C., in cities such as Babylon, Ur, and Nineveh. Archeologists have discovered that asphalt was used there as a paint, as mortar to hold bricks, and as waterproofing for jars, baths, and boats. The Phoenician traders caulked their ships with asphalt 5000 years ago. The petroleum fields at Baku, on the Caspian Sea, have apparently been known since before recorded history and are still highly productive. Prior to 1000 B.C. the burning gas vents there attracted fire-worshipping cults. Marco Polo reported that oil from Baku was widely distributed for use as lamp fuel and as medicine for animals and humans.

The Chinese knew of petroleum more than 2000 years ago. They were perhaps the first to obtain it from drilled wells, although they may originally have been seeking water or salt. Both gas and oil were used for heat and light by the Chinese.

When Europeans began exploring the New World, they discovered that many Indian tribes were using petroleum products. Asphalt was an item of commerce among the Incas, who heated oil in immense clay vessels to obtain tar of the desired thickness. The material was used for embalming and for waterproofing clay pots. Petroleum was known to Indians in widely scattered areas of North America. Oil was used for war paint, for softening leather, and especially in the northeast, for medication. The Seneca Indians had such a well-established trade with other tribes that their oil used for the treatment of coughs, sores, and bruises was called Seneca oil.

In the United States and Canada the first oil wells were drilled during the period 1858–1860. Many others followed rapidly. From that time until 1900, the chief product isolated was kerosene, sold for use in lamps and cook stoves. Smaller amounts of lubricating oils were used. About the time that electric lamps might have put the oil companies out of

Petroleum, A Source of Saturated Hydrocarbons

Fig. 22-1 A modern oil pump and crude-oil storage tanks. [Courtesy of the Shell Oil Company.]

TABLE 22-1 Saturated hydrocarbons—the first four members in the series

Name	Molecular Formula	Structural Formula
methane	CH_4	H—C(H_2)—H (one carbon with 4 H)
ethane	C_2H_6	H—C(H_2)—C(H_2)—H
propane	C_3H_8	H—C(H_2)—C(H_2)—C(H_2)—H
butane (n-butane)	C_4H_{10}	H—C(H_2)—C(H_2)—C(H_2)—C(H_2)—H

business, automobiles appeared, causing a new spurt of growth in the oil industry (Fig. 22–1). Refinery practice had to be altered for the production of gasoline.

During the period 1860–1900 chemists were learning that nearly all compounds in petroleum deposits were composed of nothing more than carbon and hydrogen. Whether from petroleum or elsewhere, compounds containing only carbon and hydrogen are called **hydrocarbons**. A compound is termed a **saturated hydrocarbon** if all its carbon atoms are saturated with hydrogen atoms; that is, hydrogens occupy all the carbon valence bonds, other than the minimum number of bonds required to link the carbons to one another in the chain. Because one chain can be composed of more carbons than are found in another chain, a whole series of saturated hydrocarbons is possible. Table 22–1 lists the first few members of the series. A complete list would show some molecules which contain as many as 10, 20, 50, or more carbon atoms linked together. Such a family of compounds can be found in deposits of petroleum. Saturated hydrocarbons usually constitute well over 90% of petroleum. The remainder often includes hydrogen sulfide and organic sulfur compounds, which have a strong odor. Small amounts of organic nitrogen or oxygen compounds may also be present. Petroleum from certain localities will contain several percent of aromatic compounds (Chapter 24), but many deposits have none. Some further remarks concerning petroleum products are found in Section 22–4.

A family of organic compounds, all of which have the same type of structure, is called a **homologous series** (Gr. *homos*, the same). In this book we will rarely have need to use the term again, but the *idea* of such a series of related compounds is important. Note (Table 22–1) that the molecular formula of any member differs from the one before it or after it by CH_2. Inspection of the structural formulas reveals the reason for this. To convert ethane to propane we need to lengthen the chain by one carbon, and in the process need to add only two new hydrogens.

```
    H  H            H              H  H  H
    |  |            |              |  |  |
H—C—C···H      —C—           H—C—C—C—H
    |  |            |              |  |  |
    H  H            H              H  H  H
```

The change represented here is of course just a mental exercise, to help us understand structural relations. We cannot pick up an individual molecule and manipulate it in this fashion. Nor can we carry out a reaction in the laboratory in quite this way.

22–2 DEVELOPMENT OF NAMES

When we get through the hydrocarbon series (Table 22–1) as far as butane, we encounter a problem of naming. A substance of molecular formula C_4H_{10} could also have this structure:

CH_3—CH—CH_3
          |
         CH_3

Since the latter is an isomer of butane, chemists years ago named it isobutane. To avoid any uncertainty, the unbranched butane (shown in Table 22–1) was designated *normal* butane, or *n*-butane.

The next member in the series is pentane. *Three* different structures, each having formula C_5H_{12} are possible. All three are real compounds; they have been isolated (from petroleum or other sources) and found to differ in their melting points, boiling points, and so forth. Therefore different names are needed for the isomeric pentanes. The unbranched one was again named *n*-pentane, and the one with the simpler branch was called isopentane. Since in this case there was yet another isomer, a "new" one, it was called neopentane (Gr. *neos*, new).

$$CH_3CH_2CH_2CH_2CH_3 \qquad CH_3CHCH_2CH_3 \qquad CH_3-\underset{\underset{CH_3}{|}}{\overset{\overset{CH_3}{|}}{C}}-CH_3$$
$$ \underset{CH_3}{|}$$

   n-pentane   isopentane   neopentane

The situation becomes rapidly more complicated. There are five different structures having formula C_6H_{14}, nine having formula C_7H_{16}, 75 having formula $C_{10}H_{22}$, and 4347 having formula $C_{15}H_{32}$! It would become increasingly difficult to think up appropriate little syllables to designate the various isomers, and even more difficult to remember which one should represent which structure. Clearly, what is needed is a relatively simple, easily remembered, logical system for naming organic compounds. It should be general and inclusive enough that it could apply to any possible organic compound, including those not yet discovered. Such a method for assigning names to organic compounds has been developed through the cooperation of chemists from many different countries. Hereafter we will call such names the **systematic names**.

By the 1880's this problem of organic nomenclature had become critical. Chemists had determined the structure of many natural products, including a few complex ones. Also, enough was known about organic reactions to permit the synthesis of completely new compounds, as well as the duplication of some of those discovered in nature. Consequently, the number of chemical structures known had surpassed any convenient method for naming them, and thereby readily communicating knowledge about them to others.

Work to fulfill the need for a systematic method of nomenclature was begun in 1889 by the International Congress of Chemists then meeting in Paris. A few members of the Congress were appointed to a study commission, to develop a set of rules for naming organic compounds. When the Congress next met in 1892 at Geneva, the recommendations of the commission were adopted by the approximately forty chemists present, and became known as the "Geneva rules." For the first time an adequate, internationally accepted system of nomenclature was available.

Thereafter, modifications had to be agreed upon occasionally, and a few new terms added to provide names for certain types of structure which had not been known previously. Some changes were accomplished at a meeting of the International Union of Chemistry (IUC) in 1930. The organization met again in 1949 under the name International Union of Pure and Applied Chemistry (IUPAC). At that time, and at subsequent meetings, the rules have been expanded. For a while the updated rules were called the IUC system and, more recently, the IUPAC system. At the present time it seems convenient to refer to the international nomenclature as *systematic names*.

22-3 THE SYSTEMATIC NAMES

In the systematic method of nomenclature, the saturated hydrocarbons, as a class, are called **alkanes**. Each individual member of the series has been assigned a name which represents the number of carbons in its chain, and ends with -*ane*. Thus, pentane is an alk*ane* (saturated hydrocarbon) having five carbons. This name, used alone, designates only the *straight* chain (unbranched) isomer having formula C_5H_{12}. The systematic names for several alkanes are listed in Table 22-2. Except for the first four, nearly all the names were derived from the Greek names for numbers; *nonane* is of Latin origin.

TABLE 22-2 Names for alkanes

Molecular Formula	Name	Molecular Formula	Name
CH_4	methane	$C_{11}H_{24}$	undecane
C_2H_6	ethane	$C_{12}H_{26}$	dodecane
C_3H_8	propane	$C_{13}H_{28}$	tridecane
C_4H_{10}	butane	$C_{14}H_{30}$	tetradecane
C_5H_{12}	pentane	$C_{15}H_{32}$	pentadecane
C_6H_{14}	hexane	$C_{16}H_{34}$	hexadecane
C_7H_{16}	heptane	$C_{17}H_{36}$	heptadecane
C_8H_{18}	octane	$C_{18}H_{38}$	octadecane
C_9H_{20}	nonane	$C_{19}H_{40}$	nonadecane
$C_{10}H_{22}$	decane	$C_{20}H_{42}$	eicosane (I-co-sane)

For a saturated hydrocarbon *ring* compound (Section 21-10) the term **cyclo** (Gr. *kyklos*, circle) is included in the name. The general name is *cycloalkane*. Individual compounds have the usual name designating the number of carbons. For example:

$H_2C\!\!-\!\!CH_2\!\!-\!\!CH_2\!\!-\!\!CH_2\!\!-\!\!CH_2$ (ring) or ⬠ cyclopentane

One more bit of terminology is necessary in order to name *branched* skeleton hydrocarbons by the systematic method. An **alkyl group** is a unit of structure derived by removing (imaginarily) one hydrogen from an alkane, so that the group has a bond available for attachment to a different atom. Illustrations are given in Table 22-3. Any particular group is named by using the ending -*yl* in place of the -*ane* ending in the name of the parent alkane. When we wish to generalize, the symbol R is used to designate any alkyl group.

The rules for writing systematic names of alkanes may be summarized as follows:
1) Name the longest carbon chain which can be found in the structure.
2) Preceding the longest-chain name, write down, in alphabetical order, the names of each of the kinds of alkyl groups which are attached to the main chain.

22-3 The Systematic Names 301

TABLE 22-3 Derivation of alkyl group names

alkane	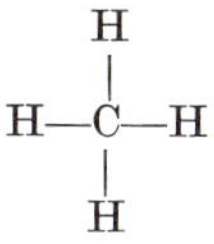 H—C(H)(H)—H CH_4 methane	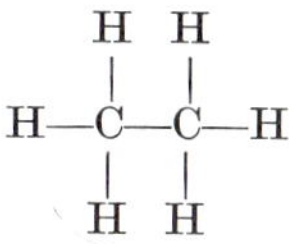 H—C(H)(H)—C(H)(H)—H CH_3CH_3 ethane	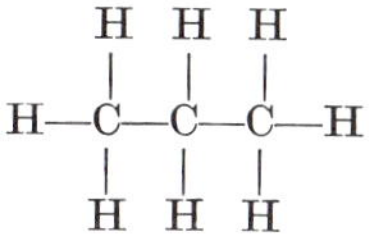 H—C(H)(H)—C(H)(H)—C(H)(H)—H $CH_3CH_2CH_3$ propane	R—H alkane (general)
alkyl group	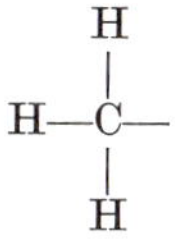 H—C(H)(H)— CH_3— methyl	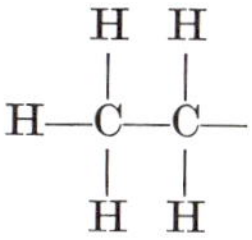 H—C(H)(H)—C(H)(H)— CH_3CH_2— ethyl	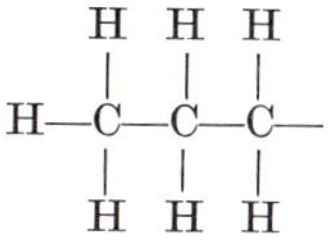 H—C(H)(H)—C(H)(H)—C(H)(H)— $CH_3CH_3CH_2$— propyl	R— alkyl (general)

3) If there are several groups of the same kind attached to the main chain, list that group name only once, using the appropriate prefix such as *di*, *tri*, *tetra*-, etc., to indicate two, three, or four groups, respectively, of that kind.

4) Assign a number, as a prefix, to *each* of the alkyl groups in the name, to indicate the position of the group on the chain. For this purpose, start numbering the carbon atoms consecutively from either end of the chain, so that the attached groups will have the lowest numbers possible.

The rules are best illustrated by applying them, step by step, to some examples.

Example 1

1) 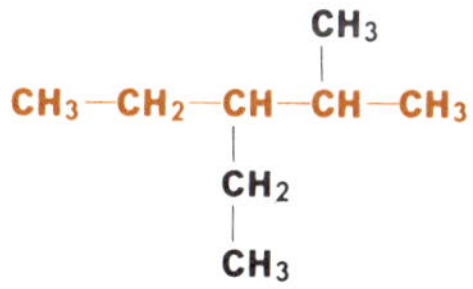 $CH_3-CH_2-CH(-CH_2-CH_3)-CH-CH_3$ with CH_3 above — pentane

2) 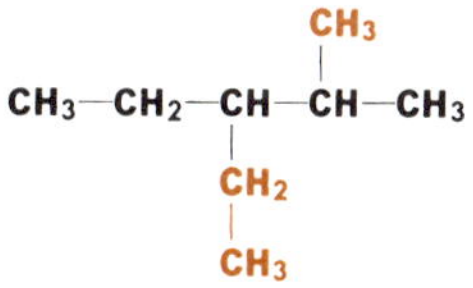 $CH_3-CH_2-CH(-CH_2-CH_3)-CH-CH_3$ with CH_3 above — ethyl methylpentane

[Rule (3) does not apply to this example.]

4) 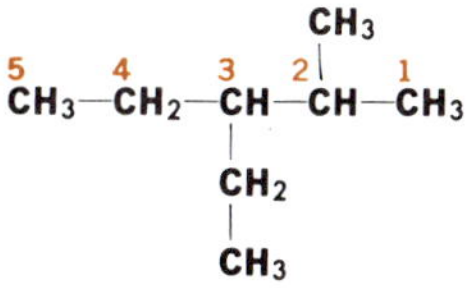 $\overset{5}{CH_3}-\overset{4}{CH_2}-\overset{3}{CH}(-CH_2-CH_3)-\overset{2}{CH}(-CH_3)-\overset{1}{CH_3}$ — 3-ethyl-2-methylpentane

(*Not* 3-ethyl-4-methylpentane, which would result if we counted from the other end of the chain.)

Example 2

1) 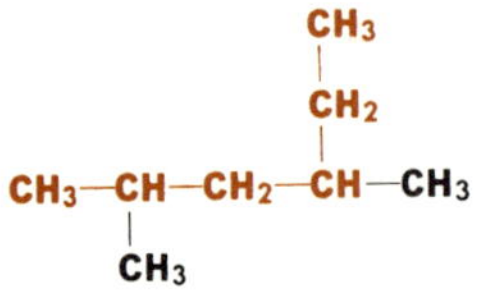
```
            CH₃
            |
            CH₂
            |
CH₃—CH—CH₂—CH—CH₃          hexane
    |
    CH₃
```

2) 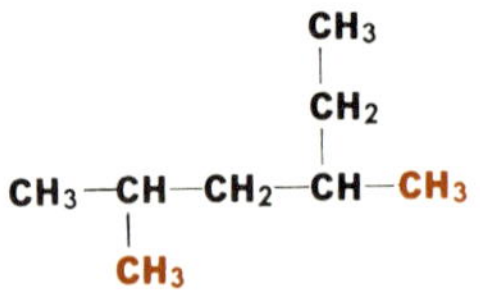
```
            CH₃
            |
            CH₂
            |
CH₃—CH—CH₂—CH—CH₃          methylhexane
    |
    CH₃
```

3) See preceding structure.   dimethylhexane

4) 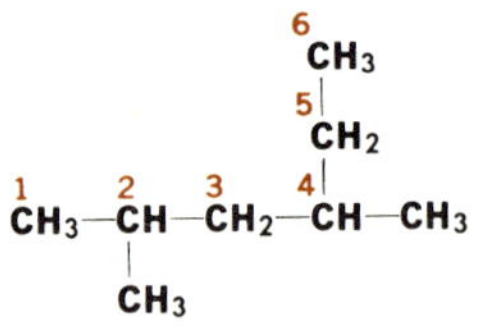
```
               6
               CH₃
              5|
               CH₂
   1   2   3  4|
   CH₃—CH—CH₂—CH—CH₃         2,4-dimethylhexane
       |
       CH₃
```

Example 3

1) 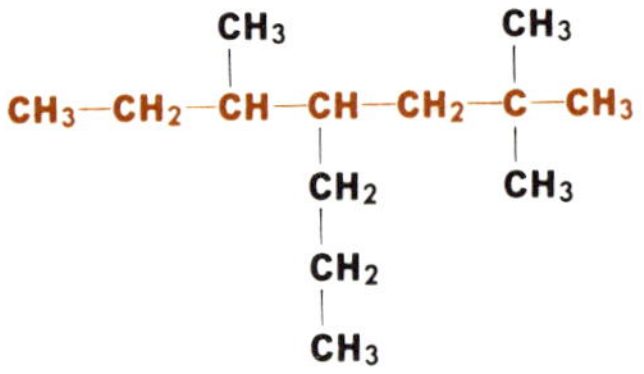
```
            CH₃         CH₃
            |           |
CH₃—CH₂—CH—CH—CH₂—C—CH₃          heptane
            |           |
            CH₂         CH₃
            |
            CH₂
            |
            CH₃
```

2) 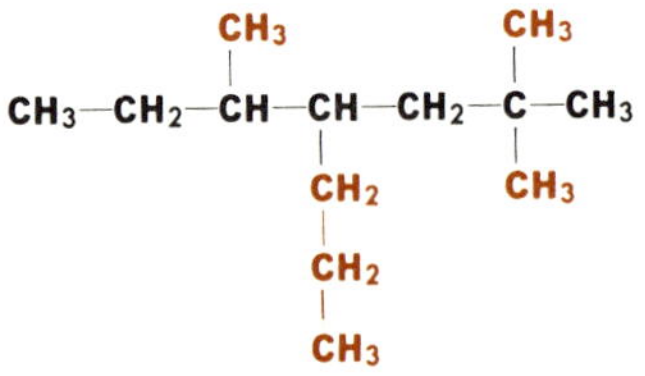
```
            CH₃         CH₃
            |           |
CH₃—CH₂—CH—CH—CH₂—C—CH₃          methyl propylheptane
            |           |
            CH₂         CH₃
            |
            CH₂
            |
            CH₃
```

3) See preceding structure.   trimethyl propylheptane

4) 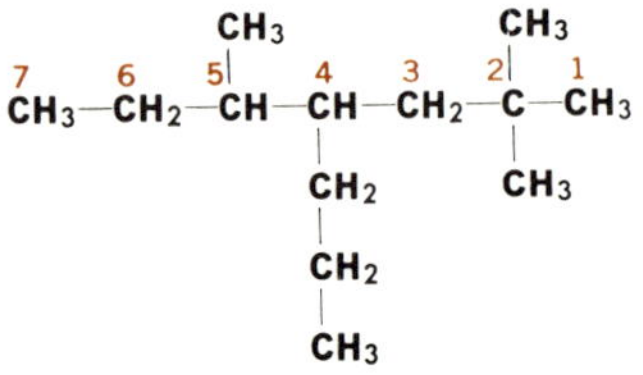
```
                CH₃         CH₃
                |           |
   7   6   5|  4   3   2|  1
   CH₃—CH₂—CH—CH—CH₂—C—CH₃     2,2,5-trimethyl-4-propylheptane
                |           |
                CH₂         CH₃
                |
                CH₂
                |
                CH₃
```

The purpose of the name is to describe accurately to another person the structure we have in mind. If the naming rules are satisfactory (and are properly used) it should be impossible for him to interpret them in such a way that he could write a *different* structure than the one we had named.

Let us now apply the rules in the opposite manner, attempting to write a structure from a name.

Example 4

Given: the name 4,4-diethyl-2-methyloctane.

First, we see at the end of the name the word *octane*. Therefore we write out a chain of eight carbons:

C—C—C—C—C—C—C—C

Next, we start reading at the beginning of the name, and as we come to each group name we write carbons to represent it in the structure, meanwhile counting down the chain to be sure we put each group in the correct position.

The *4,4-diethyl* means two ethyl groups, both on the fourth carbon:

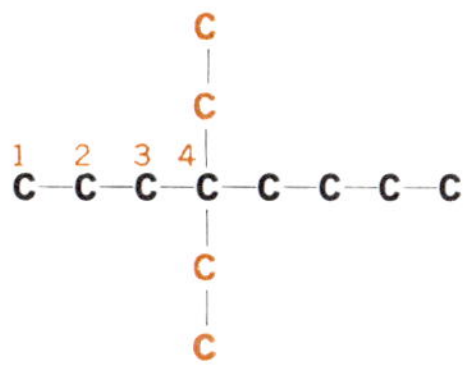

```
            C
            |
            C
  1  2  3  4|
  C—C—C—C—C—C—C—C
            |
            C
            |
            C
```

Now, *2-methyl* requires a methyl group on the second carbon; we must be sure to count in the same direction in which we counted the first time:

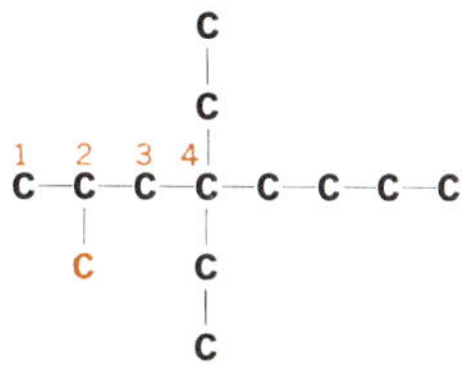

```
            C
            |
            C
  1  2  3  4|
  C—C—C—C—C—C—C—C
     |      |
     C      C
            |
            C
```

Finally, since we usually want to write a proper condensed structural formula, we can go back and put the required number of hydrogens at each carbon, keeping in mind that the covalence of carbon is 4:

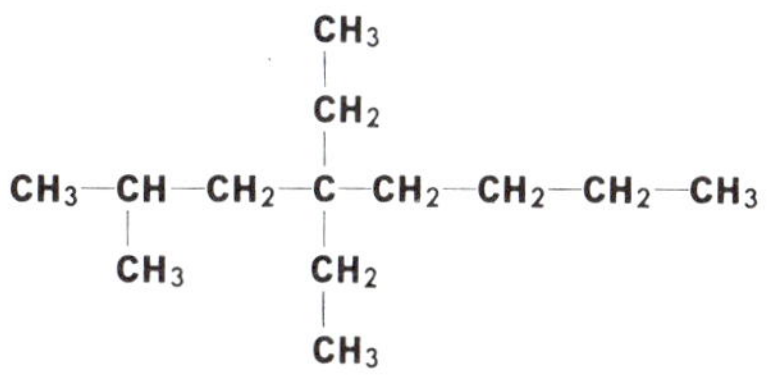

$$\begin{array}{c}
CH_3\\
|\\
CH_2\\
|\\
CH_3-CH-CH_2-C-CH_2-CH_2-CH_2-CH_3\\
||\\
CH_3CH_2\\
|\\
CH_3
\end{array}$$

Note in Example 4 that the group name *ethyl* was mentioned only once; the prefix *di-* told us that there were two such groups. However, even though both are at the same position on the chain, the number must be given for *both* of them, thus: 4,4-. Had it been written only as *4-diethyl*, we would have been uncertain when writing the structure. Did the author really mean both are at 4? Or is this a typographical omission; should it have been 3,4-diethyl?

If we count all the atoms in the structure shown in Example 1, we find the molecular formula is C_8H_{18}; the compound is one of the isomeric octanes. Although we first wrote the word *pentane* as the basis of a name for this structure, it does not mean the compound is one of the pentane isomers. If we examine the *complete* name we see that it does indeed say "eight carbons." Ethyl means two, methyl is one more, and pentane means five, for a total of eight. It is wise to check for errors by counting the total in both the name and the structure, to see whether they agree.

It is also important to realize that an alkyl group, such as any of those shown in Table 22–3, is a strictly imaginary device, convenient for naming purposes. It is not necessarily true that an alkyl group is a real compound that we might find in nature or that we might make one by some chemical reaction and attach it to some other group.

All organic compounds have a carbon skeleton of some sort. They may also have certain groups attached to this skeleton: —OH, —NH_2, —Cl, and so forth. We can name compounds containing these groups merely by learning the term used to name each particular group. This term will be incorporated in the complete name, along with a position number, just as we used a name and number for each alkyl group. Once you have learned these few simple rules, you will know the important basis for naming organic compounds by the international system of nomenclature.

22–4 GENERAL PHYSICAL AND CHEMICAL PROPERTIES

Crude oil can be separated into various fractions by distillation (Fig. 22–2). Several familiar petroleum products thus obtained are listed in Table 22–4, along with the size of alkane molecules normally found in each of the products. Evidently there is a very simple relation between the size of the molecules and some of their physical properties. The table shows that small lightweight molecules have lower boiling points, which means they are more volatile. A larger molecule requires more energy to escape from the attraction of its neighbors in the liquid state and become a separate molecule in the vapor state. The temperature at which it boils is higher, corresponding to a greater energy requirement.

We are all aware of the fact that gasoline vaporizes more easily than kerosene. Oils are even less volatile and paraffin wax is solid, at least until the melting point at 50°C is reached. Methane or butane, whose molecules are very light, are gases at room temperature. Pentane, having a boiling point of 36°C, is the first alkane in the series which is liquid at room temperature and pressure. Propane and butane (the two just lighter than pentane) can easily be liquefied if compressed and kept in a tank under pressure; they are often handled in this state and called liquefied petroleum gas (LPG).

It is also apparent from the list in Table 22–4 that the compounds consisting of heavier molecules will be more viscous; compare gasoline, kerosene, oil, and petroleum jelly.

Most of us have observed that oil floats on water; so does kerosene or gasoline. This simple observation reveals two other physical properties of the alkanes: they are insoluble in water and are less dense than water.

TABLE 22-4 Some common petroleum products

Product	Main Alkanes Present	Boiling Range, °C
natural gas	C_1	below 0°
liquefied petroleum gas (LPG); propane, butane	C_3—C_4	below 0°
petroleum ether; light naphtha (solvents and cleaning fluids)	C_5—C_7	30–100°
gasoline	C_5—C_{10}	35–175°
kerosene; jet fuel	C_{10}—C_{18}	175–275°
diesel fuel	C_{12}—C_{20}	190–330°
fuel oil	C_{14}—C_{22}	230–360°
lubricating oil	C_{20}—C_{30}	above 350°
mineral oil (refined; decolorized)	C_{20}—C_{30}	above 350°
petroleum jelly (Vaseline)	C_{22}—C_{40}	(m.p. 40–60°)
paraffin wax	C_{25}—C_{50}	(m.p. 50–65°)

volatile

boiling pt increases as °C ↑.

less volatile

"Like dissolves like" is a general rule concerning solubilities. Water is a highly polar substance which generally dissolves other polar or ionic compounds. This suggests that *an alkane*, being insoluble in water, *is nonpolar*. Nonpolarity is an important characteristic of hydrocarbon skeletons, one which influences the behavior of all organic compounds. The reason an alkane is nonpolar can be understood if we examine a portion of its structure.

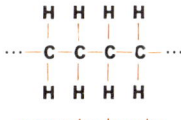

nonpolar bonds

We would expect the covalent bond between two carbon atoms to be nonpolar; since the atoms are identical, the electrons should be shared between them equally. It also happens that a hydrogen atom has just about the same tendency to attract electrons that a carbon

Fig. 22-2 Distillation tower in an oil refinery. Crude oil enters via the large pipe at the lower right. Distilled fractions are drawn off through pipes at various levels in the tower, the most volatile material rising to the top. Streams of intermediate boiling range pass to smaller columns nearby, for further separation. [Courtesy of the Shell Oil Company.]

atom does; that is, it has a similar electronegativity (Section 6–6). Consequently, electrons shared between a carbon atom and a hydrogen atom also form a nonpolar bond.

The nonpolar nature of alkanes makes them good *solvents* for other nonpolar or only slightly polar substances. However, because of their flammability, extreme caution must be observed if volatile hydrocarbons are used as solvents. Year after year, hundreds of people are seriously burned, and homes are destroyed because people were unwisely using gasoline to clean grease off a floor, an auto part, or a piece of clothing.

Alkanes are colorless. Recall that paraffin wax, kerosene, and the "white" gasoline used in camp stoves all lack color. The gasoline used in automobiles is orange because a little dye has been added as a warning that it contains tetraethyllead, a deadly poison. Common lubricating oils have a blue-green color because they still contain small amounts of colored substances originally present in the crude petroleum. They need not be removed for lubricants. The mineral oil used medically is essentially the same mixture of hydrocarbons, decolorized and refined.

Chemically, the distinctive feature of *an alkane* is that it *is generally inert*. There are only a very few chemical changes it will undergo. Most *compounds* which react (especially those studied in this book) are ionic or very polar substances. Consequently, the nonpolar alkanes do not react with them. A few examples are listed below, for it is as important to know what will *not* happen as it is to know what will.

$CH_3CH_2CH_2CH_3 + NaOH$
$CH_3CH_2CH_2CH_3 + HCl$
$CH_3CH_2CH_2CH_3 + H_2SO_4$
$CH_3CH_2CH_2CH_3 + Na$ $\longrightarrow$ no reaction
$CH_3CH_2CH_2CH_3 + KMnO_4$
$CH_3CH_2CH_2CH_3 + NaHCO_3$
$CH_3CH_2CH_2CH_3 + AgNO_3$

One important reaction of an alkane is its decomposition at high temperature, which will be discussed in Sections 23–2 and 24–9.

Alkanes do react with oxygen (combustion) and with chlorine or bromine. These reactions will be discussed in the next two sections. It is interesting to note that even these reagents do not attack an alkane under ordinary conditions at room temperature. For instance, kerosene or gasoline remain unchanged when exposed to air for a very long time. Combustion occurs only if a flame or spark provides the initial burst of high temperature to start the reaction with oxygen. The reaction, once started, is exothermic, so the high temperature is maintained and promotes the reaction of other molecules.

22–5 COMBUSTION

The combustion of methane is represented by Eq. (22–1).

$$H-\underset{\underset{H}{|}}{\overset{\overset{H}{|}}{C}}-H + 2O_2 \rightarrow CO_2 + 2H_2O + \text{heat} \qquad (22-1)$$

Energy, provided by the high temperature of a flame or spark, starts the reaction by breaking the covalent bonds of methane. Only then are the carbon and hydrogen atoms free to combine with oxygen. The formation of new bonds with oxygen releases even larger amounts of energy. The energy produced by the reaction of the first few molecules breaks covalent bonds in other methane molecules nearby, allowing them to react with oxygen. The reaction continues in this fashion until all the material is used up. At each step more energy is released than is required to initiate the combustion of adjacent molecules. The net result is the release of a tremendous amount of heat. Depending on conditions, a little of the energy may be dissipated as light.

Other alkanes react with oxygen in a similar manner. In each case the alkane molecule is completely broken up into separate hydrogen and carbon atoms which combine with oxygen to form water and carbon dioxide. This fact is used in writing a complete, balanced equation for such an oxidation. For example, if heptane burns, Eq. (22–2), we know the products will be CO_2 and H_2O. Furthermore, we can write down 7 molecules of CO_2, to account for all the carbons from heptane. Then we can write $8H_2O$ molecules to account for the 16 hydrogen atoms. Finally, we must provide enough O_2 molecules to accomplish these changes.

$$C_7H_{16} + 11O_2 \rightarrow 7CO_2 + 8H_2O \tag{22-2}$$

If the amount of oxygen available to the burning hydrocarbon is not sufficient, carbon monoxide (CO) may be produced. Or, if some carbon atoms are not oxidized at all, they form the fine black particles called soot.

The largest consumption of alkanes is for the purpose of utilizing the energy released by combustion. Various sorts of engines have been designed to convert the heat energy to mechanical energy for doing work. Frequently, however, the heat may be used directly for heating homes and public buildings or for cooking. Some furnaces and cooking ranges operate on kerosene, others on butane, and so on. A heating fuel now widely used in the United States is natural gas. It is chiefly methane, with small amounts of other light hydrocarbons such as propane. Natural gas occurs in petroleum deposits, along with the mixture of liquid alkanes constituting the crude oil. The gas is carried by huge pipelines directly from the wells to most major cities in the United States.

Since carbon chains form the skeletons of organic compounds, nearly all the organic compounds will burn. Even solid organic substances are combustible; some familiar examples are wood, fats, cotton, and paraffin wax. How *easily* the compounds burn depends largely on their volatility. Gases and easily vaporized liquids, such as methane, butane, or octane, are readily ignited and burn vigorously.

22–6 SUBSTITUTION BY CHLORINE

Under ordinary circumstances an alkane does not react with chlorine. A reaction will start, however, if a mixture of alkane and chlorine is exposed to a high temperature (400°–500°C) or to ultraviolet light from a lamp or from direct sunlight. Equation 22–3 shows the results with methane.

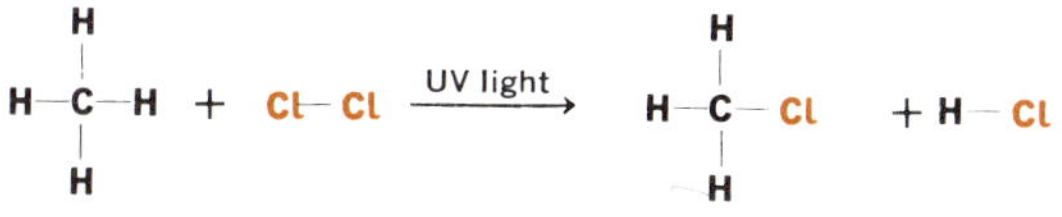

22-6 Substitution by Chlorine

$$\underset{\text{methane}}{H-\underset{\underset{H}{|}}{\overset{\overset{H}{|}}{C}}-H} + Cl-Cl \xrightarrow{\text{UV light}} \underset{\substack{\text{chloromethane}\\ \text{(methyl chloride)}}}{H-\underset{\underset{H}{|}}{\overset{\overset{H}{|}}{C}}-Cl} + H-Cl \qquad (22\text{-}3)$$

This is a **substitution reaction**. One chlorine atom has been substituted in place of one of the hydrogens originally bonded to the carbon. The displaced hydrogen atom has paired off with another chlorine atom to form hydrogen chloride.

Here again, as with combustion, we see that no reaction can occur unless some covalent bond is broken. Only after a hydrogen atom has been separated from the carbon atom can its place be taken by a chlorine atom. Indeed, it is also necessary to break the covalent bond between the two chlorine atoms in the Cl_2 molecule. The need for bond-breaking once again explains the purpose of the energy required to initiate the reaction. Ultraviolet light, which provides a larger amount of energy than does visible light, is sufficient to promote the reaction.

The reaction represented by Eq. (22–3) could be carried out by mixing methane gas and chlorine gas in a tank having a window through which we could shine ultraviolet light. As soon as a few molecules had reacted, some chloromethane would be present in the tank. It could react with the remaining chlorine in much the same way that the methane did; the result, Eq. (22–4), would be the substitution of a second chlorine atom onto the same carbon.

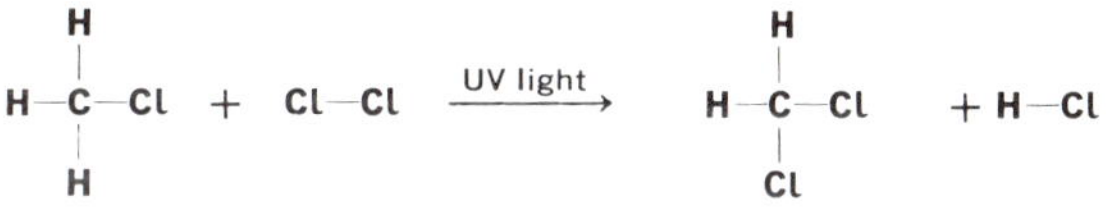

$$H-\underset{\underset{H}{|}}{\overset{\overset{H}{|}}{C}}-Cl + Cl-Cl \xrightarrow{\text{UV light}} \underset{\text{dichloromethane}}{H-\underset{\underset{Cl}{|}}{\overset{\overset{H}{|}}{C}}-Cl} + H-Cl \qquad (22\text{-}4)$$

Of course, if this can happen, the substitution of a third chlorine atom and a fourth should be possible. This is shown in Eqs. (22–5) and (22–6).

$$CH_2Cl_2 + Cl_2 \xrightarrow{\text{UV light}} \underset{\substack{\text{trichloromethane}\\ \text{(chloroform)}}}{CHCl_3} + HCl \qquad (22\text{-}5)$$

$$CHCl_3 + Cl_2 \xrightarrow{\text{UV light}} \underset{\substack{\text{tetrachloromethane}\\ \text{(carbon tetrachloride)}}}{CCl_4} + HCl \qquad (22\text{-}6)$$

These reactions occur step by step. Each one is a repetition of the same fundamental substitution reaction, and at each step one molecule of hydrogen chloride must be produced.

Consequently, when methane reacts with chlorine we obtain, in addition to the hydrogen chloride, a mixture of four different organic compounds. Fortunately the mixture can be separated by distillation, because the four compounds have distinctly different boiling points. Each time a chlorine atom is substituted on the carbon, the molecular weight increases by about 35. The heavier molecule has a higher boiling point. For example, trichloromethane boils at 61°C and tetrachloromethane at 76 °C.

Alkanes other than methane can undergo substitution by chlorine. The chief difference is that a larger alkane yields a larger number of different chlorinated products. For example, when butane reacts with chlorine, the first stage of substitution (comparable to Eq. 22–3) can yield *two* isomeric chlorobutanes.

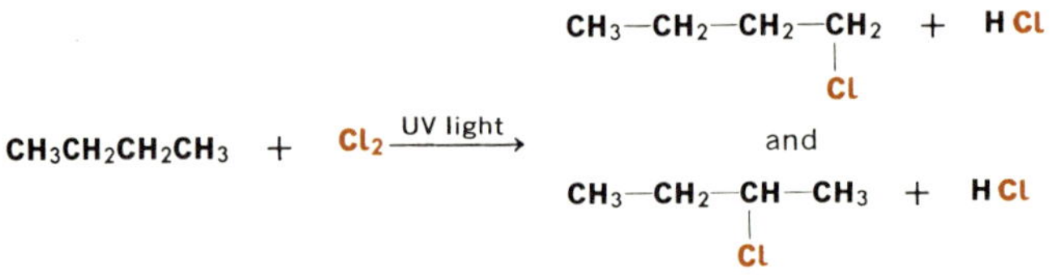

$$CH_3CH_2CH_2CH_3 + Cl_2 \xrightarrow{UV\ light} \begin{array}{c} CH_3-CH_2-CH_2-CH_2 + HCl \\ | \\ Cl \\ \text{and} \\ CH_3-CH_2-CH-CH_3 + HCl \\ | \\ Cl \end{array} \quad (22\text{--}7)$$

The chlorobutanes produced in Eq. (22–7) can react further with the chlorine in the reaction mixture to form *six* isomeric dichlorobutanes.

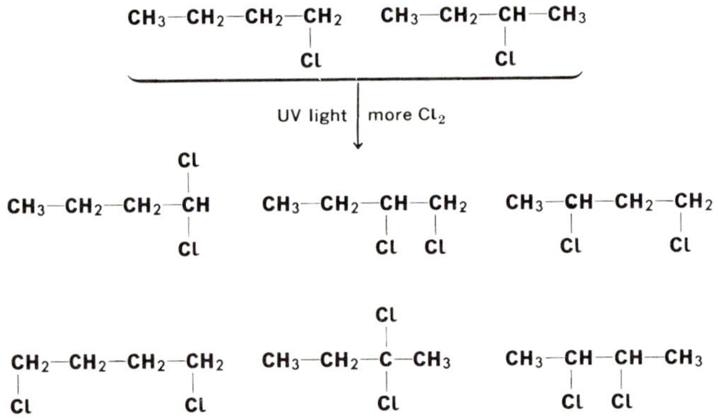

$$\underbrace{\begin{array}{cc} CH_3-CH_2-CH_2-CH_2 & CH_3-CH_2-CH-CH_3 \\ | & | \\ Cl & Cl \end{array}}_{\text{UV light} \mid \text{more } Cl_2}$$

$$\begin{array}{cccc}
Cl & & & \\
| & & & \\
CH_3-CH_2-CH_2-CH & CH_3-CH_2-CH-CH_2 & CH_3-CH-CH_2-CH_2 \\
| & \quad\ | \quad | & \quad | \qquad\ | \\
Cl & \quad Cl\ \ Cl & \quad Cl \qquad Cl
\end{array}$$

$$\begin{array}{cccc}
 & & Cl & \\
 & & | & \\
CH_2-CH_2-CH_2-CH_2 & CH_3-CH_2-C-CH_3 & CH_3-CH-CH-CH_3 \\
| & \qquad\qquad\qquad | & \quad | \quad\ | \\
Cl \qquad\qquad\qquad Cl & \qquad\qquad\qquad Cl & \quad Cl\ \ Cl
\end{array}$$

Repeated substitutions can produce trichlorobutanes, tetrachlorobutanes, and so on up to decachlorobutane, C_4Cl_{10}.

Because of differences in molecular weight, the dichlorobutanes as a group can be separated rather well by distillation from the chlorobutanes or the trichlorobutanes. The mixture of dichlorobutanes thus obtained might be useful as a solvent. For this purpose it would not matter that the various isomers present had different structures; the solvent ability of each molecule would be about the same. On the other hand, this substitution reaction would not be a good method by which to obtain, in pure condition, one particular compound, such as 1,3-dichlorobutane. It is difficult and expensive to separate one dichlorobutane isomer from another; all have the same molecular weight, so that the boiling point differences are slight.

Bromine reacts with alkanes in the same way that chlorine does, for example:

$$CH_3CH_3 + Br_2 \xrightarrow{UV\ light} CH_3CH_2Br + HBr$$

22-7 THE NATURE OF ALKANE REACTIONS

In describing the general chemical properties of alkanes, Section 22-4, we pointed out that alkanes do not participate in *ionic* reactions. Instead, when alkanes do react, the intermediate, reactive particle is a **radical**, that is, a group of atoms which has one unpaired electron to share and is in this sense comparable to an individual atom. The creation of radicals is due to the fact that when a nonpolar bond breaks, the tendency is for one electron to remain with each fragment. The appropriate energy can rupture a bond in this manner:

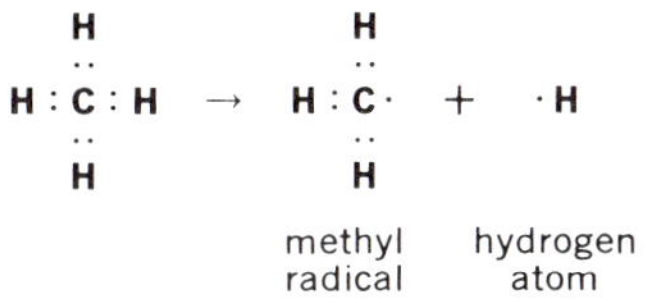

$$\begin{array}{c} H \\ H:C:H \\ H \end{array} \rightarrow \begin{array}{c} H \\ H:C\cdot \\ H \end{array} + \cdot H \qquad (22\text{-}8)$$

$$\text{methyl} \qquad \text{hydrogen}$$
$$\text{radical} \qquad \text{atom}$$

The fate of the resulting fragments then depends on the environment. If oxygen is present, the liberated hydrogen atoms combine with it to form water. Eventually the carbon atom, stripped of its hydrogens, also combines with oxygen, giving carbon dioxide and producing the familiar combustion reaction. In chlorination, the energy supplied by ultraviolet light initiates the reaction by splitting a chlorine *molecule* into two separate chlorine atoms.

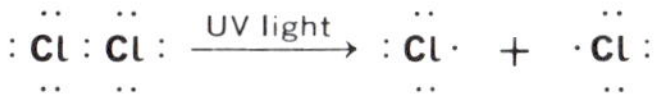

$$:\!Cl\!:\!Cl\!: \xrightarrow{\text{UV light}} :\!Cl\!\cdot + \cdot Cl\!: \qquad (22\text{-}9)$$

A chlorine atom, because of its unpaired electron, then attacks a methane molecule, picks off a hydrogen atom, and leaves a methyl radical.

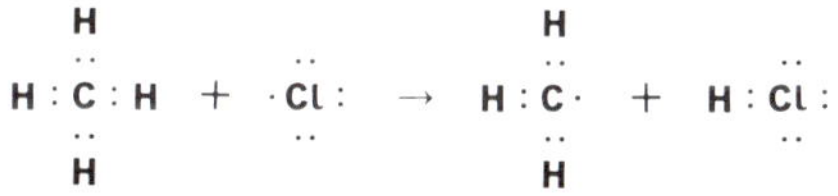

$$\begin{array}{c} H \\ H:C:H \\ H \end{array} + \cdot Cl\!: \rightarrow \begin{array}{c} H \\ H:C\cdot \\ H \end{array} + H:Cl\!: \qquad (22\text{-}10)$$

The methyl radical in turn promptly seeks an atom with which to share its unpaired electron. It picks one from a chlorine molecule, leaving a lone chlorine atom.

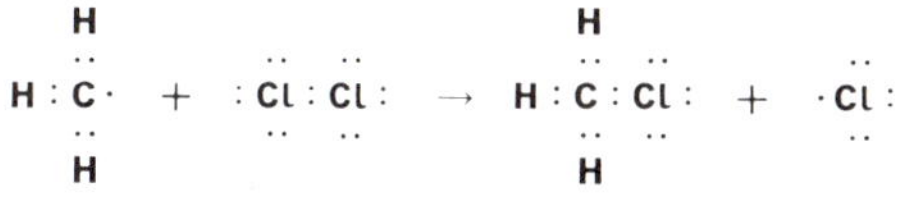

$$\begin{array}{c} H \\ H:C\cdot \\ H \end{array} + :\!Cl\!:\!Cl\!: \rightarrow \begin{array}{c} H \\ H:C:Cl\!: \\ H \end{array} + \cdot Cl\!: \qquad (22\text{-}11)$$

The chlorine atom liberated by reaction (22-11) can also attack a methane molecule as in Eq. (22-10). Thus the sequence of reactions (22-10), (22-11), (22-10), (22-11), etc., can be repeated many times. The important points are that radicals are involved, not ions, and that energy was required in Eq. (22-9) to start the reaction sequence by breaking the first bond.

If no chlorine or oxygen is present when heat energy is applied to an alkane, some of its covalent bonds break anyway. The resultant radicals simply re-pair their unshared electrons in various ways to produce new compounds. Equations (22-12), (22-13), and (22-14) show one possible fate for butane at 500°C.

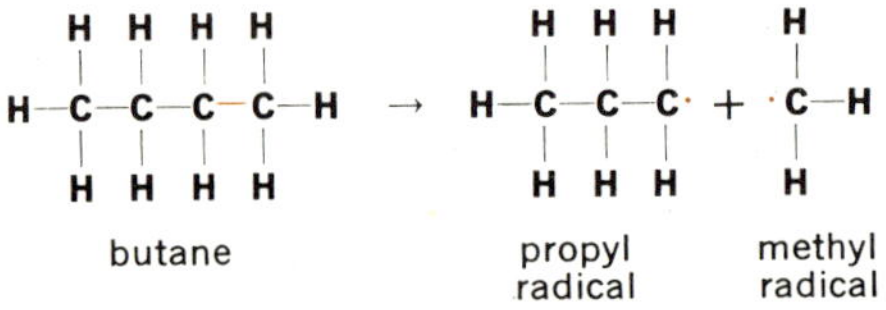

$$\underset{\text{butane}}{\text{H}_3\text{C}-\text{CH}_2-\text{CH}_2-\text{CH}_3} \rightarrow \underset{\text{propyl radical}}{\text{H}_3\text{C}-\text{CH}_2-\text{CH}_2\cdot} + \underset{\text{methyl radical}}{\cdot\text{CH}_3} \quad (22\text{-}12)$$

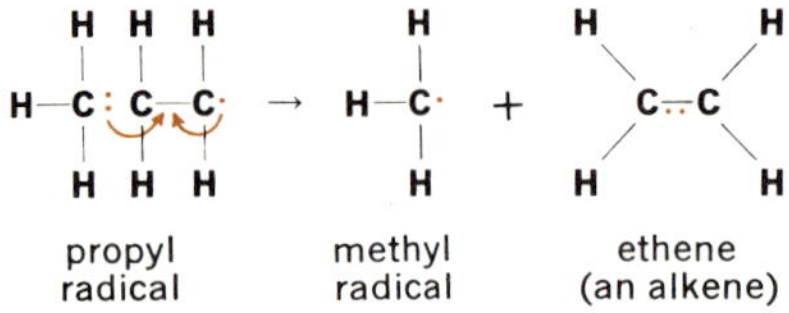

$$\underset{\text{propyl radical}}{\text{H}_3\text{C}-\text{CH}_2-\text{CH}_2\cdot} \rightarrow \underset{\text{methyl radical}}{\text{H}_3\text{C}\cdot} + \underset{\substack{\text{ethene}\\\text{(an alkene)}}}{\text{CH}_2=\text{CH}_2} \quad (22\text{-}13)$$

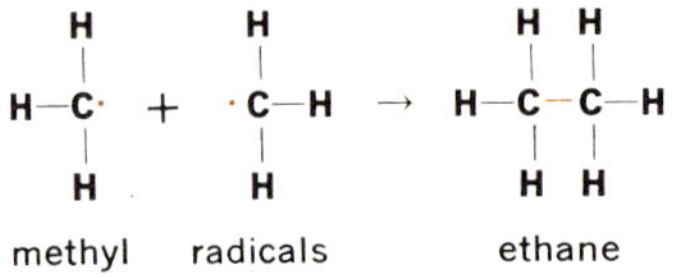

$$\underset{\text{methyl radicals}}{\text{H}_3\text{C}\cdot + \cdot\text{CH}_3} \rightarrow \underset{\text{ethane}}{\text{H}_3\text{C}-\text{CH}_3} \quad (22\text{-}14)$$

The net result of these three steps is that butane (C_4H_{10}) has been converted to ethene (C_2H_4) plus ethane (C_2H_6). The breakup of an alkane in this fashion is called **cracking**; its practical importance will be discussed in Section 23–2.

Although a radical is very reactive it is, like an atom, electrically neutral. For example, when chlorine has seven electrons in its valence shell it is a neutral atom. Only if it acquires one more electron does it become an ion having a negative charge.

$:\ddot{\text{Cl}}\cdot$ chlorine atom $:\ddot{\text{Cl}}:^-$ chloride ion

Similarly, the carbon in a methyl radical is neutral.

$\text{H}:\overset{\text{H}}{\underset{\text{H}}{\ddot{\text{C}}}}\cdot$ methyl radical

The electrons in each carbon-hydrogen bond are shared; one electron is hydrogen's share and one is carbon's. Therefore, in a methyl radical, carbon can claim a total of four electrons for its valence shell. Since this is the number carbon should normally have, it is electrically neutral.

22–8 NAMES OF ALKYL HALIDES

A compound having an alkane skeleton and a halogen in place of a hydrogen is called an **alkyl halide**. The general formula RX may be used for it. For an alkyl chloride we may use RCl; for an alkyl bromide, RBr. With the equations in Section 22–6 are given the systematic names for the four chlorine derivatives of methane; the additional names shown there in parentheses are common names.

To name an alkyl halide systematically we need only one new rule beyond those listed in Section 22–3: *an -o ending is used for a halogen group* (chloro, etc.), *and the group is given a smaller position number than is an alkyl group.*

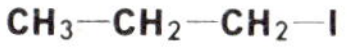

```
                      CH3                          C2H5
                      |                            |
CH3—CH2—CH2—I   CH3—CH—CH—CH3     CH3—CH2—CH—CH—CH2—CH2
                         |                 |         |
                         Cl                Br        Br

1-iodopropane   2-chloro-3-methylbutane   1,4-dibromo-3-ethylhexane
```

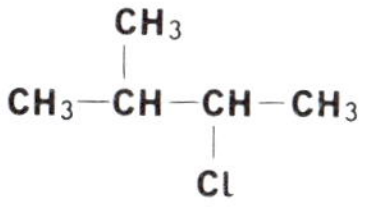

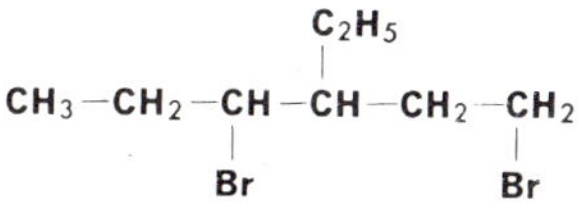

22–9 COMMON ORGANIC HALIDES

Although organic halogen compounds are very rare in nature, many can be synthesized. A few of the methods of synthesis will be mentioned in this text. Several useful halogen compounds are listed in Table 22–5; for each one the proper systematic name is listed first, followed by a common name or trademark names. Some of the examples are not *alkyl* halides because they have carbon skeletons containing double bonds or aromatic rings; these structures will be discussed in Chapters 23 and 24.

TABLE 22–5 Some organic halides

Formula	Name	Applications
CCl_4	**tetrachloromethane**; carbon tetrachloride; Carbona contains CCl_4 and benzene	dry-cleaning fluid; solvent for oils, fats, waxes; some insecticide use; in veterinary medicine, used against worms and flukes
$CHCl_3$	**trichloromethane**; chloroform	solvent; general inhalation anesthetic (safety margin between anesthetic dose and lethal dose very small; can damage liver)
CHI_3	**triiodomethane**; iodoform	antiseptic and local anesthetic applied to small wounds, especially in veterinary practice
CF_2Cl_2	**dichlorodifluoromethane**; Freon-12	fluid used in refrigerators; propellant in aerosols for cosmetics, paints, etc., but not for foods

TABLE 22–5 continued

Formula	Name	Applications
CH$_3$CH$_2$Cl	**chloroethane**; ethyl chloride	local anesthetic for minor surgery (freeze technique)
CH$_2$—CH$_2$ | | Cl Cl	**dichloroethane**	dry-cleaning fluid; solvent for fats, oils, waxes, resins, and especially rubber
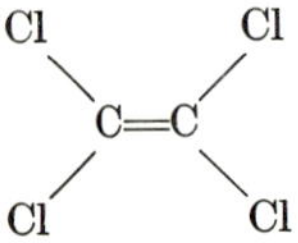 Cl\ /Cl C=C Cl/ \Cl	**tetrachloroethene**	dry-cleaning fluid, especially in coin-operated machines; metal degreasing; medically, in both humans animals, against worms and flukes
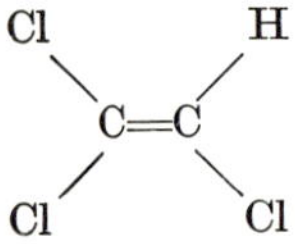 Cl\ /H C=C Cl/ \Cl	**trichloroethene**; Trilene	dry-cleaning fluid; metal degreasing; solvent; inhalation anesthetic, valuable for obstetrics and short operations
Br F | | H—C—C—F | | Cl F	**2-bromo-2-chloro-1,1,1-trifluoroethane**; halothane; Fluothane	inhalation anesthetic
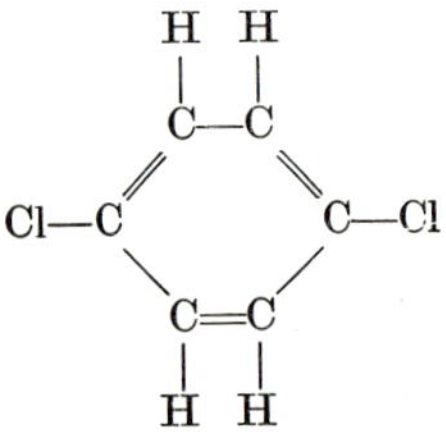 (para-dichlorobenzene ring structure)	**para-dichlorobenzene**; Dichloricide; Paramoth	kills moths in woolens (colorless crystals in flakes or lumps)
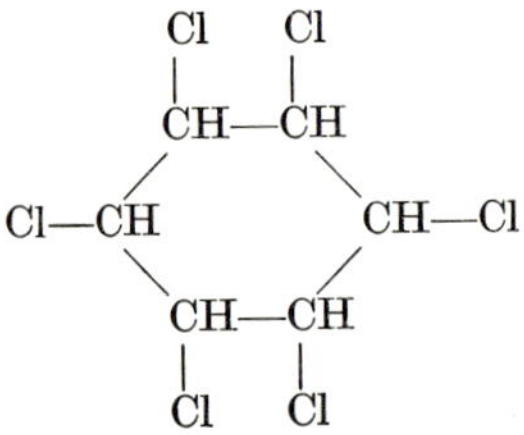 (hexachlorocyclohexane ring structure)	**hexachlorocyclohexane** (C$_6$H$_6$Cl$_6$); Lindane; 666; Gammexane; BHC, benzene-hexachloride (erroneous)	effective insecticide; has had wide agricultural use; damages the environment

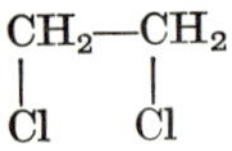

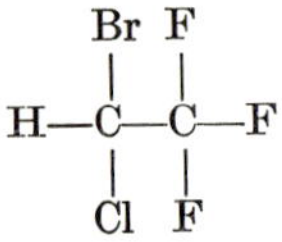

It is of interest to note that most of the organic halogen compounds are similar to each other in two characteristics. First, the liquid organic halides are good solvents for many nonpolar or moderately polar organic compounds, such as alkanes, fats, waxes, and others. Some organic halides will dissolve a greater variety of substances than the alkanes will. In water, the organic halides are insoluble. Second, their biological effects are pronounced and seem to follow a general pattern. Depending on the particular halogen compound and its concentration, it may change the activity of individual cells, especially nerve cells. It may even kill the cells or kill the whole organism. The volatile halogen compounds, which can be inhaled, affect the central nervous system, leading to unconsciousness if used in sufficient amounts. Large doses may cause permanent damage to certain tissues or may cause death due to suspension of vital functions.

A number of chlorinated hydrocarbons such as DDT and hexachlorocyclohexane proved to be very effective insecticides initially. Unfortunately, many insect populations gradually developed tolerance to these pesticides at the dosage levels at which they were being applied. As a result it became necessary to use greater and greater concentrations to achieve the same degree of insect control. This in turn greatly increased the killing or harming of other organisms. The threat to the environment is especially great because the halogenated compounds are persistent. They are destroyed only very slowly by the chemical, biochemical, and physical processes occurring in nature. DDT came into general use toward the end of World War II. It has been estimated that in the approximately 25 years between then and 1970, about 1.5 million tons of DDT were used and about 1 million tons were still present in the environment at the end of that period.

22–10 RELATION OF ALKANES TO OTHER ORGANIC COMPOUNDS

There are three reasons why it is convenient and logical to begin our study of organic chemistry with the alkanes. First, an alkane is chemically quite inert and provides a simple basis for studying the reactions of other organic compounds. For example, an *alkene* molecule, which we will discuss in the next chapter, may be regarded as a double-bond unit of structure plus an alkane skeleton.

an alkene $\quad$ $CH_3CH_2CH_2CH_2CH\!=\!CH_2$

$\qquad\qquad\quad$ alkane $\quad$ functional
$\qquad\qquad\quad$ skeleton $\quad$ group
$\qquad\qquad$ (mainly inert) (reactive)

An alkene compound is very reactive due to the presence of a double bond between two of the carbons. *The double bond does something, or functions, and therefore that part of the structure is called a* **functional group**. While reactions are occurring at a functional group, the rest of the molecule, *the skeleton, survives unchanged*. This is a very important concept, and it simplifies the study of organic chemistry. For example, any compound having a $C\!=\!C$ bond will have certain properties due to the reactivity of that bond. By learning what those properties are, we can predict the behavior of any of the thousands of different compounds in the series of alkenes. Similarly, the alcohols form a large series of compounds, each of which has an —OH group attached to an alkane skeleton. We can become acquainted with all alcohols by discovering what reactions to expect of an —OH group, and by keeping in mind that the alkane skeleton rarely reacts.

Second, the rules for naming alkanes (Section 22–3) provide the basis for naming carbon skeletons in other compounds having functional groups.

Finally, alkanes are important starting materials for synthesis because they are so abundant in petroleum. In succeeding chapters we will discuss methods for converting alkanes to other organic compounds.

22–11 CLASSES OF ORGANIC COMPOUNDS

It is very useful to classify organic compounds according to the type of structure, particularly the type of functional group, that they possess. All alcohols have similar properties because each has an —OH group covalently bonded to an alkane skeleton, and it is the structure of a molecule which determines its behavior. If we study the properties of a few typical compounds within the very large family of alcohols, we can pretty well understand the behavior of *any* alcohol. Fortunately, the hundreds of thousands of organic compounds are built up from relatively few types of carbon skeletons and functional groups. As a result, it is extremely convenient to think about organic compounds in terms of the structural units found in each. For this purpose, we tend to arrange all the compounds having one particular kind of structural unit into one class or family. At the same time, we recognize that one compound may have more than one functional group and will therefore react in several different ways, just as a human being may engage in numerous activities within one day, acting variously as pianist, baker, husband, tennis player, and so on.

In the next several chapters we will study, one by one the various classes, or families, of organic compounds. We will learn about the distinctive properties, the uses, the names, and the occurrence in nature of the compounds in each class. Some idea of the types of structure we will encounter is provided by Table 22–6, which lists the various classes of common organic compounds, with examples of each. Most of the classes listed in Table 22–6 are distinguished by their functional groups. However, three of the classes—alkane, aromatic hydrocarbon, and heterocycle—represent the three fundamental types of carbon skeleton found in organic compounds. Some compounds consist of nothing more than one of these skeletons. In the great majority of compounds, however, the skeleton carries one or several of the functional groups.

As our study proceeds, you will find it helpful to refer frequently to Table 22–6 to check on names or symbols, or to find a reference for more details. The terms used in the table are also defined in the glossary, located at the rear of the book.

IMPORTANT TERMS

Alkane	Hydrocarbon
Alkyl halide	Saturated hydrocarbon
Alkyl group	Substitution reaction
Cyclo-	Systematic name
Functional group	Unsaturated hydrocarbon

TABLE 22-6 Classes of common organic compounds

Class of Compound	Functional Group	Representative Compounds and Names	References (Section)
alkane	—	RH $CH_3CH_2CH_2CH_3$ butane 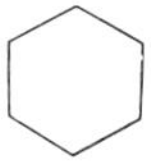 — cyclohexane (C_6H_{12})	22-1, 22-3
alkene	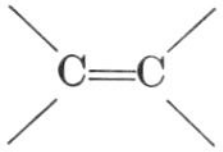 $\diagup$C=C$\diagdown$	$CH_3CH=CH_2$ propene	23-1, 23-3
alkyne	—C≡C—	$CH_3C≡CCH_3$ 2-butyne	23-6
aromatic hydrocarbon	—	ArH  benzene (C_6H_6) 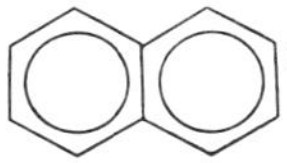 naphthalene ($C_{10}H_{14}$)	24-1, 24-2
alcohol	—OH	CH_3CH_2OH ethanol (ethyl alcohol)	25-1, 25-2
ether	—O—	ROR' $CH_3CH_2OCH_3$ ethyl methyl ether ROAr $CH_3OC_6H_5$ methyl phenyl ether	25-1, 25-7
halide	—X	RX $CH_3CH_2CH_2Br$ 1-bromopropane ArX C_6H_5Cl chlorobenzene	22-8 24-2, 24-4
carboxylic acid	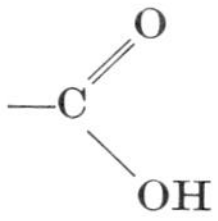 $-C(=O)-OH$	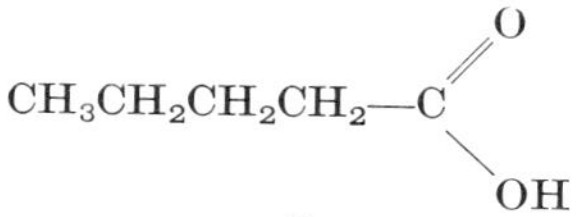 RCOOH $CH_3CH_2CH_2CH_2-C(=O)-OH$ pentanoic acid ArCOOH 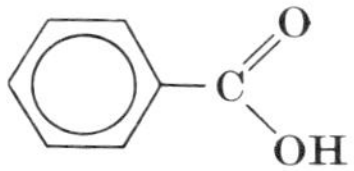 $C_6H_5-C(=O)-OH$ benzoic acid	26-1

Table 22-6 continued

Class of Compound	Functional Group	Representative Compounds and Names	References (Section)
ester	—C(=O)—OR	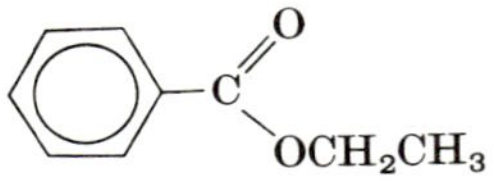 CH₃CH₂CH₂—C(=O)—OCH₃ methyl butanoate; C₆H₅—C(=O)—OCH₂CH₃ ethyl benzoate	26-4
amide	—C(=O)—NH₂	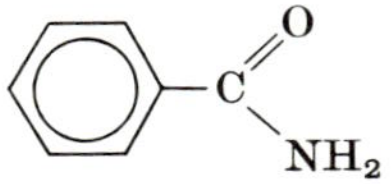 CH₃—C(=O)—NH₂ acetamide; C₆H₅—C(=O)—NH₂ benzamide	28-4
acid chloride	—C(=O)—Cl	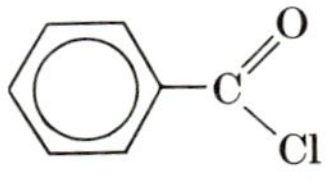 CH₃—C(=O)—Cl acetyl chloride; C₆H₅—C(=O)—Cl benzoyl chloride	26-7
aldehyde	—C(=O)—H	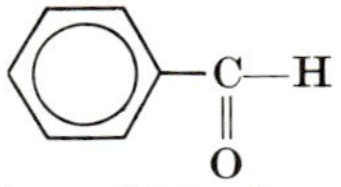 CH₃CH₂—C(=O)—H propanal; C₆H₅—C(=O)—H benzaldehyde	29-1 and Table 29-2

Table 22-6 continued

Class of Compound	Functional Group	Representative Compounds and Names	References (Section)
ketone	$-\underset{\underset{O}{\|\|}}{C}-R$	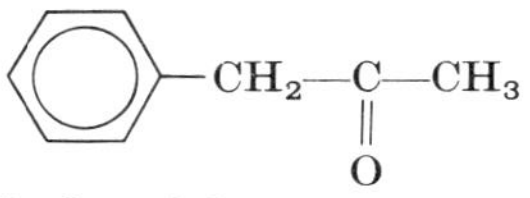 $CH_3CH_2-\underset{\underset{O}{\|\|}}{C}-CH_2CH_3$ 3-pentanone $C_6H_5-CH_2-\underset{\underset{O}{\|\|}}{C}-CH_3$ 1-phenyl-2-propanone	29-1 and Table 29-2
amine	$-NH_2$ $-NHR$ $-NR_2$	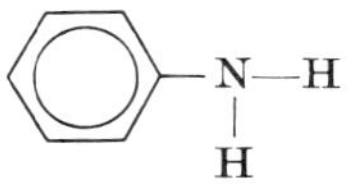 $CH_3CH_2-\underset{\underset{H}{\|}}{N}-CH_3$ ethylmethylamine $C_6H_5-\underset{\underset{H}{\|}}{N}-H$ aniline	28-1
heterocycle	—	  pyridine (C_5H_5N) thiazole (C_3H_3NS)	24-8, 28-6B

WORK EXERCISES

1. Show the condensed structural formula of
   a) 3-ethylpentane
   b) 2,2-dimethylbutane
   c) 2-methyl-3-propylhexane
   d) 2,2,4-trimethylpentane
   e) ethylcyclopentane
   f) 1,3-dibromocyclohexane
   g) 4-ethyl-2,5-dimethyloctane
   h) 3,5-dimethylheptane
   i) 2,3-dichloro-3-ethylhexane
   j) 4,6-diethyl-3,4-dimethyl-6-propyl-nonane

2. Write condensed structural formulas for all isomers having the following molecular formulas:
   a) C_6H_{14}
   b) C_4H_9Cl
   c) $C_4H_8Cl_2$
   d) $C_5H_{11}Br$

3. Write the systematic name for each structure in Problem 2. Are any of the names the same? If they are, look again to see whether the structures are really different. Is the naming system adequate—that is, does each structure have a unique name which applies to no other structure?

4. Name each structure in parts a) through j):

a) 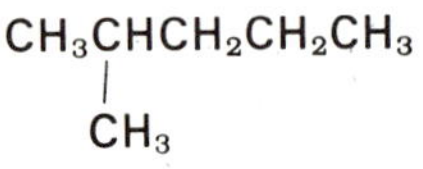 $CH_3CHCH_2CH_2CH_3$
       $|$
       CH_3

b) 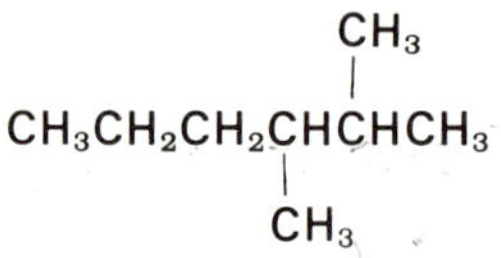 $CH_3CH_2CH_2CHCHCH_3$
                $|\ \ |$
              $CH_3\ CH_3$ (upper CH$_3$ on first labeled C)

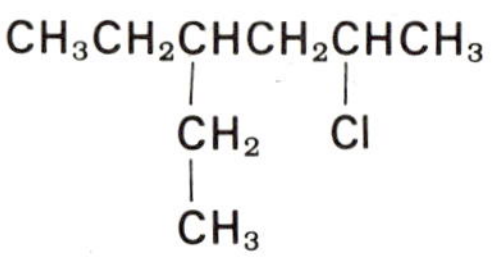

c) $CH_3CH_2CHCH_2CHCH_3$
          $|\ \ \ \ \ \ \ \ |$
        $CH_2\ \ \ Cl$
          $|$
        CH_3

d) 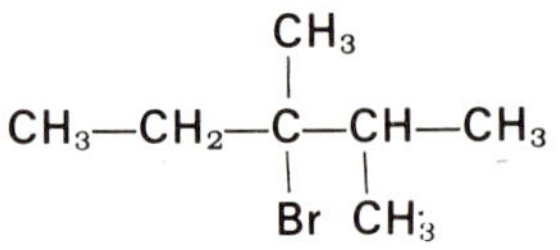 $CH_3-CH_2-C-CH-CH_3$
              $|\ \ \ \ |$
            (C has CH$_3$ above)
            $Br\ CH_3$

e) 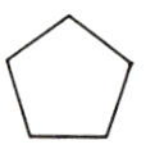 (cyclopentane)

f) 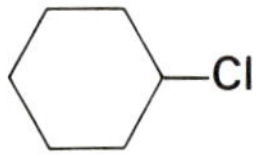 (cyclohexane)—Cl

g) 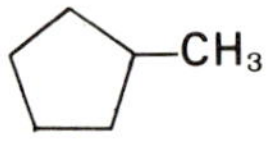 (cyclopentane)—CH$_3$

h) 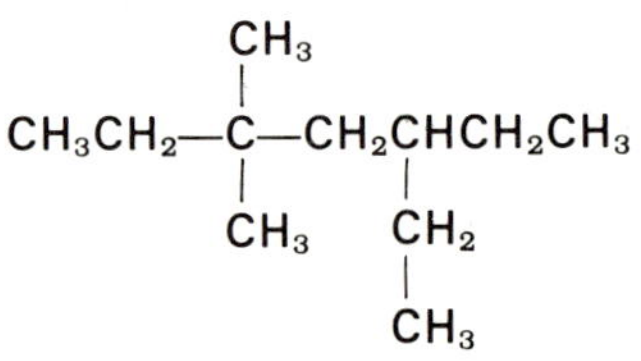 $CH_3CH_2-C-CH_2CHCH_2CH_3$
              $|\ \ \ \ \ \ \ \ \ \ \ |$
            $CH_3\ \ \ \ \ \ CH_2$
            (C also has CH$_3$ above)
                            $|$
                           CH_3

i) 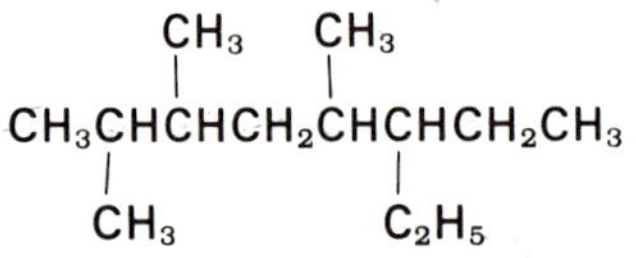 $CH_3CHCHCH_2CHCHCH_2CH_3$
        $|\ \ |\ \ \ \ \ \ \ |$
      $CH_3\ CH_3\ \ \ C_2H_5$ (with CH$_3$ groups above on first two labeled carbons)

j) 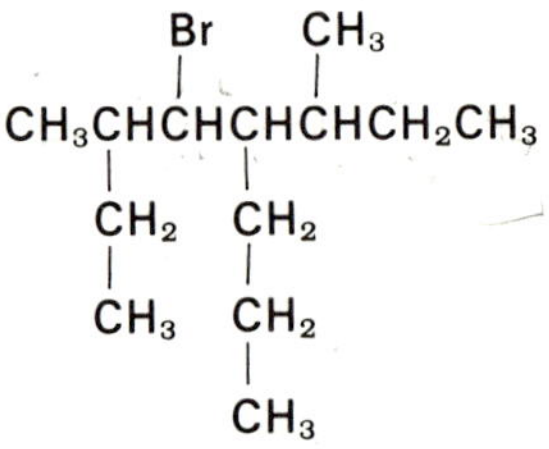 $CH_3CHCHCHCHCH_2CH_3$
        $|\ \ |\ \ \ \ \ \ \ $
      $Br\ CH_3$ (above)
        $CH_2\ CH_2$
        $|\ \ \ \ \ |$
      $CH_3\ CH_2$
                $|$
              CH_3

5. Give both a common name and a systematic name for:

   a) $CHCl_3$ b) CCl_4 c) $CH_3-CH-CH_3$
                                   $|$
                                   CH_3

   d) $CH_3-CH_2-CH-CH_3$ e) CH_3Br
              $|$
              CH_3

6. Write complete balanced equations for the reaction occurring in each mixture below, using structural formulas. If necessary, state "no reaction." If more than one reaction is possible in a given mixture, write equations for all of them.
   a) propane burning in a camp stove
   b) ethane and hydrogen chloride at 100°C
   c) methane and chlorine at high temperature
   d) isobutane and oxygen, heated
   e) cyclopentane and chlorine in ultraviolet light
   f) cyclohexane and air, heated
   g) pentane and concentrated sodium hydroxide
   h) methane and bromine in ultraviolet light
   i) dichloromethane and bromine in ultraviolet light

7. Name the compounds shown in Figs. 21–18 and 21–19.

8. Write a name and an abbreviated structural formula for the compounds shown in Fig. 21–21.

9. List all the types of structural units (both carbon skeletons and functional groups) that you can find in the compounds below. Consult Table 22–6.
   a) aspirin (Section 26–9D)
   b) citral (Table 29–2)
   c) coniine, the poison in hemlock which was used to kill Socrates. (Section 28–8B)
   d) the compound below, produced by the plant *coreopsis* (Compositae):

   $\bigcirc-C{\equiv}C-C{\equiv}C-CH{=}CH-CH_2-OH$

   e) epinephrine (adrenalin):

   $CH_3-N-CH_2-CH-\bigcirc-OH$
       $|$ $|$ $\backslash$
       H OH OH

SUGGESTED READING

DeNevers, Noel, "Liquid Natural Gas," *Scientific American*, October, 1967. When natural gas is chilled to −162°C it becomes liquid and can be transported in containers other than pipelines. This new method of handling facilitates even wider distribution and use of natural gas.

DePuy, C. H., and K. L. Rinehart, Jr., *Introduction to Organic Chemistry*, Chapters 1 and 2, John Wiley, New York, 1967.

Eglinton, G., J. R. Maxwell, and C. T. Pillinger, "The Carbon Chemistry of the Moon," *Scientific American*, Oct. 1972. The simple organic compounds found on the moon were not produced by living organisms, but they do provide additional clues to the origin of life.

Evieux, E. A., "The Geneva Congress on Organic Nomenclature, 1892," *J. Chem. Educ.* **31**, 326 (1954).

Hart, H., and R. D. Schuetz, *Organic Chemistry*, 4th ed., Chapter 2, Houghton Mifflin, Boston, 1972.

Kimberlin, C. N., Jr., "Chemistry in the Manufacture of Modern Gasoline," *J. Chem. Educ.* **34**, 569 (1957).

Lawless, J. G., C. E. Folsome, and K. A. Kvenvolden, "Organic Matter in Meteorites," *Scientific American*, June 1972.

Nelson, T. W., "The Origin of Petroleum," *J. Chem. Educ.* **31**, 399 (1954). A thorough, documented review.

Noller, C. R., *Textbook of Organic Chemistry*, 3rd ed., Chapter 5, W. B. Saunders, Philadelphia, 1966.

Rossini, F. D., "Hydrocarbons in Petroleum," *J. Chem. Educ.* **36**, 554 (1960).

Shoemaker, B. H., E. L. d'Ouville, and R. F. Marschner, "Recent Advances in Petroleum Refining," *J. Chem. Educ.* **32**, 30 (1955).

23

UNSATURATED HYDROCARBONS: Alkenes and Alkynes

23-1 STRUCTURE AND OCCURRENCE OF ALKENES

In an **unsaturated hydrocarbon** the carbon skeleton holds *less* than the maximum number of hydrogens which it might hold. "Extra" covalent bonds can exist between some carbon atoms, because not all their valence electrons are occupied in bonds with hydrogens. An **alkene** is an unsaturated hydrocarbon having a carbon–carbon double bond; a good example is propene.

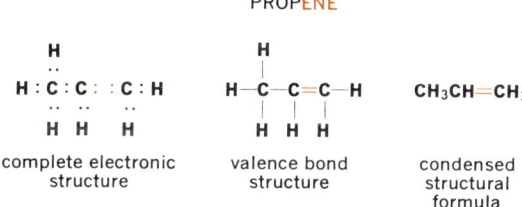

PROPENE

complete electronic structure | valence bond structure | condensed structural formula

The carbon–carbon double bond is very common in organic compounds. It is found in both chain and ring structures, in simple molecules and highly complex ones, in those produced within living cells, and in man-made substances.

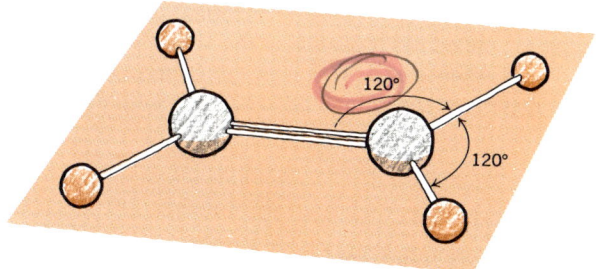

Fig. 23-1 Structure of the molecule H$_2$C=CH$_2$, ethene, showing all atoms in one plane and bond angles of 120°.

A carbon atom having a double bond holds its attached groups at different angles than does a saturated carbon. Each doubly-bonded carbon is linked to only three other atoms, which move away from each other as far as possible. For example, in the molecule C_2H_4, called ethene, one particular carbon is bonded to two hydrogens and one other carbon. When these three atoms are spread out in space, we find bond angles of 120° (Fig. 23–1). Another result of this arrangement is that the molecule is planar; that is, the two carbons and their attached hydrogens lie in one plane.

23–2 SOURCE AND USES OF ALKENES

The most economical way to obtain simple alkenes in large quantity is to **crack** alkanes from petroleum. For instance, if butane is heated to about 500°C the covalent bonds begin to break. Since any of the bonds is susceptible to fracture at this temperature, all the imaginable fragments are formed. (A more detailed explanation of the reaction is given in Section 22–7.)

$$
\begin{matrix} H & H & H & H \\ | & | & | & | \\ H-C-C-C-C-H \\ | & | & | & | \\ H & H & H & H \end{matrix} \rightarrow \begin{cases} CH_3CH=CH_2 + CH_4 \\ CH_2=CH_2 + CH_3CH_3 \\ CH_3CH_2CH=CH_2 + H_2 \\ CH_3CH=CHCH_3 + H_2 \end{cases} \quad (23-1)
$$

butane

Note in Eq. (23–1) that each time a segment is broken from the carbon chain there are insufficient hydrogens for all the carbon valences; consequently each break produces an alkene plus a saturated compound. The resultant mixture can be separated into its various components, each of which has a use. The cracking process is the industrial source of simple alkenes, because only small amounts exist originally in petroleum.

Alkenes which can be obtained inexpensively in this way are important intermediates for the synthesis of other organic compounds. For example, alkenes can be converted to alcohols, and these in turn to a wide variety of other organic materials.

Another important use of the cracking reaction is in the manufacture of gasoline. The hydrocarbons suitable for gasoline are chiefly octanes and other alkanes slightly smaller or larger (see Table 22–4). The amount of these found naturally in petroleum is seldom great enough to satisfy the demand for gasoline. Therefore an important part of any petroleum refinery is a cracking unit in which larger alkanes found in kerosene or fuel oil are broken into smaller molecules (Fig. 23–2). The cracking yields a great assortment of fragments, but the mixture can be distilled to yield several fractions, just as was the original petroleum. Fractions containing molecules too small or too large for gasoline are used for other purposes. The fraction of alkanes and alkenes in the C_5 to C_{10} size range constitutes "cracked gas." It has a superior octane rating and is blended with the natural gasoline.

23–3 ALKENE NAMES

Table 23–1 shows the names of some representative alkenes. *Systematic names* have the ending *-ene* to designate the double bond. The position of the double bond in the carbon chain is indicated by a number placed just before the name for the chain; although the

Fig. 23-2 A catalytic cracking unit in an oil refinery. The reactor is in the center. The cracked products are piped from there to the tower at the right, a fractional distillation column which separates the material into gasoline, light oil, etc. [Courtesy of the Shell Oil Company.]

---

double bond is between two carbons, we need to tell only the number of the first carbon involved as we count down the chain.

$CH_3CH_2CH=CHCH_3$ 2-pentene

If there are halogen atoms or alkyl groups attached to the chain, counting should be done from whichever end of the chain will allow *the lowest number for the double bond*.

We see in Table 23–1 several common names. These originated in the days before systematic names were developed. They are still with us because it is difficult to convince people to stop using common names for frequently encountered compounds. It is rather like trying to prevent a child's pals from calling him by a nickname. Unfortunately, some of the alkene common names are only slightly different from systematic names, which can lead to confusion and misspelling. The common names are given in the table only so that you will be aware of the problem and can look up a name here if you see it used elsewhere. As much as possible we will use proper systematic names in this book; we urge you to do the same. In later chapters it will be necessary to learn a few common names because of their wide usage.

TABLE 23-1 Names of some alkenes

Structure	Systematic Name	Common Name
$CH_2\!=\!CH_2$	ethene	ethylene
$CH_3CH\!=\!CH_2$	propene	propylene
$CH_3CH_2CH\!=\!CH_2$	1-butene	α-butylene
$CH_3CH\!=\!CHCH_3$	2-butene	β-butylene
$CH_3\!-\!\underset{\underset{CH_3}{\mid}}{C}\!=\!CH_2$	methylpropene	isobutylene
$CH_3\!-\!\underset{\underset{CH_3}{\mid}}{C}\!=\!CH\!-\!\underset{\underset{CH_3}{\mid}}{\overset{\overset{CH_3}{\mid}}{C}}\!-\!CH_3$	2,4,4-trimethyl-2-pentene	
$CH_2\!=\!\underset{\underset{CH_3}{\mid}}{C}\!-\!CH\!=\!CH_2$	2-methyl-1,3-butadiene	isoprene
 (cyclopentane ring structure)	cyclopentene	
$R_2C\!=\!CR_2$ (general)	alkene	olefin

In common names, a double-bonded compound can be called an **olefin**, as a general term.

> That is, olefin is the common name equivalent to alkene. When "ethylene" gas was discovered in 1794, it was found to react with chlorine to form a water-insoluble oil. Hence it was called an olefiant gas (L. *oleum*, oil plus *-fiant*, making). Over the years the term evolved into the shorter olefin.

23-4 CIS-TRANS ISOMERS

When two carbon atoms are joined by a double bond, one carbon cannot rotate in relation to the other, under ordinary circumstances. An interesting consequence of this property is shown in Fig. 23-3; there are *two* different compounds corresponding to the name 2-butene! If the molecule has the methyl groups held on the same side, the compound is named *cis*-2-butene; if the two methyl groups are across from each other, the compound is *trans*-2-butene. These are two distinctly different substances, each of which has its own physical properties. The double bond acts as a rigid barrier to rotation, thus restricting a methyl group in *cis*-2-butene from flipping over to produce a molecule of *trans*-2-butene.

> Strictly speaking, one should not say that rotation is *impossible* at a carbon–carbon double bond. However, because it can occur only if a bond is broken, we have said it will not take place under ordinary circumstances. If we supply enough energy (heat, for instance) to a molecule of *cis*-2-butene the extra bond in the double bond can be

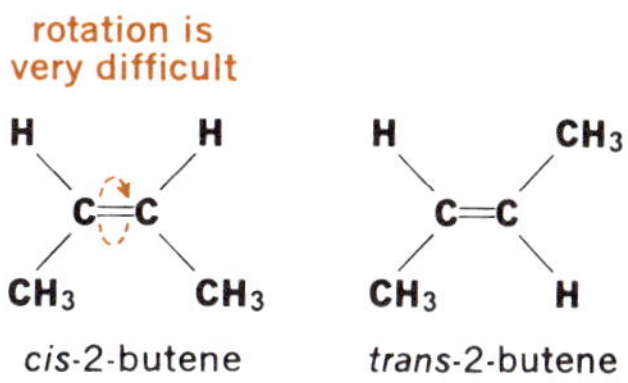

rotation is very difficult

H   H   H   CH₃
 \\   /   \\   /
  C=C   C=C
 /   \\   /   \\
CH₃   CH₃   CH₃   H

cis-2-butene   trans-2-butene

Fig. 23-3 The two isomers named 2-butene.

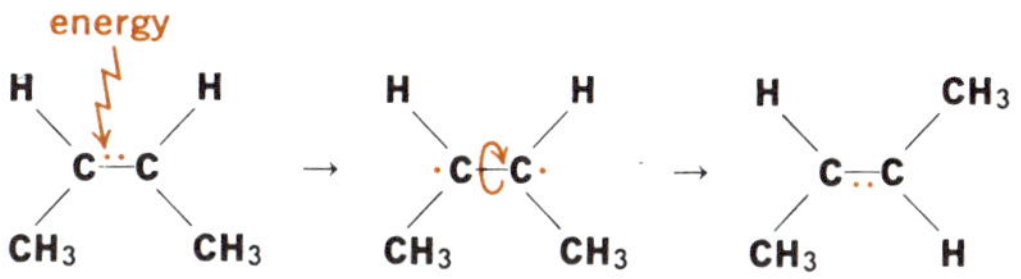

energy

H   H   H   H   H   CH₃
 \\ /   \\   /   \\   /
  C=C   →   ·C─C·   →   C=C
 / \\   /   \\   /   \\
CH₃ CH₃   CH₃   CH₃   CH₃   H

Fig. 23-4 Rotation at a double bond occurs only when a bond is broken by high energy.

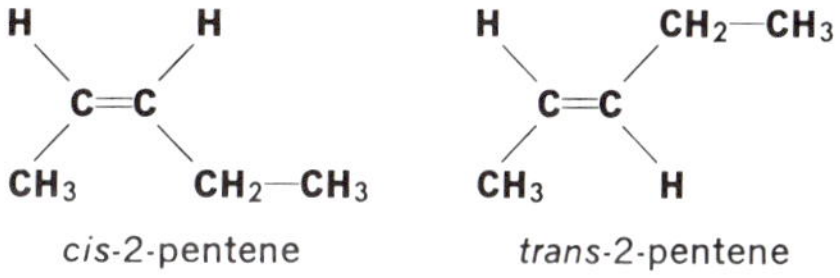

H   H   H   CH₂—CH₃   H   H
 \\   /   \\   /   \\   /
  C=C   C=C   C=C
 /   \\   /   \\   /   \\
CH₃   CH₂—CH₃   CH₃   H   H   CH₃

cis-2-pentene   trans-2-pentene

Fig. 23-5 Isomers of 2-pentene. 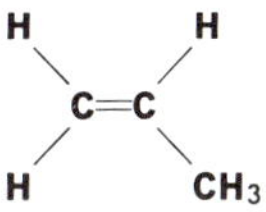   **Fig. 23-6** Propene, no *cis-trans* isomers.

---

broken, for it is somewhat weaker than a carbon–carbon single bond. The bond-breaking is essentially a matter of unpairing the electrons. Once this happens, the carbons can then rotate, on the remaining single bond, into the shape of *trans*-2-butene. When the extra bond forms again (i.e., the electrons pair off again, releasing energy to the surroundings), a molecule of the *trans* compound results (Fig. 23-4). This is just another example of a chemical reaction: a bond has been broken and a new bond has been formed, yielding a different substance.

Therefore, it is quite proper to consider *trans*-2-butene as a different compound than *cis*-2-butene, even though one can be changed *chemically* to the other. It is possible to isolate a sample of each of the two compounds. Each will retain its specific form indefinitely, unless we purposely decide to carry out a chemical reaction to change one to the other.

Cis-trans isomerism is possible whenever there are enough different groups bonded around a carbon–carbon double bond. For example, there are *cis* and *trans* isomers of 1,2-dichloroethene or of 2-pentene (Fig. 23-5). However, *cis-trans* isomers of propene are not possible because one of the carbons at the double-bond holds two hydrogens (Fig. 23-6). Consequently the methyl group, whichever side it is on, will always be across from a hydrogen and beside a hydrogen.

23-5 ALKENE REACTIONS

A. Introduction

The electrons in the "extra" bond of an alkene make it very reactive because they are loosely held and can be easily pulled into a different location to make new covalent bonds. Most reactions of alkenes fall into one category; they are **addition** reactions. Other atoms can be directly added to a double bond because valence electrons are available there: it is not necessary to displace atoms already present in order to make room for new atoms. (Recall that an alkane generally has to react by *substitution* because its valence bonds are already saturated with hydrogens.)

"extra" electrons readily available for reaction

Alkenes can also be attacked by common ionic oxidizing agents ($KMnO_4$, and so forth). This reaction, too, depends on the readily available electrons. Oxidation amounts to loss of electrons, or at least displacement of electrons toward a new atom.

In this chapter the properties of a carbon–carbon double bond will be illustrated with simple alkenes. But keep in mind that even when a large molecule contains other functional groups, a double bond reacts in its own characteristic way. This is why we can study organic compounds by examining the properties of the various units of structure separately, and later predict the behavior of complex molecules having many functional groups. A molecule of citronellol provides a simple illustration (Fig. 23–7). Each of the types of structure present (alkene, alkane, alcohol) will behave in its own distinctive manner, and the chemical properties of citronellol will be the sum of all these.

alkene reactions — alcohol reactions

$CH_3C\!=\!CHCH_2CH_2CHCH_2CH_2OH$
$\quad\ \ |\qquad\qquad\ \ \ |$
$\ \ CH_3\qquad\qquad CH_3$

alkane reactions

Fig. 23–7 Citronellol, 3,7-dimethyl-6-octene-1-ol.

B. Addition of Hydrogen (Hydrogenation)

When an alkene is treated with hydrogen gas, in the presence of certain metal catalysts, it is converted to an alkane; the process is called **hydrogenation**.

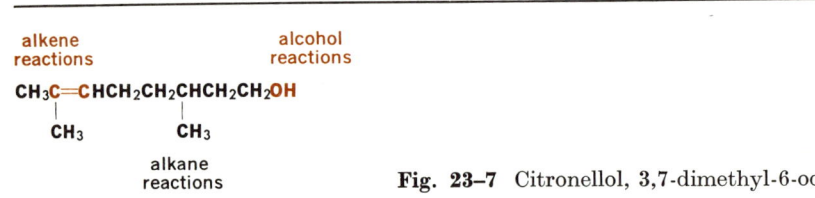

2-pentene   hydrogen   pentane   (23-2)

The metals most successful as catalysts for hydrogenation are nickel, palladium, and platinum. Reaction occurs when both the alkene and hydrogen are adsorbed onto the surface of the metal. Since the amount of surface is important, the metal is used in a finely divided form; in this condition it is a black powder resembling charcoal. The reaction shown in Eq. (23-2) is carried out by adding a very small amount of nickel powder to some liquid pentene in a bottle or tank. The rest of the vessel is filled with hydrogen gas, and the mixture is shaken to keep the nickel powder dispersed. After ten or fifteen minutes the reaction is complete.

Such a reaction would not be the most practical way to get a simple alkane such as pentane, which is easily obtained from petroleum. Furthermore, we learned in Section 23-2 that the reverse procedure is more common; that is, a simple alkene is generally obtained *from* an alkane. However, hydrogenation might be very useful if we wanted to convert a segment of alkene structure in a more complex molecule to a saturated segment. For instance, cottonseed oil can be converted to a fat (Section 27-4), or maleic acid to succinic acid (Eq. 23-3).

$$HO-\underset{\underset{O}{\|}}{C}-CH=CH-\underset{\underset{O}{\|}}{C}-OH + H_2 \xrightarrow{Pt} HO-\underset{\underset{O}{\|}}{C}-CH_2-CH_2-\underset{\underset{O}{\|}}{C}-OH \qquad (23\text{-}3)$$

maleic acid   succinic acid

C. Addition of Chlorine or Bromine

Chlorine or bromine adds readily to a carbon–carbon double bond.

$$CH_3-CH_2-CH_2-CH_2-CH=CH_2 + Cl_2 \rightarrow CH_3-CH_2-CH_2-CH_2-\underset{\underset{Cl}{|}}{CH}-\underset{\underset{Cl}{|}}{CH_2} \qquad (23\text{-}4)$$

1-hexene   1,2-dichlorohexane

Reaction (23-4) can be carried out by simply bubbling chlorine gas into the liquid 1-hexene; no catalyst or heat is required.

Since chlorine is one of the few substances which can react with an *alkane*, it is especially important to note the differences in behavior of an alkene and an alkane with chlorine. *The alkene undergoes an addition reaction* (Eq. 23-4) to give only one product. Ordinary conditions are sufficient. In contrast *an alkane reacts by substitution;* besides the organic products, HCl is always formed (Eq. 22-3). Drastic conditions are required: either ultraviolet light or a high temperature.

A reaction with bromine under ordinary conditions can be used as a simple laboratory test to detect the presence of unsaturation in a hydrocarbon.

$$R-CH=CH-R + Br_2 \rightarrow R-\underset{\underset{Br}{|}}{CH}-\underset{\underset{Br}{|}}{CH}-R \qquad (23\text{-}5)$$

an alkene   bromine   a dibromoalkane
(colorless)   (red-brown)   (colorless)

Bromine itself has a dark red-brown color. If it is used up by addition to an alkene, the color disappears. However, if even one or two drops of bromine is mixed with an alkane, the solution takes on and continues to have a red-brown color because the bromine remains unchanged. Consequently, an alkene can be distinguished from an alkane by testing with bromine.

It is convenient to use tetrachloromethane as a solvent for the test reaction, (Eq. 23–5), for several reasons. It is colorless and chemically inert to the bromine; it is a good solvent for all the substances involved; and very dilute solutions of bromine are sufficient. Thus we test a hydrocarbon by dissolving a drop of it in a few milliliters of tetrachloromethane in a test tube. To this we add, drop by drop, a solution containing 5% bromine in tetrachloromethane. If the color of the bromine solution disappears as fast as the drops enter the hydrocarbon solution, we assume an unsaturated hydrocarbon is present.

Although chlorine, too, reacts easily with an unsaturated compound, for practical reasons it is not used in a test. Chlorine is such a pale yellow-green color that in dilute solutions it is impossible to detect a color change. Also, chlorine is troublesome to handle because it is a gas.

D. Addition of Hydrogen Chloride

A number of acidic compounds will add to an alkene. We will mention the most common examples, one of which is hydrogen chloride.

$$CH_3-CH=CH_2 + HCl \quad \begin{array}{l} \nearrow CH_3-CH-CH_3 \\ | \\ Cl \\ \text{(the sole product)} \\ \\ \searrow CH_3-CH_2-CH_2 \\ | \\ Cl \\ \text{(none produced)} \end{array} \qquad (23\text{–}6)$$

It is apparent (Eq. 23–6) that hydrogen chloride could add to propene in two different ways. Curiously, only one product is formed. This fact was first reported in 1869 by the Russian chemist Markovnikov. Somewhat paraphrased, **Markovnikov's rule** *states that when an acid adds to an alkene, the hydrogen becomes attached to the carbon already bearing the most hydrogens.*

Markovnikov originally formulated this generality after observing the behavior of several alkenes. He had no idea *why* the results were thus. In 1932, Professor Frank Whitmore, of Pennsylvania State University, proposed a general theory which explains how a reaction such as (23–6) occurs, and hence why Markovnikov's rule is correct. Briefly, the explanation is as follows: A Cl^- ion would be repelled by the extra electrons in the alkene double bond. The alkene is first attacked by an H^+ ion which is seeking electrons. An approaching positive ion can distort the electron pair in the "extra" bond of the alkene, pulling it into position to form a new covalent bond.

$$\begin{array}{c} \text{H H H} \\ | | | \\ \text{H–C–C··C–H} \\ | \\ \text{H} \\ \\ \text{H}^+ \end{array} \quad \rightarrow \quad \begin{array}{c} \text{H H H} \\ | | | \\ \text{H–C–C–C–H} \\ + \\ \text{H} \text{H} \\ \text{(a carbonium ion)} \end{array} \qquad (23\text{–}7)$$

The intermediate structure produced in Eq. (23-7) is a carbonium ion. The possible existence of such ions was first suggested in 1922 by the German chemist Hans Meerwein. A carbonium ion is a highly reactive ion which promptly combines with a chloride ion, as shown in Eq. (23-8).

$$\begin{array}{c} \text{H} \;\; \text{H} \;\; \text{H} \\ | \;\;\; | \;\;\; | \\ \text{H—C—C—C—H} \\ | \;\; + \;\; | \\ \text{H} \;\;\;\;\;\; \text{H} \end{array} + \text{Cl}^- \rightarrow \begin{array}{c} \text{H} \;\; \text{H} \;\; \text{H} \\ | \;\;\; | \;\;\; | \\ \text{H—C—C—C—H} \\ | \;\;\; | \;\;\; | \\ \text{H} \;\; \text{Cl} \;\; \text{H} \end{array} \qquad (23\text{-}8)$$

In the first step, the H^+ *might* have combined with propene as shown in Eq. (23-9), producing a different carbonium ion.

$$\begin{array}{c} \text{H} \;\; \text{H} \;\; \text{H} \\ | \;\;\; | \;\;\; | \\ \text{H—C—C}\!=\!\text{C—H} \\ | \\ \text{H} \\ \;\; \\ \text{H}^+ \end{array} \rightarrow \begin{array}{c} \text{H} \;\; \text{H} \;\; \text{H} \\ | \;\;\; | \;\;\; | \\ \text{H—C—C—C—H} \\ \;\;\;\; | \;\; + | \\ \;\;\;\; \text{H} \;\;\; \text{H} \end{array} \qquad (23\text{-}9)$$

(a carbonium ion)

The nature of carbonium ions is such that the type produced in Eq. (23-7) is more stable and more easily formed than that shown in Eq. (23-9). In the competition, Eq. (23-7) wins, giving the results predicted by Markovnikov's rule.

During the years since Meerwein's first proposal and Whitmore's comprehensive theory it has become evident that carbonium ions are very common and important in certain organic reactions, especially those involving acids. As a result, chemists have studied carbonium ions intensively and have added significant details to Whitmore's original theory. Much research on the behavior of carbonium ions is still being conducted.

E. Addition of Sulfuric Acid

Sulfuric acid adds to an alkene, following Markovnikov's rule.

In order to understand the structures resulting from this addition reaction, we should first examine the structure of sulfuric acid. The formula H_2SO_4 lists the atoms but tells nothing of the manner in which they are bonded. Actually, the four oxygens are clustered around the sulfur, and the hydrogens are bonded to oxygens.

SULFURIC ACID

$$\begin{array}{c} \;\;\;\; :\!\ddot{\text{O}}\!: \\ \;\;\;\;\;\; \ddot{\;}\;\;\;\; \ddot{\;} \\ \text{H}:\!\ddot{\text{O}}\!:\!\text{S}\!:\!\ddot{\text{O}}\!:\!\text{H} \\ \;\;\;\;\;\; \ddot{\;}\;\;\;\; \ddot{\;} \\ \;\;\;\; :\!\ddot{\text{O}}\!: \end{array} \qquad \begin{array}{c} \;\;\;\; \text{O} \\ \;\;\;\; \uparrow \\ \text{H—O—S—O—H} \\ \;\;\;\; \downarrow \\ \;\;\;\; \text{O} \end{array} \qquad \text{HOSO}_2\text{OH}$$

electronic structure　　valence bond structure　　condensed structural formula

Hereafter, in writing organic reactions, we will often use the condensed structural formula of sulfuric acid.

The addition of sulfuric acid to a double bond is illustrated in Eq. (23–10).

$$CH_3\text{—}CH\!=\!CH_2 + HOSO_2OH \rightarrow \underset{\underset{OSO_2OH}{|}}{CH_3\text{—}CH\text{—}CH_3} \qquad (23\text{–}10)$$

propene   an alkyl hydrogen sulfate

The greatest use for the resulting alkyl hydrogen sulfate is to convert it to an alcohol. This is done by warming the compound in water:

$$\underset{\underset{OSO_2OH}{|}}{CH_3\text{—}CH\text{—}CH_3} + H\text{—}O\text{—}H \rightarrow \underset{\overset{OH}{|}}{CH_3\text{—}CH\text{—}CH_3} + HOSO_2OH \qquad (23\text{–}11)$$

2-propanol
(isopropyl alcohol)

Thus, reactions (23–10) and (23–11) constitute a two-step synthesis of an alcohol from an alkene.

F. Oxidation of Alkenes by Permanganate

An alkene is easily oxidized. It reacts, for example, with a dilute solution of potassium permanganate at room temperature.

$$3R\text{—}CH\!=\!CH\text{—}R + 2KMnO_4 + 4H_2O \rightarrow 3R\text{—}\underset{\overset{OH}{|}}{CH}\text{—}\underset{\overset{OH}{|}}{CH}\text{—}R + 2MnO_2\!\downarrow + 2KOH \qquad (23\text{–}12)$$

(purple)   (brown)

Reaction (23–12) is used chiefly as a test, which is possible because of the easily detected color changes. The permanganate is a *purple solution*, looking very much like grape juice. When it reacts, its color disappears, and we see in its place a *dark brown precipitate* of manganese dioxide. A positive reaction *suggests* the presence of unsaturation in the organic compound; as a check we should also try the bromine test (Section 23–5C). Certain organic compounds other than unsaturated hydrocarbons can react with potassium permanganate.

> If an alkene is treated with *hot, concentrated* potassium permanganate solution, a more drastic change occurs. The alkene is split apart at the site of the double bond.
>
> $$CH_3\text{—}CH_2\text{—}\underset{\overset{CH_3}{|}}{C}\!=\!CH\text{—}CH_3 + KMnO_4 + H_2O \rightarrow CH_3\text{—}CH_2\text{—}\underset{\overset{CH_3}{|}}{C}\!=\!O + \underset{O}{\overset{KO}{\diagdown}\!C\text{—}CH_3} + MnO_2 + KOH$$
>
>   (a ketone)   (a salt)
>
> The fragments thus obtained are often well-known compounds which can be identified rather easily. If they are, we can deduce the structure of the original alkene, knowing that the functional groups in the fragments arose from the double bond.

Until about the time of World War II, oxidative cleavage was one of the most fruitful techniques employed by chemists to determine the structure of an organic compound. Now there are analytical instruments which reveal much information about a structure with less labor by the chemist. The instruments not only provide answers to simple problems rapidly, but enable the chemist to attack more complex problems such as the structure of antibiotics, vitamin B_{12}, or insulin.

G. Combustion

If a mixture of an alkene and oxygen is ignited, combustion occurs.

$CH_3CH_2CH=CH_2 + 6O_2 \rightarrow 4CO_2 + 4H_2O$

This behavior is not unique to alkenes. In fact, alkanes and practically all other organic compounds will also burn (Section 22–5). Therefore, in the remaining chapters, when we discuss each new series of compounds, we will not bother to mention that each one is combustible.

23-6 ALKYNES

Unsaturated hydrocarbons having a *triple* bond between two carbons are called **alkynes**. The ending -*yne* is used in a systematic name; the rest of the naming rules are the same as those for alkenes.

$CH_3C \equiv CCH_3$

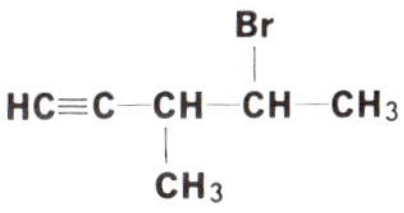

2-butyne

$HC \equiv C - CH(CH_3) - CH(Br) - CH_3$

4-bromo-3-methyl-1-pentyne

Chemical reactions of alkynes are similar to those of alkenes, the chief difference being that the alkyne has *two* "extra" bonds which can react.

$CH_3-CH_2-C \equiv CH + 2H_2 \xrightarrow{Ni} CH_3-CH_2-CH_2-CH_3$

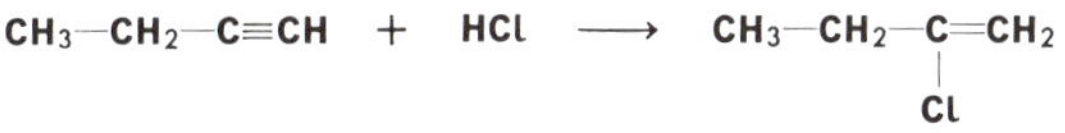

$CH_3-CH_2-C \equiv CH + HCl \longrightarrow CH_3-CH_2-C(Cl)=CH_2$ **Markovnikov rule applies**
(1 mole)   (1 mole)

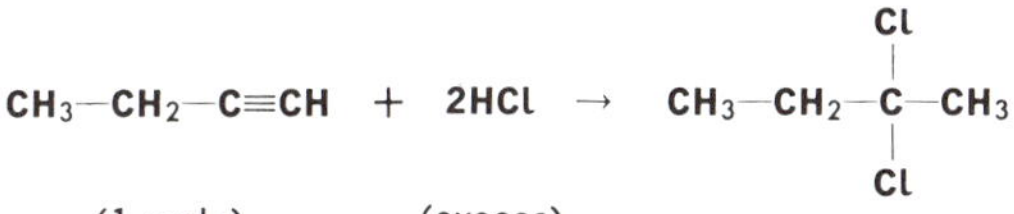

$CH_3-CH_2-C \equiv CH + 2HCl \rightarrow CH_3-CH_2-CCl_2-CH_3$
(1 mole)   (excess)

Alkynes also give positive results with the bromine test (Section 23–5C) and the permanganate test (Section 23–5F). The tests, therefore, cannot be used to distinguish an alkene from an alkyne. Instead, they are general for an unsaturated compound.

The structural formulas we have written for alkynes show quite accurately that each of the bond angles at a triple bond is 180°.

$$CH_3-C\overset{180°}{\equiv}C-CH_3$$

This angle illustrates, once again, the principle that the groups bonded to one particular carbon will get as far away from each other as possible.

23-7 ACETYLENE (Ethyne)

Ethyne, HC≡CH, requires some extra attention because of its tremendous practical importance. It is used widely and therefore is nearly always called by its common name, acetylene. The production of acetylene in the United States is now well over one billion pounds a year. The oldest method of synthesis involves the following reactions:

$$3C + CaO \xrightarrow{2500°C} CaC_2 + CO \qquad (23\text{-}13)$$
coke   lime   calcium   carbon
                         carbide   monoxide

$$(^-C\equiv C^-)Ca^{2+} + 2HOH \rightarrow HC\equiv CH + Ca(OH)_2 \qquad (23\text{-}14)$$
calcium   acetylene
carbide

In an electric furnace, coke and lime are converted to calcium carbide, a solid (Eq. 23–13). It is essentially a salt, as shown in Eq. (23–14). Treatment of calcium carbide with water replaces the calcium ions with hydrogens, yielding acetylene.

Slightly less than half of the acetylene manufactured in the United States now comes from a different raw material, natural gas (methane). Exposure to a very high temperature for less than one-tenth of a second converts the methane to acetylene.

$$2CH_4 \xrightarrow{1600°} HC\equiv CH + 3H_2 \qquad (23\text{-}15)$$

Combustion of acetylene, especially in the presence of oxygen rather than air, produces a very hot flame (near 2800°C). About 15% of the domestic production of acetylene is used in oxyacetylene torches for cutting and welding steel.

The rest of the acetylene goes into the synthesis of other organic compounds. The equations below show the most important synthetic uses of acetylene. Note that each one starts with an addition to the unsaturated bond of acetylene.

$$HCl + HC\equiv CH \xrightarrow[200°C]{HgCl} CH_2\!=\!\underset{Cl}{CH} \rightarrow \left(\!CH_2CH\!\atop Cl\right)_{\!n} \qquad (23\text{-}16)$$

vinyl   polyvinylchloride
chloride   (Table 30–1)
                        ↓
                   vinyl plastics

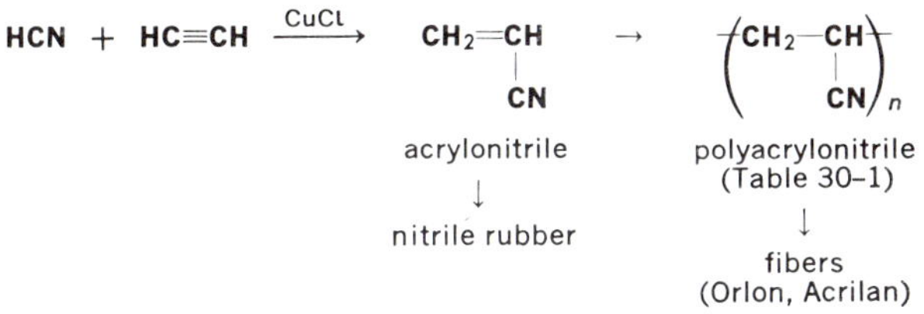

$$HCN + HC\equiv CH \xrightarrow{CuCl} \underset{\underset{CN}{|}}{CH_2=CH} \rightarrow \left(\underset{\underset{CN}{|}}{CH_2-CH}\right)_n \quad (23\text{-}17)$$

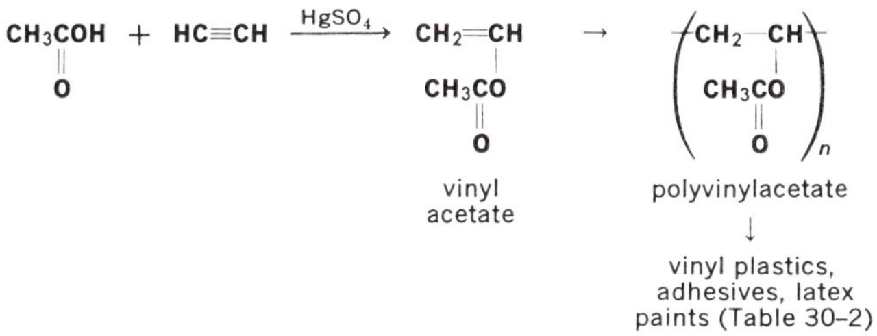

acrylonitrile   polyacrylonitrile
↓   (Table 30-1)
nitrile rubber   ↓
    fibers
    (Orlon, Acrilan)

$$\underset{\underset{O}{\|}}{CH_3COH} + HC\equiv CH \xrightarrow{HgSO_4} \underset{\underset{\underset{O}{\|}}{CH_3CO}}{CH_2=CH} \rightarrow \left(\underset{\underset{\underset{O}{\|}}{CH_3CO}}{CH_2-CH}\right)_n \quad (23\text{-}18)$$

vinyl acetate   polyvinylacetate
↓
vinyl plastics,
adhesives, latex
paints (Table 30-2)

$$HC\equiv CH + HC\equiv CH \xrightarrow{CuCl} CH_2=CH-C\equiv CH \quad (23\text{-}19)$$

(self addition)   vinylacetylene

↓ HCl

$$CH_2=CH-\underset{\underset{Cl}{|}}{C}=CH_2 \rightarrow \text{neoprene rubber (Section 30-3)}$$

chloroprene

(handwritten annotation: $CH_2=CH$) vinyl group)

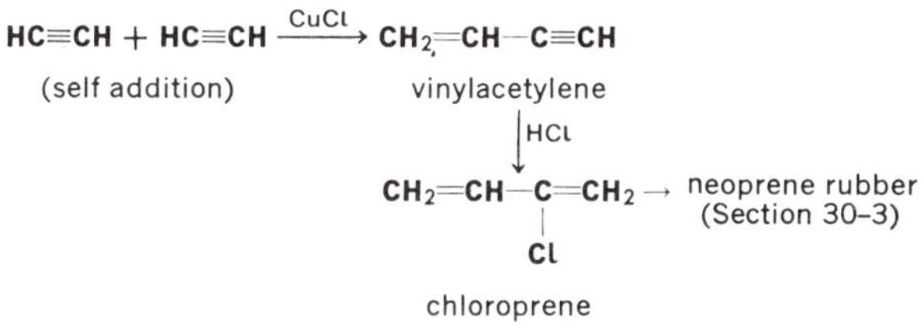

IMPORTANT TERMS

Addition reaction	Catalyst	Markovnikov rule
Alkene	Cis-trans isomers	Olefin
Alkyl hydrogen sulfate	Cracking reaction	Oxidation
Alkyne	Hydrogenation	Unsaturated hydrocarbon

WORK EXERCISES

1. Write the condensed structural formula and molecular formula for the following:

    a) hexane, 1-hexene, and cyclohexane   b) heptane, 2-heptene, and cycloheptane
    c) 2-methylpentane; 2,3-dimethyl-2-butene; methylcyclopentane

2. Which compounds in Problem 1 are isomers? What generalization can you make about the relation between alkenes and cycloalkanes?

3. Write the structural formulas of all possible open-chain compounds (i.e., those whose carbon skeletons are not rings) that have the following molecular formulas:

    a) C_4H_8 b) C_5H_{10}

4. Write the structural formulas of all possible ring compounds having molecular formulas:

    a) C_4H_8 b) C_5H_{10}

5. In view of your answers to Problems 3 and 4, give the total number of isomers of
   a) C_4H_8 b) C_5H_{10}

6. Write a systematic name for each compound in Problem 3.

7. Write the condensed structural formula for each of the following:
   a) 3-chloro-1-butene
   b) *trans*-2-butene
   c) cyclohexene
   d) 3-hexyne
   e) *cis*-4-methyl-2-pentene
   f) 1-bromo-2-butyne

8. Briefly describe the principal method for obtaining alkenes in large quantities. What is the raw material? What processes and reactions are used to obtain an alkene from it?

9. Why are alkenes more reactive than alkanes?

10. Write a complete, balanced equation for the reaction occurring in each mixture below, using structural formulas. If necessary, state "no reaction."
    a) 1-butene, hydrogen, platinum
    b) propene, bromine, 25°C
    c) 2-pentene burning in air
    d) 1-butene, hydrogen chloride
    e) cyclopentene, chlorine, 25°C
    f) cyclohexene, hydrogen, nickel
    g) 2-butene, concentrated sulfuric acid
    h) 1 mole of ethyne, 1 mole of hydrogen chloride
    i) propene, hydrogen chloride
    j) propene, chlorine
    k) propyne, excess bromine
    l) 1 mole propyne, 2 moles hydrogen chloride
    m) propyne burning in air

11. Which of the following would react with dilute permanganate solution at 25°C, causing its purple color to be replaced by a brown precipitate of MnO_2?
    a) propene
    b) butane
    c) cyclopentane
    d) butyne
    e) 2-butene
    f) cyclohexene
    g) 3-methyl-1-pentene
    h) 2-chlorobutane

12. Show structural formulas of at least six compounds, not counting *cis-trans* isomers, which would result from the cracking of pentane at 500°C.

13. Consider the synthesis of chloroethane from ethane, on an industrial scale.
    a) What would be the source of the ethane?
    b) Write an equation to show how ethane could be converted to chloroethane in one step; i.e., by one reaction.
    c) Write equations to show how ethane could be converted to chloroethane by two steps.
    d) Suggest a reason why method (c) might be more satisfactory for producing pure chloroethane than method (b), despite the fact that (c) requires two steps.

SUGGESTED READING

DePuy, C. H., and K. L. Rinehart, Jr., *Introduction to Organic Chemistry*, Chapter 3, John Wiley, New York, 1967.

Hart, H., and R. D. Schuetz, *Organic Chemistry*, 4th ed., Chapter 3, Houghton Mifflin, Boston, 1972.

Jones, J., "The Markovnikov Rule," *J. Chem. Educ.* **38,** 297 (1961). A fascinating account of the development of a scientific "law." Includes many details and special examples not described in this book.

Noller, C. R., *Textbook of Organic Chemistry*, 3rd ed., Chapter 6, W. B. Saunders, Philadelphia, 1966.

AROMATIC HYDROCARBONS: Benzene and Related Compounds

24

24–1 TYPE OF STRUCTURE

Chemists long ago became intrigued by the fragrant oils produced by certain plants, such as clove oil, wintergreen oil, and almond oil. The compounds seemed to share so many similar properties that chemists thought of them as a particular class of substances. They became known as *aromatic* compounds because of their odors.

One simple aromatic compound is **toluene,** C_7H_8, obtained by heating the balsam formed in the bark of the South American *tolu* tree. Another is produced when the fragrant *benzoin gum* is decomposed to yield a volatile liquid, of formula C_6H_6, which was named **benzene.** Gradually chemists recognized that the compounds in their "aromatic" class all had a low ratio of hydrogens to carbons, had at least six carbons, and were apparently related to benzene. They also soon learned that "aromatic" was not a very accurate description. Many of the compounds which obviously belong in this class because of their structures and properties do *not* have odors; conversely, many substances having entirely different types of structure *are* fragrant. Although the old name has persisted, we now define an **aromatic compound** *as one having a characteristic structure like that of benzene*.

Clearly, it is important to learn *what* about the structure of benzene is so distinctive. For several decades chemists were baffled about how to write the structure of this compound. We shall not have time to review all the experimental observations and the numerous attempts to write structures consistent with the facts. Instead, we shall present the formulas which now seem to be the best representation of the structure of benzene.

Benzene, C_6H_6, has its six carbons joined in a ring, with one hydrogen on each carbon. This would require only three of the four valence electrons on each carbon, leaving one electron on each unshared, as shown in formula (a). It might seem then that an unshared electron could pair off with a similar one from an adjacent carbon, forming a structure such as (b) or (c).

```
      H                H                H
      |                |                |
      C                C                C
     / \              // \              / \\
H—C·   ·C—H      H—C    C—H        H—C    C—H
     ‖                                    
H—C·   ·C—H      H—C    C—H        H—C    C—H
     \ /              \  //             \\ /
      C                C                C
      |                |                |
      H                H                H
     (a)              (b)              (c)
```

Formulas (b) and (c) raise some annoying questions, the most puzzling of which is the fact that *benzene does not have the chemical properties of an alkene*. We said that when a compound has a carbon–carbon double bond, and when we write this in its formula, we expect it to behave like a typical alkene. Since benzene does not behave in this way, neither formula (b) nor (c) seems to represent its true structure.

According to modern theory we believe that the "extra" electrons in a benzene ring do *not* pair off in specific locations, where they would be available to react as alkene double bonds. Instead, each electron interacts with the other unpaired electrons on *both* sides of it. The result is a continuous doughnut-shaped *electron-cloud* around the carbon skeleton of the benzene ring. All we know is that somewhere in the cloud there are six electrons. This is illustrated in the perspective diagram of Fig. 24-1. The ordinary covalent bonds forming the skeleton of the molecule are shown with dashed black lines; the electron cloud is in color.

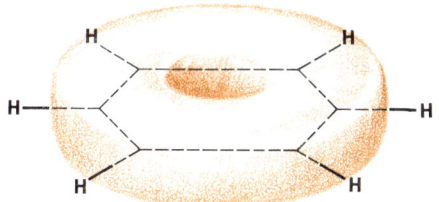

Fig. 24-1 Electronic structure of benzene. The cloud represents the space occupied by six electrons.

It is possible, by means of complicated theoretical calculations, to show that this sort of electron cloud makes the molecule especially stable. It is stable because the charge of the electrons does not have to be concentrated in certain locations. We say the electrons are "delocalized," or spread out through the cloud. *Pairs* of electrons are not accessible for the type of addition reaction which is so easy for an alkene; nor are the electrons commonly attacked by an oxidizing agent. Consequently, we consider that benzene is not an unsaturated hydrocarbon, despite the fact that it has very few hydrogens and does have some "extra" electrons.

Although Fig. 24-1 gives a satisfactory picture of the distribution of the electrons in benzene, we are still left with the problem of how to write a simple formula using our standard symbols. It is inconvenient to draw a detailed picture such as Fig. 24–1 each time we want to write a chemical equation for benzene. Organic chemists now often use formula (d) or (e), in which the circle represents the continuous cloud of six electrons.

(d) (e) (f)

Formula (e) is the abbreviated style, in which we assume there is one carbon at each corner and each is holding one hydrogen. (Compare this formula with that for cyclohexane, Fig. 21–20.) In this book we shall use formula (e) for benzene. Another style, still found in many books, is to use a formula such as (c) or its abbreviated form (f.) The disadvantage of formula (c) or (f) is that you must remember that benzene does not *really* have double bonds. The three alternating "extra" bonds shown in (f) must be thought of as representing a total of six electrons held in a special way.

24–2 NAMES

A compound having an alkyl group attached to a benzene ring is named by prefixing the alkyl group name to the word benzene. Methylbenzene, however, is ordinarily called toluene.

benzene   toluene (methylbenzene)   ethylbenzene   n-butylbenzene

The hydrocarbon names are used as the basis for naming other compounds. If two or more groups are attached to a ring, their relative positions are indicated by numbers. Since there is no "end" to the ring, we start counting at one of the groups.

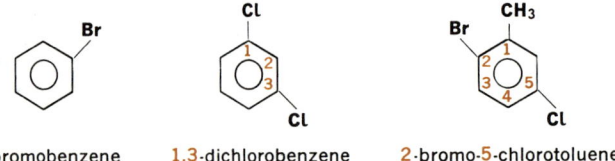

bromobenzene   1,3-dichlorobenzene   2-bromo-5-chlorotoluene

If *only* two groups are present, their relative positions can be designated by the terms *ortho*, *meta*, or *para*, abbreviated *o-*, *m-*, or *p-*:

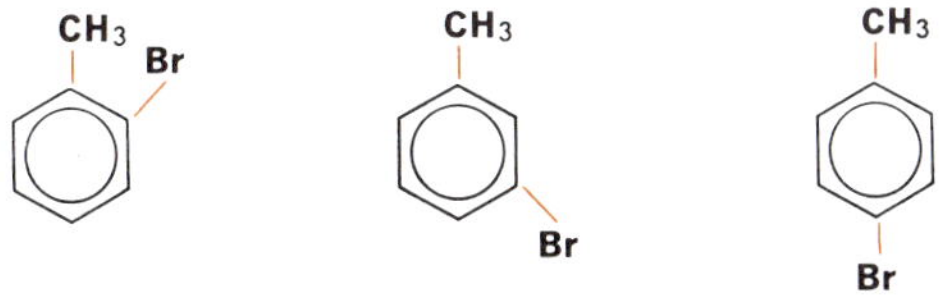

o-bromotoluene   m-bromotoluene   p-bromotoluene

It is necessary to realize that, for benzene structures, the distance between groups is the important factor in assigning names. For example, all the formulas below represent the same compound, even though the molecules are shown in different positions.

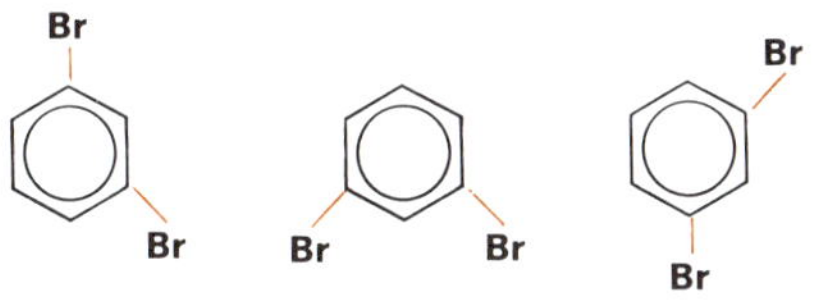

1,3-dibromobenzene or *m*-dibromobenzene

It is possible for a compound to be built from two or more fused benzene rings.

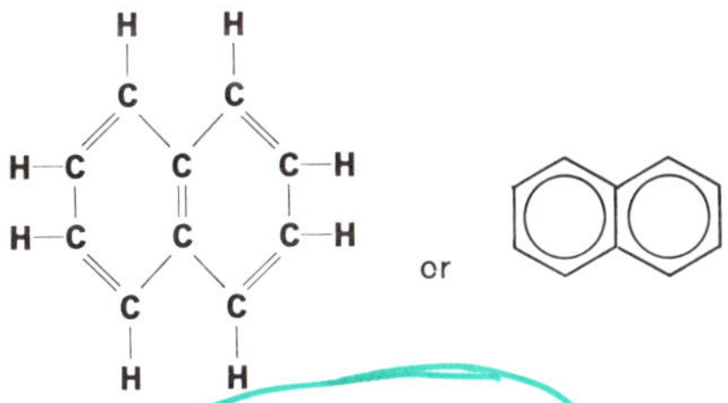

naphthalene, $C_{10}H_8$

The numbering system for naphthalene derivatives is shown below. In case only one group is present, the positions are sometimes designated by the Greek letters α and β.

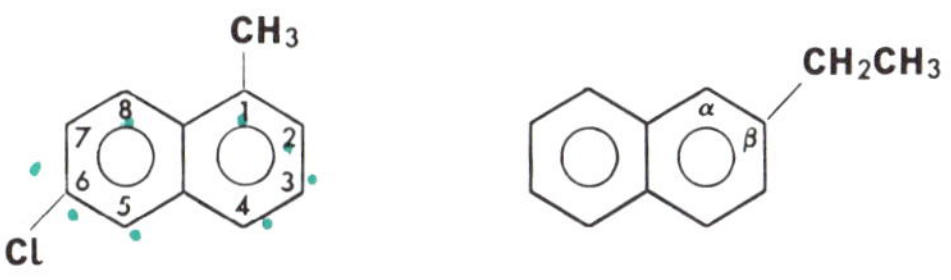

6-chloro-1-methylnaphthalene

β-ethylnaphthalene
2-ethylnaphthalene

The skeletons of a few of the other possible fused-ring aromatic hydrocarbons are shown below.

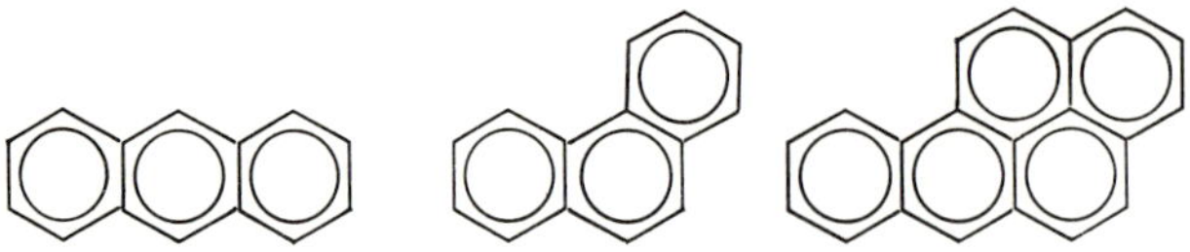

Some of the compounds composed of several rings induce the growth of cancer. One example is the last compound shown just above. Such compounds occur in coal tar and in cigarette tar.

In the examples of aromatic compounds shown so far, names have been written by naming the groups attached to the aromatic ring. Sometimes it is not convenient to do this, and it is easier to think of the ring as a group attached to some other carbon skeleton. In this situation, **phenyl** is the name used for the *group* obtained by removing one hydrogen from a benzene ring.

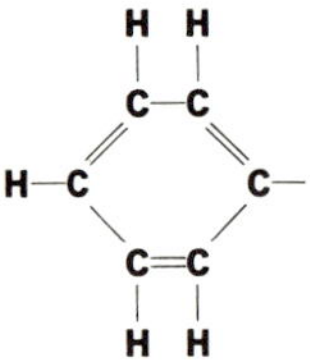

phenyl group

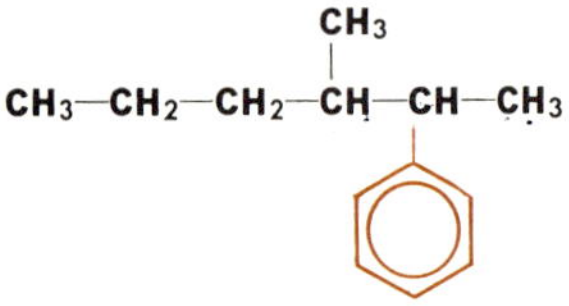

3-methyl-2-phenylhexane

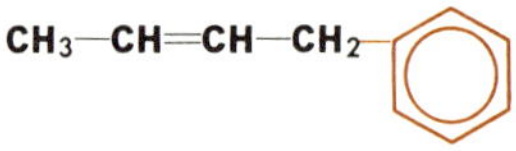

1-phenyl-2-butene

Here is a case of an illogical mixture of names surviving from the past. During the 1840's the compound now known as benzene was called *phene*. Hence it was reasonable to call the related group *phenyl*. The name benzene for the parent compound gradually became accepted, so phene unfortunately died out, but phenyl did not! The name phene had been assigned to the compound because it and other aromatic compounds were associated with the manufacture of a type of illuminating gas (Gr. *phaino-*, shining).

In general, any group of this type is called an **aryl** group; the structures below are illustrations.

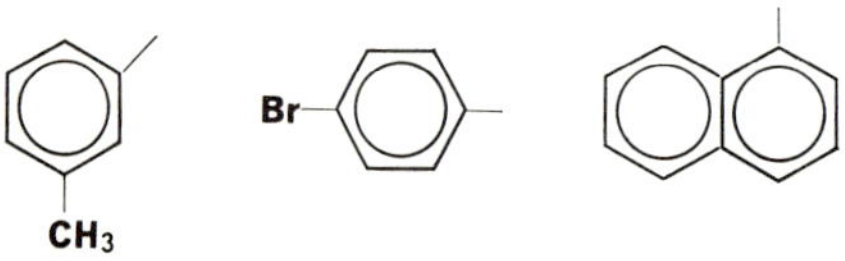

The symbol Ar- is used for an aryl group. Thus ArBr is an aryl bromide and ArH is an aromatic hydrocarbon.

24-3 GENERAL CHEMICAL PROPERTIES

We discovered in Section 24–1 that because benzene does not exhibit any of the customary alkene reactions it is necessary to develop a model, or picture, of benzene which truthfully represents these facts. Therefore we adopted the symbol (e) to portray benzene, with the understanding that the circle indicates the six "extra" electrons which are held in a unique way. The concept to keep in mind is that the six electrons are very stable in the cloud around benzene. The cloud can rarely be broken up in such a way as to make the electrons available for addition reactions.

The following list shows several of the reactions in which benzene does *not* participate.

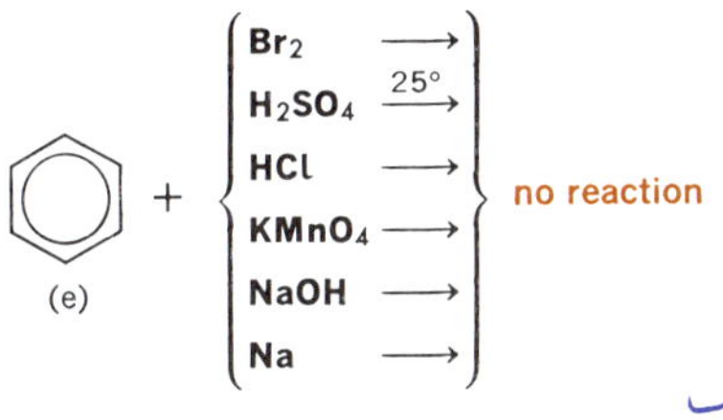

(e) + { Br_2 →, H_2SO_4 $\xrightarrow{25°}$, HCl →, $KMnO_4$ →, NaOH →, Na → } no reaction   *imp* *

There are, however, a number of substances which react quite easily with benzene. Since addition reactions involving the electrons are rarely possible, apparently the only other parts of the molecule available for reaction are the hydrogen atoms. This is exactly the case. *The reactions typical of a benzene ring are substitutions, in which a new group replaces a hydrogen.* These substitutions are distinctly different from alkane substitutions. Neither a high temperature nor ultraviolet light is necessary for benzene substitutions. If the proper catalysts are provided, the reactions occur easily at moderate temperatures. Three important examples are described in the following sections.

24-4 REACTION WITH CHLORINE OR BROMINE

If we add some bromine to benzene, the bromine dissolves nicely in the liquid. However, no chemical change takes place; this was represented by the first equation of Section 24–3. If we then add to the solution a small amount of iron (a tack will do, or a chip of iron from a machine shop), some striking changes occur promptly. The solution becomes warm. Gas bubbles out and turns to white fumes when it reaches the moist air outside the flask. These fumes have a sharp, acid taste and odor, and they turn litmus red; more thorough analysis proves that they consist of hydrogen bromide. If we also analyze the liquid remaining in the flask we find bromobenzene, C_6H_5Br. Clearly, a chemical reaction has occurred which was apparently catalyzed by the iron. The products suggest that a bromine atom was *substituted* in place of a hydrogen on the benzene ring. We can write the reaction as shown in Eq. (24–1).

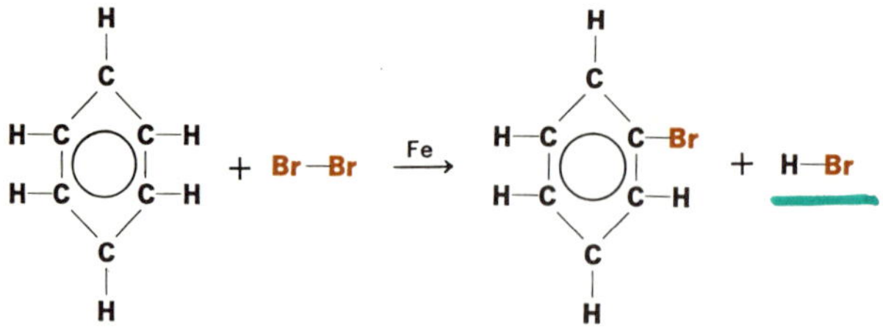

$$C_6H_6 + Br-Br \xrightarrow{Fe} C_6H_5Br + H-Br \quad \text{*imp.} \quad (24\text{–}1)$$

benzene → bromobenzene

Chlorine can react in the same way.

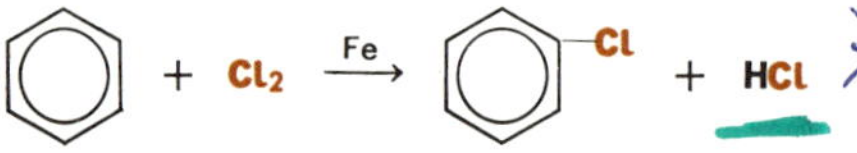

$$C_6H_6 + Cl_2 \xrightarrow{Fe} C_6H_5Cl + HCl \quad \text{*imp.} \quad (24\text{–}2)$$

Equation 24–2 is actually balanced; remember that a hydrogen is present at each corner of the benzene ring even though it is not shown in the abbreviated formula.

Reactions of this type provide a simple means to obtain aryl chlorides or bromides.

The unique behavior of benzene is well illustrated by its reaction with chlorine or bromine. As we have noted, the reactions occur by substitution, not addition. Furthermore, the substitution does not require ultraviolet light or high temperature (which is necessary for halogen substitution onto an alkane); instead, the reaction is catalyzed by another chemical substance, in this case, iron.

24–5 REACTION WITH SULFURIC ACID

Concentrated sulfuric acid does not react with benzene at room temperature; in fact, the two liquids are immiscible. However, if the mixture is heated to a moderate temperature (130–150°C) a substitution reaction, called **sulfonation**, occurs.

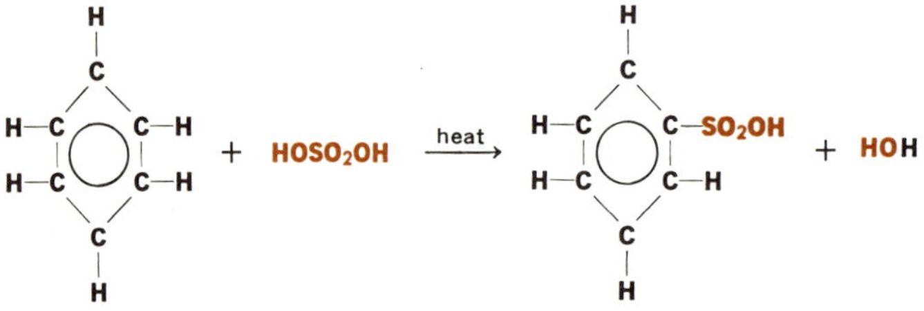

$$C_6H_6 + HOSO_2OH \xrightarrow{heat} C_6H_5\text{–}SO_2OH + HOH \quad (24\text{–}3)$$

benzenesulfonic acid

Sulfonation, like the other benzene substitutions, requires a catalyst, despite the fact that in Eq. (24–3) none is apparent. Chemists have discovered that the reaction is self-catalyzed; one molecule of sulfuric acid promotes the reaction of another molecule.

Observe carefully the structure of the **benzenesulfonic acid.** There is a bond from a carbon to the sulfur (*not* to an oxygen). The three remaining oxygen atoms are all clustered around the sulfur. Review the structure of sulfuric acid discussed in Section 23–5E.

The structure of the sulfonic acid group is the same as that of sulfuric acid, except for the organic portion. Therefore, many of the properties of benzenesulfonic acid are similar to those of sulfuric acid. Benzenesulfonic acid is very soluble in water and is a highly ionized, strong acid (Eq. 24–4).

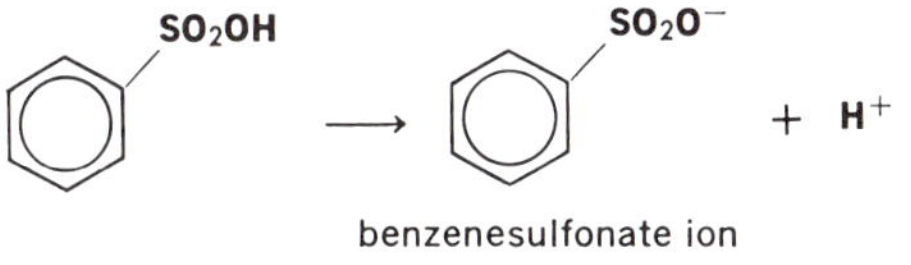

$$Ph\text{-}SO_2OH \longrightarrow Ph\text{-}SO_2O^- + H^+ \quad (24\text{–}4)$$

benzenesulfonate ion

It readily neutralizes bases (Eq. 24–5), and reacts with carbonates (Eq. 24–6).

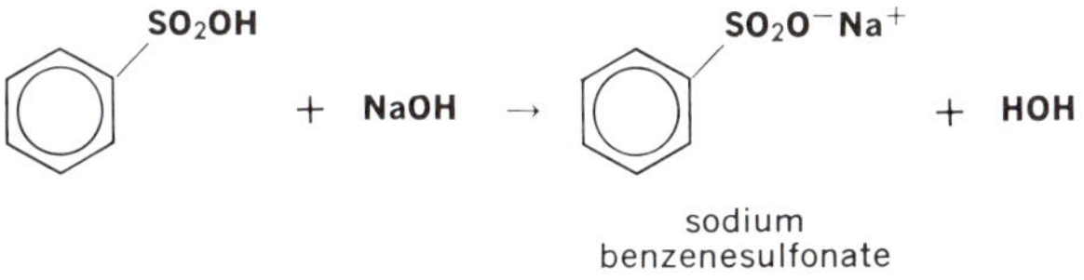

$$Ph\text{-}SO_2OH + NaOH \longrightarrow Ph\text{-}SO_2O^-Na^+ + HOH \quad (24\text{–}5)$$

sodium benzenesulfonate

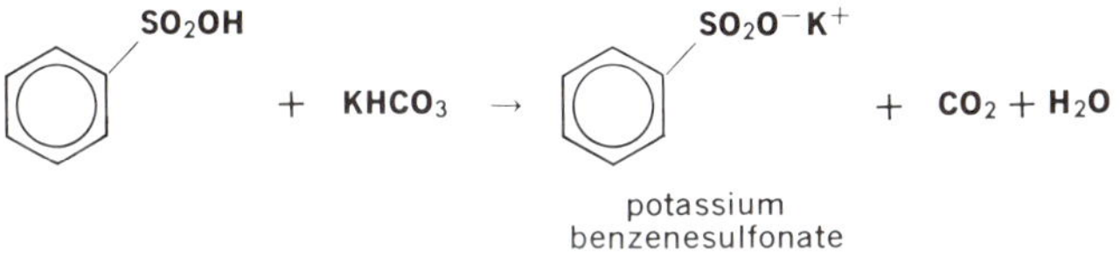

$$Ph\text{-}SO_2OH + KHCO_3 \longrightarrow Ph\text{-}SO_2O^-K^+ + CO_2 + H_2O \quad (24\text{–}6)$$

potassium benzenesulfonate

24–6 REACTION WITH NITRIC ACID

If benzene is treated with a solution containing both concentrated nitric acid and concentrated sulfuric acid, substitution begins even at room temperature. A **nitro group** from nitric acid is substituted onto the ring; the process is termed **nitration.**

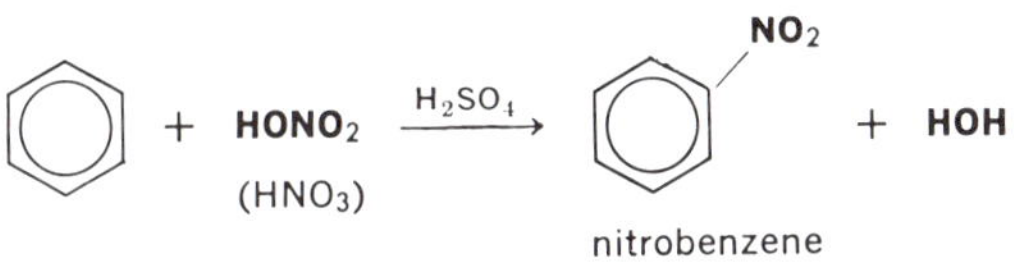

$$Ph\text{-}H + HONO_2 \xrightarrow{H_2SO_4} Ph\text{-}NO_2 + HOH \quad (24\text{–}7)$$
$$(HNO_3)$$

nitrobenzene

The sulfuric acid is written over the arrow in Eq. (24–7) because it is the catalyst. Although sulfuric acid is important to promote the reaction, substitution by nitro groups is much easier than by sulfonic acid groups; the nitrobenzene will *not* be contaminated with any benzenesulfonic acid. In this connection, note that sulfonation (Eq. 24–3) requires a higher temperature than does nitration (Eq. 24–7).

The nitration reaction is used to prepare nitro compounds and is also one of the most frequently used methods for introducing other groups, indirectly, onto an aromatic ring. It will be shown in Section 28–3B that an aryl nitro compound (ArNO$_2$) can easily be converted to an aryl amine (ArNH$_2$). From the amine a great variety of other compounds can be made.

Compounds containing several nitro groups have explosive properties. Equation (24–8) shows how TNT can be made. Persons inexperienced in chemical procedures should not attempt this reaction because of the great danger involved.

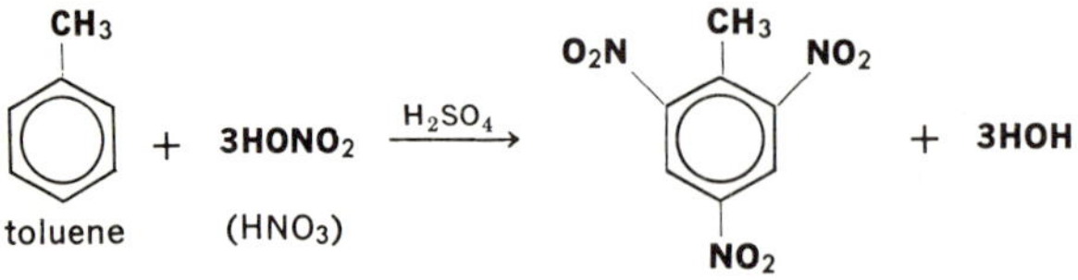

$$\text{toluene} + 3\text{HONO}_2 \xrightarrow{\text{H}_2\text{SO}_4} \text{2,4,6-trinitrotoluene (TNT)} + 3\text{HOH} \qquad (24\text{–}8)$$
(HNO$_3$)

24–7 OXIDATION OF SIDE CHAINS

An alkyl group attached to an aromatic ring is often called a **side chain**. Although the saturated hydrocarbon structure found in an alkyl group normally resists oxidation, an aromatic ring affects a side chain in such a way that it *can* be oxidized.

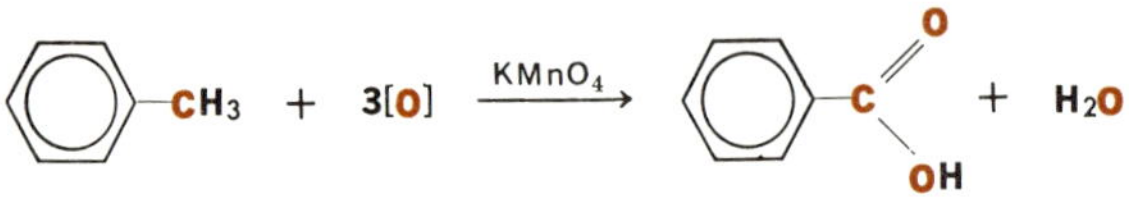

$$\text{C}_6\text{H}_5\text{–CH}_3 + 3[\text{O}] \xrightarrow{\text{KMnO}_4} \text{C}_6\text{H}_5\text{–COOH} + \text{H}_2\text{O} \qquad (24\text{–}9)$$
benzoic acid
(an organic acid)

It is often convenient to write an organic oxidation reaction as in Eq. (24–9). The oxygen in brackets indicates oxidation, but does not mean that O$_2$ is the oxidizing agent. In effect, the oxygen is supplied by the inorganic oxidizing agent, which is written above the arrow. We do not attempt to show the change in the oxidizing agent nor include it in the balanced part of the equation; instead, we focus our attention on the changes in the organic substances. The equation is balanced as far as the oxygen and organic compounds are concerned. Actually, the KMnO$_4$ taking part in reaction (24–9) would be reduced to MnO$_2$, since the toluene is being oxidized.

If there is more than one side chain attached to a benzene ring, all of them can be oxidized. Also, other oxidizing agents can be used, as shown in Eq. (24–10).

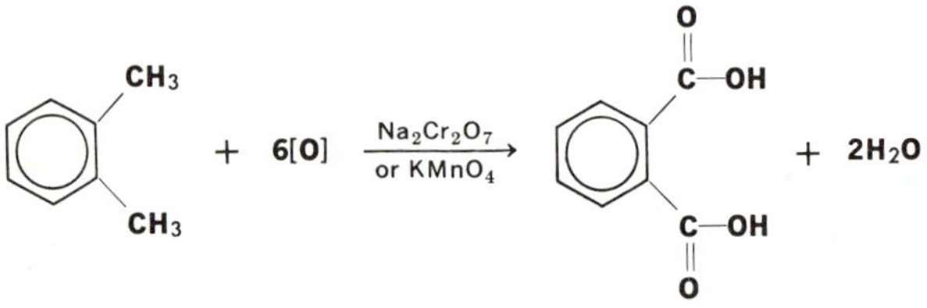

$$\text{o-C}_6\text{H}_4(\text{CH}_3)_2 + 6[\text{O}] \xrightarrow{\text{Na}_2\text{Cr}_2\text{O}_7 \text{ or KMnO}_4} \text{o-C}_6\text{H}_4(\text{COOH})_2 + 2\text{H}_2\text{O} \qquad (24\text{–}10)$$
phthalic acid

Longer side chains can likewise be oxidized (Eq. 24–11). In each case the oxidation starts at the carbon attached to the ring, emphasizing the fact that it is the ring which makes the alkyl side chain sensitive to attack. Once the oxidation begins, the rest of the carbons in the side chain are, as a rule, completely oxidized to carbon dioxide.

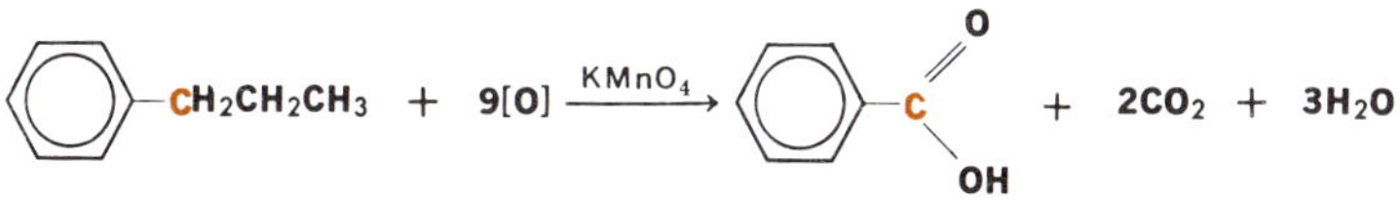

$$\text{C}_6\text{H}_5\text{-CH}_2\text{CH}_2\text{CH}_3 + 9[\text{O}] \xrightarrow{\text{KMnO}_4} \text{C}_6\text{H}_5\text{-COOH} + 2\text{CO}_2 + 3\text{H}_2\text{O} \quad (24\text{-}11)$$

Side-chain oxidation is useful for the preparation of an aromatic acid and for the identification of the original hydrocarbon.

24–8 HETEROCYCLIC COMPOUNDS

Many important compounds, both natural and synthetic, have ring structures in which there are some atoms other than carbons; hence they are called **heterocyclic** compounds (Gr. *hetero-*, other, different). The *hetero* atoms commonly found are nitrogen, sulfur, or oxygen. The ring may be saturated, unsaturated, or aromatic. Although heterocyclic compounds will be encountered occasionally later in the book, we will not attempt to discuss them in detail. It is important, however, to be aware that structures of this kind exist. Examples are shown below.

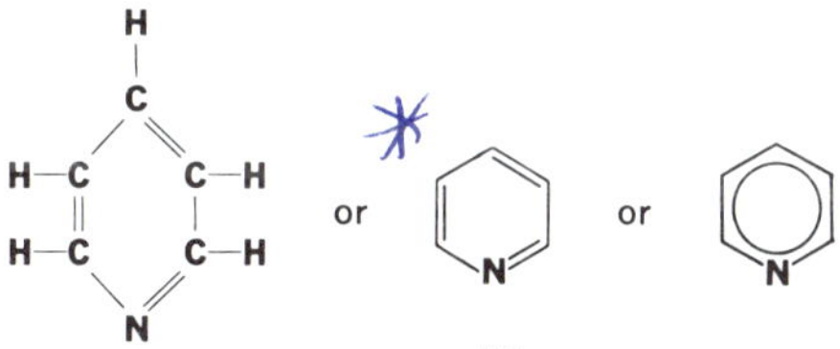

pyridine

quinoline

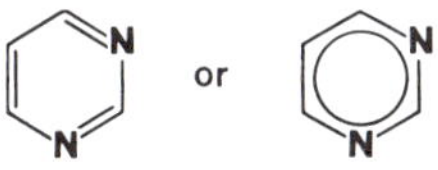

pyrimidine

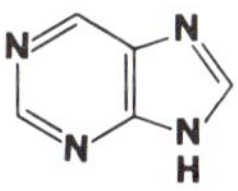

purine

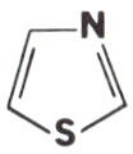

thiazole

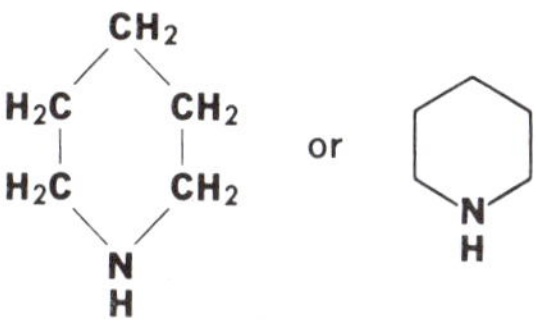

piperidine

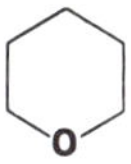

tetrahydropyran

thiophene

24–9 SOURCES OF AROMATIC COMPOUNDS

If coal is heated in the absence of air (so that it is not just burned up), a portion of the material vaporizes. The residue, called **coke,** is carbon plus small amounts of mineral impurities, mainly clay.

$$\text{coal} \xrightarrow{\text{heat}} \underset{\text{(residue)}}{\text{coke}} + \underset{\text{(volatile)}}{\text{coal tar}} \tag{24-12}$$

The hot vapors which escape are led to a cool chamber where they condense to become a thick liquid called **coal tar.** Coal tar is a valuable mixture composed of many different aromatic compounds, a few of which are listed in Table 24–1 to provide some idea of the variety present. Although living plants and animals produce aromatic substances, they rarely yield large quantities of them. Consequently coal tar is the richest natural source of aromatic compounds.* (See Fig. 24–2.)

Fig. 24–2 A modern plant for the conversion of coal to coke and coal tar. Coal is stored in the concrete bin at the right. To its left is a battery of many coke ovens, each sixteen feet high. At the opposite end of the oven battery is an elevated pipe carrying the volatile coal gas and tar to the cooling towers at the left. At front center are two distillation columns for separation of the oils. [Courtesy of the Allied Chemical Corporation.]

---

* Recall that petroleum has been mentioned as the ultimate source of *alkanes* (Section 22–1).

TABLE 24-1 Types of compounds in coal tar

Hydrocarbons

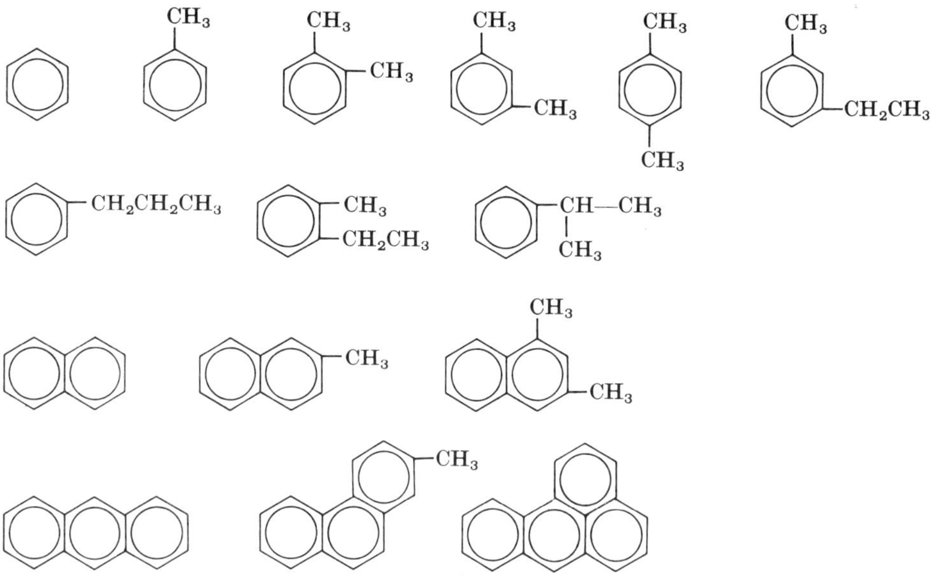

Phenols

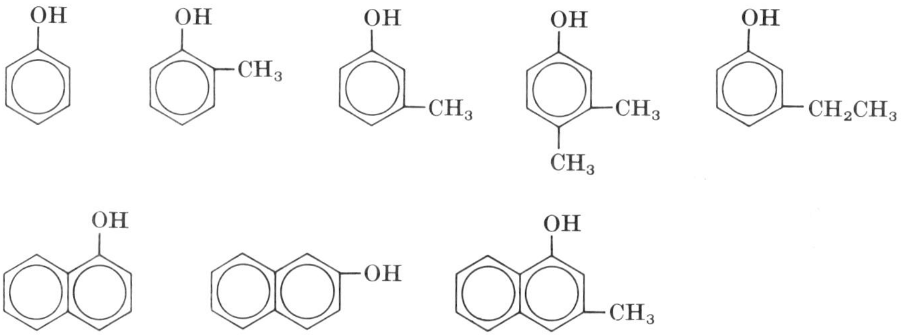

Nitrogen Heterocycles

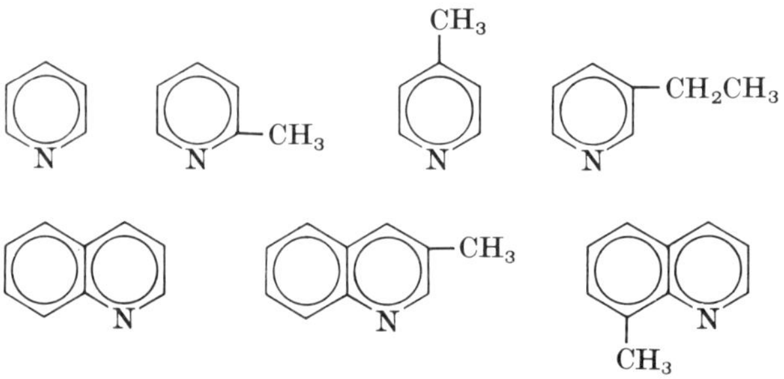

Fig. 24-3 A catalytic reforming unit. Alkanes pass from a heater at the left into four reactors (right) containing platinum catalyst. The aromatic molecules produced there go to the tall distillation columns (center, rear) for separation into benzene, xylene, etc. [Courtesy of the Shell Oil Company.]

Of course, the coal tar mixture must be separated into various fractions before individual compounds can be obtained. Both chemical and physical methods of separation are used. The phenols, which are weakly acidic (Section 25-6), can be converted to water soluble salts with sodium hydroxide, and hence can be washed away from the rest of the coal tar. In a similar fashion the weakly basic nitrogen heterocycles (Section 28-2) can be extracted with sulfuric acid.

Each of the three groups—hydrocarbons, phenols, heterocycles—is then further separated by distillation into various fractions. Depending on the boiling range and the variety of compounds originally present, a particular fraction may be a single, relatively pure compound, or it may still contain several compounds of similar weight and physical properties. For instance, a fraction boiling at 136–144°C will contain ethylbenzene and the three dimethylbenzenes (xylenes). Even a fraction having a boiling range as narrow as 136–139°C will include all but one of these. Since they have similar structures and all have formula C_8H_{10}, their boiling points are nearly the same; it is difficult to separate one from the others.

Conversion of 1000 pounds of coal to coke (Eq. 24–12) yields only about 50 pounds of coal tar as a by-product. Since coke is used mainly for steel making, the demands of that industry determine the amount of coal tar available from this source. For many years the conversion of coal to coke has not supplied enough benzene and other simple aromatics to supply the great quantities used in the manufacture of dyes, drugs, plastics, and fabrics. It has been necessary to seek additional sources of aromatic hydrocarbons.

Petroleum is a natural resource from which simple aromatic hydrocarbons can be *manufactured*. Most petroleum deposits contain only very small quantities of aromatic compounds. However, certain of the petroleum alkanes can be chemically transformed to aromatic hydrocarbons by the **catalytic reforming** process. Alkane vapors are passed over a platinum catalyst at about 500°C. (See Fig. 24–3.) For example, a petroleum fraction containing primarily C_6 compounds is converted to benzene.

$$CH_3CH_2CH_2CH_2CH_2CH_3 \xrightarrow{\text{Pt}, 500°C} \text{benzene} + 4H_2$$

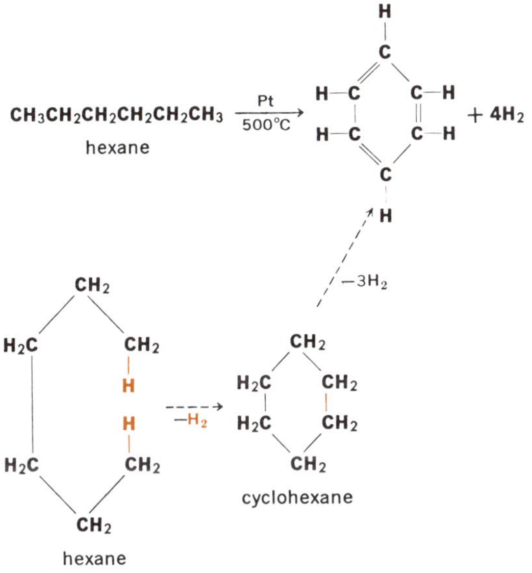

(24-13)

The reaction (Eq. 24–13) involves loss of hydrogens, causing both ring closure and aromatization of the newly-formed ring. [The broken arrows in Eq. (24–13) show the intermediate steps.] The C_6 petroleum fraction initially contains substantial amounts of cyclohexane. However, we can see from Eq. (24–13) that all the cyclohexane present is converted to benzene as easily as is hexane.

In an analogous manner, toluene can be formed from a C_7 petroleum fraction containing both heptane and methylcyclohexane.

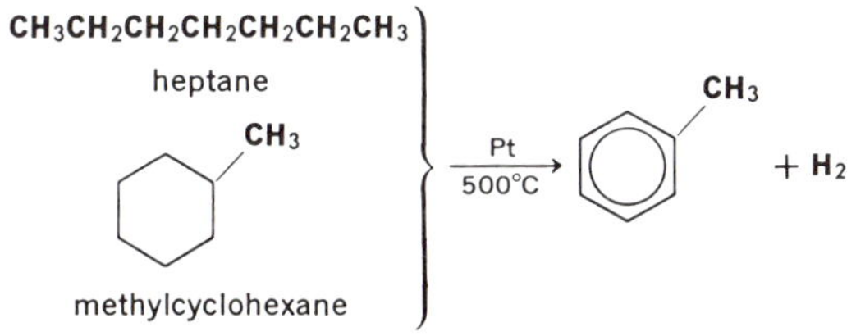

$$\left. \begin{array}{l} CH_3CH_2CH_2CH_2CH_2CH_2CH_3 \\ \text{heptane} \\ \text{methylcyclohexane} \end{array} \right\} \xrightarrow{\text{Pt}, 500°C} \text{toluene} + H_2$$

(24-14)

Equation 24–14 indicates that an aromatic ring results even from heptane. This reaction occurs because a six-membered ring closes more easily than does a seven-membered ring, and once it does, the very stable benzene ring structure can be formed.

IMPORTANT TERMS

Aromatic compound
Aryl group
Benzenesulfonic acid
Catalytic reforming

Coal tar
Heterocyclic compound
Meta
Nitration

Nitro group
Ortho
Para
Phenyl group
Sulfonation

WORK EXERCISES

1. Write a structural formula for
   a) toluene
   b) ethylbenzene
   c) *para*-dichlorobenzene
   d) 1,2-dinitrobenzene
   e) 3-bromonitrobenzene
   f) pyridine
   g) sodium benzenesulfonate

2. Give the structure and name of the products formed by the reaction of benzene with the reagents shown. If necessary, state "no reaction."
   a) concentrated sulfuric acid at 150°C b) hot concentrated hydrochloric acid
   c) a mixture of concentrated nitric and sulfuric acids
   d) chlorine and iron   e) bromine at 25°C

3. Write the structure of the organic acid produced by the reaction of
   a) toluene with sodium dichromate
   b) ethylbenzene with potassium permanganate
   c) 1,2,4-trimethylbenzene with potassium permanganate
   d) *m*-bromotoluene with sodium dichromate

4. Describe briefly the following, and give an example of each.
   a) aromatic compound   b) heterocyclic compound

5. Briefly discuss the two chief sources of aromatic hydrocarbons.

6. Write the structural formulas to show how many trichlorobenzene ($C_6H_3Cl_3$) isomers are possible.

7. What are the three principal types of compounds found in coal tar?

SUGGESTED READING

Breslow, R., "The Nature of Aromatic Molecules," *Scientific American*, August, 1972.

DePuy, C. H., and K. L. Rinehart, Jr., *Introduction to Organic Chemistry*, Chapter 5, John Wiley, New York, 1967.

Hart, H., and R. D. Schuetz, *Organic Chemistry*, 4th ed., Chapter 5, Houghton Mifflin, Boston, 1972.

Noller, C. R., *Textbook of Organic Chemistry*, 3rd ed., Chapter 20, W. B. Saunders, Philadelphia, 1966.

ALCOHOLS, PHENOLS, AND ETHERS

25-1 INTRODUCTION

Nearly every primitive tribe has stumbled onto the fact that overripe fruits may undergo a pleasant type of spoilage (fermentation) which produces an intoxicating liquid. The more advanced societies discovered that by distillation they could isolate the intoxicating substance from the original liquid. During the Middle Ages this essence or spirit became known as alcohol in some European languages. (The word was derived from the Arabic *al kuh'l*, but with considerable alteration in meaning.) Scientists eventually learned that this organic compound contained not only carbon and hydrogen, but also oxygen. Furthermore, many other oxygen-containing substances, having different amounts of carbon, were found to be structurally related to alcohol. Consequently the term alcohol was used to designate all compounds having this type of structure, and also to name the one particular compound. Two other classes of compounds, the ethers and phenols, proved to be structurally related, each in a different way, to the alcohols.

In the water molecule (H—O—H) the two covalent bonds of oxygen hold hydrogen atoms. Either one or both of these bonds may instead be linked to organic groups. In an **alcohol** (R—O—H), the oxygen holds one alkyl group. In a **phenol** (Ar—O—H), the oxygen holds one aryl group. In an **ether**, the oxygen holds two alkyl or aryl groups which may be the same or different (R—O—R, R—O—R', R—O—Ar).

These oxygen structures may be found in such diverse materials as sugars, pain-relievers, disinfectants, plastics, and solvents. Vanillin, whose structure is shown on the next page, is the pleasant flavoring agent extracted from vanilla beans. Note that the compound has both an ether group and a phenol structure. Thymol, produced by the herb thyme, is used medically as a fungicide and also as a preservative for anatomical specimens. A few representative compounds are shown on the following page.

CH₃—CH₂—O—H
ethanol
("alcohol," ethyl alcohol)

CH₃—CH₂—O—CH₂—CH₃
ethyl ether
(anesthetic "ether")

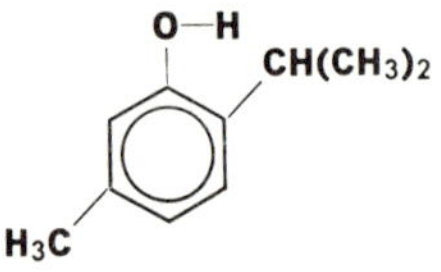

2-isopropyl-5-methylphenol
(thymol)

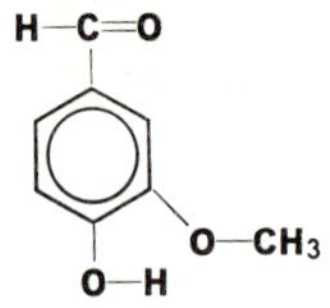

4-hydroxy-3-methoxybenzaldehyde
(vanillin)

25-2 ALCOHOL NAMES

In *systematic nomenclature* an alcohol is named by using the ending *-ol* in place of the final *-e* in the hydrocarbon name. Thus, the two-carbon alcohol is called ethanol. In most cases it is necessary to include a number to indicate the location of the hydroxyl group on the carbon chain. The hydroxyl, an important functional group, is assigned the lowest number possible for it, in preference to halogen or alkyl groups which may also be present. These rules are illustrated in the examples below.

Common names are also still used for many of the familiar alcohols. For these, a group name for the carbon skeleton is used, together with the word alcohol. In the examples, the common name is shown in parentheses.

CH₃CH₂CH₂OH
1-propanol
(*n*-propyl alcohol)

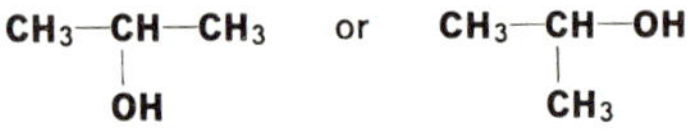

CH₃—CH—CH₃   or   CH₃—CH—OH
    | |
    OH   CH₃
2-propanol
(isopropyl alcohol)

CH₃CH₂CH₂CH₂OH
1-butanol
(*n*-butyl alcohol)

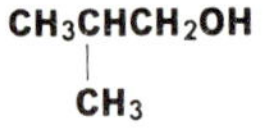

CH₃CHCH₂OH
     |
     CH₃
2-methyl-1-propanol
(isobutyl alcohol)

CH₃CH₂CH₂CH₂CH₂CH₂OH
1-hexanol
(*n*-hexyl alcohol)

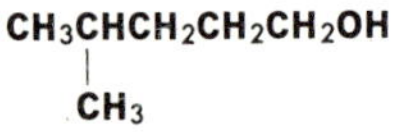

CH₃CHCH₂CH₂CH₂OH
     |
     CH₃
4-methyl-1-pentanol
(isohexyl alcohol)

Note that the term *iso* is used to indicate a methyl branch in the chain at the end farthest from the functional group; it must not be used for groups branched elsewhere.

If we observe the reactions of the different butanol isomers, we find that their rates of reactions and kinds of products depend upon the position of the hydroxyl group in the molecule. Similar behavior of other isomeric alcohols leads to the generalization that their reactions depend not only upon the presence of the hydroxyl group but also upon the amount of branching in the carbon skeleton adjacent to the hydroxyl group. Consequently, we classify an alcohol as *primary*, *secondary*, or *tertiary* if there are, respectively, one, two, or three carbons directly bonded to the carbon holding the OH group.

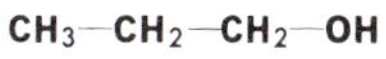

CH₃—CH₂—CH₂—OH

a primary alcohol

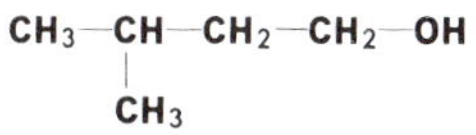

CH₃—CH—CH₂—CH₂—OH
          |
         CH₃

a primary alcohol

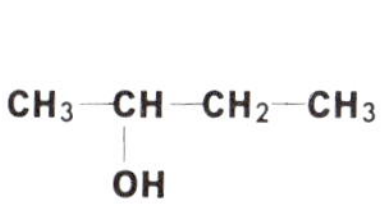

CH₃—CH—CH₂—CH₃
     |
     OH

a secondary alcohol

         CH₃
          |
CH₃—C—CH₂—CH₃
     |
     OH

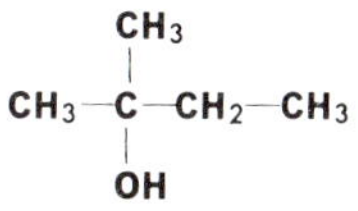

a tertiary alcohol

Note that the total number of carbons in the molecule, or branching elsewhere in the carbon skeleton, is not considered when we classify an alcohol as primary, secondary, or tertiary.

Some names for alkyl groups include the terms secondary (abbreviated *sec-*), or tertiary (abbreviated *tert-* or *t-*).

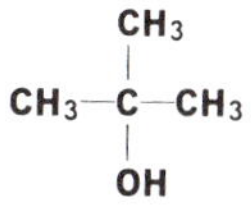

     CH₃
      |
CH₃—C—CH₃
      |
     OH

(tert-butyl alcohol)
2-methyl-2-propanol

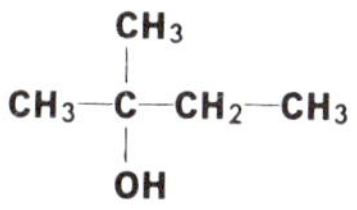

     CH₃
      |
CH₃—C—CH₂—CH₃
      |
     OH

(t-pentyl alcohol)
2-methyl-2-butanol

CH₃—CH₂—CH—CH₃   or   CH₃—CH₂—CH—OH
          | |
          OH   CH₃

(sec-butyl alcohol)
2-butanol

Note carefully the difference between the sec-butyl group (shown just above) and the isobutyl group (shown on p. 354). Also note that a common group name, such as isohexyl, designates in this one name the total number of carbons and that there are no additional names for alkyl substituents.

Some molecules may have more than one hydroxyl group. The most familiar of these are shown below. Their systematic names are included for illustration, but their common names are nearly always used. (The term *glycol* is a general name for those alcohols having two OH groups, usually on adjacent carbons. The name ethylene glycol is derived from the fact that this compound can be made from ethylene.)

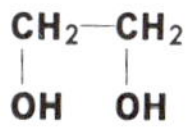

CH₂—CH₂
 | |
OH   OH

(ethylene glycol)
1,2-ethanediol

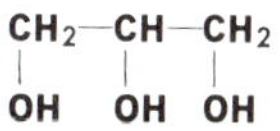

CH₂—CH—CH₂
 | | |
OH OH OH

(glycerol)
1,2,3-propanetriol

25-3 COMMON ALCOHOLS

Methanol, also called *methyl alcohol,* or sometimes *wood alcohol,* was formerly obtained by destructive distillation* of wood. The product now manufactured synthetically is less expensive and so much purer that its industrial use has greatly increased. Each year tons of methanol are dehydrogenated (Section 25–5) to formaldehyde, which in turn is converted to phenol-formaldehyde, urea-formaldehyde, and melamine-formaldehyde polymers (Bakelite, Melmac, etc., Table 30–3) and to numerous other compounds. A nearly equal amount of methanol is used in automotive antifreezes. Methanol is also used as a solvent, and to build the structure of many dyes, drugs, and perfumes.

Methanol is a severe poison which can cause blindness and death. Even prolonged breathing of its vapor may be damaging; factory workers must be protected from the vapor by adequate ventilation.

Ethanol is the correct chemical name for the substance popularly known as alcohol. Some other common names for it are spirits of wine, spirits, grain alcohol, and ethyl alcohol.

Physiologically, alcohol induces relaxation, a feeling of well-being, poor coordination, a dulled sense of judgment, and if consumed in sufficient quantities, causes drowsiness and unconsciousness. A large quantity consumed within a short time can even cause death. However, because of the initial effects of alcohol, a fatal concentration in the body is rarely established.

Grapes, honey, milk, cactus sap, berries, or almost any other imaginable source of sugar can be fermented to alcohol by yeasts. Furthermore, the starch in rice, corn, oats, potatoes, rye, etc., can be biochemically broken down to sugar and then fermented to alcohol. Fermentation rarely produces an alcohol content in excess of 13%. Natural table wines are 11–13% alcohol. The higher percentages in strong liquors can be obtained only by distilling to concentrate the alcohol. *Proof* is a term used in the United States to designate the strength of alcoholic beverages. Numerically, proof is twice the percentage; for example, a 90-proof gin contains 45% alcohol; the remainder is water and trace amounts of flavoring substances.

In almost every nation the production of alcohol is closely regulated and is taxed, often heavily. Enough beverage alcohol is consumed to make this tax a very important source of revenue. Unfortunately, alcohol also frequently creates medical and sociological problems.

Great quantities of alcohol are used industrially, in the form of **denatured alcohol.** This is alcohol to which a small amount of a poisonous substance has been added to make it unfit for drinking. The additive also usually has a disagreeable odor and taste. Some of the common denaturing agents are methanol, benzene, gasoline, and isopropyl alcohol. Denatured alcohol is used to avoid the heavy beverage tax. In the United States, the current price for a gallon of 95% (190 proof) denatured alcohol is about 70 cents, when purchased in large quantity. The same amount of alcohol with the beverage tax included costs $20.70!

In both laboratory and industry, 95% alcohol is used extensively. This is the highest concentration which can be obtained by ordinary distillation; it is pure enough for many purposes. If, by special means, the last 5% of water is removed, the pure, water-free alcohol is called **absolute alcohol.**

---

* In destructive distillation a material is heated (in the absence of air) to a very high temperature, so that it decomposes and releases volatile substances.

Part of the industrial alcohol is obtained by fermentation of molasses, or sometimes of potatoes. Most is synthesized from ethene, obtained as usual from petroleum sources. The two-step process described in Section 23–5E is utilized in some factories. In others, the conversion from ethene to ethanol is accomplished in *one* operation by passing steam and ethene through a hot tube containing phosphoric acid adsorbed on an inert material.

Ethanol is used for the synthesis of other organic compounds, as a component of lotions, perfumes, and cosmetics, and as a solvent.

Isopropyl alcohol is most familiar as the compound in "rubbing alcohol." Other applications are in hand lotions, after-shave lotions, or cosmetics, and in quick-drying inks and paints. The largest quantities of isopropyl alcohol are converted to acetone by the dehydrogenation reaction (Section 25–5). The isopropyl alcohol required for all these needs is synthesized from propene by the two-step hydration process, utilizing sulfuric acid (Section 23–5E).

Ethylene glycol is an excellent automotive antifreeze because it is nonvolatile. It is also used in hydraulic brake fluids, printer's inks, stamp-pad inks, ball-point pens, and as a solvent for certain paints, plastics, and other materials. Large quantities are converted to the polymers constituting Dacron fibers, Mylar film, and alkyd paints.

Glycerol, also called glycerin, is a constituent of all fatty foods and is readily used by the human body. This is interesting in view of the fact that other small alcohols, including ethylene glycol, are toxic.

Most countries obtain glycerol entirely as a by-product of soap-making. In the United States additional quantities are made synthetically because of the great demands for it. (One authority has reported more than 1500 uses for glycerol!) The synthesis, developed by the Shell companies, uses propene from petroleum as the starting material.

Glycerol is a good humectant (moisture-retaining agent). This property makes it useful in the manufacture of tobacco, candy, cosmetics, skin lotions, inks, dentifrices, and pharmaceuticals. Eye corneas, blood cells, and other live tissues are often treated with glycerol for frozen storage. A few of the materials synthesized from glycerol are alkyd paints and other polymers, the explosive nitroglycerin, and monoglycerides and di-glycerides used as emulsifiers and softening agents.

25–4 PHYSICAL PROPERTIES OF ALCOHOLS, PHENOLS, AND ETHERS

The volatility of an organic compound is directly related to its molecular weight. We found that alkanes of low molecular weight are gases at room temperature, those of higher weight are liquids, and the heaviest are solids (Section 22–4). We also know that within the group of alcohols a larger molecule has a higher boiling point than does a smaller one. We might expect, then, that methanol and ethanol would be gases, and would have very low boiling points. Instead, both are liquids. Evidently some factor other than weight is also affecting the volatility of alcohols. We now believe that most physical properties of alcohols are influenced by hydrogen bonds.

Hydrogen bonds may be formed from one alcohol molecule to another, because of the highly polar covalent bond between oxygen and hydrogen (Fig. 25–1). Just as hydrogen bonds do in the case of water (Section 10–7), these strong intermolecular attractions have a profound effect on the physical properties of **alcohols.**

First, an alcohol has a higher boiling point than does an alkane hydrocarbon of about the same weight and size. This is illustrated by the data for 1-butanol and *n*-pentane, given in Table 25–1. Because of the hydrogen bond attractions between alcohol

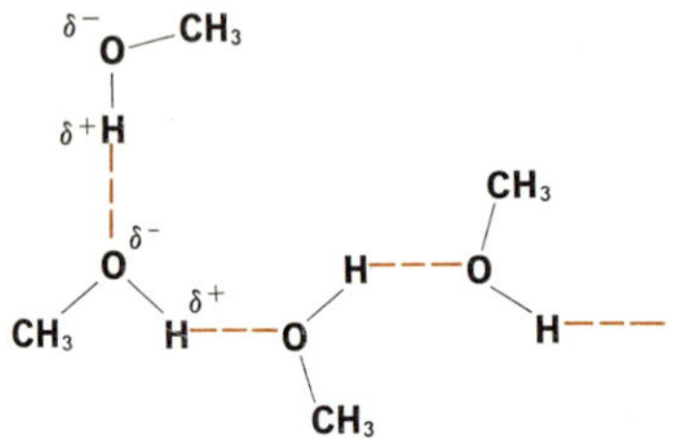

Fig. 25-1 Hydrogen bonds (color) create strong attractions between alcohol molecules.

molecules, it is difficult for a molecule to break away from its neighbors in the liquid and exist as an individual molecule in the gas state. Therefore the alcohol requires more energy (higher temperature) for vaporization than does the hydrocarbon.

TABLE 25-1 Relation of boiling point to structure

Compound	Molecular Weight	Structure	Boiling Point, °C
n-pentane	72	$CH_3CH_2CH_2CH_2CH_3$	36
1-butanol	74	$CH_3CH_2CH_2CH_2OH$	118
ethyl ether	74	$CH_3CH_2OCH_2CH_3$	35

Second, the simple alcohols are much *more soluble in water* than are the hydrocarbons of similar size. This solubility is due to the fact that when an alcohol is mixed with water it can form hydrogen bonds to water molecules, and hence be pulled into the water, as indicated in Fig. 25-2.

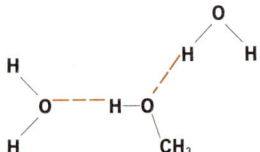

Fig. 25-2 Methanol dissolves in water because water molecules are attracted to it by hydrogen bonds (color).

We find, however, that a long-chain alcohol such as 1-octanol,

$CH_3CH_2CH_2CH_2CH_2CH_2CH_2CH_2OH$,

is practically insoluble in water. We explain this by saying that the attraction of the polar OH group toward water molecules is not sufficient to pull such a large hydrocarbon chain into solution; the 1-octanol behaves nearly the same as does nonpolar octane. Alcohols of medium size, of course, have intermediate solubilities. The data in Table 25-2 show that the longer the carbon chain, the less soluble the alcohol is in water.

TABLE 25-2 Solubility of alcohols in water

Alcohol	Solubility at 20°C (g ROH/100 g H$_2$O)
methanol	completely miscible
ethanol	completely miscible
1-propanol	completely miscible
1-butanol	8
1-pentanol	2.7
1-hexanol	0.6

Although long-chain alcohols are insoluble in water, they do dissolve in octane or other hydrocarbons. The bulk of the long-chain alcohol molecule is the nonpolar hydrocarbon group, which readily associates with solvents similar to it in structure. Alcohol solubility is nicely summarized by the statement "like dissolves like."

In hot, dry locations tremendous amounts of water are lost from reservoirs through evaporation. Because of its solubility properties, octadecanol (C$_{18}$H$_{37}$OH) has been used successfully to reduce this loss. When octadecanol is thrown on the water, the OH group is attracted to the water but is unable to pull such a long hydrocarbon group into solution. Consequently, the octadecanol molecules line up side by side on the surface, with the OH groups in the water and the long hydrocarbon "tails" sticking out of the water vertically. A continuous film (just one molecule thick!) forms on the surface, greatly hindering the evaporation of water molecules.

Alcohols can, in turn, act as solvents for *other* compounds; their ability to do this depends on the principles just discussed. Ethanol is a good solvent for alkanes and aromatic hydrocarbons as well as for polar compounds such as benzoic acid, acetone, and other alcohols. Methanol is a poorer solvent for alkanes, but dissolves the polar compounds and even some ionic salts.

Phenols follow the same solubility principles as do alcohols. Phenols are more water-soluble and have higher boiling points than do aromatic hydrocarbons of similar size. However, since even the smallest phenol has a skeleton of six carbons, phenols are only moderately soluble in water. Those having large carbon-skeletons are more soluble in hydrocarbons such as benzene. Most phenols are quite soluble in ethanol.

We note in Table 25–1 that the boiling point of **ethyl ether** is almost identical to that of *n*-pentane. This similarity is due to the fact that hydrogen bonding cannot exist between two ether molecules. In an ether, all the hydrogens are held on carbon atoms (Fig. 25–3). Since the carbon-hydrogen covalent bonds are *nonpolar*, a hydrogen atom of one molecule does not attract an oxygen atom in another molecule. Hydrogen bonding is important only when a hydrogen atom is caught between two very negative atoms, such as two oxygens (Fig. 25–1). Since hydrogen bonds do not exist between ethyl ether molecules, their volatility is very similar to that of *n*-pentane.

Ethyl ether, the most common ether, forms a separate layer when mixed with water. In a practical sense it is, therefore, often regarded as insoluble. Actually, some of the ether dissolves (about 7 g/100 g of water), but the amount is not easily observed by the eye. Ether is completely miscible with alkanes.

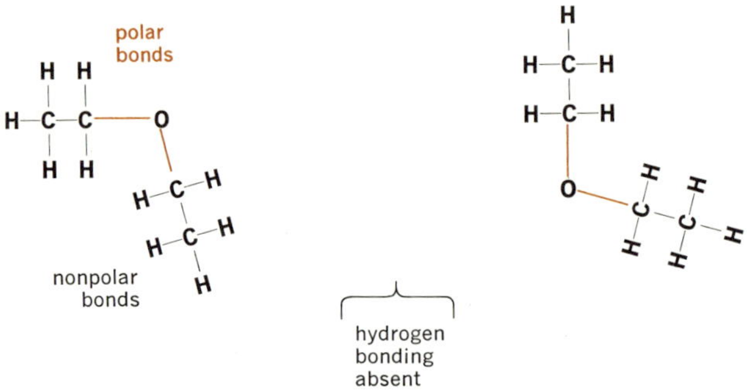

Fig. 25-3 There are no hydrogen-bond attractions between ether molecules because their hydrogen atoms are held by nonpolar bonds.

Ether can dissolve a wide variety of organic structures because of its own intermediate character; the alkyl portion of the molecule is typically nonpolar, but the oxygen bonds are moderately polar (Fig. 25–3). Consequently, ether associates with and dissolves nonpolar compounds such as alkanes, benzene, and fats, as well as moderately polar compounds such as phenol or acetic acid. The compounds which fail to dissolve in ether are chiefly those having several very polar groups, such as sugars, glycerol, or ethylene glycol. The chemical inertness of ether (Section 25–7) is another factor making it a desirable solvent for organic compounds. On the other hand, the great flammability of ether makes it very dangerous.

25-5 CHEMICAL PROPERTIES OF ALCOHOLS

A. Covalent Functional Group

First of all, we should keep in mind that in an alcohol molecule such as H—O—CH$_3$, all the bonds, including those from oxygen to carbon and from oxygen to hydrogen, are *covalent*. Although the oxygen bonds are quite *polar* covalent bonds, they are not ionic. Therefore, the alcohol is *not* a strong base like NaOH which releases separate hydroxide ions, HO$^-$, into solution. The functional group of an alcohol may be called a *hydroxyl group*, but not a hydrox*ide ion*. Although the alcohol functional group does not ionize, in many other respects it is quite reactive, and hence an alcohol may be converted to a variety of other substances. Several important chemical properties of alcohols are discussed in the following paragraphs.

B. Dehydrogenation; Oxidation

When vapors of an appropriate alcohol are passed over a hot copper catalyst (Eqs. 25–1 and 25–2), hydrogen is lost, and an aldehyde or ketone is produced.

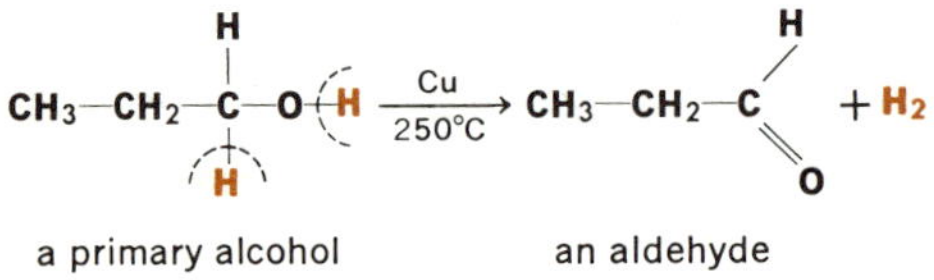

(25-1)

a primary alcohol   an aldehyde

$$CH_3-CH_2-\underset{\underset{H}{|}}{\overset{\overset{CH_3}{|}}{C}}-O-H \xrightarrow{\text{Cu}, 250°C} CH_3-CH_2-\underset{\underset{O}{\|}}{\overset{\overset{CH_3}{|}}{C}} + H_2 \qquad (25\text{-}2)$$

a secondary alcohol → a ketone

$$CH_3-\underset{\underset{CH_3}{|}}{\overset{\overset{CH_3}{|}}{C}}-O-H \xrightarrow{\text{Cu}, 250°C} \text{no reaction} \qquad (25\text{-}3)$$

a tertiary alcohol

The reaction is termed **dehydrogenation** because it involves *removal of hydrogens* from the molecule; in this case, one from the oxygen and one from the carbon adjacent. Since the adjacent carbon in a tertiary alcohol does not hold a hydrogen, the reaction is not possible, Eq. (25-3).

The dehydrogenation reaction is important for at least three reasons:

1) It demonstrates the structural relations among the organic compounds involved. Note that a primary alcohol is converted to an aldehyde, a secondary alcohol becomes a ketone, and a tertiary alcohol is unchanged.

2) Dehydrogenation also occurs in vital biochemical reactions. In such cases, the reaction must take place at body temperature, and the catalyst is an enzyme. For example, an enzyme called alcohol dehydrogenase can catalyze the conversion of ethanol to acetaldehyde.

3) This reaction provides a valuable synthetic method for making aldehydes or ketones from the readily available alcohols. It is adaptable to either laboratory or factory use. Aldehydes and ketones are discussed further in Chapter 29.

Dehydrogenation is in a sense equivalent to *oxidation:* the aldehyde or ketone product is in a state of higher oxidation than was the alcohol. This fact can be demonstrated experimentally by showing that the results are similar when alcohols are treated with the ionic oxidizing agents commonly used in the laboratory. In Eq. (25-4) below, the oxygen in brackets, [O], represents oxygen supplied by an ionic oxidizing agent such as $KMnO_4$ or $Na_2Cr_2O_7$. Note that this oxygen combines with the two hydrogens removed from the alcohol. (The actual mechanism for donation of the oxygen by the oxidizing agent is quite complex.)

$$CH_3-CH_2-\underset{\underset{H}{|}}{\overset{\overset{CH_3}{|}}{C}}-O-H + [O] \rightarrow CH_3-CH_2-\underset{\underset{O}{\|}}{\overset{\overset{CH_3}{|}}{C}} + H_2O \qquad (25\text{-}4)$$

Such a reagent is rarely useful for preparing an aldehyde from a primary alcohol, because the aldehyde is easily oxidized further to an organic acid:

$$CH_3-\underset{\underset{H}{|}}{\overset{\overset{H}{|}}{C}}-O-H \xrightarrow{[O]} CH_3-\underset{\underset{O}{\|}}{\overset{\overset{H}{\diagup}}{C}} + H_2O \xrightarrow[{[O]}]{\text{more}} CH_3-\underset{\underset{O}{\|}}{\overset{\overset{OH}{\diagup}}{C}} \qquad (25\text{-}5)$$

(difficult to isolate)   organic acid

In contrast, the dehydrogenation reaction is generally an excellent method for synthesizing aldehydes as well as ketones.

Combustion occurs with alcohols, as it does with most other organic compounds, when they are exposed to oxygen at a sufficiently high temperature. The resultant complete oxidation yields the usual products, carbon dioxide and water:

$$CH_3CH_2OH + 3O_2 \xrightarrow{heat} 2CO_2 + 3H_2O$$

C. Reactions with Acids

1. Dehydration. Alcohols are dehydrated when they are treated with strong acids such as sulfuric or phosphoric. The water lost may be split out from one alcohol molecule, in which case an alkene results (Eq. 25–6); or it may be split out between two molecules, so than an ether results (Eq. 25–7).

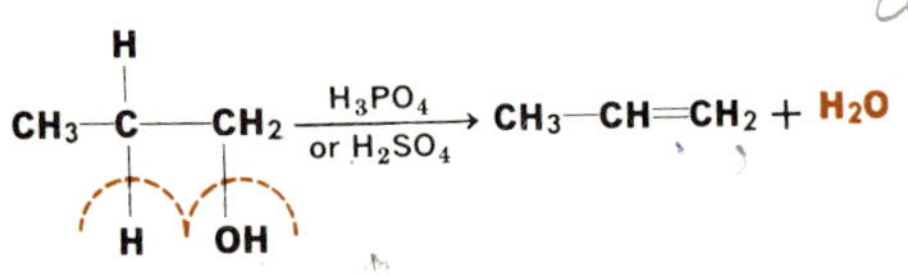

alcohol + strong acid → alkene + water

$$CH_3-\underset{\underset{H}{|}}{\overset{\overset{H}{|}}{C}}-\underset{\underset{OH}{|}}{CH_2} \xrightarrow[\text{or } H_2SO_4]{H_3PO_4} CH_3-CH=CH_2 + H_2O \quad (25\text{–}6)$$

1-propanol → propene

(adjacent) 2 alcohols + strong acids → ether

$$CH_3CH_2CH_2-O-H \quad H-O-CH_2CH_2CH_3 \xrightarrow[\text{or } H_2SO_4]{H_3PO_4} CH_3CH_2CH_2-O-CH_2CH_2CH_3 + H_2O \quad (25\text{–}7)$$

1-propanol → propyl ether

Which product results depends on the reaction conditions and the particular alcohol used. A number of both alkenes and ethers have been synthesized in this fashion. Ethyl ether is manufactured by this reaction or a modification of it.

2. Halide Formation. Concentrated halogen acids convert alcohols to organic halides, Eq. (25–8).

$$ROH + HBr \rightarrow RBr + H_2O \quad (25\text{–}8)$$

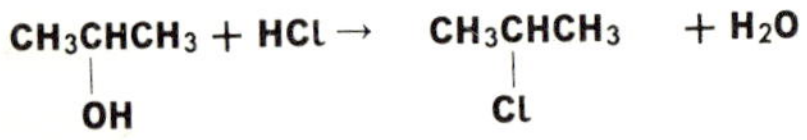

$$CH_3\underset{\underset{OH}{|}}{CH}CH_3 + HCl \rightarrow CH_3\underset{\underset{Cl}{|}}{CH}CH_3 + H_2O$$

2-propanol → 2-chloropropane

3. Ester Formation. Alcohols react with organic acids to form compounds called esters.

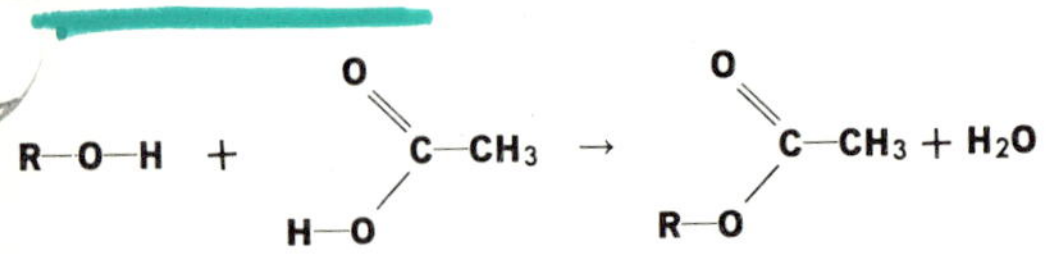

$$R-O-H + \underset{H-O}{\overset{O}{\overset{\|}{C}}-CH_3} \rightarrow \underset{R-O}{\overset{O}{\overset{\|}{C}}-CH_3} + H_2O \quad (25\text{–}9)$$

an ester

The esters, of interest both chemically and biologically, will be discussed in Chapters 26 and 27.

Under the proper conditions, concentrated nitric acid changes an alcohol to an alkyl nitrate. The most familiar example of this reaction, Eq. (25–10), occurs in the manufacture of glyceryl trinitrate, an explosive used as an ingredient of blasting mixtures and of propellants in ammunition. Glyceryl trinitrate is also used for treatment of coronary attacks.

$$\begin{array}{c} CH_2OH \\ | \\ CHOH \\ | \\ CH_2OH \end{array} + 3HONO_2 \rightarrow \begin{array}{c} CH_2ONO_2 \\ | \\ CHONO_2 \\ | \\ CH_2ONO_2 \end{array} + 3H_2O \qquad (25\text{–}10)$$

glycerol   nitric acid   glyceryl trinitrate ("nitroglycerin")

The product in Eq. (25–10) is often called *nitroglycerin*, a confusing and erroneous name. Structurally, it is a nitrate compound; there are three oxygens around the nitrogen, and the carbon is bonded to an oxygen. Compare this structure with that of a true nitro group such as that in nitrobenzene or trinitrotoluene (Section 24–6), in which there are two oxygens around nitrogen, and the carbon is directly bonded to the nitrogen.

Another explosive, cellulose nitrate (nitrocellulose) is obtained when the cellulose of cotton or wood pulp is treated with nitric acid. There are alcoholic OH groups in cellulose (Section 33–5) so the reaction with nitric acid is the same as that shown in Eq. (25–10). Working with either cellulose nitrate or glyceryl trinitrate is extremely hazardous; early attempts to use them ended in disaster. Finally, in 1867, the Swedish chemist and industrialist Alfred Nobel patented dynamite, a mixture of glyceryl trinitrate with diatomaceous earth. He had found that glyceryl trinitrate could be safely handled in this condition. A decade later Nobel discovered that a mixture of glyceryl trinitrate and cellulose nitrate formed a colloidal mass which, curiously, stabilized both of them. This gelatin could be used for blasting purposes or, in a modified form, in guns. We can also stabilize cellulose nitrate by treating it with alcohol-ether solvent to render it colloidal, then forming it into grains, and finally removing the solvent. Nobel's invention of a detonator for these stabilized explosives was equally important. Explosives are of little use if they cannot be set off when desired.

Either cellulose nitrate alone, or a mixture of glyceryl and cellulose nitrates is called *smokeless powder*. (Ancient gunpowder was a mixture of potassium nitrate, charcoal, and sulfur.) Nitrate materials of this kind are widely used in guns, rockets, and missiles. The explosives are tremendously useful in mining, road building, etc.; they are cruelly destructive in military use. Advances in technology very often present mankind with dilemmas.

Alfred Nobel acquired a large fortune from oil fields and from manufacturing explosives. He willed funds to the establishment of the Nobel Prizes for peace, chemistry, physics, literature, and physiology or medicine. The first prizes were awarded in 1901, five years after his death.

25–6 PHENOLS

The term **phenol** is used to name this whole class of compounds, as well as the simplest example in the group. (The term phen*ol* implies a hydroxyl derivative of phene; the origin of the word *phene* is mentioned in Section 24–2.)

ArOH

general formula for a phenol

phenol (carbolic acid)

All phenols are weakly acidic and able to react with sodium hydroxide, Eq. (25–11). Most, however, are unable to react with sodium hydrogen carbonate. (In contrast, alcohols do not react even with sodium hydroxide.)

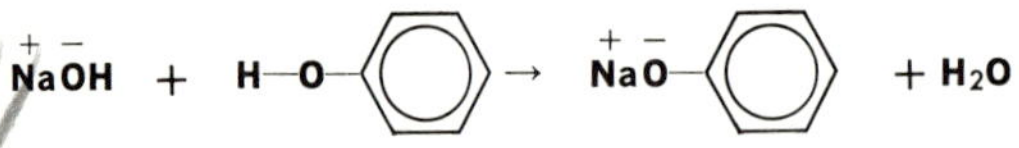

$$\text{NaOH} + \text{H-O-}\underset{\text{phenol}}{\bigcirc} \rightarrow \text{Na-O-}\underset{\text{sodium phenoxide}}{\bigcirc} + \text{H}_2\text{O} \qquad (25\text{–}11)$$

This acidic property is reflected in some of the names. An aqueous solution of phenol is still occasionally called "carbolic acid." Picric acid is an especially acidic phenol.

The structures and names of several phenols are shown below.

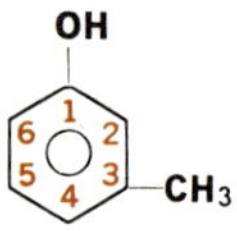

3-methylphenol
(*m*-cresol)

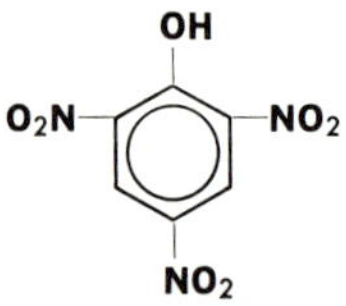

2,4,6-trinitrophenol
(picric acid)

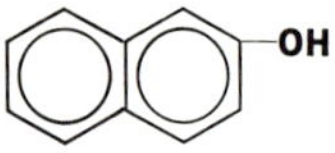

2-naphthol
(β-naphthol)

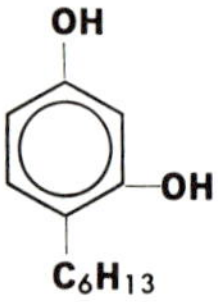

(hexylresorcinol)

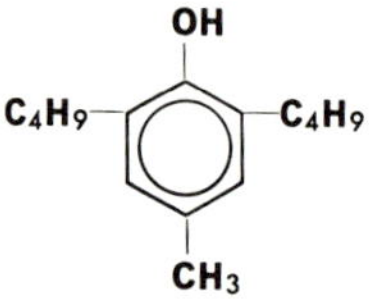

2,6-di-t-butyl-4-methylphenol
BHT

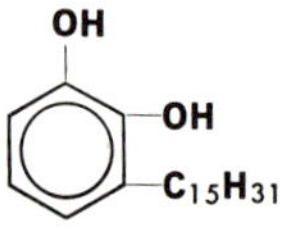

(urushiol)
the irritant in poison ivy
and poison oak*

Phenols have interesting physiological properties. Phenol itself is a disinfectant used on floors and apparatus. If strong solutions are spilled on the skin they will kill (burn) some of the tissue. Similarly, the germicidal properties of creosote (a crude mixture, obtained from coal tar, of aromatic hydrocarbons, cresols, and other phenols) make it a good wood preservative. Hexylresorcinol is used in humans and animals to combat intestinal worms and urinary infections. Butylated hydroxytoluene, BHT, is now used

---

* Urushiol is actually a mixture of closely related compounds in which the C_{15} side-chains have varying numbers of double bonds.

extensively as an antioxidant in foods, rubber, plastics, petroleum products, and soaps. It prevents deterioration caused by exposure to air.

Phenol is manufactured into important polymers, chiefly the phenol-formaldehyde and epoxy types. Thousands of pounds of dyes are made annually from naphthol. Alizarin, a phenolic compound found in madder root, has been used as a dye since ancient times. It is now made synthetically.

The phenols used in quantity are often obtained from coal tar (Section 24–9); for some, the supply is augmented by synthesis from petroleum materials.

25–7 ETHERS

An ether group occurs as a unit of structure in a variety of compounds produced by living organisms. However, the role, if any, that the ether unit plays in crucial biochemical reactions is not apparent. Possibly its chief contribution is to provide a molecule with the proper physical characteristics, such as size or solubility. This situation may be related to the fact that ethers are chemically very inert, in much the same way that alkanes are. Combustion occurs with oxygen, and chlorine will substitute onto the alkyl portion of an ether in the presence of ultraviolet light. However, hydrogen, strong bases, reactive metals, or strong reducing agents have no chemical effect on an ether. Some strong acids and oxidizing agents may attack an ether at high temperatures, but these, too, are without effect at moderate temperatures.

Therefore, except when studying the structure of natural products, we are usually concerned with only a very few simple ethers. By far the most common of these is ethyl ether, sometimes called diethyl ether, but most often simply ether. It is used mainly as a solvent (Section 25–3) and to some extent as an anesthetic.

In 1846, William Morton, a dentist, tested ether as an anesthetic while extracting a tooth. About two weeks later he was the first to make public its effectiveness, by administering it to a patient during surgery. The use of ether had been suggested to Morton by a chemist, Charles Jackson, who had experimented with inhalation of ether. Before this time surgery was an agonizing procedure. The patient was strapped to a table and (if he was lucky) soon lost consciousness due to pain. Ethanol had some value as a relaxant, but only by the use of ether and other anesthetics could surgery develop to its present state, involving kidney and heart transplants.

Ether is a very useful anesthetic. A safe level of unconsciousness can be achieved without depressing respiration or circulation. The amount of ether required is not toxic. Some disadvantages of using ether are that most patients experience prolonged and unpleasant recovery, and that ether is highly flammable. A number of other agents, or combinations of agents, have now replaced ether in many cases.

Ethene, cyclopropane, and vinyl ether ($CH_2=CH-O-CH=CH_2$) are also successful anesthetics, but share with ether a high flammability. Trichloromethane, commonly called chloroform (Table 22–5) is effective and nonflammable. It must be administered with great care, however, to avoid overdosage or liver damage, but is used in many parts of the world. Another halogen compound, halothane (Table 22–5), was developed in 1956 specifically for anesthesia, and by 1966 had become widely accepted. The inorganic compound nitrous oxide (N_2O) is a weak anesthetic often used as a supplement to others. It is also effective in conjunction with intravenous administration of thiopental sodium, a heterocyclic compound (Section 24–8).

IMPORTANT TERMS

Alcohol
Alkoxide
Dehydrogenation
Ether
Hydrogen bond

Hydroxyl group
Primary alcohol
Phenol
Secondary alcohol
Tertiary alcohol

WORK EXERCISES

1. Show the structure of all possible alcohols having the molecular formula $C_4H_{10}O$.
2. Write a systematic name for each isomer shown in Question 1.
3. Label each isomer in Question 1 as primary, secondary, or tertiary.
4. Show the product that each alcohol in Question 1 would give with copper at 250°C.
5. Show the structure of:
   a) 2-propanol
   b) phenol
   c) isobutyl alcohol
   d) 4-methyl-2-pentanol
   e) potassium methoxide
   f) sec-butyl alcohol
   g) m-nitrophenol
   h) sodium ethoxide
   i) 5-bromo-2-methyl-3-hexanol
   j) two different secondary alcohols having formula $C_5H_{12}O$
   k) 3-ethyl-2-phenyl-1-pentanol
   l) 3-chloro-4-isopropyl-2-heptanol
   m) cyclohexanol
6. Methyl ethyl ether, $CH_3OCH_2CH_3$, has mol. wt. 60 and is a gas (b.p. 10°C), whereas ethylene glycol, $HOCH_2CH_2OH$, has about the same mol. wt. (62), but has b.p. 197°C.
   a) Explain why ethylene glycol has such a high boiling point.
   b) Why is the high boiling point an advantage when ethylene glycol is used as antifreeze in an automobile engine?
7. Write complete equations for these reactions. If necessary, state "no reaction."
   a) 1-butene and cold concentrated sulfuric acid
   b) product from (a) plus warm water
   c) 3-pentanol and copper at 250°C
   d) ethanol and copper at 250°C
   e) 2-butanol and warm potassium permanganate solution
   f) phenol and sodium hydroxide
8. Show how 2-propanol (rubbing alcohol) could be manufactured, starting with petroleum as the raw material.
9. List all the types of structural units (both carbon skeletons and functional groups) which you can find in each of the mood-altering drugs shown below. (Consider the groups discussed in this chapter and consult Table 22–6.)
   a) mescaline

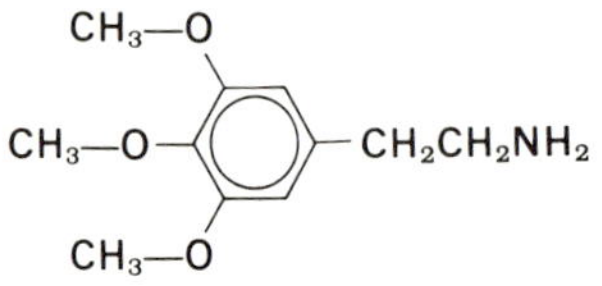

b) tetrahydrocannabinol

[Structure of tetrahydrocannabinol showing CH₃, OH, (CH₂)₄CH₃, CH₃, O, CH₃ groups]

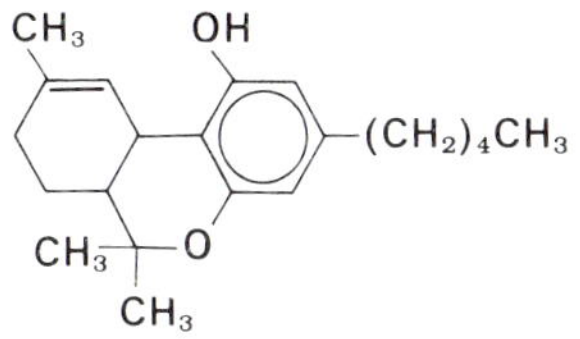

10. Write a structural formula for:
    a) 1-propanol
    b) 3-pentanol
    c) 2-methyl-2-butanol
    d) 3-methyl-2-butanol
11. Which alcohol in Exercise 10 is not dehydrogenated by hot copper?
12. Write an equation for the dehydrogenation of 1-butanol. In this reaction, is the alcohol reduced or oxidized? Discuss briefly.
13. Write an equation for one example of a biochemical dehydrogenation.
14. What type of compound is formed by the dehydrogenation of:
    a) a primary alcohol?
    b) a secondary alcohol?
    c) a tertiary alcohol?

SUGGESTED READING

Beecher, H. K., "Anesthesia," *Scientific American*, Jan. 1957. A fascinating review of the history of anesthesia, from early to recent times.

Clevenger, S., "Flower Pigments," *Scientific American*, June 1964 (Offprint #186). Despite the wide range in color from red to blue, these pigments are phenolic compounds with many similarities in structure.

DePuy, C. H., and K. L. Rinehart, Jr., *Introduction to Organic Chemistry*, Chapter 6, John Wiley, New York, 1967.

Greenberg, L. A., "Alcohol in the Body," *Scientific American*, Dec. 1953. A thorough discussion of the biochemical reactions and physiological effects of alcohol.

Hart, H., and R. D. Schuetz, *Organic Chemistry*, 4th ed., Chapters 6 and 7, Houghton Mifflin, Boston, 1972.

Kermode, G. O., "Food Additives," *Scientific American*, March 1972.

Lesser, M. A., "Glycerin—Man's Most Versatile Chemical Servant," *J. Chem. Educ.* **26,** 327 (1949). A survey of some of the history and uses of glycerol, but out-of-date in terms of new applications in recent years.

26 ORGANIC ACIDS

26–1 INTRODUCTION

Some of the most abundant and well-known organic compounds are acids. Of equal importance and familiarity are the fats, which are made up of organic acids chemically combined with the alcohol glycerol. For centuries man used tart fruits, sour milk, and vinegar, even though he was not acquainted with the individual compounds responsible for the sour tastes. When men did finally begin to isolate and identify separate acids (starting mostly in the eighteenth century), they usually named them after some familiar source, as shown in Table 26–1. Formic acid is the irritant in the sting of red ants, bees, and nettle plants. Butyric acid occurs in rancid butter, aged cheese, and human perspiration; it is the chief cause of their strong, offensive odor.

As a class, the acids shown in Table 26–1 are most properly called **carboxylic acids.** The functional group characteristic of the class is a **carboxyl group**; in this group the four covalent bonds of carbon may be represented as follows:

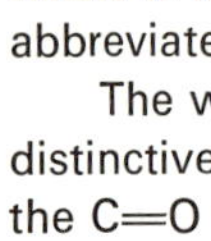

$$-\overset{\displaystyle O}{\underset{\displaystyle O-H}{C}} \quad \text{carboxyl group}$$

In most cases the structure should be carefully written out, as shown above. For the sake of brevity, the group is sometimes written —COOH (especially in the printing of books); when it is represented thus, be sure to remember the actual pattern of bonding. In abbreviated form, the acids themselves may be represented by RCOOH or ArCOOH.

The whole carboxyl group should be regarded as *one* functional group with its own distinctive properties. These unique properties are due to the fact that the OH portion and the C=O portion of the carboxyl group interact with each other strongly. Consequently, the carboxyl group has few chemical properties in common with a hydroxyl group, OH (Chapter 25), or with a carbonyl group, C=O (Chapter 29).

The systematic names for carboxylic acids follow the usual pattern: To the name for a carbon skeleton we add an ending for the carboxyl group by replacing the final *-e* with *-oic acid*. Thus, the five-carbon acid is called pentanoic acid. In the systematic nomenclature *benzoic acid* is used as a parent name for aromatic carboxylic acids (Table 26–1).

TABLE 26–1 Some common organic acids (carboxylic acids)

$-C\overset{=O}{\underset{O-H}{}}$ carboxyl group.

Common Name	Origin of Name	Structure	Systematic Name
formic acid	L. *formica*, ant	HCOOH	methanoic acid
acetic acid	L. *acetum*, vinegar	CH_3COOH	ethanoic acid
propionic acid	Gr. *pro(tos)*, first + *pion*, fat	CH_3CH_2COOH	propanoic acid
butyric acid	L. *butyrum*, butter	$CH_3CH_2CH_2COOH$	butanoic acid
caproic acid	L. *caper*, goat	$CH_3CH_2CH_2CH_2CH_2COOH$	hexanoic acid
lactic acid	L. *lactis*, milk	$CH_3-CH-COOH$ $\quad\quad\;\; |$ $\quad\quad\;\; OH$	2-hydroxy-propanoic acid
benzoic acid	obtained from benzoin, a plant gum	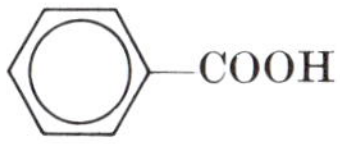 C₆H₅—COOH (phenyl-COOH)	benzoic acid
salicylic acid	obtained from the willow tree L. *salix*; Fr. *salicine*	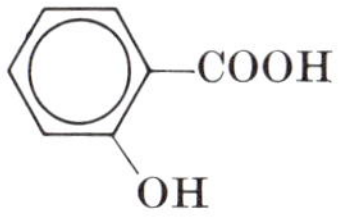 (2-OH phenyl)—COOH	2-hydroxy-benzoic acid

26–2 ACIDIC PROPERTIES; Salt Formation

Acetic acid is a good example to use in beginning our discussion of chemical properties. By the mid-1700's chemists could isolate acetic acid in fairly pure condition from the distillation of vinegar. By the mid-1800's many of the properties of acetic acid had been carefully observed. The customary analysis indicated that the molecular formula was $C_2H_4O_2$. The substance had the sour taste thought to be typical of acids, and it neutralized caustic soda (sodium hydroxide). The resultant salt had the formula $C_2H_3O_2Na$. Even if an excess of caustic soda was used, only one of the four hydrogens of acetic acid could be replaced by sodium. Various other observations were made, including some which suggested that the two oxygens were bound to one carbon. To be consistent with all the experimental observations, and with the ideas of valence becoming accepted at that time (with C = 4, O = 2, H = 1), acetic acid was written this way:

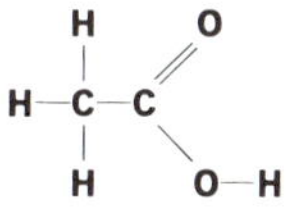

$$\begin{array}{c} \text{H} \quad\;\; \text{O} \\ | \quad\;\; \| \\ \text{H}-\text{C}-\text{C} \\ | \quad\;\;\; \backslash \\ \text{H} \quad\; \text{O}-\text{H} \end{array}$$

Acetic acid

Over the years this has proven to be a satisfactory structural formula for acetic acid; that is, none of the more recent experimental facts have indicated that the atoms are bonded in some different pattern.

370 Organic Acids

All the bonds of acetic acid are covalent. However, the nature of the carboxyl group is such that the oxygen–hydrogen bond is even more intensely polar than is the oxygen–hydrogen bond of water or of an alcohol. In fact, when acetic acid is dissolved in water an occasional acid molecule is able to ionize; the hydrogen escapes from the oxygen, to become an H⁺ ion in solution. At any one time in a 0.1 M solution, only about one acetic acid molecule out of every thousand exists in the ionized condition. The arrows in Eq. (26–1) are meant to show that most acetic acid molecules are in the nonionized form.

$$CH_3-C\begin{matrix}\nearrow O \\ \searrow O-H\end{matrix} \rightleftarrows CH_3-C\begin{matrix}\nearrow O \\ \searrow O^-\end{matrix} + H^+ \qquad (26\text{–}1)$$

acetic acid acetate ion

Recall that when HCl dissolves in water, *all* the molecules become ionized, yielding separate H⁺ and Cl⁻ ions. By comparison, *acetic acid is termed a weak acid* because its water solution contains only a few hydrogen ions at any one time. However, the presence of even such a small number of hydrogen ions is enough to give the solution *typical acid properties:* it changes the color of litmus paper, etc.

If sodium hydroxide is added to acetic acid, the hydroxide ions pull hydrogen ions away from nearly all the acetic acid molecules (rather than from just one out of a thousand, as water does). As usual, this acid-base reaction Eq. (26–2) produces a salt.

$$CH_3-C\begin{matrix}\nearrow O \\ \searrow O-H\end{matrix} + Na^+OH^- \rightarrow CH_3-C\begin{matrix}\nearrow O \\ \searrow O^-Na^+\end{matrix} + HOH \qquad (26\text{–}2)$$

sodium acetate

The reaction occurs very rapidly, actually as fast as one can stir the solutions together. A second, very important feature of the reaction is that one gram molecular weight of acetic acid can convert one gram molecular weight of sodium hydroxide to a salt. In other words, molecule by molecule, a weak acid (e.g., acetic) has as much *capacity* to neutralize a base as does a strong acid (e.g., hydrochloric).

However, acetic acid, when alone in a dilute water solution, releases only a few hydrogen ions at any moment. Since the typical acidic properties are due to the hydrogen ions, the solution of acetic acid in water is not strongly acidic. This property of organic acids is important in biological systems. The tissues may need to have available a large supply of some acid, but at any given time a high concentration of hydrogen ions cannot be tolerated. The organic acids, being only slightly ionized, satisfy these conditions very nicely.

For the same reason, if we wish to halt the caustic effects of a strong base which has been spilled, vinegar or some other weak acid is a desirable neutralizing agent. If we used a strong acid, it might cause as much damage as the base we were trying to neutralize.

This whole situation can be dramatically summarized by considering another example. A 1 M solution of hydrochloric acid is much more strongly acid than is a 10 M solution of acetic acid. (In fact, the hydrochloric acid solution is about 100 times as acidic!) Yet, a given volume of the acetic acid solution has the *capacity* to neutralize ten times as much sodium hydroxide as does the hydrochloric acid solution.

Acidic Properties: Salt Formation

※ Salts of Carboxylic acids are formed:
1. carboxylic acid + base → carboxylate salt + H₂O
2. " " + Carbonate or Hydrogen carbonate → salt + H₂CO₃ → H₂O + CO₂

The reaction of carboxylic acids with bases is quite general. Other bases such as KOH, $Mg(OH)_2$, or $Al(OH)_3$ could be involved. Also, the acidity of nearly all carboxylic acids is about the same, despite rather wide structural variations elsewhere in the molecule. Benzoic acid has almost exactly the same acid strength as does acetic acid. Here are some further examples of acid-base reactions:

① b.

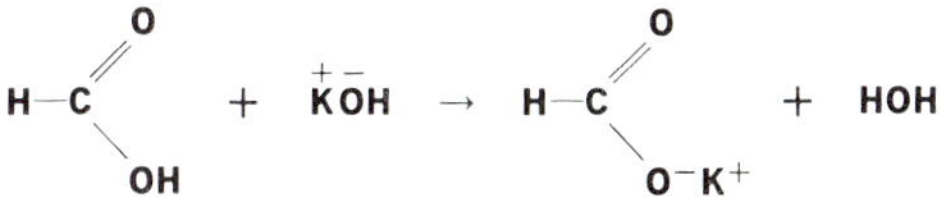

$$H-C(=O)OH + KOH \rightarrow H-C(=O)O^-K^+ + HOH \qquad (26\text{-}3)$$

formic acid   potassium formate

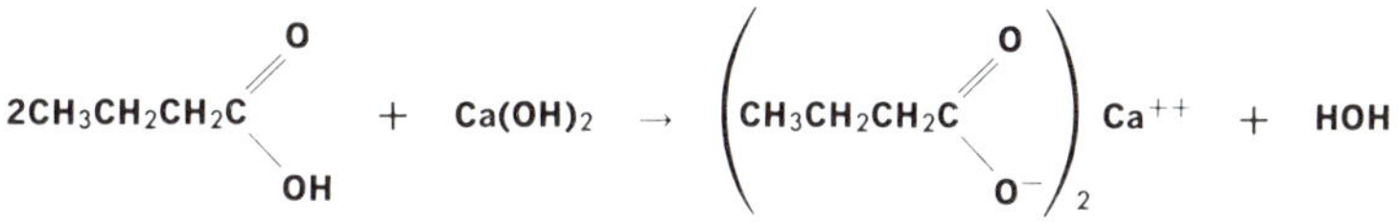

$$2 CH_3CH_2CH_2C(=O)OH + Ca(OH)_2 \rightarrow (CH_3CH_2CH_2C(=O)O^-)_2 Ca^{++} + HOH \qquad (26\text{-}4)$$

butanoic acid (butyric acid)   calcium butanoate (calcium butyrate)

As shown above, the salts of carboxylic acids are named by changing the *-ic* ending of the acid name to *-ate* for the salt. This applies whether we are using common names or systematic names.

Still another type of base is ammonia, NH_3. It reacts with HCl to form the salt $NH_4^+Cl^-$, as was shown in Section 12–5. In a similar manner, ammonia reacts readily with a carboxylic acid to form a salt.

c.

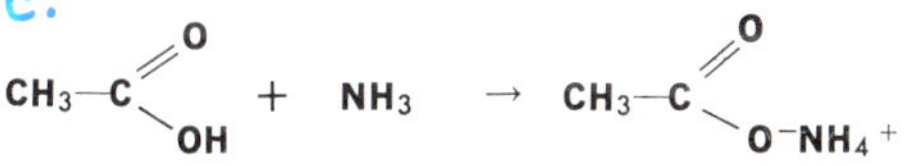

$$CH_3-C(=O)OH + NH_3 \rightarrow CH_3-C(=O)O^-NH_4^+ \qquad (26\text{-}5)$$

acetic acid   ammonia   ammonium acetate

In a carboxylate salt, such as $RCOO^-Na^+$, the functional group has an ionic bond. Therefore the compound has properties similar in many ways to those of an inorganic salt. The carboxylate salts are often quite soluble in water and conduct an electric current in solution. They form relatively hard crystals and have high melting points.

When a piece of limestone ($CaCO_3$) or a little baking soda ($NaHCO_3$) is dropped into vinegar, a distinct fizzing occurs. Water solutions of other carboxylic acids give the same results. Proper testing reveals that the gas being evolved is carbon dioxide. Evidently a carboxylic acid is capable of donating a hydrogen ion to a carbonate or hydrogen carbonate ion, Eqs. (26–6) and (26–7). (Incidentally, this means that the carboxylic acid is a considerably stronger acid than is a phenol—see Section 25–6.)

② a.

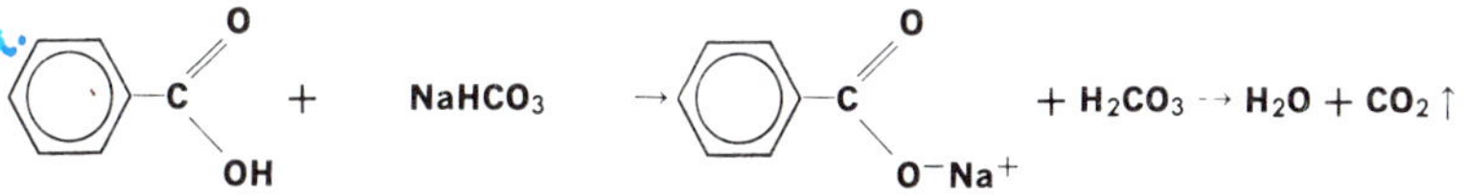

$$C_6H_5-C(=O)OH + NaHCO_3 \rightarrow C_6H_5-C(=O)O^-Na^+ + H_2CO_3 \rightarrow H_2O + CO_2\uparrow \qquad (26\text{-}6)$$

benzoic acid   sodium hydrogen carbonate   sodium benzoate

Organic Acids

b. $2CH_3CH_2COOH + K_2CO_3 \rightarrow 2CH_3CH_2COO^-K^+ + H_2CO_3 \rightarrow H_2O + CO_2\uparrow$ (26-7)

propanoic acid / potassium carbonate / potassium propanoate

Thus, the reaction of carboxylic acids with carbonates or hydrogen carbonates is another means by which salts are created.

If we add hydrochloric acid to an aqueous solution of sodium benzoate, a white precipitate of benzoic acid appears. (This acid happens to be a solid compound.) Or, if we add sulfuric acid to a solution of sodium butyrate, we promptly note the strong odor of butyric acid.

⌬—$COO^-Na^+ + H^+Cl^- \rightarrow$ ⌬—$COOH + Na^+Cl^-$ (26-8)

sodium benzoate / benzoic acid

$2CH_3CH_2CH_2COO^-Na^+ + (H^+)_2SO_4^{2-} \rightarrow 2CH_3CH_2CH_2COOH + (Na^+)_2SO_4^{2-}$ (26-9)

sodium butyrate / butyric acid

In general, we find that mineral acids will convert carboxylate salts to the corresponding carboxylic acids. Our explanation of this is simply that the carboxylic acids are weak acids, so that the strong mineral acids are capable of forcing hydrogen ions onto carboxylate ions, Eqs. (26-8) and (26-9).

To summarize, carboxylic acids have acidic properties for the same reason that inorganic acids do: they release hydrogen ions. The difference is one of degree; organic acids are weak acids because they are only slightly ionized. Like the inorganic acids, they form salts by reaction with metallic hydroxides, oxides, or carbonates.

26-3 WATER SOLUBILITY OF CARBOXYLIC ACIDS AND SALTS

In the preceding section we mentioned sodium benzoate and benzoic acid, Eq. (26-8). Sodium benzoate is quite soluble in water (61 g will dissolve in 100 g of water at 25°C). On the other hand, benzoic acid is only slightly soluble (about 0.25 g of benzoic acid will dissolve in 100 g of water at 25°C), so most of it is precipitated from the solution. In general, this relationship exists for carboxylic acids and their alkali metal salts; although the salt may be either fairly soluble or very soluble in water, the acid is less soluble, often much less so. The alkali salts are more water-soluble because they are ionic; the fully developed charges on ions strongly attract polar water molecules, causing solution. The carboxylic acids, however, are chiefly nonionized. In order to dissolve, they must depend upon hydrogen-bond attractions between their carboxyl groups, which are polar, and the water molecules (Fig. 26-1). These attractions due to *partially* developed charges in polar molecules are less strong than the attractions due to ions.

The result is that the solubility of carboxylic acids in water is about like that of alcohols; the smallest acids are very soluble, those of five and six carbons are slightly soluble, and the larger molecules are "insoluble." In contrast, most sodium or potassium carboxylate salts are quite soluble in water.

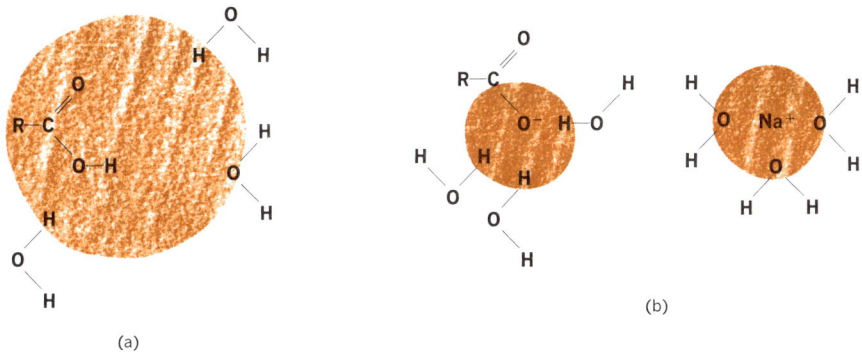

Fig. 26-1 Comparison of solvent action of water on polar and ionic substances: (a) moderate attraction between water molecules and a polar acid molecule, RCOOH; (b) strong attraction between water molecules and the ionic salt, RCOO⁻Na⁺.

26-4 REACTION WITH ALCOHOLS: ESTERIFICATION

Before 1800, a few chemists had discovered that if they heated an organic acid with alcohol, in the presence of sulfuric acid, two products resulted: water and an oily, water-insoluble liquid usually having a pleasant, fruitlike odor. During the same era it was also observed that by proper treatment the fragrant oil could be converted back to the original acid and alcohol. Clearly, this new kind of compound seemed to be some combination of alcohol with an organic acid, achieved by splitting out water. During the 1800's, this reaction was found to be very generally applicable to a great variety of acids in combination with many different alcohols. Eventually the products were called **esters** (apparently a made-up name), and this method of preparing them was termed **esterification** (literally, ester-formation).

The necessity for logically relating the structures of the alcohol, the acid, and the ester played an important part in the development of the concepts of organic structural formulas (Section 21-4). We are now able to write the esterification reaction as in Eq. (26-10).

$$\text{R-O-H} + \underset{\text{H-O}}{\overset{\text{O}}{\underset{\|}{\text{C}}}}\text{-R'} \xrightarrow[\text{heat}]{H_2SO_4} \underset{\text{R-O}}{\overset{\text{O}}{\underset{\|}{\text{C}}}}\text{-R'} + \text{H-O-H} \qquad (26\text{-}10)$$

alcohol   carboxylic acid   ester   water

$$\text{CH}_3\text{CH}_2\text{-O-H} + \underset{\text{H-O}}{\overset{\text{O}}{\underset{\|}{\text{C}}}}\text{-CH}_3 \xrightarrow[\text{heat}]{H_2SO_4} \underset{\text{CH}_3\text{CH}_2\text{-O}}{\overset{\text{O}}{\underset{\|}{\text{C}}}}\text{-CH}_3 + \text{H-O-H} \qquad (26\text{-}11)$$

ethyl alcohol   acetic acid   ethyl acetate

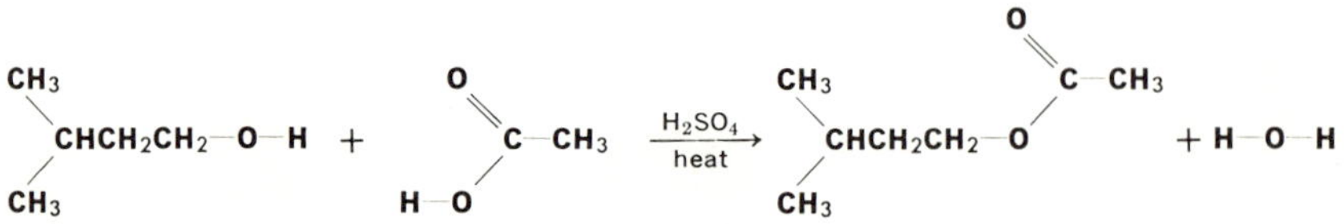

$$\underset{\text{isopentyl alcohol}}{\begin{array}{c}CH_3\\ \diagdown\\ CHCH_2CH_2-O-H\\ \diagup\\ CH_3\end{array}} + \underset{\text{acetic acid}}{\begin{array}{c}O\\ \|\\ C-CH_3\\ |\\ H-O\end{array}} \xrightarrow[\text{heat}]{H_2SO_4} \underset{\text{isopentyl acetate}}{\begin{array}{c}CH_3O\\ \diagdown\|\\ CHCH_2CH_2-O-C-CH_3\\ \diagup\\ CH_3\end{array}} + H-O-H \quad (26\text{-}12)$$

Note that the structure of an ester is described in its name. The group name for the alcohol portion is stated first, followed by the name of the acid component, with its *-ic* ending changed to *-ate*.

The carboxyl group of an ester holds an alkyl group, whereas a carboxylic acid holds a hydrogen. As shown above, the ester name ends with *-ate*. Recall that this is the same system used to name an organic *salt*, in which a metal ion has replaced the hydrogen. Thus, acet*ic* acid (CH_3COOH) yields sodium acet*ate* (CH_3COONa). The same is true for *inorganic* oxygen-containing acids and salts: sulfur*ic* acid (H_2SO_4) gives sodium sulf*ate* (Na_2SO_4). Of course, in chemical *structure* an ester differs from a salt in one very important manner; the alkyl group of the ester is *covalently* bound to the oxygen, whereas the metal in a salt exists as an *ion*. Also, the structural *changes* occurring when a carboxylic acid is converted to an ester are entirely different from those occurring in salt formation. When an acid is converted to a salt there is a direct replacement of the carboxylic hydrogen by a metal ion; see Eqs. (26–2) and (26–6). However, there is good evidence that when an acid is converted to an ester, as in Eq. (26–10), the whole OH group of the acid is replaced by the OR group from the alcohol. Correspondingly, the water produced by the reaction is formed by loss of the OH from the carboxylic acid and the H from the alcohol.

Aromatic carboxylic acids undergo the esterification reaction as readily as do the alkanoic acids, for example,

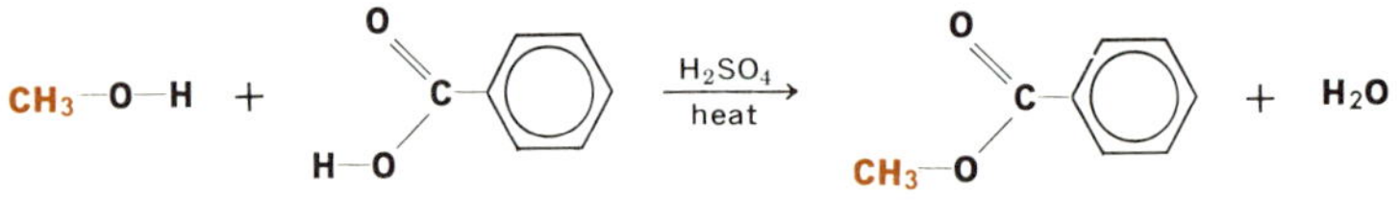

$$\underset{\text{methanol}}{CH_3-O-H} + \underset{\text{benzoic acid}}{\begin{array}{c}O\\ \|\\ C-\bigcirc\\ |\\ H-O\end{array}} \xrightarrow[\text{heat}]{H_2SO_4} \underset{\text{methyl benzoate}}{\begin{array}{c}O\\ \|\\ C-\bigcirc\\ |\\ CH_3-O\end{array}} + H_2O \quad (26\text{-}13)$$

26–5 COMMON ESTERS, NATURAL AND SYNTHETIC

The odors of flowers and fruits are due to mixtures of several organic compounds. It was not until chemists developed skillful techniques that they were able to separate such mixtures into individual compounds and determine their structures. (Some of the more complicated mixtures are still being investigated.) Eventually, chemists found in plant materials a few esters which were identical with those synthesized in the laboratory by the esterification reaction, Eq. (26–10). **Isopentyl acetate,** Eq. (26–12), was detected in bananas, apples, and some other fruits. It is often called *banana oil* (or *amyl* acetate, an old common name). Since it is a good solvent for certain lacquers, cements, and plastics, large quantities are made industrially by the esterification reaction. **Isopentyl butanoate** occurs in cocoa oil, along with several other compounds; by itself it has a distinctly sweet, fruity odor resembling pears. **Ethyl acetate,** Eq. (26–11), is found in wine, pineapples, and other fruits. It is a valuable solvent used in nail polish removers and in the manufacture of perfumes, plastics, paints, and photographic films.

Methyl benzoate, Eq. (26–13), occurs in a few plant oils; its odor is somewhat medicinal rather than sweet. Another example of an ester derived from an aromatic acid is methyl salicylate (Section 26–9D).

Several other simple esters, not known to occur in fruits or flowers, have nevertheless been prepared and used for decades by manufacturers of artificial flavors and scents. **Methyl butanoate** resembles the scent of apples. **Butyl butanoate** has a fine odor useful in pineapple essence. **Ethyl butanoate** is used in artificial peach flavor, as well as in pineapple and apricot. **Pentyl propanoate** resembles apricot essence.

The effects of vapors on the human nose are intriguing, but not well understood. For instance, the sweet, fruity odors of the simple esters present a marked contrast to the odors of the acids from which they are derived. Methanoic (formic) acid and ethanoic (acetic) acid have pungent, stinging odors. The larger acids, butanoic, hexanoic, and octanoic, have goatlike or rancid odors. Propanoic acid, of intermediate size, has some of both the pungent and the rancid qualities. Acids larger than about decanoic are so slightly volatile that we detect very little odor in their presence.

The esters discussed so far have been derived from alcohols. Esters of phenols are also of frequent occurrence in nature and in the laboratory. One example of a phenolic ester is **phenyl acetate;** note very carefully the structural difference between this ester and methyl benzoate, Eq. (26–13).

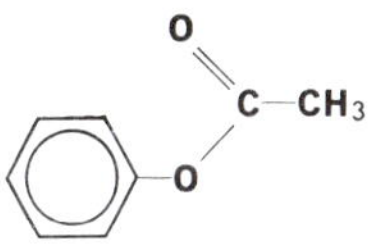

phenyl acetate

The phenols will not react directly with a carboxylic acid to form an ester (another case in which phenols differ from alcohols). However, esters of phenols are easily synthesized by an indirect method discussed in Section 26–7.

All told, the ester group is very common in nature. It occurs in fats, to be discussed in Chapter 27, and in many simple compounds, as we have seen in this chapter. The ester group may also appear in a complex molecule, along with several other functional groups, as shown by the structures below.

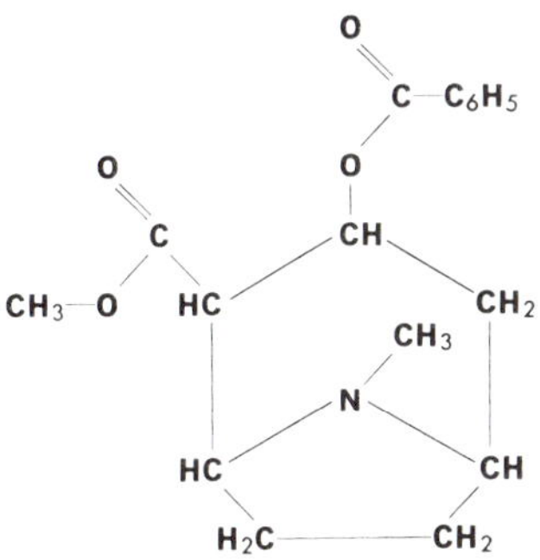

cocaine

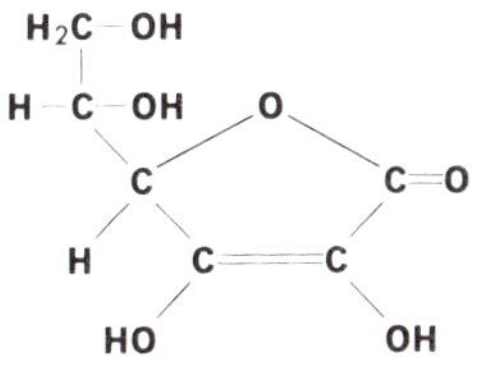

vitamin C

Cocaine, produced by the coca bush of Bolivia and Peru, is a habit-forming narcotic if taken internally. In medicine it is used as a surface anesthetic. Note that cocaine has two different ester groups and a basic amine group (Section 28–2). Note also the interesting location of the ester group in **vitamin C.**

26–6 HYDROLYSIS AND SAPONIFICATION OF ESTERS

In Section 26–4, we learned that after chemists discovered esters they also discovered how to split them into the component alcohol and acid. From the beginning, this has been a useful, frequently-applied manipulation. One method for accomplishing the cleavage is now called hydrolysis; another is saponification.

An ester can be cleaved by heating it in water containing a little strong acid.

$$\underset{\text{an ester}}{CH_3CH_2-O-\overset{O}{\underset{\|}{C}}-CH_3} + \underset{\text{water}}{H-O-H} \xrightarrow[\text{heat}]{H^+} \underset{\text{an alcohol}}{CH_3CH_2-O-H} + \underset{\text{a carboxylic acid}}{\overset{O}{\underset{\|}{C}}-CH_3 \atop H-O} \quad (26\text{–}14)$$

A chemical decomposition, such as this one, due to reaction with water is termed **hydrolysis** (Gr. *hydro*, water plus *lysis*, loosen, break away). In this particular case, the catalyst is a strong acid. Any of the common mineral acids are effective, for example, hydrochloric, sulfuric, or phosphoric acid.

The ester-hydrolysis reaction just written, Eq. (26–14), looks suspiciously like the equation for esterification written backward, Eq. (26–10). Furthermore, we find experimentally that *complete* hydrolysis is not achieved in the reaction mixture; some of the alcohol and the carboxylic acid molecules thus produced combine with each other to recreate some ester molecules and water. We often show in one equation, Eq. (26–15), the relation between these two easily reversible reactions:

$$CH_3CH_2-O-\overset{O}{\underset{\|}{C}}-CH_3 + H-O-H \underset{B}{\overset{A}{\underset{H^+ \text{ heat}}{\rightleftarrows}}} CH_3CH_2-O-H + \overset{O}{\underset{\|}{C}}-CH_3 \atop H-O \quad (26\text{–}15)$$

As written here, the forward reaction, A, would be hydrolysis; the reverse reaction, B, would be esterification. The heat and catalyst have the same influence on both reactions, merely speeding up the action. This observation, and many others, causes us to believe that exactly the same events of bond-breaking and bond-making occur as the reactions proceed in either direction.

On a practical basis, we can carry out either a hydrolysis or an esterification, depending on which materials we put into the flask. However, because the products can react with each other to give back the starting materials, we never get complete conversion. That is, if we attempt to hydrolyze an ester to an alcohol plus a carboxylic acid, we will obtain *chiefly* those products; but we must expect that there still will be *some* ester in our reaction mixture. Similarly, if we attempt an esterification reaction, Eqs. (26–15B) or (26–10) the conversion to ester will not be a complete one.

A second method for cleaving an ester is to treat it with an aqueous solution of a metallic hydroxide, most often sodium hydroxide. The reaction can be conducted at room temperature within a reasonable time, but as usual can be hastened by heat.

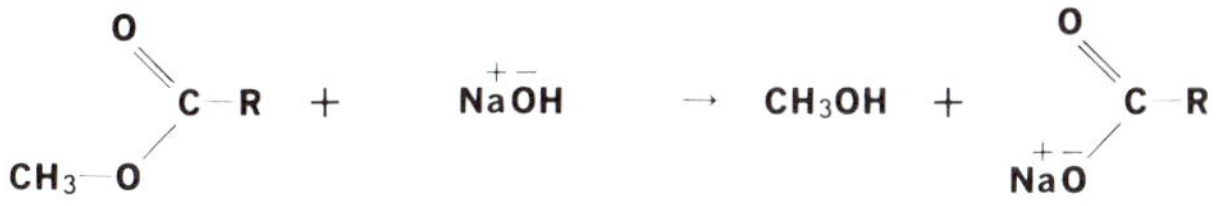

$$\underset{\text{an ester}}{\underset{CH_3-O}{\overset{O}{\overset{\|}{C}}}-R} + \underset{\substack{\text{sodium hydroxide}\\\text{(in water)}}}{Na^+OH^-} \rightarrow \underset{\substack{\text{an}\\\text{alcohol}}}{CH_3OH} + \underset{\substack{\text{sodium salt}\\\text{of an acid}}}{\underset{Na^+O^-}{\overset{O}{\overset{\|}{C}}}-R} \qquad (26\text{-}16)$$

Because of the base used, one of the products of this reaction is a salt of an acid rather than the acid itself; the other cleavage product is an alcohol, as in the hydrolysis reaction, Eq. (26–14).

If the R-group in this example happens to have a long, straight chain (12 to 18 carbons), the salt produced is a soap. Indeed, we will see in Section 27–5 that treatment with sodium hydroxide solution is used to produce soap from fats (a special class of esters). The term **saponification** (literally, soap making) is now applied quite broadly to the reactions of hydroxides with any kind of ester (Eq. 26–16).

Although the *salt* of an acid is produced by saponification, mere acidification of the reaction mixture (for example, with hydrochloric acid) immediately converts the salt to the corresponding acid.

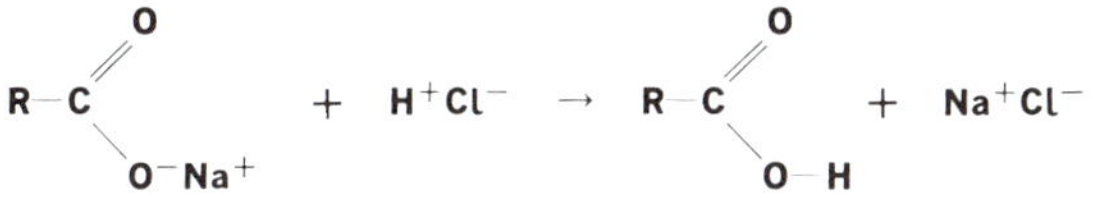

$$\underset{O^-Na^+}{R-\overset{O}{\overset{\|}{C}}} + H^+Cl^- \rightarrow \underset{O-H}{R-\overset{O}{\overset{\|}{C}}} + Na^+Cl^- \qquad (26\text{-}17)$$

With this simple after-treatment, saponification can give the same net result as hydrolysis; an ester can be split into its component alcohol and carboxylic acid. Indeed, saponification may often be the preferred method because the cleavage is complete; the salt of the carboxylic acid cannot recombine with the alcohol to give back some of the ester.

26–7 FORMATION AND REACTIONS OF ACID CHLORIDES

A carboxylic acid can be converted to an acid chloride by treatment with phosphorous trichloride, for example:

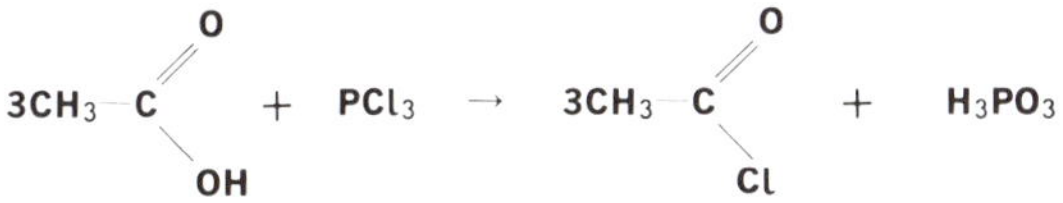

$$\underset{\text{acetic acid}}{3CH_3-\underset{OH}{\overset{O}{\overset{\|}{C}}}} + PCl_3 \rightarrow \underset{\substack{\text{acetyl chloride}\\\text{(an acid chloride)}}}{3CH_3-\underset{Cl}{\overset{O}{\overset{\|}{C}}}} + \underset{\substack{\text{phosphorous}\\\text{acid}}}{H_3PO_3} \qquad (26\text{-}18)$$

The acid chloride is a highly reactive substance which can be converted to still other acid derivatives. For example, it reacts with an alcohol to produce an ester:

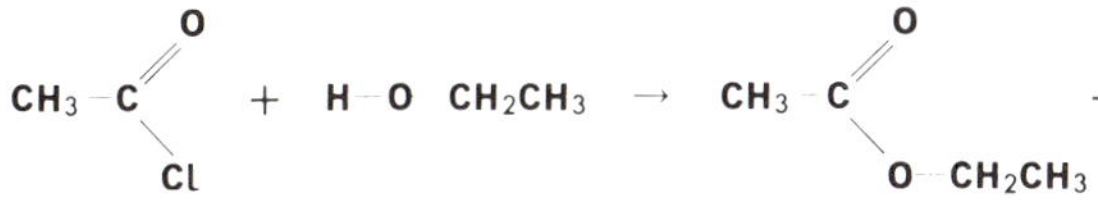

$$\underset{\text{acetyl chloride}}{CH_3-\underset{Cl}{\overset{O}{\overset{\|}{C}}}} + \underset{\text{ethanol}}{H-O-CH_2CH_3} \rightarrow \underset{\text{ethyl acetate}}{CH_3-\underset{O-CH_2CH_3}{\overset{O}{\overset{\|}{C}}}} + HCl \qquad (26\text{-}19)$$

An advantage of this indirect, two-step pathway from the acid to the ester, Eqs. (26–18) and (26–19), is that both steps give complete conversion to the desired products.

378 Organic Acids

This may be important if either the alcohol or the acid is expensive. Of course, the ester could also have been produced in one step by direct reaction of the alcohol with acetic acid (esterification—see Eq. 26–11). However, the conversion to ethyl acetate would not be very high because esterification is a reversible reaction (Section 26–6).

The reactive acid chloride also provides a method for synthesizing an ester of a phenol.

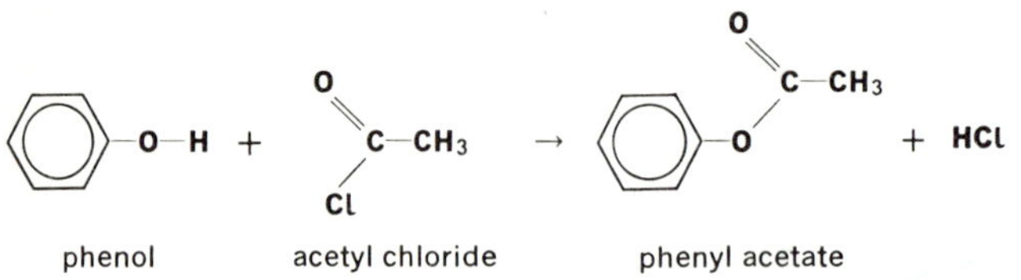

$$\text{phenol} + \text{acetyl chloride} \longrightarrow \text{phenyl acetate} + HCl \qquad (26\text{–}20)$$

Direct esterification of a phenol with a carboxylic acid is not possible.

Amides are another type of acid derivative which can be synthesized via acid chlorides (Section 28–4).

26–8 SOURCES OF ACIDS — Know

Carboxylic acids may be synthesized by a great variety of methods. A few of the common procedures will be discussed in this section. Other methods are of interest primarily to professional chemists.

A. By Oxidation

Often the acids may be obtained by oxidation of other organic compounds, on either a laboratory or an industrial scale. Primary alcohols, as well as aldehydes, may be oxidized to acids (Eq. 25–5). In the laboratory these oxidations are usually accomplished by reagents such as sodium dichromate or potassium permanganate. For commercial production it is desirable to use an inexpensive oxidizing agent such as oxygen from air.

We learned in Section 24–7 that an alkylbenzene can be oxidized to benzoic acid. For industrial production, a vanadium pentoxide or manganese acetate catalyst makes possible the use of air for the oxidation. Nearly all benzoic acid is obtained in this way.

B. From Organic Cyanides

Hydrolysis of an organic cyanide provides an important laboratory method for the preparation of a carboxylic acid.

$$R\text{–}C\equiv N + 2H_2O + HCl \xrightarrow{\Delta} R\text{–}C\underset{O}{\overset{OH}{\diagup}} + NH_4{}^+Cl^- \qquad (26\text{–}21)$$

an organic cyanide (a nitrile)   a carboxylic acid

This reaction of an organic cyanide with water requires a catalyst. Either acid or base is an effective catalyst. However, if acid is used, as in Eq. (26–21), the carboxylic acid is obtained directly, without the necessity of acidifying a basic mixture after the hydrolysis reaction is complete.

The organic cyanide is also sometimes called a *nitrile*. It is important to note that the cyanide group is *covalently* bonded to the rest of the carbon skeleton in an organic compound, whereas in inorganic salts a cyanide group commonly exists as an ion.

The organic cyanide required for reaction (26–21) can be obtained in a variety of ways. One common starting material is an alkyl halide, which can undergo displacement by a cyanide ion:

$$\overset{+\ -}{\text{NaCN}} + \text{CH}_3\text{CH}_2\text{CH}_2\text{Br} \rightarrow \text{CH}_3\text{CH}_2\text{CH}_2\text{CN} + \overset{+\ -}{\text{NaBr}} \tag{26-22}$$

The reaction shown in Eq. (26–22) does not work for an *aryl* halide (e.g., chlorobenzene or bromobenzene), because in an aryl compound the halogen is held too tightly to allow displacement. However, a different route is possible; aryl amines (Section 28–3C) can be converted to aryl cyanides by the following series of reactions:

$$\text{ArNH}_2 \xrightarrow[\text{HCl}]{\text{HNO}_2} \text{ArN}_2{}^+\text{Cl}^- \xrightarrow{\text{CuCN}} \text{ArCN} \tag{26-23}$$

C. From Fats

By hydrolyzing a natural fat, one may obtain a mixture of octadecanoic acid (stearic acid) and certain other long-chain acids (Sections 27–1 and 27–2).

26–9 OTHER ACIDS, ESTERS, AND SALTS OF INTEREST

A. Acetic Acid

Dilute solutions of acetic acid (vinegar) have been used from antiquity and are still being used to preserve meat, fish, pickles, and other foods. Acetic acid arrests the growth of various microorganisms, preventing spoilage of the food to which it is added.

Acetic acid which is nearly pure, and undiluted by water, is called *glacial* acetic acid, so named because it freezes at about 16°C. Industrially, is it as important among organic compounds as sulfuric acid is among the inorganics. A weak acid, it is a useful acidulant in the dyeing and processing of textiles and in the coagulation of rubber latex. Acetic acid is one of the intermediates essential for the manufacture of cellulose acetate film, acetate textile fibers, and innumerable pharmaceuticals. The insecticide Paris green is a mixture of copper acetate, $Cu(CH_3COO)_2$, and copper arsenite, $Cu(AsO_2)_2$.

B. Benzoic Acid

Benzoic acid is an intermediate for the synthesis of many drugs and dyes. Increasing amounts are being used to improve the quality of alkyd enamels. Sodium benzoate, C_6H_5COONa, has long been used as a food preservative. It is most satisfactory in somewhat acid foods (pH 4 or lower), particularly in fruit juices, catsup, pickles, pie fillings, jams, and margarine. Sodium benzoate is an effective bactericide at the customary concentration of 0.1%, and is tasteless and nontoxic. It is also a preservative for drugs, cosmetics, toothpastes, gum, and starch.

C. Dicarboxylic Acids

In a number of organic compounds we may find two or more carboxyl groups in one molecule. The simplest example is **oxalic acid,** HOOC—COOH, which is just two carboxyl groups bonded together. Its name arises from the fact that it is the sour material in wood sorrel, botanically called oxalis (Gr. *oxys*, acid). Since both the hydrogens in oxalic acid can be neutralized by a base, it can form either normal salts such as NaOOC—COONa, sodium oxalate, or acidic salts such as HOOC—COOK, potassium hydrogen oxalate. It is the latter form which occurs in sorrel plants, rhubarb, and spinach. Apparently the amount of rhubarb or spinach we eat, or the manner in which we eat it, does not provide a harmful dose of oxalic acid. However, the acid and its salt are definitely toxic. Livestock have died because of its presence in the poisonous weed halogeton. Oxalic acid can be used in some circumstances to remove ink stains and rust. It is quite effective for cleaning the rusty scale from the insides of auto radiators.

Although many dicarboxylic acids are well known, we will mention just one more for now. **Succinic acid,** HOOC—CH_2—CH_2—COOH, apparently occurs in almost all plants and animals. It was first mentioned in 1550 by Agricola, who isolated some from amber (L. *succinum*). The acid is an important intermediate for the manufacture of numerous pharmaceuticals and polymers. The salt sodium succinate is an antidote for poisoning by heavy metals or by barbiturate drugs.

D. Hydroxy Acids

Another structural possibility is that a molecule may have both hydroxyl and carboxyl functional groups. We have already mentioned (Section 26–1) that sour milk contains **lactic acid** (2-hydroxypropanoic acid). This acid may also occur in fatigued muscle tissue.* Lactic acid is a common acidulant for food products because its mild acid taste does not overpower other flavors; a few examples of its use are in soups, olives, beer, soft drinks, cheese, and sherbets. The lactic acid for these purposes is commercially obtained by the action of *lactobacillus* organisms on molasses or starch.

Tartaric acid, a by-product of wine-making, is structurally dihydroxysuccinic acid:

$$\text{HOOC—CH—CH—COOH} \atop \phantom{\text{HOOC—}} \text{OH} \text{OH}$$

During the aging of wine, a substance, originally present in the grape juice, crystallizes on the inside of the barrel. The medieval Greeks called this hard crust tartaron; in modern English it is tartar. The purified material, being white, is called *cream of tartar*. We now know that this is the potassium hydrogen salt of tartaric acid. Cream of tartar is widely used in baking powders. It is only slightly soluble at room temperature. However, at baking temperatures it dissolves and reacts with the sodium hydrogen carbonate also present in the baking powder, releasing bubbles of carbon dioxide. Tartaric acid itself may be obtained when desired by treating crude tartar with sulfuric acid. Tartaric acid is used in some foods, soft drinks, and metal polishes.

---

* The breakdown of glucose, through a complex series of reactions, releases the energy needed for movement and body heat. Under certain circumstances, lactic acid may be one of the end products of glucose metabolism.

Citric acid (L. *citrus*) is a more complex substance having three carboxyl groups and one alcohol group.

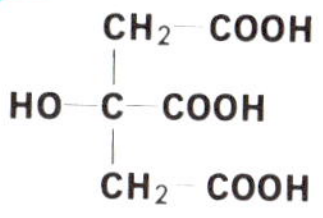

$$\begin{array}{l} CH_2-COOH \\ | \\ HO-C-COOH \\ | \\ CH_2-COOH \end{array} \quad \text{citric acid}$$

This acid occurs in several berries and other fruits, but especially in lemons, oranges, etc. In fact, it is present in very small amounts in all living cells which derive energy from the metabolism of carbon compounds. In food processing, citric acid is used more than any other solid organic acid because it is nontoxic, very soluble in water, and has a pleasant, mildly sour taste. A few of its applications are in fruit and vegetable juices, candies, desserts, jellies, frozen fruits, soft drinks, and effervescent tablets; others are in cosmetics, hair rinses, rust and scale removers, and bottle-washing mixtures. It is usually the most satisfactory acidulant for drug preparations.

Although citric acid was formerly isolated from citrus wastes, since about 1925 it has been produced more economically by growing a fungus, *aspergillus niger*, in a glucose solution. Citric acid is a metabolic product elaborated by the fungus as it consumes the glucose. The mat of fungus is then filtered off, and the citric acid is crystallized from the solution.

The salt sodium citrate, in conjunction with citric acid, is valuable for its buffering ability in jellies, ice cream, candy, gelatin desserts, and whipping cream; the setting of these foods depends on the proper pH. The sodium citrate-citric acid mixture is likewise the most desirable buffer for medicines. In samples of human blood collected for transfusion, the mixture acts as a buffer and anticoagulant.

Salicylic acid and its derivatives are an interesting family. They constitute about half of all the "coal tar" drugs (i.e., benzenoid and related compounds) manufactured in the United States.

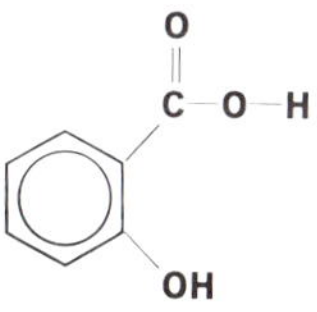

salicylic acid	methyl salicylate "oil of wintergreen"	acetyl salicylic acid "aspirin"

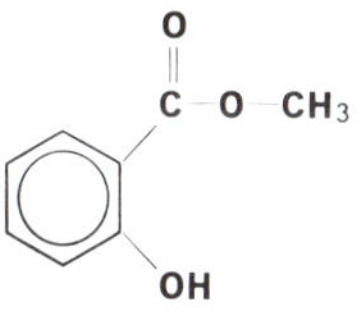

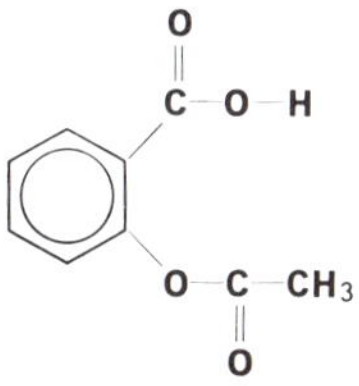

Salicylic acid, a better disinfectant than phenol, is a common ingredient of topical ointments for skin diseases. It causes the outer layer of skin to flake off, but does not kill the underlying tissue.

The ester methyl salicylate is a fragrant oil occurring in numerous plants, especially wintergreen. It is a flavoring agent for candy, gum, foods, and dentifrices, and a constituent of some antiseptics and perfumes. Methyl salicylate is used in liniments as a counterirritant. The mild surface inflammation it creates affects the circulation in a manner which relieves sore muscles. Commercial quantities of methyl salicylate are produced by esterification of salicylic acid with methanol.

Aspirin is the most widely used synthetic drug because it is cheap, relatively safe, easily available, and quite effective in reducing fever and in relieving headaches or similar discomforts. Recently, enough aspirin has been produced in the United States each year to provide every person in the country with about 200 of the standard 5-grain tablets; this amounts to about 40 tons per day! Note in the structure of aspirin that the phenolic group has been converted to an acetate ester.

E. Waxes

Waxes are produced by a number of plants and animals. The wax in our ears and beeswax are typical of those produced by animals. Carnuba wax, from the leaves of certain Brazilian palm trees, makes an excellent polish for floors and automobiles. These waxes, although often mixtures of compounds, are chiefly esters in which both the acid portion and the alcohol portion have very long chains. A common example is $C_{27}H_{55}COOC_{30}H_{61}$. The simple term *wax* is usually reserved for these ester compounds. Modified names are used for other substances having waxy properties. For instance, recall that paraffin wax is a mixture of long-chain hydrocarbons.

IMPORTANT TERMS

Acid chloride	Carboxylate salt	Ester	Hydrolysis
Alkanoic acid	Carboxylic acid	Esterification	Saponification

WORK EXERCISES

1. Give the structural formula of
   - a) propanoic acid
   - b) hexanoic acid
   - c) benzoic acid
   - d) sodium acetate
   - e) potassium benzoate
   - f) ethyl benzoate
   - g) isopropyl acetate
   - h) 3-chlorobenzoic acid
   - i) a wax

2. Write complete equations for the following reactions:
   - a) formic acid, sodium hydroxide
   - b) acetic acid, sodium hydrogen carbonate
   - c) benzoic acid, ethanol, H_2SO_4, heat
   - d) sodium benzoate, HBr
   - e) ethyl butanoate, H_2O, H_2SO_4, heat
   - f) butanoic acid, isopentyl alcohol, H_2SO_4, heat
   - g) methyl benzoate, dilute NaOH, heat
   - h) CH_3CH_2CN, H_2O, heat

3. Show how
   - a) hexanoic acid could be obtained from 1-hexanol
   - b) benzoic acid could be obtained from toluene

4. Write the equation for the reaction of methanol with *p*-bromobenzoic acid to produce an ester.
   - a) Which would you estimate would be less expensive, the methanol or the *p*-bromobenzoic acid?
   - b) If you were to prepare this ester in the laboratory, what conditions, concentrations, etc., would you use to make maximum use of the more expensive component?

5. Write a systematic name for each compound:

a) 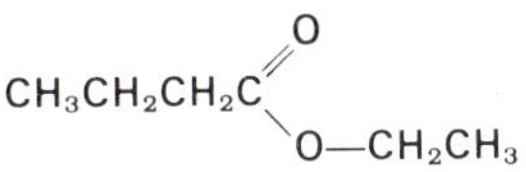 $CH_3CH_2CH_2\overset{\displaystyle O}{\underset{\displaystyle O-CH_2CH_3}{C}}$

b) 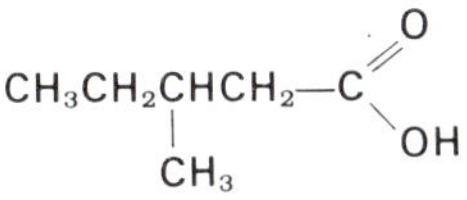 $CH_3CH_2\underset{\displaystyle CH_3}{CH}CH_2-\overset{\displaystyle O}{\underset{\displaystyle OH}{C}}$

c) $\text{Ph}-\overset{\displaystyle O}{\underset{\displaystyle O^-NH_4^+}{C}}$

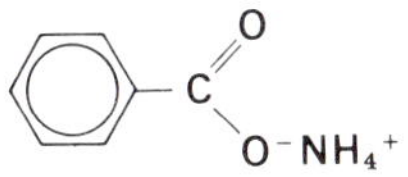

d) 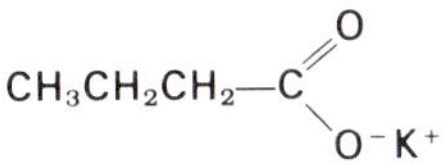 $CH_3CH_2CH_2-\overset{\displaystyle O}{\underset{\displaystyle O^-K^+}{C}}$

6. Large amounts of phthalic acid are required for the manufacture of glyptal enamels (Section 30–7) and a variety of other substances. Show an industrial method for economically converting o-xylene (1,2-dimethylbenzene) to phthalic acid (benzene-1,2-dicarboxylic acid).

7. Show reactions that could be used to convert 1-butanol to:
   a) butanoic acid.
   b) pentanoic acid. Hint: This will require more than one step and is a more difficult problem than part (a). If you get stuck, consult Eq. (25–8) for a reaction that might give a useful intermediate.

8. List all the types of structural units (both carbon skeletons and functional groups) which you can find in the compounds below.
   a) oil of wintergreen (Section 26–9D)
   b) ferulic acid, an intermediate formed during the biosynthetic processes of many plants:

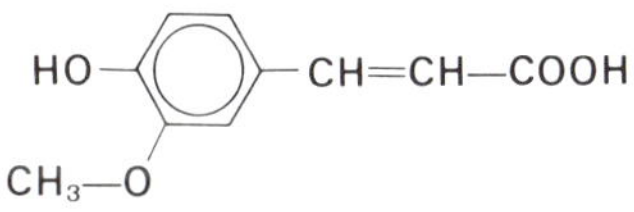

   HO—C$_6$H$_3$(OCH$_3$)—CH=CH—COOH

SUGGESTED READING

Collier, H. O. J., "Aspirin," *Scientific American*, Nov. 1963 (Offprint #169). There is now a better understanding of the reasons for the dramatic effectiveness of the most widely used drug.

DePuy, C. H., and K. L. Rinehart, Jr., *Introduction to Organic Chemistry*, Chapter 11, John Wiley, New York, 1967.

Hart, H., and R. D. Schuetz, *Organic Chemistry*, 4th ed., Chapter 10, Houghton Mifflin, Boston, 1972.

Jacobson, M., and M. Beroza, "Insect Attractants," *Scientific American*, Aug. 1964 (Offprint #189). The sex attractants excreted by insects are being isolated so that the chemical structure can be determined. If an attractant can be synthesized, it can be used to lure insects to traps. The structures discovered so far include alcohols, esters, and other types.

Noller, C. R., *Textbook of Organic Chemistry*, 3rd ed., Chapters 11 and 27, W. B. Saunders, Philadelphia, 1966.

THE CHEMISTRY OF FATS

27-1 STRUCTURE AND HYDROLYSIS OF FATS

Typical fats, such as bacon fat or beef tallow, can be hydrolyzed to glycerol and a mixture of fatty acids. From detailed studies of the hydrolysis fragments and of the other properties of fats, it is certain that fats *are esters of glycerol with fatty acids.*

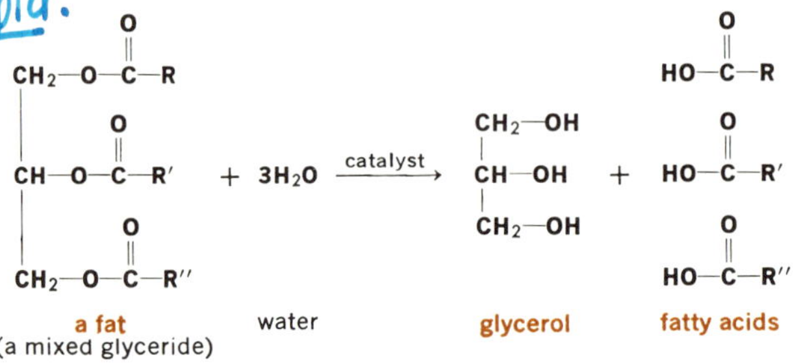

$$\begin{array}{c} CH_2-O-\overset{O}{\overset{\|}{C}}-R \\ | \\ CH-O-\overset{O}{\overset{\|}{C}}-R' \\ | \\ CH_2-O-\overset{O}{\overset{\|}{C}}-R'' \end{array} + 3H_2O \xrightarrow{catalyst} \begin{array}{c} CH_2-OH \\ | \\ CH-OH \\ | \\ CH_2-OH \end{array} + \begin{array}{c} HO-\overset{O}{\overset{\|}{C}}-R \\ HO-\overset{O}{\overset{\|}{C}}-R' \\ HO-\overset{O}{\overset{\|}{C}}-R'' \end{array} \quad (27\text{-}1)$$

a fat   water   glycerol   fatty acids
(a mixed glyceride)

The term **glyceride** is a special name designating an ester of glycerol; the term triglyceride is also frequently used (Section 35–3). Since there is a variety of different acids combined in a natural fat, it is often called a **mixed glyceride**. The arrangement of R-groups in the fat structure shown above is just one typical example. In another molecule from the same sample of fat the arrangement might be R′, R, R″ or R′, R, R or R, R″; R, etc. Therefore, a fat is not a pure compound by the usual chemical definition that all molecules in a sample are identical. (See Table 27–2.)

In the laboratory, hydrolysis of a fat can be accomplished with a mineral acid, such as hydrochloric or phosphoric, acting as catalyst.

The chemical reaction taking place during the digestion of fatty food in the intestines of man or an animal is another example of the hydrolysis reaction, Eq. (27–1). The catalyst for the reaction is an enzyme.

In certain tissues of plants and animals, *biosynthesis* of fats occurs.

$$\text{fatty acids} + \text{glycerol} \xrightarrow{\text{enzyme}} \text{fat} + \text{water} \quad (27\text{-}2)$$

TABLE 27-1 Common fatty acids

Structural Formula and Common Name	Abbreviated Formula
$CH_3(CH_2)_{10}COOH$ lauric acid	$C_{11}H_{23}COOH$
$CH_3(CH_2)_{12}COOH$ myristic acid	$C_{13}H_{27}COOH$
$CH_3(CH_2)_{14}COOH$ palmitic acid	$C_{15}H_{31}COOH$
$CH_3(CH_2)_5CH{=}CH(CH_2)_7COOH$ palmitoleic acid	$C_{15}H_{29}COOH$
$CH_3(CH_2)_{16}COOH$ stearic acid	$C_{17}H_{35}COOH$
$CH_3(CH_2)_7CH{=}CH(CH_2)_7COOH$ oleic acid	$C_{17}H_{33}COOH$
$CH_3(CH_2)_4CH{=}CHCH_2CH{=}CH(CH_2)_7COOH$ linoleic acid	$C_{17}H_{31}COOH$
$CH_3CH_2CH{=}CHCH_2CH{=}CHCH_2CH{=}CH(CH_2)_7COOH$ linolenic acid	$C_{17}H_{29}COOH$

This synthesis is the reverse of the hydrolysis reaction, Eq. (27-1). Plants first synthesize fatty acids and glycerol from very simple molecules and then combine them, Eq. (27-2), to produce a fat. Animals can do the same, but they also obtain large amounts of fatty acids and glycerol from digested food.

27-2 FATTY ACID CONSTITUENTS

Since the glyceride structure is common to all fats, the variations in properties from one type of fat to another must be due to differences in their fatty-acid components. It is therefore convenient to describe fats in terms of the fatty acids present.

In general, a natural fatty acid has a skeleton composed of an even number of carbon atoms in a long, unbranched chain. A chain length of sixteen or eighteen carbons is predominant. Carbon–carbon double bonds are common. The most typical acids found in fats are listed in Table 27-1. Common names, rather than systematic names, are often used for these acids; origins of the names are discussed in Section 35-2.

Note that three of the common unsaturated acids, oleic, linoleic, and linolenic, have eighteen-carbon skeletons like that of stearic acid, but have a certain number of carbon–carbon double bonds. Consequently, in this family one formula differs from the next by two hydrogens (see the last four formulas in Table 27-1). Two sixteen-carbon acids, palmitic and palmitoleic, are similarly related to each other.

The curious fact that nearly all natural fatty acids possess an even number of carbon atoms was explained by experiments in the 1950's (Section 40-5). It was found that most organisms capable of synthesizing fatty acids do so by using acetate groups (CH_3COO^-), which are two-carbon building blocks. There are a few exceptions. It

TABLE 27-2 Composition of some fats and oils

Fat or Oil	Fatty Acid Present, % by Weight[a]						
	lauric C_{12} sat'd	myristic C_{14} sat'd	palmitic C_{16} sat'd	stearic C_{18} sat'd	oleic C_{18} one C=C	linoleic C_{18} two C=C	linolenic C_{18} three C=C
lard		1–2	25–30	12–16	40–50	5–10	1
beef tallow		3–5	25–30	20–30	40–50	1–5	
mutton tallow		1–5	20–25	25–30	35–45	3–6	
butterfat (cow)[b]	2–5	8–14	25–30	9–12	25–35	2–5	
coconut fat[c]	45–48	16–18	8–10	2–4	5–8	1–2	
palm kernel fat[d]	43–47	15–20	8–9	2–5	10–18	1–3	
palm oil or fat		1–3	35–45	4–6	40–50	8–11	
peanut oil			8–10	3–5	55–60	25–30	
sardine oil		5–6	12–16	2–3	(75–82)[e]		
olive oil			8–16	2–3	70–85	5–15	
cottonseed oil		1	20–25	1–2	20–30	45–50	
soybean oil			10	3	25–30	50–55	4–8
safflower oil			6	3	13–15	75–78	
linseed oil					20–35	15–25	40–60
tung oil			4–6	5–10	9–12	(76–82)[f]	

[a] The approximate range of typical values for each fat is indicated by these figures which were compiled from various sources.
[b] Also 3–4% butyric acid and 1–3% each of C_6, C_8, C_{10} acids.
[c] Also 5–9% each of C_8 and C_{10} acids.
[d] Also 4% C_8 and 4–8% C_{10} acids.
[e] Total unsaturated acids, 75–82%, including 10–15% palmitoleic; some as large as C_{24} with up to six C=C.
[f] Tung oil contains eleostearic acid rather than linolenic; the three C=C are in different locations.

has been discovered that certain bacteria can produce fatty acids having methyl branches and an odd number of carbons. One such organism is *mycobacterium tuberculosis*. Branched-chain or odd-numbered acids are apparently rare in higher plants and animals.

Acids containing fewer than twelve carbons, or more than twenty, occur in some fats but are less common.

The composition of a particular fat can be determined by hydrolyzing it, Eq. (27–1), and then analyzing the resultant mixture of fatty acids. The percent, by weight, of each acid present in various fats is listed in Table 27–2.

27–3 FATS AND OILS

It has become customary to call a glyceride a fat if it is solid, or an oil if it is liquid, at ordinary temperatures. This differentiation of course is arbitrary and depends on the climate. The differences in melting points of the glycerides are due chiefly to the varying numbers of double bonds present; the more of these there are, the lower the melting point. For example, olive oil and cottonseed oil have considerably higher percentages of unsaturated acids than do lard and tallow (Table 27–2).

Palm oil or fat represents an interesting borderline case. It is usually called an oil because it is a liquid as it originates in the tropics. However, in the temperate regions it may be a semisolid. In palm oil the total percentage of unsaturates (oleic plus linoleic) is only slightly higher than that of lard or tallow (Table 27–2).

The *length* of the fatty acid chains can also influence the melting point. As in other series of organic compounds, shorter chains mean lower melting points. Although coconut fat and palm kernel fat are highly saturated, they soften at a lower temperature than does lard because of their high percentage of lauric acid. Another interesting example is butter. Although it is quite saturated, it is not as stiff as tallow because butter has a moderate number of short chain acids. There are few other examples of this effect among the glycerides because most of them are constituted of C_{16} and C_{18} acids.

The data in Table 27–2 suggest that animal glycerides are usually fats, while plant glycerides are usually oils. However, this differentiation does not always hold since the environmental temperature of the organism does have an effect on the form of the glyceride. Thus, for example, the sardine glycerides are oils. On the other hand, the tropical plants (coconut, palm) produce higher-melting glycerides. Experiments have shown that when two groups of the same plants or animals are raised at different temperatures, the group raised at the lower temperature produces more unsaturated glycerides (Section 40–4). In general, however, the oils are found in plants or in animals living in cold environments.

Animal fats may slowly become rancid if exposed to air at room temperature. The attack of oxygen on unsaturated groups produces smaller acids and aldehydes, many of which have foul odors. Although plant oils are more unsaturated, they are less susceptible to rancidity. They normally contain significant amounts of tocopherols (vitamin E) which inhibit the oxidation. Animal fats may be protected by the addition of an anti-oxidant such as BHT (Section 25–6).

When the term *oil* is used in connection with plant products, it refers to the glycerides which are structurally esters. Petroleum oil, however, consists of hydrocarbons (Section 22-4).

27-4 HYDROGENATION OF OILS

For some purposes, solid fats rather than liquid oils are preferred. Oils, which are more unsaturated, can be "hardened" to higher-melting solids by the addition of hydrogen to their double bonds. The hydrogenation reaction, first described for alkenes (Section 23-5B), can be applied to oils:

An oil
(lower melting; liquid at room temperature)

$$\begin{array}{l} CH_2-O-\overset{O}{\underset{\parallel}{C}}-(CH_2)_7CH=CHCH_2CH=CH(CH_2)_4CH_3 \\ | \\ CH-O-\overset{O}{\underset{\parallel}{C}}-(CH_2)_{16}CH_3 \\ | \\ CH_2-O-\overset{O}{\underset{\parallel}{C}}-(CH_2)_7CH=CH(CH_2)_7CH_3 \end{array} \quad + \; 3H_2 \;\xrightarrow[\text{heat}]{\text{Ni}}\;$$

A fat
(higher melting; solid at room temperature)

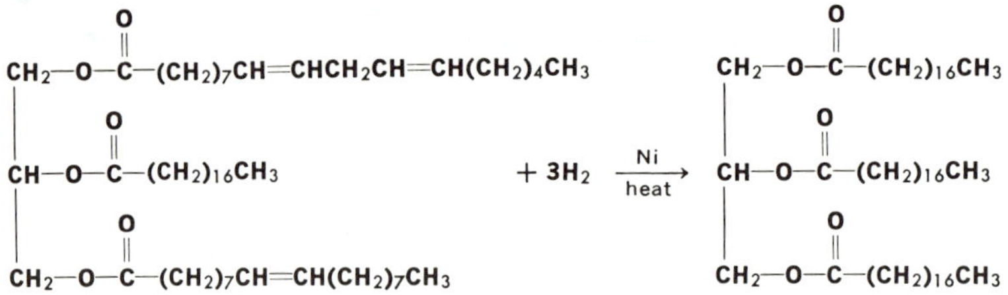

$$\begin{array}{l} CH_2-O-\overset{O}{\underset{\parallel}{C}}-(CH_2)_{16}CH_3 \\ | \\ CH-O-\overset{O}{\underset{\parallel}{C}}-(CH_2)_{16}CH_3 \\ | \\ CH_2-O-\overset{O}{\underset{\parallel}{C}}-(CH_2)_{16}CH_3 \end{array} \quad (27\text{-}3)$$

glyceryl tristearate

Note that in this example of an oil molecule there are two unsaturated side chains, one oleic and one linoleic. In the resultant fat, all three groups are stearic, so the product is called glyceryl tristearate.

The reaction is carried out by bubbling hydrogen into a tank of the hot oil. A little powdered nickel is dispersed in the oil to catalyze the hydrogenation—the nickel can be removed later.

In this manner cheap, abundant vegetable oils, such as cottonseed or soybean oils, can be converted to oleomargarine, cooking greases (Crisco, Spry, etc.), or stocks to be further processed into soap (Section 27-5). The hydrogenation can be controlled to provide the degree of firmness most suitable for the product desired. For oleomargarine a certain amount of milk, vitamins A and D, emulsifying agents, flavors, and yellow food colors are added to the hardened oil. It should be noted that the same coloring agents are frequently added to butter to improve its appearance. When cows are not on fresh pasture the color of the butter is pale and the vitamin A content decreases by about half.

27-5 SAPONIFICATION

It was shown in Section 26-6 that a typical ester can be cleaved either by water or by a sodium hydroxide solution. Since fats are esters, they display the same behavior. Cleavage of a fat by water was described in Section 27-1; cleavage by hydroxide, also called saponification, is illustrated in Eq. (27-4).

Know

$$\begin{array}{c} CH_2-O-\overset{O}{\underset{\|}{C}}-C_{17}H_{35} \\ | \\ CH-O-\overset{O}{\underset{\|}{C}}-C_{17}H_{35} \\ | \\ CH_2-O-\overset{O}{\underset{\|}{C}}-C_{17}H_{35} \end{array} + 3NaOH \rightarrow \begin{array}{c} CH_2-OH \\ | \\ CH-OH \\ | \\ CH_2-OH \end{array} + 3 \overset{O}{\underset{NaO}{\diagdown C}}-C_{17}H_{35} \qquad (27\text{-}4)$$

a fat   glycerol   (sodium stearate)
                                 a soap

An alkali metal carboxylate salt having a chain of 12 to 18 carbons has cleansing properties and is called a **soap**; the most common example is sodium stearate.

Ordinary hand soap has for many years been manufactured by the chemical reaction shown in Eq. (27-4). The ingredients are sodium hydroxide (caustic soda) and a natural fat such as tallow or a hardened vegetable oil; these are heated together with water in a large vessel called a soap kettle (Fig. 27-1). When the reaction is complete, salt is added to the mixture, causing precipitation of the soap. After the soap has been collected, the valuable glycerol (Section 25-3) can be separated from the saltwater solution.

Soap is still manufactured in kettles by the saponification reaction (Eq. 27-4) in most parts of the world. In the United States a new process utilizing the hydrolysis reaction (Eq.

Fig. 27-1 A soap kettle used for the traditional saponification reaction. This is a view at the top of the kettle, which may be 20–30 ft high. [Courtesy of Procter and Gamble Company.]

Fig. 27-2 The tower (foreground) is a hydrolyzer used in the continuous process for soapmaking. [Courtesy of Procter and Gamble Company.]

27-1) has been developed. Fats are treated in a tall tower (Fig. 27-2) under pressure, with very hot water. The resultant fatty acids pass out through a pipe at the top of the tower and glycerol, in water, is drawn off at the bottom. After purification, the fatty acids are carried to a neutralizer tank (Fig. 27-3) where treatment with exactly the correct amount of alkali converts the acids to soap (Eq. 27-5).

$$\text{R—COOH} + \text{NaOH} \rightarrow \text{RCOONa} + \text{HOH} \qquad (27\text{-}5)$$
fatty acids   soap

The advantages of this process for soap-making are that it is rapid, it can be run continuously, and it permits more efficient separation and purification of glycerol and the fatty acids.

By using potassium hydroxide in the manufacturing process, a potassium soap can be prepared. It is more expensive but is desirable for a liquid soap or shaving cream because it is more soluble and produces a softer lather. Sodium stearate ($C_{17}H_{35}$COONa) produces a firm lather and has excellent cleansing properties, but dissolves best in hot water. Coconut and palm kernel fats, having a higher percentage of C_{12} and C_{14} acids (Table

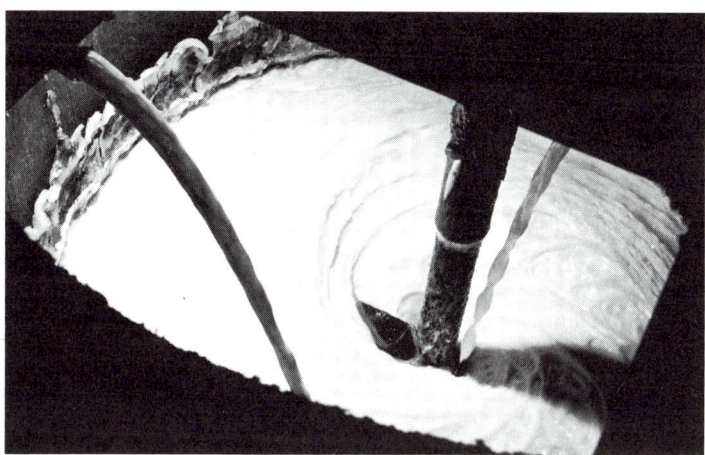

Fig. 27-3 Soap is formed in this neutralizer tank when a stream of fatty acids from the hydrolyzer (Fig. 27-2) is mixed with a stream of alkali. [Courtesy of Procter and Gamble Company.]

27-2), yield soaps which are more soluble. These are often blended with soap mixtures made from lard or tallow.

> The Romans of the first century A.D. were well acquainted with the art of soap-making. It is believed that they learned it from the Greeks. During the Middle Ages the process became known throughout the Mediterranean countries. Animal fats or olive oil were boiled in a kettle with water and ashes from wood or seaweed. The ashes provided potassium and sodium bases derived from the minerals originally present in the plants. It was not until about 1850 that people learned that glycerol was present as a by-product and was useful for other purposes.

27-6 THE SOLUBILITY AND CLEANSING ACTION OF SOAPS

A sodium carboxylate salt (R—COO$^-$Na$^+$) is highly ionic and usually quite water-soluble because of the strong attraction of water molecules to the charges on the ions (Section 26-3). Let us now consider how this behavior might be modified in the case of a soap, in which the R-group is a *long* hydrocarbon chain. The structure of sodium stearate, a typical soap, is written out in Fig. 27-4. Beneath it is a diagram symbolizing the two important features of the structure. The circle represents the ionic carboxylate end of the molecule and the long line represents the nonpolar hydrocarbon chain. The nonpolar hydrocarbon group is not soluble in water, but water would be attracted to the ionic groups and hence *tend* to dissolve the molecules.

The outcome of these conflicting tendencies is illustrated in Fig. 27-5. When soap molecules are placed in water, the hydrocarbon portions will not permit themselves to be exposed to water. Instead, they are attracted to each other, forming a cluster in which they are literally dissolved in each other. This grouping allows the ionic groups at the ends of the soap molecules to be attracted to the surrounding water molecules. The result is that

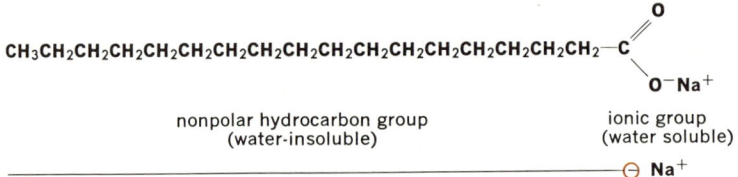

nonpolar hydrocarbon group (water-insoluble) ionic group (water soluble)

Fig. 27-4 Structural features of a soap molecule.

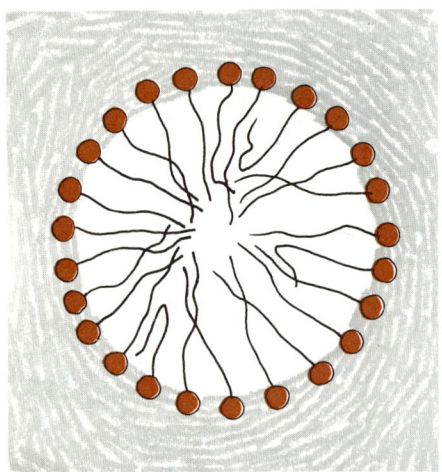

Fig. 27-5 A soap micelle; about 100 soap molecules clustered together and surrounded by water.

the soap molecules are, in a sense, able to "dissolve" in the water. However, this is a colloidal solution, not a true solution.

The droplet of soap molecules surrounded by water, illustrated in Fig. 27-5, is called a *soap micelle*. Experimental measurements indicate that there are about 100 soap molecules in each cluster. The alignment of the hydrocarbon groups is random and frequently changing, as is typical in a liquid. In particular, the hydrocarbon chains must be bent, because of the crowding that would result at the center of the cluster if all were straight. The liquid soap micelles dissolved in liquid water constitute one example of an emulsion, one of the types of colloidal solutions (Section 14-3).

Recall that in a true solution each dissolved ion or molecule is separately and completely surrounded by solvent molecules. (Examples would be potassium bromide ions in water, hexane molecules in ether, or iodine molecules in ethyl alcohol.) However, the "solution" of soap in water is designated as a colloidal solution because each bit of dissolved material is a cluster of *many* solute molecules. Although each cluster or micelle is fairly large, they do not coalesce and settle out. The most important factor in keeping the micelles dispersed throughout the water is the negative electrical charge, from carboxylate ions, on the surface of each micelle. When two of the micelles approach each other they are repelled by their like charges and hence do not coalesce into a larger globule.

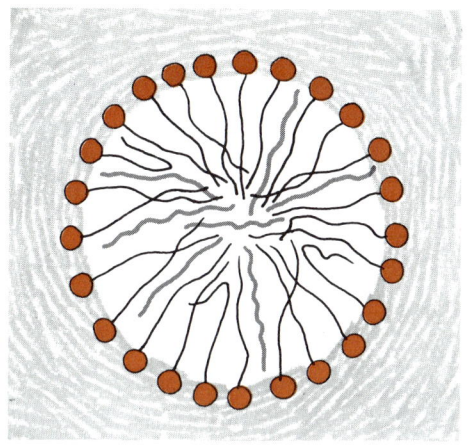

Fig. 27-6 Grease molecules dissolved inside soap micelle.

Figure 27-6 shows how soap can pull into solution (emulsify) oil and dirt particles which alone would not be soluble in water. The interior of the soap micelle is essentially the same as a liquid alkane mixture, such as kerosene. It is therefore a good solvent for materials similar to it in structure, namely nonpolar or slightly polar substances.

When a soap solution is splashed over a dirty dish, an automobile, or a fabric, soap molecules spread over the surface of the object, loosening any oil material. Agitation helps jar the oil from the surface and disperse it into tiny droplets. More and more soap molecules surround the oil, until it becomes incorporated within a soap micelle, as illustrated in Fig. 27-6. Particles of dirt often cling to an object because of an oily film. Once the oil is emulsified and removed by the soap, the dirt particles are also loosened. The very small particles may even be trapped inside the soap micelle.

From the standpoint of human civilization, it is interesting to note that the carboxylate salts obtainable from natural fats (Eq. 27-4) have just the right properties to make them good cleansing agents. If the carbon chain is longer than eighteen carbons, the salt is not sufficiently water-soluble to be useful. If it contains less than ten carbons it does not emulsify oil.

27-7 SYNTHETIC DETERGENTS

It was shown in the previous section that a soap has cleansing properties because it can emulsify an oil in water. This action depends on the structural features shown in Fig. 27-4. In more general terms, any molecule will show cleansing and emulsifying behavior if it possesses the necessary combination of two different structural units: a large nonpolar, hydrocarbon group and a water-soluble end group. In addition to the soaps, there are numerous other compounds, both natural and synthetic, which fulfill these structural requirements.

The term **detergent** (L. *detergere*) means simply cleansing agent. Soap is just one of many compounds in this group. However, in everyday language when a person speaks

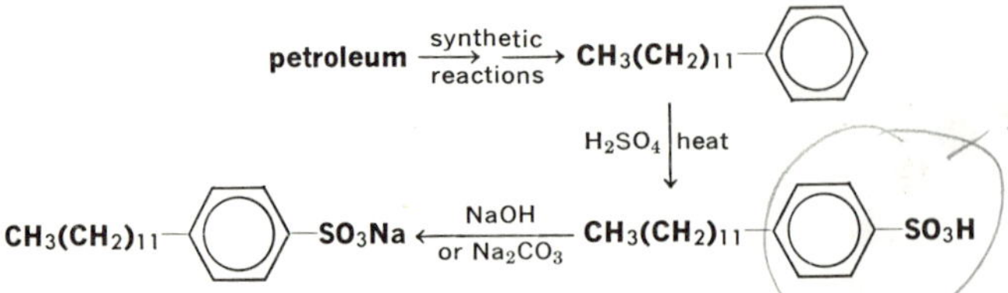

The Chemistry of Fats

petroleum $\xrightarrow[\text{reactions}]{\text{synthetic}}$ CH$_3$(CH$_2$)$_{11}$—◯

$\downarrow$ H$_2$SO$_4$ | heat

CH$_3$(CH$_2$)$_{11}$—◯—SO$_3$Na $\xleftarrow[\text{or Na}_2\text{CO}_3]{\text{NaOH}}$ CH$_3$(CH$_2$)$_{11}$—◯—SO$_3$H

Fig. 27-7. Synthetic route to a typical sulfonate detergent.

---

of a detergent he is usually thinking of a modern synthetic detergent. The old familiar soap is still often regarded as being in a separate class.

The sulfonate salt shown below is a synthetic detergent which has excellent cleansing ability and is therefore widely distributed for both household and commercial use. Some common brand names are Tide, Fab, Cheer, and Trend. (In general, synthetic detergents of this type are called alkylbenzenesulfonates.)

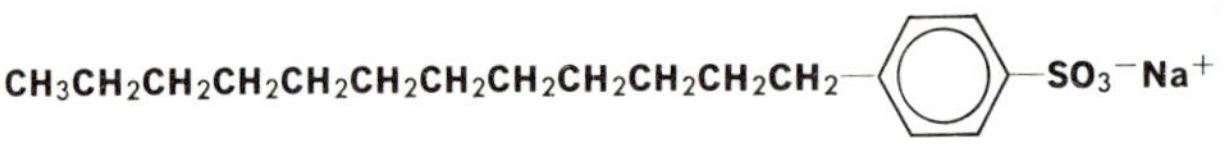

CH$_3$CH$_2$CH$_2$CH$_2$CH$_2$CH$_2$CH$_2$CH$_2$CH$_2$CH$_2$CH$_2$CH$_2$—◯—SO$_3^-$ Na$^+$ a common synthetic detergent

sodium *n*-dodecylbenzenesulfonate

This sulfonate salt is synthesized commercially by the route outlined in Fig. 27-7, starting from petroleum raw materials. Appropriate chemical changes yield *n*-dodecylbenzene. A typical sulfonation reaction (Eq. 24-3) then substitutes a sulfonic acid group onto the benzene ring *para* to the alkyl group. The resultant sulfonic acid is finally neutralized with a base to obtain the detergent.

The alkylbenzenesulfonate detergent shown above is said to be *biodegradable*; that is, it can be decomposed by microorganisms in a sewage plant or septic tank. Before 1965 the alkylbenzenesulfonates being manufactured had highly *branched* alkyl groups which were not easily degraded. The metabolism of microorganisms is adapted to straight-chain alkyl groups, such as those found in soap or natural fats. Consequently, the nondegraded detergents caused serious problems of foaming and contamination in the water released from sewage plants. Research chemists provided a solution to the problem. The manufacturing process has been changed so that a straight-chain alkyl group is built into the detergent molecule (Fig. 27-7).

A problem entirely separate from the question of biodegradability is the practice of mixing phosphates with the detergent, which improves its cleansing ability. Much of this phosphate ends up in lakes and streams where it promotes rampant growth of algae. When the algae die and decay, nearly all the available oxygen in the water is consumed, killing off most other living organisms there. Now that we realize the ecological destruction that this process represents, there is a serious question as to whether any cleansing benefits derived from adding phosphates to detergents are really worthwhile.

Another useful detergent is the sodium alkyl sulfate shown below. Note that it has the two features of structure previously described as being essential for cleansing action:

CH$_3$(CH$_2$)$_{10}$CH$_2$OSO$_2$O$^-$ Na$^+$

It is also possible for a detergent molecule to have a water-soluble end with a *positive* charge (rather than a negative charge as in the previous examples). Such a compound is therefore sometimes called an *invert soap*. Here is one example:

$$\underset{Cl^-}{\underset{}{}}\text{C}_6\text{H}_5-\text{CH}_2-\overset{+}{\underset{\underset{CH_3}{|}}{\overset{\overset{CH_3}{|}}{N}}}-\text{R} \quad \text{R} = \text{C}_8 \text{ to } \text{C}_{18}$$

This type of compound is known as a quaternary ammonium salt (Section 28–3A). Since it is not only a good washing agent but also has germicidal properties, it is especially suitable for use in hospitals, etc.

It is not even necessary that the end group of a detergent be ionic, so long as it is water soluble. Polar covalent structures, if there are enough of them, can provide the required solubility. This is illustrated in the examples below.

$$\text{C}_8\text{H}_{17}-\text{C}_6\text{H}_4-\text{O}-(\text{CH}_2-\text{CH}_2-\text{O})_8-\text{CH}_2-\text{CH}_2-\text{OH}$$

$$\underset{O}{\overset{\|}{\text{R}-\text{C}}}-\text{O}-(\text{CH}_2-\text{CH}_2-\text{O})_8-\text{CH}_2-\text{CH}_2-\text{OH} \quad \text{the R group is from a typical fatty acid}$$

Compounds of this type, known as *nonionic detergents*, are used extensively in dishwashing liquids, in low-sudsing products for clothes washers, and in special applications in which the absence of inorganic ions is desirable.

27–8 HARD WATER

Much of the water used in homes and factories is *hard water*, containing appreciable amounts of iron, magnesium, and calcium ions. These accumulated in the water as it percolated through soils and rock strata. Ions of this type interfere with the washing action of a soap because they form an insoluble precipitate with the soap; for example,

$$2\text{CH}_3(\text{CH}_2)_{16}\text{COO}^-\text{Na}^+ + \text{Mg}^{2+} \rightarrow [\text{CH}_3(\text{CH}_2)_{16}\text{COO}]_2\text{Mg}\downarrow + 2\text{Na}^+ \quad (27\text{-}6)$$

The resultant precipitate appears as sticky curds in the water; some of it forms a ring as it clings to the side of the washbowl or bathtub. This film of course represents waste. Enough soap must be used to precipitate all the magnesium ions before any dissolved soap is available for washing action. Furthermore, the presence of the curdy precipitate interferes with the cleansing ability of the additional soap which is used.

An important advantage of a synthetic detergent such as an alkylbenzenesulfonate is that it does *not* form an insoluble precipitate with magnesium or calcium ions. It is therefore a very good washing agent even in hard water.

Naturally *soft* water can be obtained by collecting fresh rain water before it runs through the soil. Hard water may be softened by treating it in some way to remove the undesirable ions.

27-9 DRYING OILS

not imp-

A drying oil is the essential ingredient causing a paint or varnish to become a tough, protective coating. A good drying oil must be highly unsaturated. Tung oil and linseed oil are excellent examples (Table 27–2). When such oil is spread out in a thin layer, air attacks the unsaturated groups, causing the formation of covalent bonds from one carbon chain to another. A vast network of crosslinks develops, transforming the oil into a tough film. (See Section 30–7 for other examples of cross-linked materials.) In this sense, the "drying" is actually a chemical change, not merely an evaporation of solvent.

Paint consists of a drying oil, a pigment, a drier, and a thinner.

> The **pigment** contributes the color and enough opaqueness for good hiding quality. The most important white pigments (because of their hiding power) are oxides of zinc, titanium, or antimony, and lead oxide, carbonate, or sulfate. A desired color is achieved by adding to these varying amounts of other inorganic compounds: chromium oxide (green), cadmium sulfide (yellow, orange, red), iron oxides (yellow, red, maroon, brown), lead chromate (yellow), or complex iron cyanides (blue). Colored organic compounds are used in certain paints, but do not withstand sunlight well. The principal black pigment is carbon.
>
> A **drier** is a catalyst which hastens the reaction of the drying oil with air. Carboxylate salts of cobalt, manganese, and lead are effective for this purpose.
>
> The **thinner** is a solvent used in sufficient amounts to make the paint flow evenly over the surface. Paint thinners usually consist of *mineral spirits* (a mixture of C_9 and C_{10} hydrocarbon isomers from petroleum) and turpentine (C_{10} hydrocarbons produced by pine trees). The thinner evaporates soon after the paint is applied to the surface. Subsequent drying of the paint involves the reaction with air, described above.

IMPORTANT TERMS

Detergent	Hard water	Mixed glyceride	Triglyceride
Drying oil	Hydrogenation	Oils	Unsaturated chains
Fats	Hydrolysis	Saponification	
Glyceride	Micelle	Soap	

WORK EXERCISES

1. Write the structure of
    a) an ester containing oleic, palmitic, and stearic acids combined with glycerol. With these components is more than one structure possible? What general name is given to such a compound?
    b) a fat   c) an oil   d) a soap
    e) a synthetic detergent   f) a drying oil

2. Write a complete equation, using structural formulas, for each reaction and tell the name used for that type of reaction.
    a) glyceryl tripalmitate, sodium hydroxide solution, heat
    b) glyceryl trioleate, hydrogen, nickel, heat
    c) glyceryl tristearate, water, hydrochloric acid, heat

3. Which of these compounds would you expect to be solid, and which liquid, at room temperature? Check your answers in a chemical handbook.

    a) glyceryl trioleate   b) glyceryl tristearate   c) glyceryl tributyrate
4. What type of structure must an organic compound have to be a good cleansing agent?
5. Discuss briefly the process for manufacturing oleomargarine. In general, how do you think its nutritional value compares with that of butter?

SUGGESTED READING

DePuy, C. H., and K. L. Rinehart, Jr., *Introduction to Organic Chemistry*, Chapter 11, John Wiley, New York, 1967.

Hart, H., and R. D. Schuetz, *Organic Chemistry*, 4th ed., Chapter 11, Houghton Mifflin, Boston, 1972.

Levey M. "The Early History of Detergent Substances," *J. Chem. Educ.* **31,** 521 (1954).

Noller C. R. *Textbook of Organic Chemistry*, 3rd ed., Chapter 12, W. B. Saunders, Philadelphia, 1966.

Snell, F. D., "Soap and Glycerol," *J. Chem. Educ.* **19,** 172 (1942). An excellent discussion although it lacks a description of the modern continuous process of soap making.

Snell, F. D., and C. T. Snell, "Syndets and Surfactants " *J. Chem. Educ.* **35,** 271 (1958). A good, survey of the types, properties, and manufacturing of synthetic detergents.

28 AMINES

28–1 THE RELATIVES OF AMMONIA

The amines are structurally related to ammonia. Whereas in ammonia the nitrogen holds three hydrogens, in an amine it is bonded to at least one organic group.

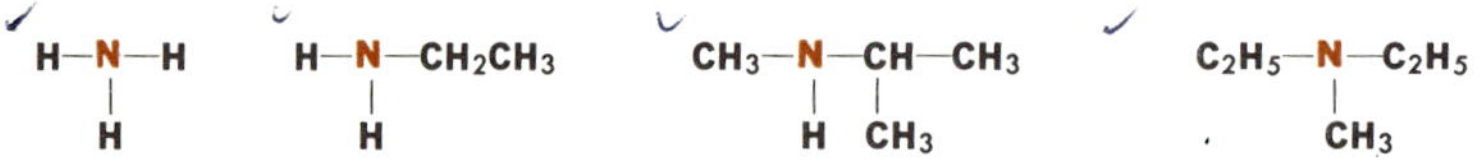

H—N—H   H—N—CH$_2$CH$_3$ CH$_3$—N—CH—CH$_3$ C$_2$H$_5$—N—C$_2$H$_5$
  | | | | |
  H   H   H   CH$_3$ CH$_3$

ammonia   ethylamine   methylisopropylamine   diethylmethylamine

The easiest method for naming the amines, shown above, is simply to name the organic groups bonded to the nitrogen. However, doing so may not be convenient if the organic group is complex. In such cases, illustrated below, an —NH$_2$ can be regarded as a *group* attached to a carbon skeleton. The name **amino** is used to indicate that the —NH$_2$ group is present.

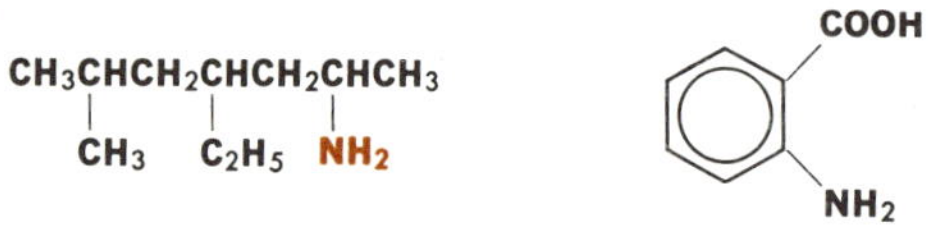

CH$_3$CHCH$_2$CHCH$_2$CHCH$_3$
    | | |
    CH$_3$ C$_2$H$_5$ NH$_2$

2-amino-4-ethyl-6-methylheptane   o-aminobenzoic acid

The aryl amines are named as derivatives of aniline, the simplest amine in the aromatic class.

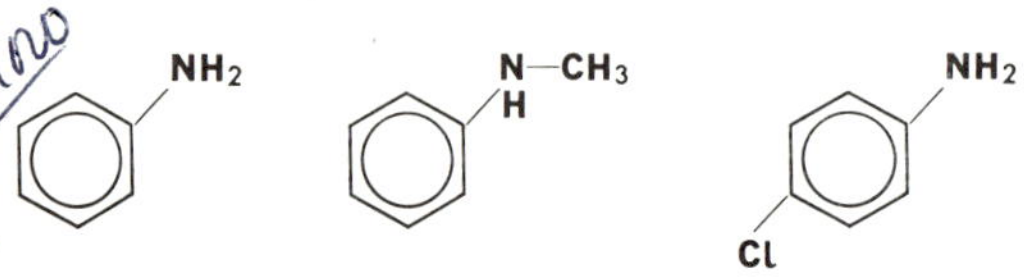

aniline   methylaniliné*   p-chloroaniline

---

*Another, more precise, name for this compound is N-methylaniline. The capital N indicates that the methyl group is definitely on the nitrogen atom, not on the ring.

As might be expected, several of the properties of amines are similar to those of ammonia. The smaller amines, such as methylamine, dimethylamine, or ethylamine, are gases which are quite soluble in water and have a pungent odor very much like the odor of ammonia. The larger the carbon skeleton of an amine, the less soluble it is in water. Medium-sized amines (butyl, pentyl, etc.) have fishy odors. Two of the decomposition products found in decaying flesh are amines having foul, putrid odors; they have been given the descriptive names putrescine and cadaverine.

$H_2NCH_2CH_2CH_2CH_2NH_2$ $\quad$ $H_2NCH_2CH_2CH_2CH_2CH_2NH_2$
putrescine $\quad\quad\quad\quad\quad\quad\quad\quad$ cadaverine

Fairly small amounts of these compounds, formed by bacterial action, can cause ptomaine poisoning if one eats spoiled meat, fish, eggs, etc. The biochemical reaction by which the bacteria produce cadaverine and putrescine is discussed in Section 28–7C.

28–2 BASIC PROPERTIES

Any amine reacts w/ acid to form acid base reaction.

It can be demonstrated that an ammonia molecule behaves as a base because it has an electron pair which it readily shares with a hydrogen ion. Either an organic or an inorganic acid may provide the hydrogen ion. These acid-base reactions occur rapidly at room temperature.

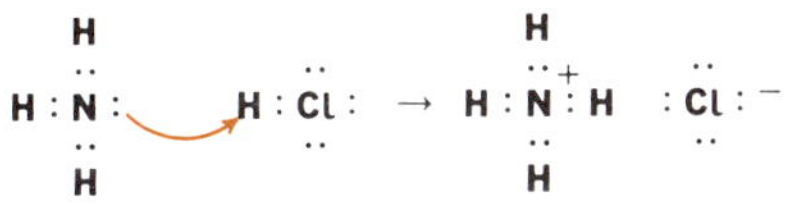

$$H:\underset{H}{\overset{H}{N}}: \quad H:Cl: \rightarrow H:\underset{H}{\overset{H}{\overset{+}{N}}}:H \quad :Cl:^- \qquad (28\text{-}1)$$

ammonia $\quad$ hydrogen chloride $\quad$ ammonium chloride

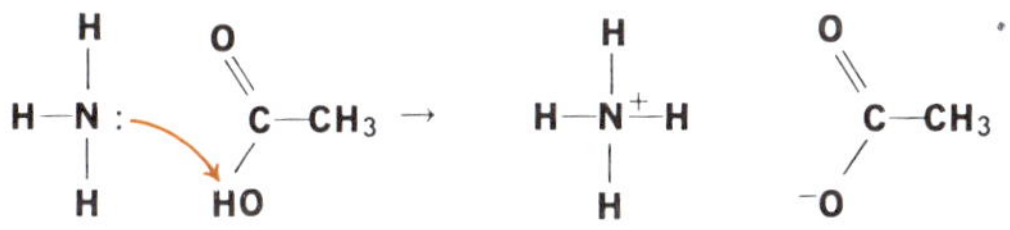

$$H-\underset{H}{\overset{H}{N}}: \quad \underset{HO}{\overset{O}{\underset{\|}{C}}}-CH_3 \rightarrow H-\underset{H}{\overset{H}{\overset{+}{N}}}-H \quad \underset{-O}{\overset{O}{\underset{\|}{C}}}-CH_3 \qquad (28\text{-}2)$$

ammonia $\quad$ acetic acid $\quad\quad\quad$ ammonium acetate

Any amine also is a base; it, too, has an electron pair available which can attract a hydrogen ion:

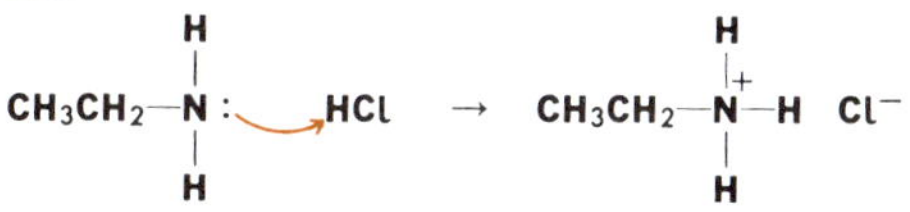

$$CH_3CH_2-\underset{H}{\overset{H}{N}}: \quad HCl \rightarrow CH_3CH_2-\underset{H}{\overset{H}{\overset{+}{N}}}-H \quad Cl^- \qquad (28\text{-}3)$$

ethylamine $\quad$ hydrogen chloride $\quad$ ethylammonium chloride

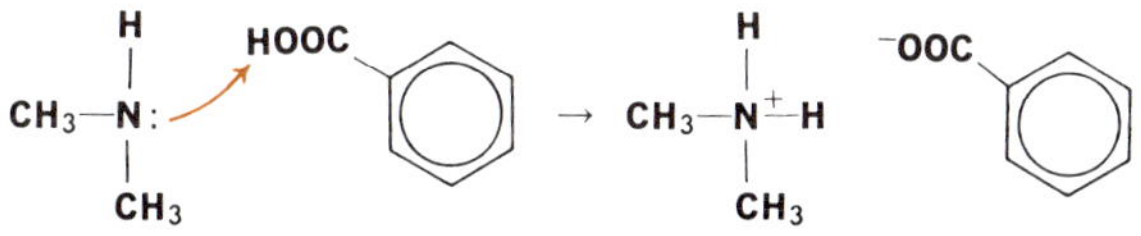

$$CH_3-\underset{CH_3}{\overset{H}{N}}: \quad HOOC-C_6H_5 \rightarrow CH_3-\underset{CH_3}{\overset{H}{\overset{+}{N}}}-H \quad ^-OOC-C_6H_5 \qquad (28\text{-}4)$$

dimethylamine $\quad$ benzoic acid $\quad\quad\quad$ dimethylammonium benzoate

The salts which result from these acid-base reactions are named as **substituted ammonium salts,** as shown above. Since the basic property of an amine is due to the pair of available electrons, an amine of the type $R_3N:$ reacts as a base, too (Eq. 28–5).

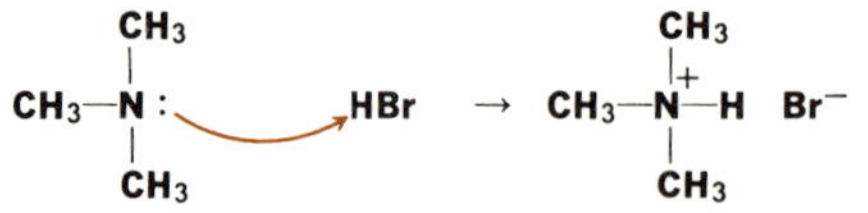

$$CH_3-\underset{\underset{CH_3}{|}}{\overset{\overset{CH_3}{|}}{N:}} + HBr \rightarrow CH_3-\underset{\underset{CH_3}{|}}{\overset{\overset{CH_3}{|+}}{N}}-H \ \ Br^- \qquad (28\text{–}5)$$

trimethylamine   hydrogen bromide   trimethylammonium bromide

For the same reason, the heterocyclic nitrogen compounds, Section 24–8, are likewise basic, despite the fact that the nitrogen is held in the ring.

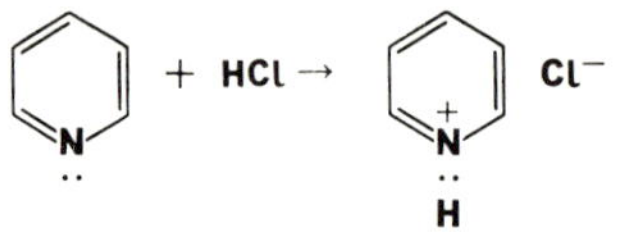

$$\text{pyridine} + HCl \rightarrow \text{pyridinium chloride} \quad Cl^- \qquad (28\text{–}6)$$

28–3 PREPARATION OF AMINES

A. Alkyl Amines, From Halides

A common method for the synthesis of an *alkyl* amine involves the reaction of an alkyl halide with ammonia, often in the presence of a stronger base such as sodium hydroxide.

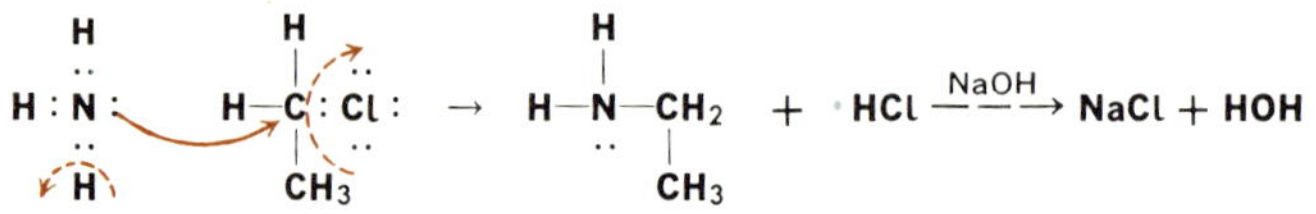

$$H:\underset{H}{\overset{H}{N:}} + H-\underset{CH_3}{\overset{H}{C}}:Cl: \rightarrow H-\underset{CH_3}{\overset{H}{N}}-CH_2 + HCl \xrightarrow{NaOH} NaCl + HOH \qquad (28\text{–}7)$$

ammonia   ethyl chloride   ethylamine

This is another reaction which depends upon the electron pair of nitrogen. When the pair of electrons attacks the alkyl group, a halide ion (for example, Cl^-) is displaced off the other side of the carbon being attacked (Eq. 28–7). Following the substitution of the alkyl group onto nitrogen, a hydrogen ion can be lost from the nitrogen. Indeed, the reaction is promoted if a strong base is present to pull the hydrogen ion off the nitrogen. From this method of formation we see that an amine can truly be regarded as a substitution product of ammonia.

Further substitution occurs if an amine is allowed to react with more alkyl halide (Eqs. 28–8 and 28–9). The nitrogen of an amine still has a pair of electrons available, so it can react in the same way that ammonia does.

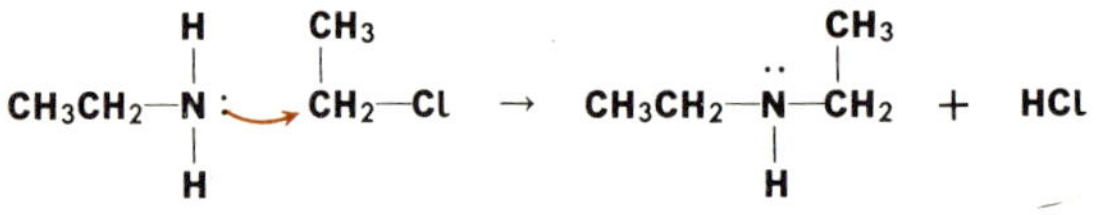

$$CH_3CH_2-\underset{H}{\overset{H}{N:}} + \overset{CH_3}{\underset{}{CH_2-Cl}} \rightarrow CH_3CH_2-\underset{H}{\overset{CH_3}{N}}-CH_2 + HCl \qquad (28\text{–}8)$$

ethylamine   diethylamine

$$\underset{\text{diethylamine}}{\overset{\overset{\text{H}}{|}}{\underset{\underset{\text{C}_2\text{H}_5}{|}}{\text{C}_2\text{H}_5-\text{N}:}}} \quad \overset{\overset{\text{CH}_3}{|}}{\text{CH}_2-\text{Cl}} \quad \rightarrow \quad \underset{\text{triethylamine}}{\overset{\overset{\text{CH}_3}{|}}{\underset{\underset{\text{C}_2\text{H}_5}{|}}{\text{C}_2\text{H}_5-\text{N}-\text{C}_2\text{H}_5}}} \quad + \quad \text{HCl} \qquad (28\text{-}9)$$

In fact, even a trialkylamine can react, because it too has an electron pair available (Eq. 28–10). However, in this case, there is no hydrogen to be lost from the nitrogen so the product is an ionic salt:

$$\underset{\text{triethylamine}}{\overset{\overset{\text{C}_2\text{H}_5}{|}}{\underset{\underset{\text{C}_2\text{H}_5}{|}}{\text{C}_2\text{H}_5-\text{N}:}}} \quad \overset{\overset{\text{CH}_3}{|}}{\text{CH}_2-\text{Cl}:} \quad \rightarrow \quad \underset{\text{tetraethylammonium chloride}}{\overset{\overset{\text{C}_2\text{H}_5}{|}}{\underset{\underset{\text{C}_2\text{H}_5}{|}}{\text{C}_2\text{H}_5-\overset{+}{\text{N}}-\text{C}_2\text{H}_5}}} \quad :\text{Cl}:^- \qquad (28\text{-}10)$$

Since this final product ($\overset{+}{\text{NR}_4}\overset{-}{\text{Cl}}$) is a completely substituted ammonium salt (compare $\overset{+}{\text{NH}_4}\overset{-}{\text{Cl}}$), it is named accordingly. A compound of the type $\overset{+}{\text{NR}_4}\overset{-}{\text{Cl}}$ is also called, in general, a **quaternary ammonium salt** because of the four organic groups present.

Another variation of this synthetic method is to treat some amine with an alkyl halide having a different group.

$$\underset{\text{aniline}}{\text{C}_6\text{H}_5\text{-N(H)-H}} \;+\; 2\text{CH}_3\text{I} \;\xrightarrow{\text{NaOH}}\; \underset{\text{dimethylaniline}}{\text{C}_6\text{H}_5\text{-N(CH}_3\text{)-CH}_3} \;+\; 2\text{HI} \qquad (28\text{-}11)$$

B. Alkyl Amines, From Carbonyl Compounds

Another quite general method for preparing an alkyl amine consists of treating some carbonyl compound (either aldehyde or ketone) with ammonia and hydrogen in the presence of a nickel catalyst. The process is called **reductive amination**.

$$\text{NH}_3 \;+\; \underset{R}{\overset{O}{\underset{\|}{C}}}-\text{R}' \;+\; \text{H}_2 \;\xrightarrow{\text{Ni}}\; \text{H}_2\text{N}-\underset{R}{\overset{H}{\underset{|}{C}}}-\text{R}' \;+\; \text{H}_2\text{O} \qquad (28\text{-}12)$$

$$\text{NH}_3 \;+\; \underset{H}{\overset{O}{\underset{\|}{C}}}-\text{R}' \;+\; \text{H}_2 \;\xrightarrow{\text{Ni}}\; \text{H}_2\text{N}-\underset{H}{\overset{H}{\underset{|}{C}}}-\text{R}' \;+\; \text{H}_2\text{O} \qquad (28\text{-}13)$$

Once an amine has been synthesized, for example by Eq. (28–13), it can be made to react with another carbonyl compound (either the same or different from the first) in the same manner that ammonia did, thus producing a *di*alkyl amine.

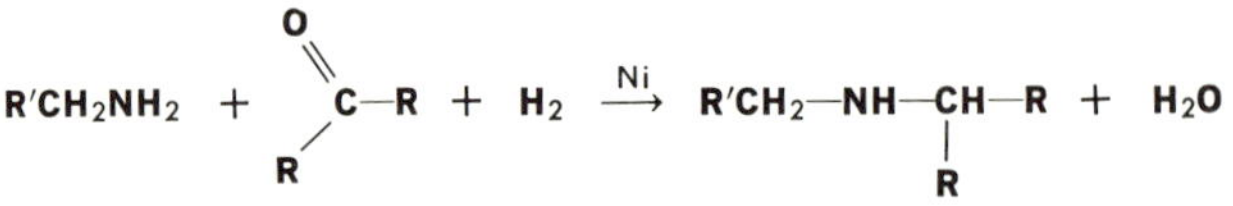

$$\text{R'CH}_2\text{NH}_2 + \underset{R}{\overset{\overset{O}{\|}}{C}}-R + H_2 \xrightarrow{Ni} \text{R'CH}_2-\text{NH}-\underset{R}{\text{CH}}-R + H_2O \qquad (28\text{-}14)$$

Further details of these reductive amination reactions are discussed in Section 29–9, including some related biological reactions.

C. Aryl Amines, From Nitro Compounds

A different synthetic method is usually employed to prepare *aryl* amines. A nitro group can be reduced to an amino group by hydrogens produced from the action of an acid on a metal. Iron or tin are commonly used.

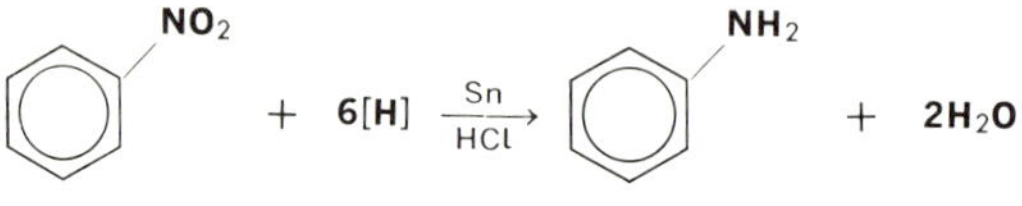

$$\text{nitrobenzene} + 6[H] \xrightarrow{\text{Sn}/\text{HCl}} \text{aniline} + 2H_2O \qquad (28\text{-}15)$$

$$\text{1-nitronaphthalene} + 6[H] \xrightarrow{\text{Fe}/\text{HCl}} \text{1-aminonaphthalene} + 2H_2O \qquad (28\text{-}16)$$

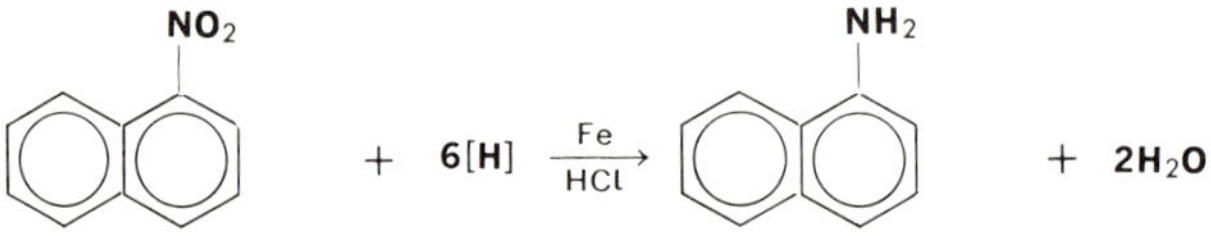

The reactions shown in Eqs. (28–15) and (28–16) are particularly useful because the nitro group is so easily substituted onto an aromatic ring (Section 24–6). This in turn means that a great variety of aryl amines can be synthesized. For example, consider the following synthetic route:

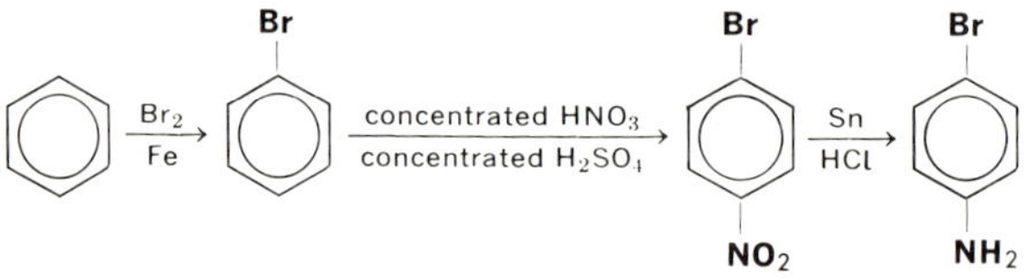

$$\text{C}_6\text{H}_6 \xrightarrow{\text{Br}_2/\text{Fe}} \text{C}_6\text{H}_5\text{Br} \xrightarrow[\text{concentrated H}_2\text{SO}_4]{\text{concentrated HNO}_3} \text{Br-C}_6\text{H}_4\text{-NO}_2 \xrightarrow{\text{Sn}/\text{HCl}} \text{Br-C}_6\text{H}_4\text{-NH}_2$$

28–4 CONVERSION OF AMINES TO AMIDES

An acid chloride reacts rapidly with ammonia to form an amide. A particular amide is named according to the acid group from which it is derived.

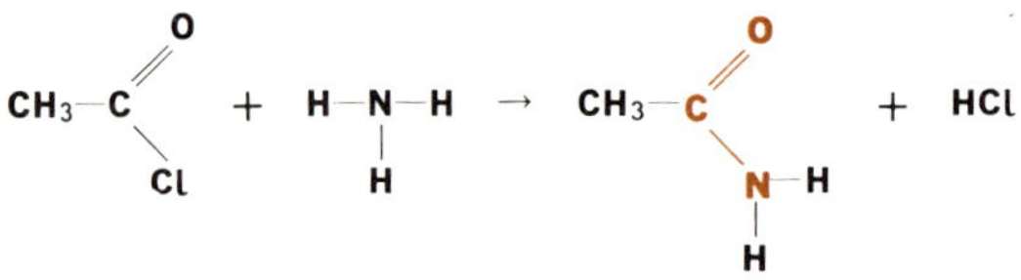

$$\underset{\text{acetyl chloride}}{\text{CH}_3-\overset{\overset{O}{\|}}{C}-\text{Cl}} + \text{H}-\underset{H}{\text{N}}-\text{H} \rightarrow \underset{\text{acetamide}}{\text{CH}_3-\overset{\overset{O}{\|}}{C}-\underset{H}{\text{N}}-\text{H}} + \text{HCl} \qquad (28\text{-}17)$$

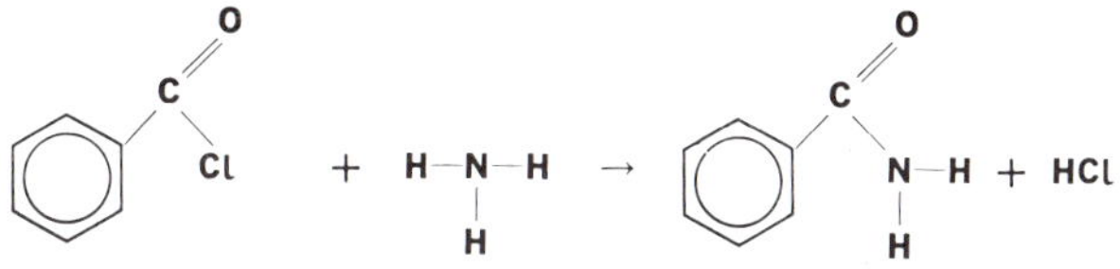

Conversion of Amines to Amides

$$\text{benzoyl chloride} + H-\underset{H}{\underset{|}{N}}-H \longrightarrow \text{benzamide} + HCl \tag{28-18}$$

Acid chlorides react also with amines, provided at least one hydrogen is available for displacement; the trialkylamines therefore do not react. When amines react with acid chlorides, the products are N-substituted amides.

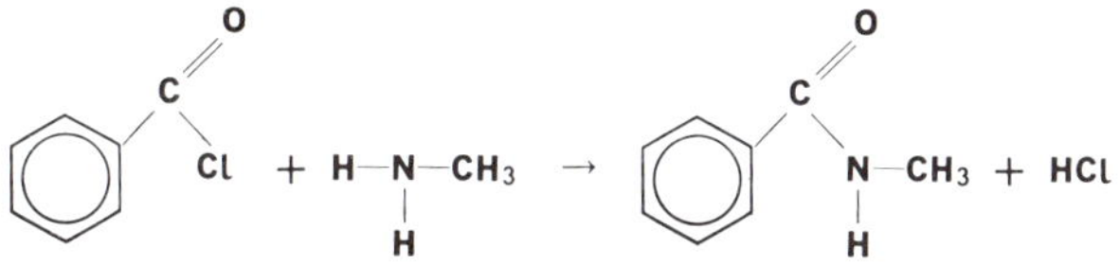

$$\text{PhCOCl} + H-\underset{H}{\underset{|}{N}}-CH_3 \longrightarrow \text{PhCO-N(H)-}CH_3 + HCl \tag{28-19}$$

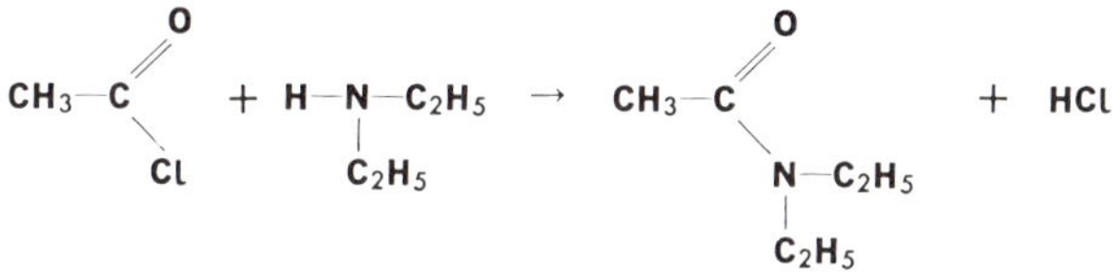

N-methylbenzamide

$$CH_3-\underset{Cl}{\overset{O}{C}} + H-\underset{C_2H_5}{\underset{|}{N}}-C_2H_5 \longrightarrow CH_3-\underset{\underset{C_2H_5}{|}}{\overset{O}{C}}-\underset{}{N}-C_2H_5 + HCl \tag{28-20}$$

N,N-diethylacetamide

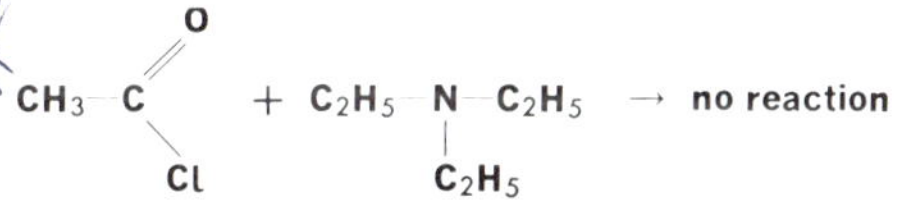

$$CH_3-\underset{Cl}{\overset{O}{C}} + C_2H_5-\underset{C_2H_5}{\underset{|}{N}}-C_2H_5 \longrightarrow \text{no reaction} \tag{28-21}$$

$$CH_3-\underset{Cl}{\overset{O}{C}} + H-\underset{}{N}(H)(C_6H_5) \longrightarrow CH_3-\overset{O}{\underset{}{C}}-\underset{}{N}(H)(C_6H_5) + HCl \tag{28-22}$$

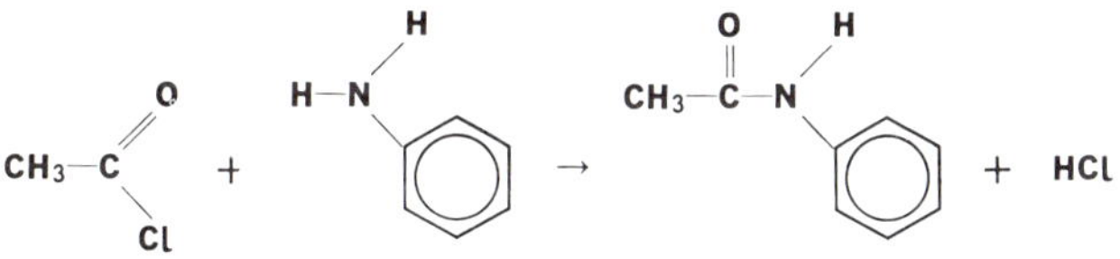

acetyl chloride   aniline   acetanilide

These reactions of amines with acid chlorides are analogous to the reactions of alcohols with an acid chloride (Section 26–7).

Amides can also be formed from the carboxylic acids themselves (Eq. 28–23). However, heat is required because a carboxylic acid is less reactive than an acid chloride.

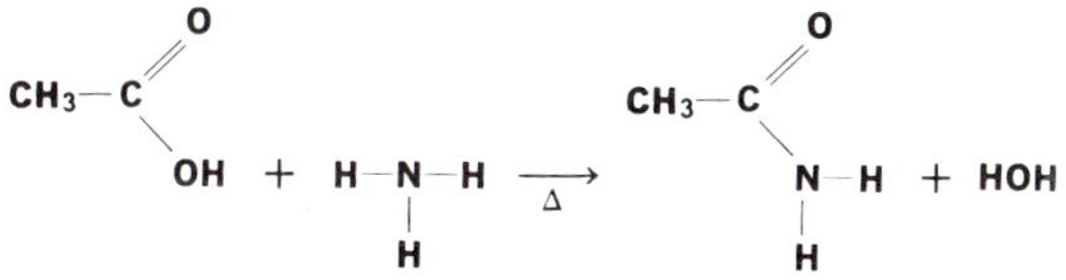

$$CH_3-\underset{OH}{\overset{O}{C}} + H-\underset{H}{\underset{|}{N}}-H \xrightarrow{\Delta} CH_3-\underset{\underset{H}{|}}{\overset{O}{C}}-N-H + HOH \tag{28-23}$$

acetic acid   ammonia   acetamide

$$R-\overset{O}{\underset{OH}{C}} + H-\underset{H}{\overset{R'}{N}} \xrightarrow{\Delta} R-\overset{O}{\underset{N-R'}{C}} + HOH \quad (28-24)$$

carboxylic acid   amine   N-substituted amide

Reaction 28–24 is comparable to the esterification reaction of an acid with an alcohol (Section 26–4) and, like it, is very slow unless heat is used. Furthermore, the reverse reaction is possible (Eq. 28–25); an amide can be hydrolyzed (a mineral acid catalyst helps), just as can an ester (Section 26–6).

Hydrolysis reaction

$$CH_3-\overset{O}{\underset{N-C_2H_5}{C}} + HOH \xrightarrow[\Delta]{H^+} CH_3-\overset{O}{\underset{OH}{C}} + H-\underset{H}{\overset{C_2H_5}{N}} \quad (28-25)$$

Amides can be hydrolyzed to give back Am. or Am. sub + Acid

The formation and hydrolysis of amide bonds are not only frequent laboratory procedures but are also essential to the chemistry of amino acids and proteins (Section 28–5). In the latter case, enzymes in the living cells catalyze the reactions so that they occur easily at body temperature, provided energy sources are available.

    The alert reader may have noticed that in equations such as (28–2) and (28–4) we showed ammonia, or an amine, reacting with a carboxylic acid to produce an ammonium salt, whereas in Eq. (28–23) we showed the same reactants producing an amide. The product obtained depends on the temperature. Reaction (28–2), which leads to ammonium salt, occurs rapidly at room temperature or below. In contrast, the amide-formation reaction (Eq. 28–23) is slow and difficult. It takes place only if the reactants are heated.

    The amides are essentially "neutral" substances, in contrast to the basic amines. Although the nitrogen in an amide still has an extra pair of electrons, the rest of the structure affects it in such a way that the electrons are not readily shared with a hydrogen ion. Since basicity depends on donation of an electron pair (to a hydrogen ion, for example), the amide is not significantly basic. Thus a typical amide, such as acetanilide, does not form a salt when placed in dilute hydrochloric acid.

28–5 THE STRUCTURE OF PROTEINS AND AMINO ACIDS

The tissues of animal bodies are composed chiefly of protein material, aside from the water present within the tissues or the bones supporting them. Muscle, skin, hair, cartilage, nails, and hooves all consist of proteins. In addition, a number of specialized compounds such as enzymes and certain hormones are composed mainly of protein. Experiments show that proteins are very large molecules. Many have molecular weights between about fifty thousand and several hundred thousand. Some have molecular weights as high as a few million. A clue to the nature of these gigantic molecules is the observation that they can be broken up into much smaller organic molecules. One way to do this is to boil a

※ know - as illustration of amino acid

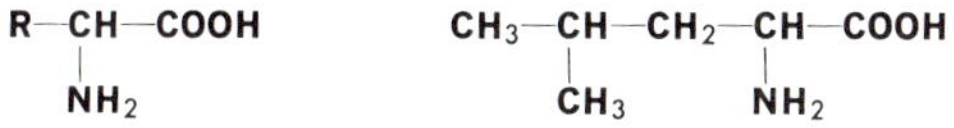

R—CH—COOH   CH₃—CH—CH₂—CH—COOH
   | | |
   NH₂   CH₃   NH₂
   (a)   (b)

Fig. 28-1 The structure of natural amino acids. (a) General formula for any α-amino acid. (b) The structure of leucine.

---

sample of protein in water containing some hydrochloric acid. After several hours, one finds that this dilute acid solution converts the protein to a mixture of amino acids.

※ know

$$\text{protein} + H_2O \xrightarrow[\Delta]{HCl} \text{mixture of amino acids} \tag{28-26}$$

This result suggests that the amino acid molecules are building-units which can somehow be linked together to form a very large protein molecule.

If one separates the mixture from reaction (28-26) into the different amino acids, each type present can be identified. (Prior to about 1965, this was a tedious and difficult job! Since then, the availability of "amino acid analyzer" instruments has greatly reduced the time and labor required for such a study.) It turns out that there are as many as 20 to 25 different kinds of amino acids in the mixture and that because of the size of the original protein, many molecules of each of the different kinds of amino acids are present. Despite the variety of amino acids in the mixture, one finds that all are α-amino acids, having the structure shown in Fig. 28-1(a). They differ from each other only in the nature of their R groups. One specific example is leucine, shown in Fig. 28-1(b). The symbol α in the name α-amino acid specifies that the amino group is bonded to the carbon adjacent to the carboxyl group. It is possible, of course, for the amino group to be at some other position in the molecule. However, since all natural amino acids derived from proteins are α-amino acids, the remainder of our discussion will be confined to this type.

Greek letters are sometimes used to designate locations on the carbon skeleton, relative to a certain functional group. The letter α is assigned to the carbon nearest the functional group, β to the next carbon, and so on. Thus in leucine (Fig. 28-1b) the amino group is at the α position and the methyl group is at the γ position relative to the carboxyl group. The name "leucine" is a common name. To write a correct systematic name for this compound one should, as usual, use numbers to designate positions in the carbon skeleton, rather than α, β, γ, δ, and so on. Thus, the systematic name for leucine would be 2-amino-4-methylpentanoic acid.

Some representative examples of natural amino acids occurring in proteins are shown in Fig. 28-2. It is worth noting some of the structural types present. The R groups in alanine and valine are merely alkyl groups. Serine possesses an alcohol group as well as the amino and carboxyl functional groups. Phenylalanine has an aromatic ring attached to the alkyl skeleton, and histidine has a heterocyclic ring (see Section 24-8). Many amino acids have one basic group, the amino group, and one acidic group, the carboxyl group. However, some have *two* carboxyl groups and only one amino group, and are therefore often called "acidic" amino acids. Glutamic acid is an example. Conversely, some, such as

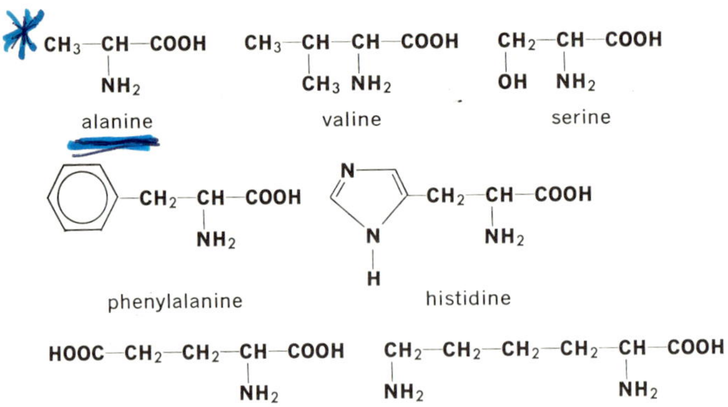

406   Amines

$CH_3-CH-COOH$ $CH_3-CH-CH-COOH$ $CH_2-CH-COOH$
       | | | | |
      NH_2 CH_3 NH_2 OH NH_2
    alanine   valine   serine

⬡$-CH_2-CH-COOH$ (imidazole)$-CH_2-CH-COOH$
             | |
           NH_2 NH_2
     phenylalanine   histidine

$HOOC-CH_2-CH_2-CH-COOH$ $CH_2-CH_2-CH_2-CH_2-CH-COOH$
                    | | |
                  NH_2 NH_2 NH_2
      glutamic acid   lysine

Fig. 28–2 The structure of several natural amino acids. These examples illustrate most of the types found in proteins.

---

lysine, have one carboxyl group but two of the basic amino groups. This type is called a "basic" amino acid. A more extensive list of amino acids is given in Chapter 34.

Reaction (28–26) indicated that a protein is composed of amino acid units. Other experiments show further that the protein is built up by linking the amino acids to each other by means of amide bonds. In Eq. (28–24) we saw that an amine can react with a carboxylic acid by splitting out water to form an amide. Now an amino acid has a very interesting type of structure which we have not previously encountered; it has, within one molecule, both an amino group and a carboxylic acid group. Consequently, if the carboxyl group of one amino acid molecule encounters the amino group of some other amino acid molecule, water could be split out and an amide bond could be formed, thus linking together the two molecules. Furthermore, since one of the units would still have a carboxyl group available, it could form another amide bond with the amino group of yet a third molecule of amino acid. Since the newly attached amino acid would also have a carboxyl group to react, it could bond to a fourth unit, and so on until many amino acids were linked to each other. Each time an amide linkage is formed, one molecule of water is split out.

This process of combining many amino acid units, via amide bonds, to form a protein is represented in Fig. 28–3. Each kind of protein is unique, depending on the particular amino acids it contains and the sequence in which they are arranged. Therefore, the build-up or synthesis of a protein is exceedingly complex. Biologically it is accomplished by enzymes acting as catalysts, and it is controlled by ribonucleic acids, so that only the exact kind of protein required for a particular function is assembled. The protein of a hormone will be quite different than that found in a muscle fiber or in the membrane of a nerve cell. Also, the protein of beef muscle differs from that of human muscle. To duplicate such a synthesis in the laboratory is extremely difficult and has been accomplished only for relatively small proteins containing about one hundred amino acids or less. Some of the details concerning the biosynthesis of proteins are discussed in Chapter 39.

Whenever one amide bond is formed by the reaction of an amino group with a carboxyl group, one molecule of water is split out. The reverse reaction, hydrolysis, is also

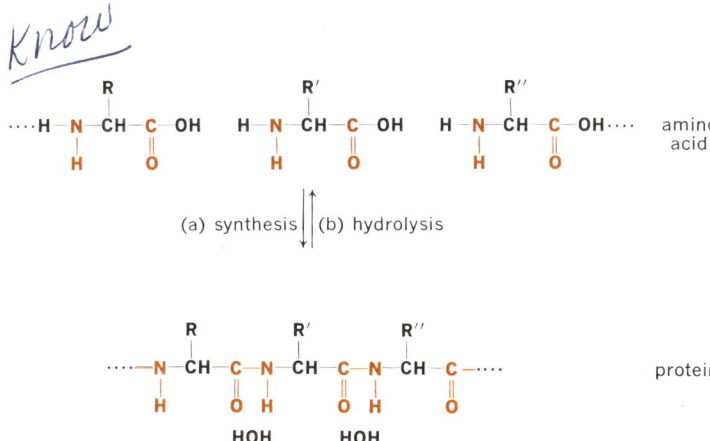

Fig. 28–3 Formation of a protein from amino acids, and the reverse. Reaction in direction (a) represents the buildup of protein from amino acids. This complex biosynthesis occurs in living organisms. In the reverse reaction (b), water can break up the protein, provided a catalyst is present. In living organisms, enzymes are the catalysts; in the laboratory, strong acid can catalyze the hydrolysis.

possible. That is, one molecule of water can cleave an amide group to give back an amine and a carboxylic acid, as shown in Eq. (28–25). In the case of a protein, hydrolysis is represented by reaction (b) in Fig. 28–3. Cleavage of each amide linkage requires one molecule of water and the aid of a catalyst. In the laboratory, hydrochloric acid is a satisfactory catalyst. Since hydrolysis involves breakdown of the protein molecule, it is much easier to carry out in the laboratory than is the synthesis of the protein. Even large, complex proteins can be hydrolyzed if heated for a few hours in the presence of hydrochloric acid. Figure 28–3(b) shows in greater detail the protein hydrolysis reaction that was first represented by Eq. (28–26).

Biologically, the hydrolysis reaction of Fig. 28–3(b) can occur in a variety of different tissues which have appropriate enzymes to catalyze the reaction. For example, the chemical reaction taking place during digestion is simply hydrolysis. In the intestines, a piece of beefsteak reacts with water, under the influence of enzymes, and is broken up into its constituent amino acids. The resultant amino acid molecules are relatively small and water-soluble and, therefore, can be absorbed through the intestinal wall, dissolved in the blood, and carried to various tissues of the body. There, other enzymes can promote a synthesis reaction (Fig. 28–3a), reassembling the amino acids into new protein molecules. Because of the very specific nature of the enzymes and ribonucleic acids controlling the biosynthesis, the collection of amino acids which originated from the beefsteak will be arranged differently in the protein of human muscle or cartilage.

We have seen that it is the formation of amide groups which holds amino acids together in a protein and that hydrolysis of the amide bonds causes the protein to revert to amino acids. *The amide bonds in a protein are called* **peptide bonds.** This is merely a special term used to designate the particular kind of amide group that results from α-amino acids linking to each other to form a protein. However, as we have seen, the chemical properties of a peptide linkage are the same as those of any other amide group. Each peptide group in the protein structure of Fig. 28–3 is shown in color.

408 Amines

The protein structure shown in Fig. 28–3 indicates that the chain can be continued in either direction as more and more amino acid units are attached. This process does, in fact, continue until a very large number of units are linked together. Since each protein molecule is composed of a discrete number of amino acids, the chain *does* have definite ending points. At one end of the protein molecule there will be a "free" amino group (i.e., one not involved in a peptide bond), and at the opposite end there will be a "free" carboxyl group, as shown below:

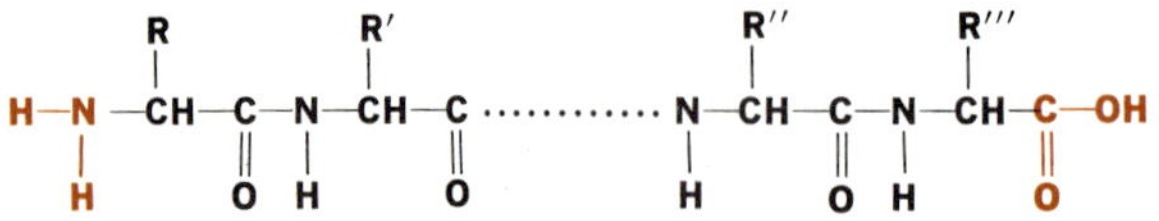

$$\text{H—N—CH—C—N—CH—C} \cdots \cdots \text{N—CH—C—N—CH—C—OH}$$

with R, R', R'', R''' side chains and the appropriate H, O atoms on N and C.

28–6 REACTIONS OF AMINO ACIDS

A. Amide Formation

In Section 28–5 we observed that the amino group and the carboxyl group of an amino acid can be involved in the formation of an amide structure. When the amide structure links α-amino acids together to yield a protein, the amide group is called a peptide group. The formation of peptide bonds is undoubtedly the most common type of amide bond produced by a natural α-amino acid. However, the amino and carboxyl groups will always display their characteristic properties, and so there is always the possibility that an amide bond could be formed with some other molecule which was not an amino acid. One example of amide formation between an amino acid and some different type of compound will be discussed.

Benzoic acid (or its salt) is sometimes ingested as a constituent of our meals. It occurs naturally in certain foods. Other foods may contain larger molecules which are broken down to benzoic acid during metabolism. Rather large amounts of such compounds are found in prunes and cranberries. Some foods contain benzoic acid or sodium benzoate which have been added as preservatives. Since benzoic acid from any of these sources is not further broken down, one way the body eliminates it is by combining it with glycine, which is the simplest amino acid and one usually present in the body in good supply. The benzoic acid and the glycine become linked together by an amide bond.

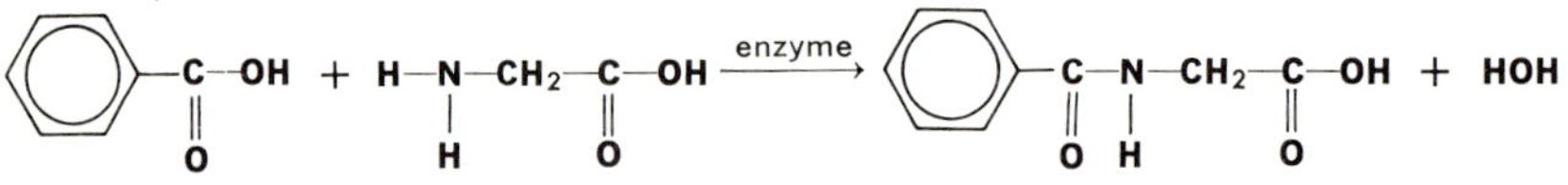

$$\text{C}_6\text{H}_5\text{—C(=O)—OH} + \text{H—N(H)—CH}_2\text{—C(=O)—OH} \xrightarrow{\text{enzyme}} \text{C}_6\text{H}_5\text{—C(=O)—N(H)—CH}_2\text{—C(=O)—OH} + \text{HOH} \quad (28\text{–}27)$$

benzoic acid   glycine   hippuric acid

The hippuric acid thus formed is subsequently excreted via the urine. Certain related compounds can be eliminated by the same route. If benzaldehyde is ingested, for instance in almond flavoring, the body can oxidize it to benzoic acid, which is then converted to hippuric acid.

B. Salt Formation

Thus far most of our discussion of amino acids has been, quite properly, devoted to their ability to form peptide bonds and thus build proteins. However, it is important to consider

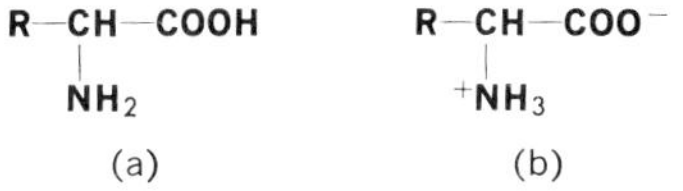

```
R—CH—COOH        R—CH—COO⁻
   |                |
   NH₂              ⁺NH₃
   (a)              (b)
```

Fig. 28-4 The structure of an amino acid. (a) Simplified structure. (b) Dipolar ion form, the more accurate representation.

---

also the properties of an individual amino acid as it might exist in a solution in the laboratory, in intestinal fluid, or in the bloodstream. We should expect that the amino group and the carboxyl group will have the typical properties we have come to expect of such groups. The only question will be to see how these groups might affect each other when both are in the same molecule.

So far we have written the structure of an amino acid as shown in Fig. 28-4(a) because this portrays the amino and carboxyl groups in the same way that they exist in many other compounds. Let us now consider acid and base properties. An amine is a fairly basic compound (Section 28-2), and it readily combines with a proton released from either a strong inorganic acid or from an organic carboxylic acid; the result is the formation of a substituted ammonium salt. An example would be the transfer of a proton that occurs when methylamine and propanoic acid are mixed:

$$CH_3\text{—}NH_2 + CH_3CH_2COOH \rightarrow CH_3\text{—}\overset{+}{N}H_3 \quad CH_3CH_2COO^- \tag{28-28}$$

methylamine   propanoic   methylammonium
               acid   propanoate

An amino acid will behave in the same way; a proton will leave the carboxyl group and reside instead on the amino group. The amino acid will actually exist as shown in Fig. 28-4(b) rather than as the simplified structure in Fig. 28-4(a). It is merely coincidence that the amino and carboxyl groups happened to be attached to the same carbon skeleton and that in the resultant salt (b) the positive and negative ions are held together in the same molecule.

In Fig. 28-4(b) the organic skeleton is held together by covalent bonds as usual but in addition, part of the structure of (b) is ionic; it is often described as being an **inner salt**. This ionic structure is also called a **dipolar ion,** since there are two poles, positive and negative, within the same molecule.

When an amino acid is dissolved in water, nearly all the molecules will exist in the dipolar-ion form (b). In fact, even without being in water solution, an amino acid by itself will exist in the dipolar-ion form. Crystals of a pure amino acid actually contain the dipolar ions. When the solid dissolves in water these dipolar ions, already formed, are merely released into the solution.

We have just described the condition of an amino acid in a water solution—that is, near neutrality. However, if a moderately strong acid or base is present, the amino acid can react with either, which means that an amino acid is *amphoteric*. An amino acid would react with a strong acid as follows:

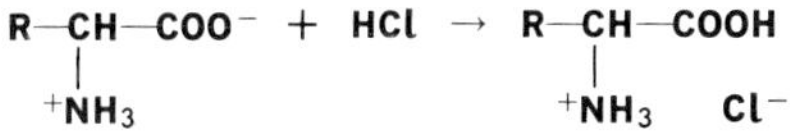

```
R—CH—COO⁻ + HCl  →  R—CH—COOH                                    (28-29)
   |                    |
   ⁺NH₃                 ⁺NH₃     Cl⁻
```

In this reaction the hydrochloric acid is able to donate a proton to the negative ion. Since the solution is acidic, the amino group remains as a positive ion; the only difference is that the negative ion associated with it is now a chloride ion rather than the original carboxylate ion.

In a basic solution this reaction would occur:

$$\text{R—CH—COO}^- + \text{NaOH} \rightarrow \text{R—CH—COO}^-\text{Na}^+ + \text{HOH} \quad (28\text{-}30)$$
$$\quad\;\;|\qquad\qquad\qquad\qquad\quad\;|$$
$$\;\;^+\text{NH}_3\qquad\qquad\qquad\qquad\text{NH}_2$$

Here the strongly basic hydroxide ion removes the proton from the amino group, a weaker base, and forms a molecule of water. The carboxylate group remains a negative ion, unchanged. However, its charge is now balanced by the sodium ion rather than by the ammonium ion group originally present.

C. Decarboxylation

In the presence of the proper enzyme, the carboxyl group of an amino acid can be split out, leaving an amine. Such a reaction is called **decarboxylation**. A reaction of this type converts the amino acid lysine to the evil-smelling amine cadaverine. This is one of the many biochemical reactions taking place when microorganisms attack dead flesh.

$$\underset{\text{lysine}}{\text{CH}_2\text{CH}_2\text{CH}_2\text{CH}_2\text{CHCOOH}} \xrightarrow{\text{enzyme}} \underset{\text{cadaverine}}{\text{CH}_2\text{CH}_2\text{CH}_2\text{CH}_2\text{CH}_2} + \text{CO}_2 \quad (28\text{-}31)$$

(with NH$_2$ groups on the terminal carbons and on the α-carbon of lysine)

Although there is no simple method for promoting decarboxylation of an α-amino acid as a chemical reaction in the laboratory, it is of some importance as a biochemical reaction in living systems. Another interesting example is the conversion of histidine, one of the common natural amino acids, to histamine.

$$\underset{\text{histidine}}{\text{[imidazole]—CH}_2\text{—CH—COOH}} \xrightarrow{\text{enzyme}} \underset{\text{histamine}}{\text{[imidazole]—CH}_2\text{—CH}_2} + \text{CO}_2 \quad (28\text{-}32)$$

Histamine is produced in the body during an allergic reaction. Anti-histamine drugs are intended to neutralize its effects and provide relief to the person suffering from the allergy.

28-7 MORE NITROGEN COMPOUNDS OF INTEREST

A. Vitamins

Know that vitamins are nitrogen types — amides or amino acids

Several (but not all) of the vitamins have nitrogen atoms in either amide groups or amine groups. Quite often the basic amino nitrogen is part of a heterocyclic ring. **Niacin** is a very important vitamin but is the simplest in structure. It has one basic nitrogen in a pyridine ring, plus a neutral nitrogen in an amide group. Most vitamins have a more complex structure, such as that of **thiamine** (vitamin B_1).

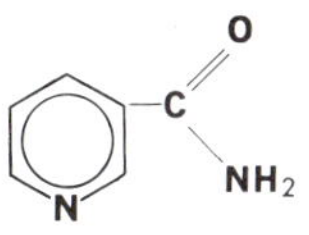

More Nitrogen Compounds of Interest 411

niacin

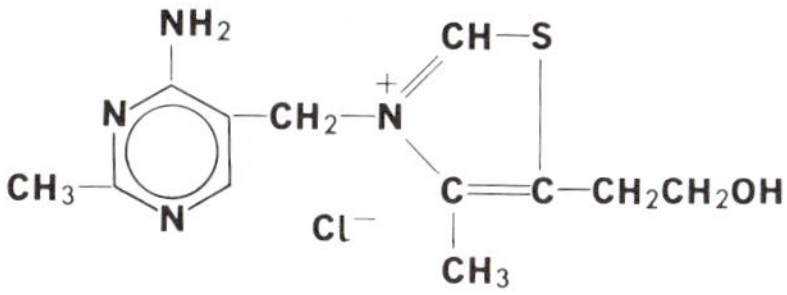

thiamine (vitamin B$_1$)

Note in thiamine that the five-membered heterocyclic ring contains both a nitrogen and a sulfur, with the nitrogen in the form of a quaternary ammonium salt (Section 28–3A).

B. Alkaloids — *nitrogen containing compds — either amides or*

A number of plants produce nitrogen-containing substances which are moderately basic and are therefore called alkaloids (alkali-like). Most of these compounds have striking effects on the nervous systems of animals; some may even be fatal. Structurally, the nitrogens present in alkaloids are most often in heterocyclic rings. The simplest alkaloid is coniine, which occurs in the poison hemlock plant (see Fig. 28–5). Cocaine is also an alkaloid. Two other familiar examples, caffeine and nicotine, are shown in Fig. 28–5. Caffeine is present in tea, coffee, cola nuts, and a number of other plants. The compound is a stimulant. Apparently rather large amounts would be necessary to cause death; no fatalities from caffeine have been reported.

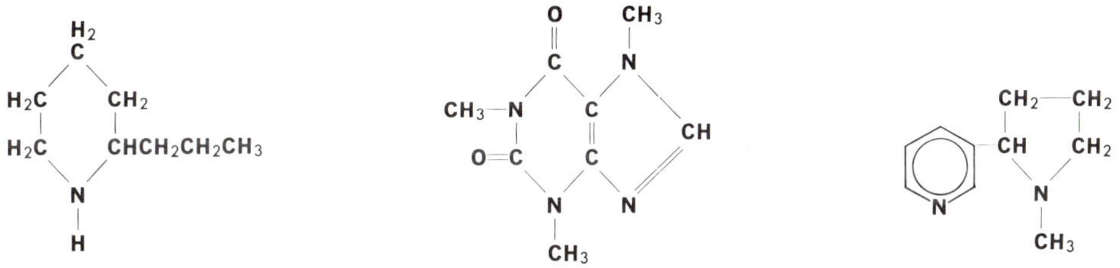

coniine

caffeine

nicotine

Fig. 28–5 Some common alkaloids.

More than ten different alkaloids have been found in tobacco leaves, but **nicotine** constitutes three-fourths of the weight of alkaloids present. Nicotine is highly toxic to animals and has considerable use as an insecticide. In very small amounts, as might be inhaled from smoking tobacco, it causes brief stimulation.

Ergot, a fungus disease of rye, produces several different amides of **lysergic acid.**

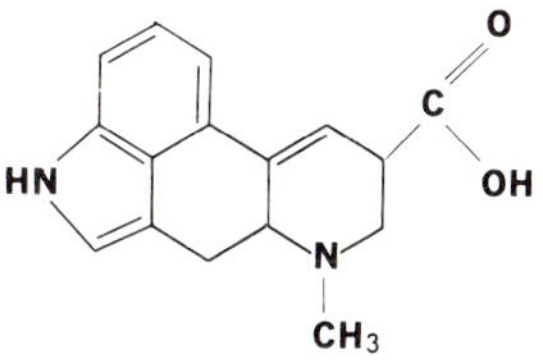

lysergic acid

A synthetic modification of these alkaloids is a hallucinogenic drug called LSD, which represents *lysergic acid diethylamide*. Appropriate treatment of the natural materials can yield lysergic acid. Its carboxyl group, by standard procedures, can react with diethylamine to become a diethylamide structure.

> Many alkaloids have even more complicated structures than those shown above. One example is **quinine**, which is moderately effective in suppressing malaria infections. **Strychnine** is a poison causing convulsions and death. A certain species of poppy plant produces the drug **morphine**. **Codeine** is a slight modification of morphine. Both are habit-forming, but when properly administered in controlled amounts are medically useful as pain relievers and sedatives. Another modification of morphine is **heroin**, which is so addicting that its use even for medical purposes is illegal in the United States.

C. Urea *Know structure*

Economically and biologically, urea is important. It is a colorless, odorless, crystalline, water-soluble compound playing a significant part in the balance of life between plants and animals. Chemically, it is the amide of carbonic acid; the structural relations are shown in the scheme below.

$$H_2O + CO_2 \updownarrow$$
$$HO-\underset{\underset{O}{\|}}{C}-OH + 2NH_3 \rightleftarrows H_2N-\underset{\underset{O}{\|}}{C}-NH_2 + 2H_2O \qquad (28\text{--}33)$$

carbonic acid　　ammonia　　　carbamide
　　　　　　　　　　　　　　　　urea

Like other amides, urea can be *hydrolyzed* to yield ammonia plus the parent acid; the hydrolysis reaction is represented by Eq. (28–33) if it is read from right to left. Since carbonic acid is unstable, it decomposes to carbon dioxide and water. In the opposite direction, urea can be *formed* from carbon dioxide and ammonia, as represented by Eq. (28–33) when read from left to right.

Metabolism of proteins by an animal could potentially release ammonia into his body. The ammonia would be harmful if it accumulated. To avoid this, many animals dispose of the amino groups from proteins or amino acids by combining them with carbon dioxide to form urea. [They carry out Eq. (28–33), in effect, in several steps promoted by enzymes. See Section 39–5 for details of the conversion.] The urea is then excreted. In the soil the enzymes of certain bacteria hydrolyze the urea, thus reversing Eq. (28–33) and releasing the ammonia, which is then available for plants to rebuild into new proteins.

Urea can also be obtained by conducting in a factory the reaction shown in Eq. (28–33). Ammonia, carbon dioxide, and a little water are mixed under pressure at about 180°C. The resultant urea can be isolated as a granular solid by evaporating the water present. Recent production of urea in the United States has exceeded one million tons per year. About 80% of this is used for fertilizers; the rest goes into urea-formaldehyde plastics and adhesives, a few pharmaceuticals, and miscellaneous other products.

> Urea is one of the essential ingredients for the synthesis of the **barbiturate** drugs. The other component is an ethyl malonate ester bearing two substituents. (Ethyl malonate itself is $C_2H_5OOC-CH_2-COOC_2H_5$, the ester of a diacid, malonic acid.) During the reaction (Eq. 28–34), two new amide bonds are formed, closing a ring.

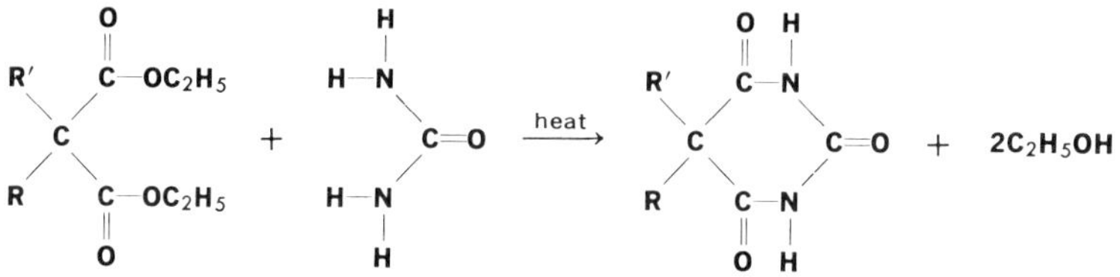

$$\begin{matrix} R' & \overset{O}{\underset{\parallel}{C}}-OC_2H_5 \\ C & \\ R & \underset{\parallel}{C}-OC_2H_5 \\ & O \end{matrix} \quad + \quad \begin{matrix} H & \\ | & \\ H-N & \\ & C=O \\ H-N & \\ | & \\ H & \end{matrix} \quad \xrightarrow{\text{heat}} \quad \begin{matrix} & O & H \\ R' & \overset{\parallel}{C}-N & \\ C & & C=O \\ R & \underset{\parallel}{C}-N & \\ & O & H \end{matrix} \quad + \quad 2C_2H_5OH \qquad (28\text{-}34)$$

substituted ethyl malonate   urea   a barbiturate

In general, the barbiturates are effective sedatives and soporifics. They can be used to induce sleep, calm epileptic patients, provide sufficient unconsciousness for minor surgery, etc. Many variations of the drugs are possible because different substituents (R, R') can be built into the compounds. These variations provide differences in depth of relaxation, effective time, and so forth. Of the dozens which have been synthesized and tested, a few of the most common are listed in Table 28-1.

TABLE 28-1 Some common barbiturates

Substituents R	R'	Generic Name	Some Brand Names
CH_3CH_2-	CH_3CH_2-	barbital	Veronal, Barbitone, Dormonal
 C_6H_5-	CH_3CH_2-	phenobarbital	Luminal, Gardenal, Barbenyl

$CH_3CHCH_2CH_2-$
    $\vert$
    CH_3
	CH_3CH_2-	amobarbital	Amytal, Somnal, Isomytal

D. Aniline and Related Compounds

Thousands of tons of aniline and naphthylamine are manufactured annually by the type of reduction shown in Eq. (28-16). For commercial purposes, scrap cast iron is the least expensive metal to use as the source of hydrogen. Aniline and its derivatives are used largely for the manufacture of dyes and compounds for rubber processing. Smaller quantities are converted to photographic chemicals and important pharmaceuticals.

Aniline has always been closely associated with important dyes. Its name is derived from *anil*, a term used in southern Europe for indigo, because in 1826 aniline was obtained from the decomposition of the natural dye indigo. For many years this source supplied most of the aniline, which remained a rare compound until about 1845 when chemists became aware that aniline could be prepared by reducing nitrobenzene. Aniline was soon to become associated with not only certain natural dyes, but also synthetic dyes.

In 1856 the English chemist William Perkin, then only eighteen years old, attempted to synthesize quinine. In the process he treated aniline with sodium dichromate, a strong oxidizing agent. The result at first appeared to be a useless, tarry mess. However, Perkin discovered (while trying to clean the flask) that alcohol dissolved out of the mess a beautiful purple substance. This proved to be a satisfactory dye for fabrics and became known as *mauve*. With his father and brother, Perkin went into the business of manufacturing the dye. Mauve became so popular and so successful commercially that it set off a rash of trial-and-error experimentation; people mixed this and that with aniline or related amines, hoping to discover an equally lucrative dye. Surprisingly, a number of fairly useful dyes were developed in this way.

It was nearly 1860 before the first structural formulas for organic compounds were written. It was during the 1860's (while mauve was in vogue) that structural formulas began to be applied with great success to a wide variety of organic compounds. Several German chemists insisted that the dyes too should be examined structurally. They reasoned that if a person could learn what types of structure were present in good dyes, then he could more sensibly plan reactions which might lead to similar structures. This attitude was eventually justified. By the end of the century the synthetic dye industry in Germany far surpassed that in England. Indeed, there had been a vigorous growth in all aspects of the German chemical industry.

Thus the development of compounds derived from aniline, in particular some of the dyes, greatly stimulated the discovery of fundamental knowledge about organic compounds. This development also gave organic chemists confidence to attempt the synthesis of organic structures, whether they were duplicates of molecules found naturally or entirely new molecules. Incidentally, the synthesis of quinine, which Perkin had originally attacked, proved to be a very difficult one. This synthesis was not achieved until nearly a century later, in 1944.

It was inevitable that sooner or later chemists would attempt to duplicate synthetically the lovely blue indigo dye produced by *Indigofera* plants. For aniline had been obtained as a decomposition product from indigo, and many of the new dyes had been built up from aniline or closely related compounds. Note the place of aniline in the structure of indigo, shown below.

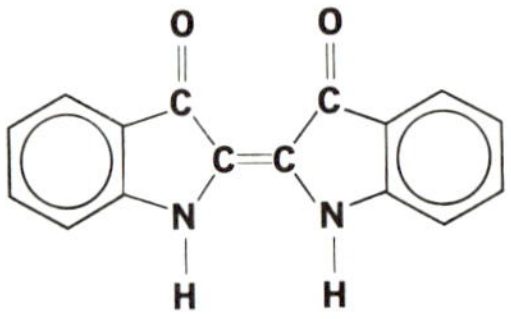

indigo

In 1880 the German chemist Baeyer reported a laboratory synthesis of indigo from relatively simple organic compounds. Unfortunately, some of the intermediate steps required expensive reactants. For factory production it was necessary to find a different series of reactions utilizing cheap materials. Not until 1897 was synthetic indigo put on the market at a price lower than that of the natural product. The economic repercussions were immediate. Frequently a technological change of this sort has a devastating effect on certain groups in society; in this case it was the indigo farmers of India. Indigo is still the most used blue dye. It is cheap and is a "fast" color; that is, it does not fade easily during washing or upon exposure to light.

Aniline is now essential for the manufacture of a great variety of dyes, produced at a rate of thousands of tons per year.

IMPORTANT TERMS

Amide	Dipolar ion	Quaternary ammonium salt
Amine	Peptide	Reductive amination
α-Amino acid	Protein	Substituted ammonium salt
Amino group		

WORK EXERCISES

1. Name these compounds:

   a) $CH_3CH_2NHCH_2CH_3$

   b) 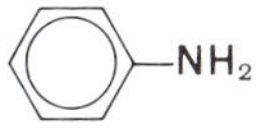 ⬡—NH_2

   c) $CH_3CHCH_2NH_2$
      |
      CH_3

   d) $CH_3NH_3^+ Br^-$

2. Give the structural formula of
   a) trimethylamine
   b) ammonia
   c) ethyl-*n*-propylamine
   d) *o*-nitroaniline
   e) 3-amino-2,4-dimethylpentane
   f) *p*-aminophenol
   g) triethylammonium chloride
   h) benzamide

3. Using structural formulas, show clearly how methylamine can behave as a base in reacting with hydrogen bromide.

4. Write complete equations for the following reaction mixtures. If necessary, state "no reaction."
   a) one mole of methyl bromide, one mole of ammonia
   b) trimethylamine, hydrogen chloride
   c) propanoic acid, ammonia, heat
   d) methyl chloride, trimethylamine
   e) *p*-nitrotoluene, tin, hydrochloric acid
   f) acetyl chloride, trimethylamine
   g) benzoic acid, ethylamine, heat
   h) methylamine, acetyl chloride

5. Using RCOOH to represent lysergic acid, show the structure of the diethylamide derived from it.

6. Write a systematic name for each of the following amino acids. (See Section 28–1 and Table 26–1, and for structures, see Section 28–5.)
   a) alanine
   b) valine
   c) phenylalanine
   d) serine

7. Show the structure of the amino acid alanine as it would exist:
   a) in water
   b) in stomach fluid, which is quite acidic (pH 1).
   c) in intestinal fluid, which is quite basic (pH 9).

8. List all the types of structural units (both carbon skeletons and functional groups) which you can find in the compounds below. (Consider the groups discussed in this chapter, as well as your previous study. Check Table 22–6.)

   a) procaine (Novocaine, Neocaine, etc.), a local and spinal anesthetic:

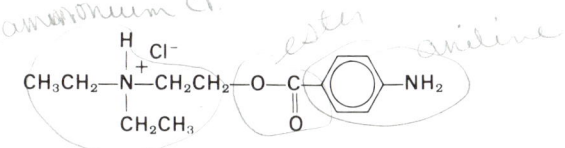

b) The insect repellent called Delphene or Off:

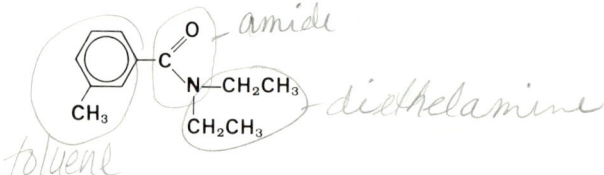

SUGGESTED READING

Asimov, I., *The New Intelligent Man's Guide to Science*, Basic Books, New York, 1965, pp. 447–462. A fascinating account of the contributions from organic synthesis.

Asimov, I., *A Short History of Chemistry*, Anchor Books, Doubleday, New York, 1965, pp. 168–182. Description of the rise of synthetic and structural organic chemistry, including dyes, drugs, explosives, and proteins.

DePuy, C. H., and K. L. Rinehart, Jr., *Introduction to Organic Chemistry*, Chapter 8, John Wiley, New York, 1967.

Dickerson, R. E., "The Structure and History of an Ancient Protein," *Scientific American*, April 1972.

Gates, M., "Analgesic Drugs," *Scientific American*, Nov. 1966 (Offprint #304). A discussion of morphine and the search for related drugs which would provide the same benefits without the undesirable effects.

Guillemin, R., and R. Burgess, "The Hormones of the Hypothalamus," *Scientific American*, Nov. 1972. Recently two of these hormones have been isolated and characterized. Both are peptides, one being composed of three amino acid units and the other of ten amino acid units.

Hart, H., and R. D. Schuetz, *Organic Chemistry*, 4th ed., Chapter 12, Houghton Mifflin, Boston, 1972.

Kurzer, F., and P. M. Sanderson, "Urea in the History of Organic Chemistry," *J. Chem. Educ.* **33**, 452 (1956). The authors state, "The early history of urea provides a vivid impression of the slow and difficult beginnings of chemical science . . ."

Nunes, F., "LSD—An Historical Reevaluation," *J. Chem. Educ.* **45**, 688 (1968).

Robinson, T., "Alkaloids," *Scientific American*, July 1959. Most alkaloids have a profound effect on animals and humans, but their function in plants is obscure.

CARBONYL COMPOUNDS

29–1 NAMES AND STRUCTURES

A carbonyl group has a carbon with a double bond to oxygen.

$$-\underset{\underset{O}{\|}}{C}-\quad \text{carbonyl group}$$

If both of the other bonds from the carbon hold organic groups, the compound is a **ketone**; if one of the bonds holds a hydrogen, the compound is an **aldehyde**.

$$\underset{\text{ketones}}{\underset{O}{\underset{\|}{R-C-R'}}\quad \underset{O}{\underset{\|}{R-C-Ar}}\quad \underset{O}{\underset{\|}{Ar-C-Ar}}}\quad \underset{\text{aldehydes}}{\underset{O}{\underset{\|}{R-C-H}}\quad \underset{O}{\underset{\|}{Ar-C-H}}}$$

> The functional group present in a carboxylic acid, RCOOH, was discussed in Section 26–1. The carboxylic acids are not generally included in the class of carbonyl compounds because a carboxyl group rarely gives the same reaction products that a carbonyl group does.

Names for typical carbonyl compounds are shown in Table 29–1. In a systematic name, the ending -*one* is used for a ketone. As usual, a number is assigned to indicate where in the carbon skeleton this functional group is located. (Note that numbers are not really necessary for propanone or butanone.) The systematic ending for aldehyde is -*al;* it is unnecessary to assign the number *1* to the aldehyde group because if the carbonyl group were at a different position in the chain (say, 2 or 3), the compound would be a ketone, not an aldehyde.

29–2 SYNTHESIS FROM ALCOHOLS

Chemists have devised a great number of methods for building aldehyde or ketone groups into a variety of molecules. We will consider the one most general method of synthesis.

418   Carbonyl Compounds

TABLE 29-1 Nomenclature of aldehydes and ketones

Structure	Common Name	Systematic Name
$CH_3-C(=O)-CH_3$	acetone	propanone
$CH_3-C(=O)-CH_2-CH_3$	methyl ethyl ketone	butanone
$CH_3-CH_2-C(=O)-CH(CH_3)-CH_2-CH_3$	ethyl sec-butyl ketone	4-methyl-3-hexanone
$H-C(=O)-H$	formaldehyde	methanal
$CH_3-C(=O)-H$	acetaldehyde	ethanal
$CH_3-CH_2-C(=O)-H$	propionaldehyde	propanal
$CH_3-CH(CH_3)-C(=O)-H$	isobutyraldehyde	2-methylpropanal

Primary or secondary alcohols can be dehydrogenated to aldehydes and ketones, respectively. This reaction is summarized in Eq. (29–1) and was discussed in detail in Section 25–5B.

$$\overset{|}{\underset{H}{C}}-O-H \xrightarrow[250°C]{Cu} \overset{\diagdown}{\underset{\diagup}{C}}=O + H_2 \tag{29-1}$$

A specific example of the dehydrogenation reaction is the following:

$$CH_3-CH_2-\underset{OH}{\overset{H}{\underset{|}{C}}}-CH_2-CH_3 \xrightarrow[250°C]{Cu} CH_3-CH_2-\underset{O}{\overset{\|}{C}}-CH_2-CH_3 + H_2 \tag{29-2}$$

3-pentanol → 3-pentanone (diethyl ketone)

Many of the aldehydes and ketones which are manufactured in large amounts are made by this procedure, or a slight variation of it (e.g., silver catalyst is sometimes used in place of copper). Examples of compounds prepared in this fashion are acetone, methyl ethyl ketone, formaldehyde, and acetaldehyde.

TABLE 29-2 Some carbonyl compounds found in nature

not respm. for names

Structure	Name	Comments
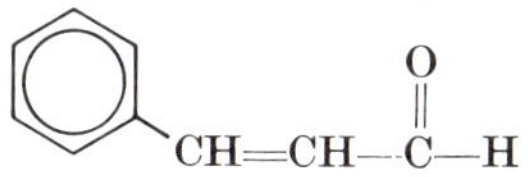 C₆H₅–CH=CH–C(=O)–H	cinnamaldehyde	in cinnamon bark; causes the odor and flavor
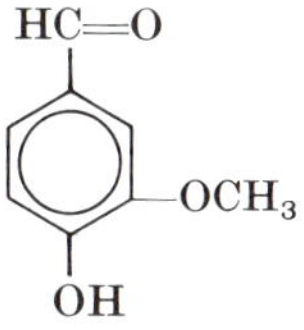 4-hydroxy-3-methoxybenzaldehyde (HC=O, OCH₃, OH on benzene ring)	vanillin	in vanilla beans
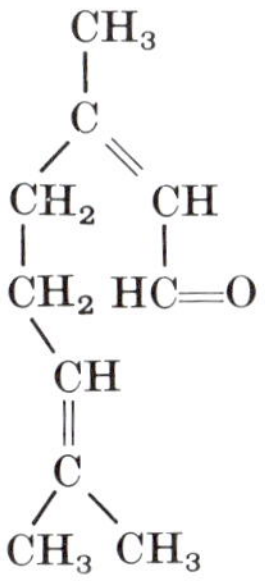 (CH₃)₂C=CH–CH₂–CH₂–C(CH₃)=CH–CHO	citral	in the rinds of citrus fruits and several other plants; characteristic lemon odor
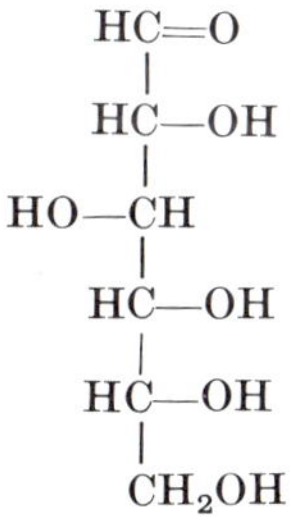 HC=O–HC–OH–HO–CH–HC–OH–HC–OH–CH₂OH	glucose	a sugar found in nearly all living organisms
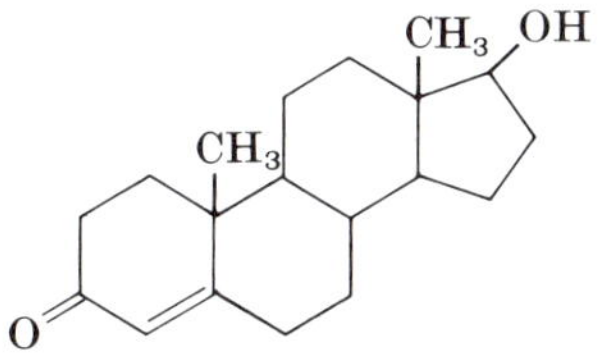 (steroid structure with CH₃, CH₃, OH, and O=)	testosterone	a male sex hormone
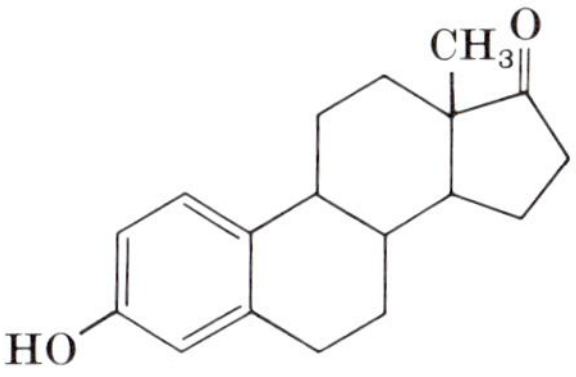 (steroid structure with CH₃, =O, and HO–)	estrone	a female sex hormone

29-3 COMPOUNDS OF INTEREST

Many aldehydes and ketones can be manufactured inexpensively by the method shown in the previous section. They are, in turn, frequently used to synthesize still other organic materials. Thus, **acetaldehyde** is converted to acetic acid, polymers, chemicals for the processing of rubber, and baits to kill snails and slugs. Approximately a million tons of **formaldehyde** are manufactured in the United States annually; nearly all is made into polymers for adhesives or plastic molding materials (Bakelite, Melmac). A solution of formaldehyde in water, called *formalin*, is used to preserve biological specimens. It reacts with the protein in such a way that decay is inhibited, and the tissue becomes firmer. Formaldehyde is also used in embalming. **Acetone** is the starting material for a variety of organic intermediates and solvents, including those manufactured into epoxy polymers and polymethacrylates (Lucite, Plexiglass). Another ketone, **cyclohexanone,** is an intermediate in the synthesis of nylon.

In addition, numerous ketones are effective solvents for materials such as lacquers, fingernail polish, waxes, adhesives, and plastics. **Acetone** (propanone) and **methyl ethyl ketone** (*MEK;* butanone) are the two used most extensively as solvents.

Acetone is also biologically important. Abnormal metabolism in individuals having diabetes causes the production of acetone (Section 38–10); it is then excreted in the urine, or in severe cases even exhaled in the breath.

Many compounds produced by plants or animals have aldehyde or ketone functional groups. The examples listed in Table 29–2 give some idea of the wide variety possible and the complexity of some of them.

29-4 GENERAL CHEMICAL PROPERTIES *of Aldehydes+Ketones*

Aldehydes and ketones are reactive compounds which can be transformed into a wide variety of other substances having either practical or theoretical importance. Many of the possible carbonyl reactions are of interest primarily to the professional chemist. In this chapter we consider three general types of reactions: hydrogenation, oxidation, and addition of polar molecules. Following that, we take a brief look at carbohydrates, which constitute a unique group of carbonyl compounds.

29-5 HYDROGENATION

Hydrogen will add to a carbonyl group in much the same fashion that it adds to an alkene double bond (Section 23–4B). Catalysts such as nickel, palladium, or platinum are necessary to promote the addition reaction, another example of hydrogenation. If we consider specific examples, we note once again the structural relation of aldehydes to primary alcohols and of ketones to secondary alcohols:

$$\begin{array}{c}\diagdown\\ \diagup\end{array}\!\!C\!=\!O + H_2 \xrightarrow{Ni} -\!\!\underset{H}{\overset{|}{C}}\!\!-\!O\!-\!H \tag{29-3}$$

$$CH_3CH_2\underset{\underset{O}{\|}}{C}H + H_2 \longrightarrow CH_3CH_2\underset{\underset{OH}{|}}{C}H_2 \tag{29-4}$$

$$CH_3CCH_3 + H_2 \xrightarrow{Ni} CH_3CHCH_3 \qquad (29\text{-}5)$$
$$\quad\;\; \overset{\|}{O} \qquad\qquad\qquad\qquad \underset{OH}{|}$$

Further, we see that the hydrogenation reaction (Eq. 29–3) is just the reverse of dehydrogenation (Eq. 29–1). In Section 25–5B we saw that dehydrogenation is equivalent to oxidation. Analogously, a hydrogenation reaction, such as that shown in Eq. (29–3), is equivalent to *reduction*. Indeed, we sometimes refer to the addition of two hydrogen atoms to a carbonyl group or to some similar structure as a reduction reaction. The reducing agent may not always be hydrogen gas. Certain other compounds can supply the two hydrogen atoms and thus act as reducing agents.

Hydrogenation of carbonyl compounds (Eq. 29–3), is seldom used to synthesize common alcohols. Instead, the alcohols are usually the source of the carbonyl compounds (Section 29–2). However, the hydrogenation reaction may often be important in preparing a more unusual alcohol or in establishing structural relations. An example of the latter is the hydrogenation of menthone to menthol (Eq. 29–6). The fragrant, flavorful oil produced by mint leaves contains both menthol and menthone.

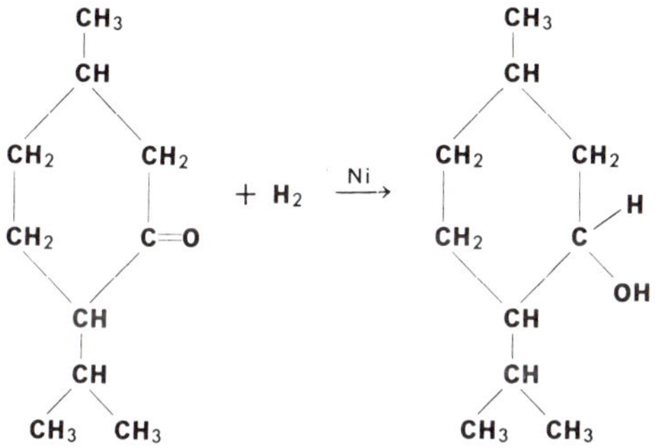

menthone + H₂ →(Ni) menthol   (29-6)

Note that although menthone and citral (Table 29–2) have different functional groups, both have a ten-carbon skeleton with similar branching. However, in menthone the carbon skeleton has been closed to form a six-membered alkane ring. Both citral and menthone belong to the class of compounds called terpenes. Such compounds are widely distributed in nature, particularly in plants. All terpenes have a ten-carbon skeleton consisting of eight carbons plus two methyl branches. Terpenes are discussed further in Section 35–7.

29–6 OXIDATION of aldehyde

Under ordinary conditions ketones resist attack by the common ionic oxidizing agents. Aldehydes, on the other hand, are very easily oxidized to carboxylic acids.

$$RCH_2-\underset{\underset{O}{\|}}{C}-CH_2R + [O] \xrightarrow{KMnO_4} \text{no reaction} \qquad RCH_2-\underset{\underset{O}{\|}}{C}-H + [O] \xrightarrow{KMnO_4} RCH_2-\underset{\underset{O}{\|}}{C}-OH$$

The oxidation involves attack at the *hydrogen* bonded to the carbonyl group. However, *carbons* bonded to the carbonyl group are resistant to attack, which explains why the ketone is unchanged.

This difference in behavior has been used as the basis of qualitative tests to distinguish an aldehyde from a ketone. One such test is provided by **Tollen's reagent**, a solution of silver nitrate in aqueous ammonia:

[margin note: Oxid. reduct. reactions →]

$$R-\underset{\underset{O}{\|}}{C}-H + 2Ag^+ + 3NH_3 + H_2O \rightarrow R-\underset{\underset{O}{\|}}{C}-O^- + 2Ag\downarrow + 3NH_4^+ \qquad (29\text{-}7)$$

While the aldehyde group is being oxidized to a carboxyl group, the silver ion is reduced to silver metal, thus providing a change which can be seen. The metallic silver precipitates in very fine black particles or coats the surface of the glass vessel with a bright silver mirror. A carbonyl compound which behaves in this way is presumed to be an aldehyde; if it does not cause the formation of metallic silver, a ketone is suspected.

Tollen's reagent is used not only as a qualitative test but also for the manufacture of mirrors. For the latter purpose, the aldehyde required (Eq. 29-7) may be either formaldehyde or a sugar such as glucose (Table 29-2).

Glucose is an aldehyde and like all simple sugars (Section 29-11A) has an OH group on the carbon adjacent to the carbonyl group. This structure is especially reactive toward Tollen's reagent. Two similar reagents, originally developed to test for sugars, are **Fehling's solution** and **Benedict's solution**. Both contain copper(II) ions (bright blue solution) which are reduced to copper(I) oxide (brick red precipitate).

$$R-\underset{\underset{O}{\|}}{C}-H + 2Cu^{2+} + 5OH^- \rightarrow R-\underset{\underset{O}{\|}}{C}-O^- + Cu_2O + 3H_2O \qquad (29\text{-}8)$$

Benedict's solution, for example, is often used to detect sugar in urine and is quite satisfactory for this purpose. The Fehling and Benedict solutions are also frequently applied to simple aldehydes or ketones; however, the results are apt to be misleading. For instance, butanal does *not* give positive results with Fehling's solution.

The alert reader may wonder why OH^- ions in the same solution with Cu^{2+} ions do not cause precipitation of insoluble $Cu(OH)_2$. This reaction is prevented by including citrate ions or tartrate ions (not shown in Eq. 29-8) in the solution. These ions form a "complex" with Cu^{2+} which protects it from combination with OH^-.

29-7 ADDITION OF HYDROGEN CYANIDE

[handwritten: aldehyde + HC≡N; Notice — CN attaches to carbonyl C. H of HCN goes to O.]

A. Cyanohydrin Formation

In Section 29-5 the hydrogenation of carbonyl compounds was discussed. This is just one example of an addition reaction; hydrogen, the substance being added, is nonpolar. There is a great variety of *polar* compounds which will also add to a carbonyl group. Reaction with these polar molecules occurs quite readily, because the carbonyl group itself is polar. A good example of addition to a carbonyl group by a polar molecule, HCN, will be shown in this section. Other important examples will appear in the succeeding sections.

Addition of Hydrogen Cyanide

If a solution of hydrogen cyanide is mixed with ethanal, the hydrogen cyanide will add to the ethanal as shown in Eq. (29–9). The reaction goes readily to completion and the product, a cyanohydrin, is stable and easily isolated.

$$CH_3-\overset{\delta+\ H}{\underset{\underset{\delta-}{O}}{C}} \quad HCN \rightarrow CH_3-\overset{H}{\underset{OH}{\overset{|}{C}}}-CN \qquad \textit{Hydroxy cyanide} \qquad (29\text{–}9)$$

a cyanohydrin

This manner of addition is simple and logical. The positive fragment, the H^+, is attracted to the oxygen, the negative end of the carbonyl group. The other portion adding, the CN^-, bonds to the carbon, which is the positive end of the carbonyl group. All other reactions involving addition of polar molecules to a carbonyl compound follow this same pattern. In each case, the hydrogen from the adding molecule becomes bonded to the oxygen, and the other fragment adding, the negative portion, becomes bonded to the carbon.

The nature of this type of addition to a carbonyl group can be better understood if we examine it in more detail. In particular, we should notice the fate of all the electrons involved, for it is pairs of shared electrons which will form the covalent bonds in the resultant product.

$$CH_3-\overset{\delta+\ H}{\underset{\underset{\delta-}{\overset{..}{\underset{..}{O}}}}{C}} \quad \overset{+\ \ -}{H\ :C\equiv N} \rightarrow CH_3-\overset{H}{\underset{\overset{..}{\underset{..}{O}}-H}{\overset{|}{C}}}-C\equiv N \qquad (29\text{–}10a)$$

Since oxygen is much more electronegative than carbon, the two covalent bonds in the carbonyl group are distinctly polar; that is, the electrons tend to be attracted more toward the oxygen atom. The electron pair forming the extra bond in the carbonyl group is especially susceptible to being displaced toward oxygen. Now, to consider the hydrogen cyanide: it is not a highly ionized compound, but it *is* definitely polar, with the electrons between the hydrogen and the cyanide group *tending* to be pulled toward the cyanide to make it a negative ion and tending to release an H^+. Consequently, when the HCN collides with the carbonyl group, the hydrogen, which needs a pair of electrons, is attracted to the oxygen. Because of the mutual attraction, the approach of the H^+ to the negative oxygen causes the electron pair from the extra bond in the carbonyl group to move entirely over to oxygen, where the electrons are shared with hydrogen to form a new covalent bond. The result of this first stage of the reaction would be this:

$$CH_3-\overset{\delta+\ H}{\underset{\underset{\delta-}{\overset{..}{\underset{..}{O}}}}{C}} \quad \overset{+\ \ -}{H\ :C\equiv N} \rightarrow CH_3-\overset{H}{\underset{\overset{..}{\underset{..}{O}}-H}{\overset{|}{C^+}}} \quad :\overset{-}{C}\equiv N \qquad (29\text{–}10b)$$

Since the carbonyl carbon has lost a pair of electrons, it now has a complete positive charge. Notice that it has only three covalent bonds (i.e., three electron pairs). It needs one more electron pair. Consequently, the reaction can be completed by the negative cyanide ion moving in with its electron pair:

$$\underset{\substack{|\\ :O-H}}{\overset{\substack{H\\|}}{CH_3-C^+}} \curvearrowleft :C\equiv N| \rightarrow \underset{\substack{|\\:O-H}}{\overset{\substack{H\\|}}{CH_3-C-C\equiv N}} \tag{29-10c}$$

Among the various carbonyl addition reactions we will consider, there may be minor variations from the pattern of reaction described above, depending on the kind of polar molecule adding and the other reaction conditions. For example, we could imagine the possibility that the two stages for adding the fragments could occur in the reverse order. That is, the negative portion (e.g., NC:⁻) might attack the carbonyl carbon *first*, forcing the electrons of the extra bond to move out to the oxygen, where they would become available to the hydrogen. But the net result would be the same. The new structure produced is always the result of the electron pair in the carbonyl group moving away from the carbon (leaving it positive) onto the oxygen (making it negative).

The product obtained from the addition of hydrogen cyanide to a carbonyl (Eq. 29–9) is called a **cyanohydrin**. It has two functional groups, a hydroxyl group (an alcohol) and an organic cyanide. Each will display the reactions typical for such a group. In Eq. (29–9) hydrogen cyanide was shown adding to an aldehyde. However, the reaction is general for carbonyl compounds; ketones react in a similar fashion, also yielding cyanohydrins.

B. Conversion of Cyanohydrin to Hydroxy Acid

The cyanide functional group in a cyanohydrin can be hydrolyzed to yield a carboxylic acid group, just as can a simple organic cyanide (Eq. 26–24). Applying this hydrolysis reaction to the cyanohydrin obtained from reaction (29–9) results in the formation of lactic acid:

$$\underset{\substack{|\\OH}}{CH_3-CH-C\equiv N} + 2H_2O + HCl \xrightarrow{\Delta} \underset{\substack{|\\OH\\ \text{lactic acid}}}{CH_3-CH-COOH} + NH_4Cl \tag{29-11}$$

Hydrolysis of a cyanohydrin will always yield an α-hydroxy acid. This series of reactions is of considerable interest as a synthetic route, because quite a few α-hydroxy acids occur in nature. Lactic acid, discussed in Section 26–9D, is just one example.

C. Conversion of Cyanohydrin to Hydroxy Aldehyde

Another possible alteration of the cyanohydrin can convert it to a still different compound. Catalytic hydrogenation of an organic cyanide in the presence of water changes it to an aldehyde:

$$R-C\equiv N + H_2 + H_2O \xrightarrow{Pt} R-\overset{\substack{H\\|}}{\underset{\substack{\|\\O}}{C}} + NH_3 \tag{29-12}$$

Treatment of a cyanohydrin in this manner would correspondingly yield a hydroxy aldehyde:

$$\underset{\substack{|\\OH}}{CH_3-CH-C\equiv N} + H_2 + H_2O \xrightarrow{Pt} \underset{\substack{|\\OH}}{CH_3-CH-\overset{\substack{H\\|}}{\underset{\substack{\|\\O}}{C}}} + NH_3 \tag{29-13}$$

The use of reaction (29-9) followed by (29-13) is a synthetic sequence of particular value, for it provides a means to build, step-by-step, a whole series of carbohydrate compounds. This method of synthesis has been widely applied to study carbohydrates and to duplicate those found in nature. Let us discuss an example of this technique for carbohydrate synthesis.

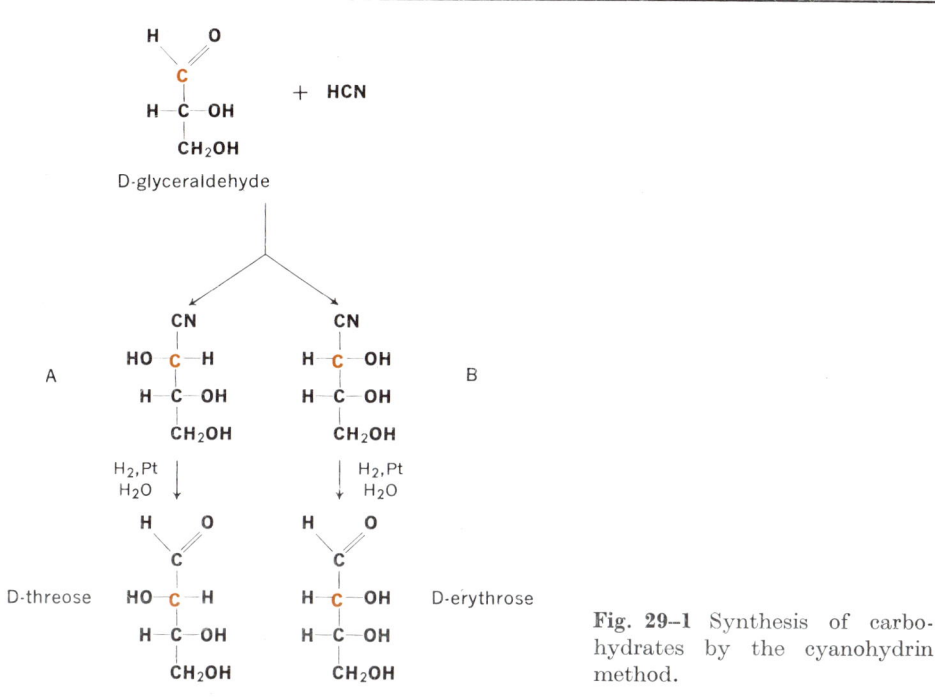

Fig. 29-1 Synthesis of carbohydrates by the cyanohydrin method.

D-glyceraldehyde, shown in Fig. 29-1, is one of the simplest carbohydrates; it is the aldehyde structurally related to glycerol. D-glyceraldehyde is one of the key compounds formed during the process of photosynthesis and in the metabolism of carbohydrates by living organisms. In the laboratory, if D-glyceraldehyde is treated with HCN it is converted to a mixture of two cyanohydrins, labeled A and B in Fig. 29-1. The only difference between A and B is the arrangement in space of their OH groups. In B the new OH group created by the addition of HCN is oriented toward the original OH group, while in A the two OH groups are directed away from each other. Now if A and B are separated and each is treated with hydrogen, platinum, and water, each will be converted to the corresponding aldehyde which has a skeleton one carbon longer than did the starting material, D-glyceraldehyde. Cyanohydrin A will yield the carbohydrate D-threose, and B will yield D-erythrose.

Since D-threose and D-erythrose are aldehydes, we could select either one and repeat the two-step synthetic scheme, first adding hydrogen cyanide and then reducing catalytically in the presence of water. By this process, D-erythrose is converted to D-arabinose and D-ribose. The formulas of these two compounds are shown in Fig.

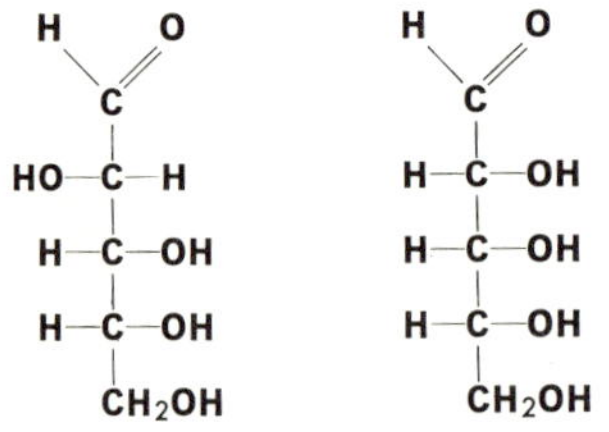

426 Carbonyl Compounds

```
   H   O              H   O
    \ //               \ //
     C                  C
     |                  |
HO—C—H              H—C—OH
     |                  |
 H—C—OH             H—C—OH
     |                  |
 H—C—OH             H—C—OH
     |                  |
   CH₂OH              CH₂OH
 D-arabinose         D-ribose
```

Fig. 29-2 Formulas of two well-known five-carbon carbohydrates.

29-2. Each of these has one carbon, together with a new hydroxyl group, more than did its precursor, D-erythrose.

D-ribose is another substance which is a component of a number of crucial biochemical compounds; for instance, it is the carbohydrate component of the ribonucleic acids which control protein synthesis in the cells (Section 29–11G).

By applying one more cycle of this synthesis, we can transform D-arabinose, with its five-carbon skeleton, to D-mannose and D-glucose, each with a six-carbon skeleton. The detailed formula for D-glucose is shown in Table 29–2.

The symbol D in the name D-glyceraldehyde indicates that the hydroxyl group on the central carbon is oriented toward the right, as shown below and in Fig. 29–1. D is derived from the Latin word *dextro*, meaning right. The other possibility is that the hydroxyl group could be toward the left, in which case the designation L would be used, from the Latin *laevus*, left.

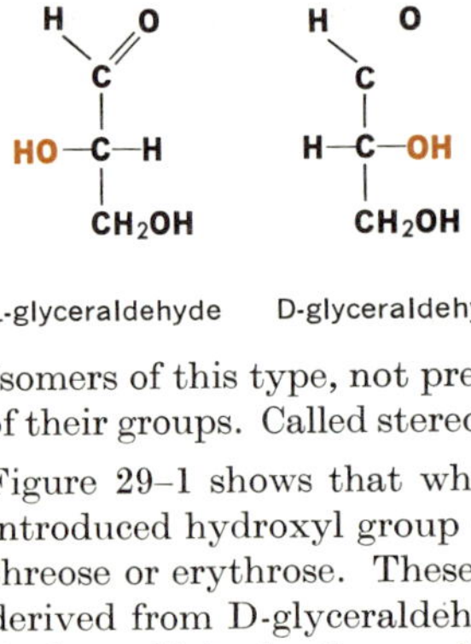

```
   H   O              H   O
    \ //               \ //
     C                  C
     |                  |
HO—C—H              H—C—OH
     |                  |
   CH₂OH              CH₂OH
 L-glyceraldehyde    D-glyceraldehyde
```

Isomers of this type, not previously discussed, differ only in the arrangement in space of their groups. Called stereoisomers, they are discussed in greater detail in Chapter 31.

Figure 29–1 shows that when a D-glyceraldehyde molecule is lengthened, the newly introduced hydroxyl group can be oriented toward either the left or the right, giving threose or erythrose. These two products are still designated as D because they were derived from D-glyceraldehyde. If the synthesis had been carried out starting with L-glyceraldehyde, the products would be L-threose and L-erythrose.

29-8 ADDITION OF AN ALCOHOL

In the presence of an acid catalyst, an alcohol will add to a carbonyl group, with the oxygen from the alcohol bonding to the carbon of the carbonyl group and the carbonyl oxygen acquiring a hydrogen:

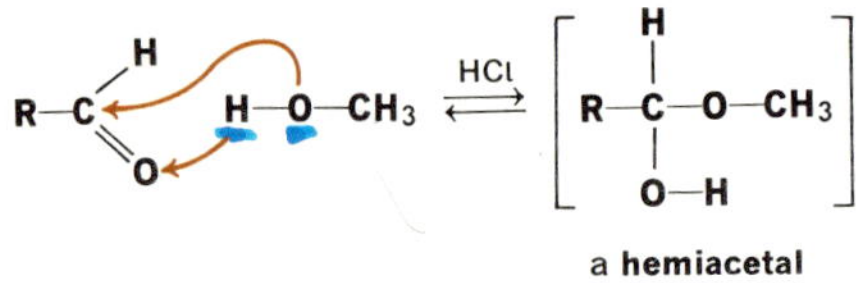

$$R-C(H)(=O) + H-O-CH_3 \underset{}{\overset{HCl}{\rightleftharpoons}} \left[R-\underset{O-H}{\overset{H}{\underset{|}{\overset{|}{C}}}}-O-CH_3 \right] \qquad (29\text{-}14)$$

a hemiacetal

This mode of addition by the alcohol is clearly similar to the pattern followed during the addition of hydrogen cyanide, Eq. (29–9). However, reaction (29–14) is readily reversible. The equilibrium does not especially favor the product, a hemiacetal, and the hemiacetal is not very stable. In fact, with most of the ordinary carbonyl compounds it is not possible to get the hemiacetal out of the reaction mixture. In Eq. (29–14), brackets are placed around the formula of the hemiacetal to indicate that it is unstable.

Although the hemiacetal is usually too stable to isolate, it can react further in this same solution to form an acetal, a product which *is* stable:

$$\left[\begin{array}{c} H \\ | \\ R-C-O-CH_3 \\ | \\ O-H \end{array} \right] + H-O-CH_3 \xrightleftharpoons{HCl} \begin{array}{c} H \\ | \\ R-C-O-CH_3 \\ | \\ O-CH_3 \end{array} + HOH \quad (29\text{–}15)$$

a hemiacetal   an **acetal** *(di-ether)*

This second stage of reaction, Eq. (29–15), will be favored if we make it a point to include an excess of the alcohol in the reaction mixture. This stage is not an addition reaction; rather, it is a condensation, with water being split out. This reaction is also reversible.

Because of the reversibility of both Eq. (29–15) and Eq. (29–14), it is possible to *hydrolyze* an acetal. By treating an acetal with a large excess of water and an acid catalyst, we can convert it back to an aldehyde plus two molecules of alcohol. Of course, the conversion takes place by passing through the intermediate hemiacetal stage. The hydrolysis of an acetal functional group facilitates the study of those natural compounds which happen to contain such a structure; quite a few important compounds do. Hydrolysis will reveal which aldehyde and alcohol components were originally combined in the compound.

Since reactions (29–14) and (29–15) are both reversible, in order to get good conversion of the aldehyde to the acetal, we need to do whatever is possible to shift the equilibrium. We should use a large excess of alcohol. We notice that water is a *product* of the reactions. Therefore, we should be sure none is present to start with; we should use alcohol which is free of water, and for catalyst we should dissolve in the alcohol some dry HCl gas, *not* an aqueous solution of HCl. With these precautions, it is possible to carry the reaction through both stages, Eqs. (29–14) and (29–15), and obtain a good yield of the acetal.

By a similar line of reasoning, we can see how to convert the acetal back to the aldehyde by reversing the equilibrium in the two reactions. To promote reaction in the reverse direction, the acetal should be treated with a large excess of water. Acid catalyst should again be included, because any substance which catalyzes a forward reaction will similarly catalyze the reverse reaction. The reverse reaction, starting with attack by water, will first split one molecule of alcohol off the acetal, converting it to a hemiacetal. Subsequent loss of a second molecule of alcohol will finally produce the aldehyde. Altogether the reverse reaction would appear as follows:

$$\begin{array}{c} H \\ | \\ R-C-OCH_3 \\ | \\ OCH_3 \end{array} + HOH \xrightleftharpoons{HCl} \left[\begin{array}{c} H \\ | \\ R-C-OCH_3 \\ | \\ OH \end{array} \right] + HOCH_3 \xrightleftharpoons{HCl} R-C\begin{array}{c} H \\ \diagdown \\ O \end{array} + 2HOCH_3 \quad (29\text{–}16)$$

acetal   hemiacetal   aldehyde

Since this conversion of an acetal back to an aldehyde, Eq. (29–16), involves reaction with water, it would be called a hydrolysis reaction.

In summary, it should be emphasized that the reaction of a carbonyl compound with an alcohol begins, as shown in Eq. (29–14), with an *addition* reaction which is very similar to many other additions to carbonyls. Although an aldehyde was used for illustration in Eq. (29–14), ketones will react in a similar manner with alcohols, in which case the final product is called a **ketal**.

The formation and hydrolysis of hemiacetals and acetals are frequently observed in the laboratory and are important as well among natural products. In particular, a knowledge of these reactions is essential for an understanding of the properties of carbohydrates, which will be discussed soon in Section 29–11 and again, later, in Chapter 33.

29–9 ADDITION OF AMMONIA

Ketone + → imine + H₂O

A. The Nature of the Reaction

The addition of ammonia to a carbonyl group takes place as follows:

$$R-C(R)=O + H-N(H)-H \rightleftharpoons \left[R-C(R)(OH)-N(H)-H \right] \rightleftharpoons \left[R-C(R)=NH \right] + HOH \quad (29-17)$$

AA   an **imine**

In several ways, this reaction is similar to the addition of an alcohol to a carbonyl group. Reaction (29–17) is reversible and the addition product, AA, which forms in the solution is unstable, just like a hemiacetal. Again we enclose in brackets the formulas of unstable compounds. As in the previous examples of carbonyl additions, when ammonia reacts the hydrogen will bond to the oxygen, and the electronegative atom, in this case nitrogen, will bond to the carbon of the carbonyl group. An aldehyde participates in a reaction such as (29–17) just as well as the ketone illustrated.

Although a substantial amount of the ammonia addition product, structure AA, is formed, it cannot be isolated. It can, however, break down by splitting out ammonia and reverting to the carbonyl compound. Alternatively, it can split out water to give an imine. This reaction is also reversible and the imine is unstable.

Equation (29–17) is an important reaction despite the fact that neither structure AA nor the imine can be isolated. There is ample evidence that they *are* present in the reaction mixture and under just slightly different conditions stable compounds *can* result from such a reaction. One way that this can happen is if the compound adding is a derivative of ammonia, G—NH$_2$, bearing a group G in place of one of the hydrogens. There are several kinds of structure which G may be that will make the resultant imine, R$_2$C=N—G a stable, easily isolated substance. Products of this type will be discussed in Section 29–10.

B. Reductive Amination

imine + H₂ —cat.→ amine

Another way in which a significant product can be isolated from a reaction such as (29–17) is by reducing (hydrogenating) the imine existing in the mixture. This process converts it to an amine, a stable and useful compound:

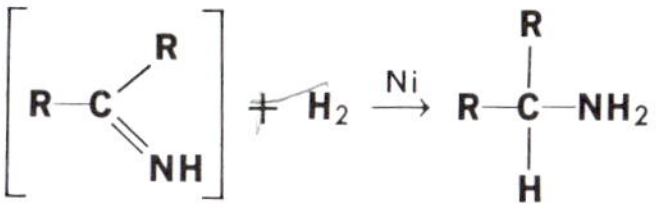

$$\left[R-C \begin{array}{c} R \\ \diagdown \\ NH \end{array} \right] + H_2 \xrightarrow{Ni} R-\underset{\underset{H}{|}}{\overset{\overset{R}{|}}{C}}-NH_2 \qquad (29\text{-}18)$$

an imine an amine

It is evident that this reaction is structurally similar to the addition of hydrogen to C=C or C=O groups. To transform the carbonyl compound all the way through to the amine simply requires placing in the same vessel all the necessary reactants: the carbonyl compound, ammonia, hydrogen, and nickel catalyst. The net reaction occurring in this mixture may be represented thus:

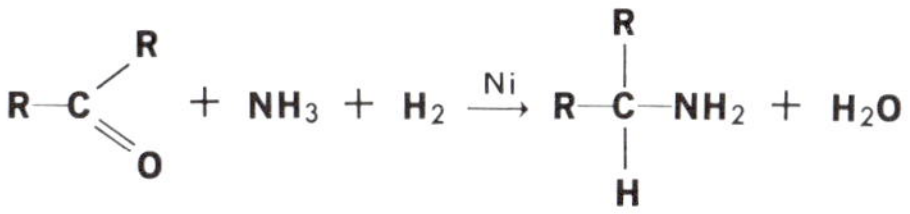

$$R-C \begin{array}{c} R \\ \diagup \\ \diagdown \\ O \end{array} + NH_3 + H_2 \xrightarrow{Ni} R-\underset{\underset{H}{|}}{\overset{\overset{R}{|}}{C}}-NH_2 + H_2O \qquad (29\text{-}19)$$

What takes place, of course, is that reaction (29–17) forms the imine. As fast as any imine appears in the mixture, it is converted by hydrogen, as in reaction (29–18), to the amine. In fact, by adding Eqs. (29–17) and (29–18) algebraically, one arrives at the net equation (29–19). The process represented by Eq. (29–19) is called **reductive amination**. It was first shown in Section 28–3B as a general method for the synthesis of amines. In our present discussion, a ketone has been used in the examples, Eq. (29–17) and so on. However, these are very general reactions of carbonyl compounds; aldehydes react every bit as well.

C. Reductive Amination in Living Organisms

Reductive amination has just been discussed as a chemical process. It is also biologically important. In living systems, all the same steps occur, the chief differences being that the catalysts are enzymes and that the pair of hydrogen atoms required for the reduction step comparable to Eq. (29–18) come not from hydrogen gas but from some organic reducing agent in the cell.

One biochemical example of reductive animation is the production of the important amino acid, glutamic acid. A great variety of plants and animals are capable of carrying out this biosynthesis, represented by Eq. (29–20).

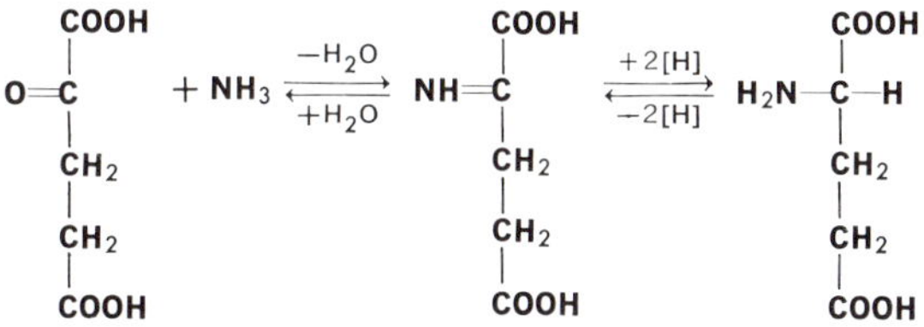

$$\underset{\substack{\text{COOH} \\ | \\ O=C \\ | \\ CH_2 \\ | \\ CH_2 \\ | \\ COOH}}{} + NH_3 \underset{+H_2O}{\overset{-H_2O}{\rightleftharpoons}} \underset{\substack{\text{COOH} \\ | \\ NH=C \\ | \\ CH_2 \\ | \\ CH_2 \\ | \\ COOH}}{} \underset{-2[H]}{\overset{+2[H]}{\rightleftharpoons}} \underset{\substack{\text{COOH} \\ | \\ H_2N-C-H \\ | \\ CH_2 \\ | \\ CH_2 \\ | \\ COOH}}{} \qquad (29\text{-}20)$$

α-keto glutaric acid α-imino glutaric acid glutamic acid

An enzyme called glutamate dehydrogenase is required as a catalyst for reaction (29–20). The necessary ammonia is obtained from cellular constituents. The two hydrogen atoms required for addition to the imine structure in the second stage are provided by a coenzyme, nicotinamide adenine dinucleotide. The α-ketoglutaric acid required for the carbon skeleton in reaction (29–20) can be derived from carbohydrates, through a complex series of metabolic reactions.

29–10 ADDITION OF AMMONIA DERIVATIVES

When one of the hydrogens of an ammonia molecule has been replaced by one of certain organic groups, the ammonia derivative will react with a carbonyl compound to give a stable product. For this reaction to be successful, the substituted group (which will be shown in a general way as G) must be of particular types. The reaction of these ammonia derivatives occurs as follows:

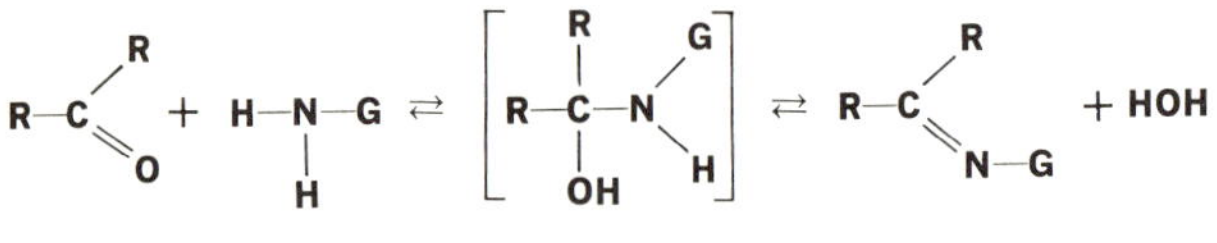

$$R-C(R)(=O) + H-N(H)-G \rightleftharpoons \left[R-C(R)(OH)-N(G)(H) \right] \rightleftharpoons R-C(R)(=N-G) + HOH \quad (29\text{-}21)$$

GA

Notice that the initial reaction involving *addition* to the carbonyl group once again gives an unstable product, just as it did with ammonia; compare structure AA in Eq. (29–17). The important difference in Eq. (29–21) is that when the intermediate structure, GA, splits out water in the second stage of the reaction, the final product is stable.

Shown below are the reactions of some of the important types of GNH_2 compounds. In each case only a net equation has been written, showing the final, stable product which one obtains. Since the unstable intermediate cannot be isolated, it has not been shown. However, in each case the reaction will take place as shown in Eq. (29–21), going by way of the intermediate, GA.

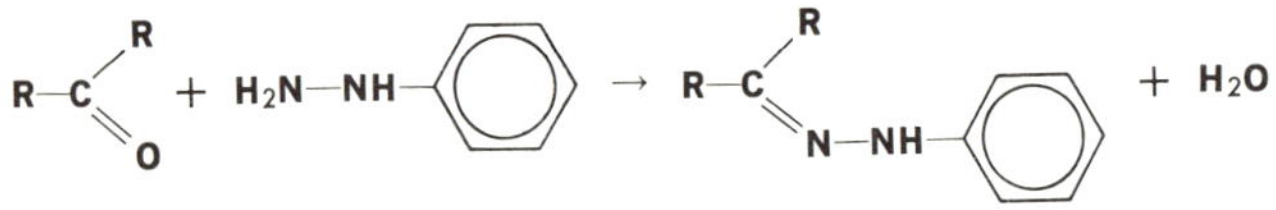

$$R-C(R)(=O) + H_2N-NH-C_6H_5 \rightarrow R-C(R)(=N-NH-C_6H_5) + H_2O \quad (29\text{-}22)$$

phenylhydrazine → a **phenylhydrazone**

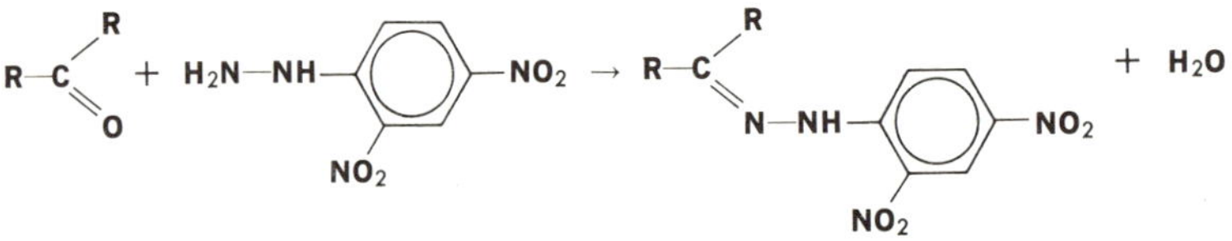

$$R-C(R)(=O) + H_2N-NH-C_6H_3(NO_2)_2 \rightarrow R-C(R)(=N-NH-C_6H_3(NO_2)_2) + H_2O \quad (29\text{-}23)$$

2,4-dinitrophenylhydrazine → a **2,4-dinitrophenylhydrazone**

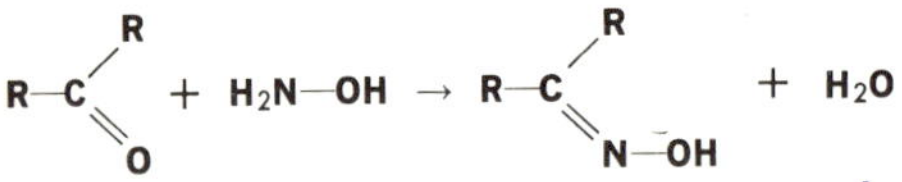

$$R-C(R)(=O) + H_2N-OH \rightarrow R-C(R)(=N-OH) + H_2O \quad (29\text{-}24)$$

hydroxylamine → an **oxime**

Each of the reactions just shown is *very general for all kinds of aldehydes and ketones*. The products, often called carbonyl *derivatives*, are nearly always solids which can be filtered off and purified by crystallization. Since each solid derivative has its own characteristic melting point, it provides important information for identifying the original carbonyl compound. Very often three or four such bits of information are sufficient for positive identification of a carbonyl compound. Since many carbonyl compounds are liquids, the solid derivatives such as phenylhydrazones and oximes provide a great advantage. Solids can be purified with more certainty and greater ease than can liquids. And once the solid is obtained, its melting point is a much more reliable property for identification than is the boiling point of a liquid.

29-11 CARBOHYDRATES

A. Introduction

The simplest carbohydrates, to which all others are related, are often called **monosaccharides**. These compounds are also called glycoses, a more systematic chemical name. A **glycose** *is a polyhydroxy aldehyde or ketone*. A great variety of such structures are found in nature, a few being produced in large quantities. Three of the common glycoses are shown in Fig. 29-3. One of the simplest, and also very common, is D-glyceraldehyde, which was mentioned previously in Fig. 29-1. Glucose is found in all living organisms. Many plants accumulate large quantities of glucose combined with itself in the form of starch and cellulose. Galactose is one of the components of milk sugar. Fructose is found free in honey and in numerous fruit juices (the basis of its name), particularly in apples and tomatoes. Ordinary table sugar, sucrose, consists of fructose combined with glucose.

Notice that all the glycoses in Fig. 29-3 are from the D-family; that is, each is related to D-glyceraldehyde in that the hydroxyl group on the next-to-last carbon (shown in color in Fig. 29-3) is on the right. Fructose is a ketone type of glycose, often called a *ketose*. Glucose and galactose are *aldoses*. In fructose, the configurations (i.e., spatial arrangement) of the OH groups beyond position 2 are the same as for the corresponding positions in glucose. Galactose and glucose differ in configuration only at position 4.

The carbonyl group and the hydroxyl groups of a glycose respond in most chemical reactions as expected for such functional groups. The aldehyde group of glucose is easily converted by mild oxidizing agents to a carboxyl group, and its hydroxyl groups, like any alcohols, can be esterified. Functional derivatives such as these exist among

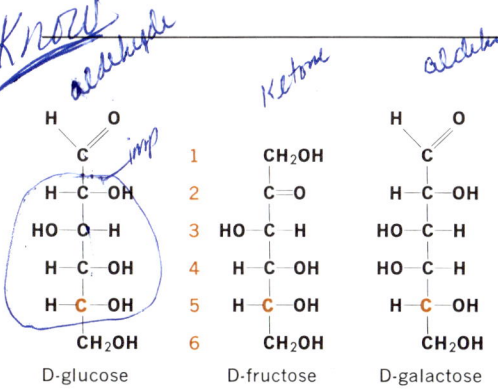

Fig. 29-3 Some common glycoses, or monosaccharides.

natural glycoses, and so the derivatives are usually included within the class of carbohydrates. Another characteristic of glycoses, or monosaccharides, is that two molecules can link to each other, by splitting out water, forming a **disaccharide**. Or, the process can continue, with more and more glycose units bonding to those previously linked, giving a **polysaccharide** (*poly* meaning many). Trisaccharides and tetrasaccharides are also possible, but considerably less common.

We have defined glycoses and have described some of the typical modifications in their structures. Now we can state that the class of organic compounds called **carbohydrates** *includes glycoses, disaccharides, polysaccharides, and other closely related derivatives of glycoses*. The word *sugar* simply means sweet substance and has no precise chemical meaning. Although a number of both organic and inorganic compounds have a sweet taste, the term sugar is most often applied to those which are carbohydrates. Since polysaccharides do not taste sweet, the compounds we think of as being **sugars** *are monosaccharides and disaccharides*, in general.

B. Ring Structures

We found in Section 29–8 that an alcohol could participate in an addition reaction with an aldehyde to form a hemiacetal. Interestingly, a glycose molecule possesses both alcohol and carbonyl functional groups. Therefore, hemiacetal formation can occur *within* one molecule, provided that the glycose carbon skeleton is long enough so that

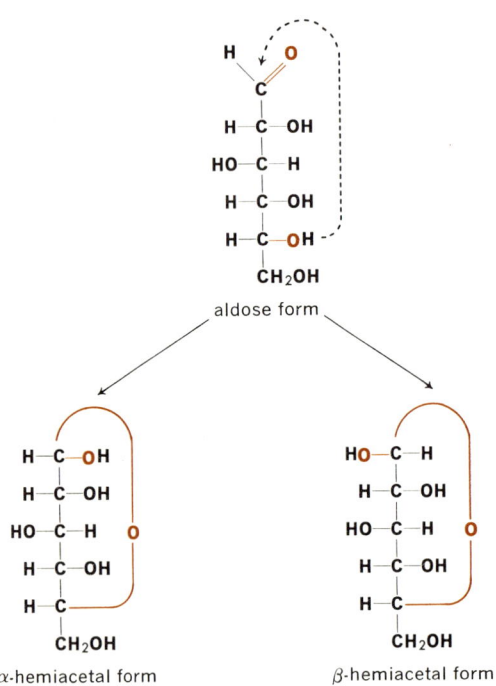

Fig. 29–4 Hemiacetal formation in D-glucose, resulting in ring closure.

during its normal twistings (arising from rotations of the single bonds; see Chapter 21) one of the hydroxyl groups will reach the carbonyl group. The reaction of the two groups within one molecule results in a ring structure. Figure 29-4 illustrates this behavior for D-glucose. The oxygen atoms and bonds involved in the addition reaction are shown in color.

Notice that the only difference between the two hemiacetal forms, called alpha and beta, is the orientation of the newly formed OH group in each. Either one is possible, its formation depending merely on which direction the C=O group happened to be pointing at the moment the OH group added to it to form the hemiacetal structure. There can be rotation on the single bond between the first and second carbons, allowing the C=O group to be oriented to either the right or the left. If reaction occurs while it is to the right, the α form of the hemiacetal ring structure results.

In terms of chemical reactivity, it is entirely reasonable that a glycose hydroxyl group should add to a carbonyl group within the same molecule, and in terms of size or distance it is extremely probable. The ring resulting from this intramolecular addition reaction is of a size which is very stable and easily formed; it is, altogether, a six-member ring, for it consists of five carbon atoms and one oxygen atom. The type of structural formula for the hemiacetal ring shown in Fig. 29-4 is convenient to write. It is easy to see its relation to the structure of the open-chain aldose. Although this sort of hemiacetal formula shows which atoms are bonded to which, it is very unrealistic in its dimensions. The oxygen bonds could never be so long and so distorted as they appear in Fig. 29-4. Actually, C—O bond lengths are very nearly the same as C—C bond lengths. Therefore, all the

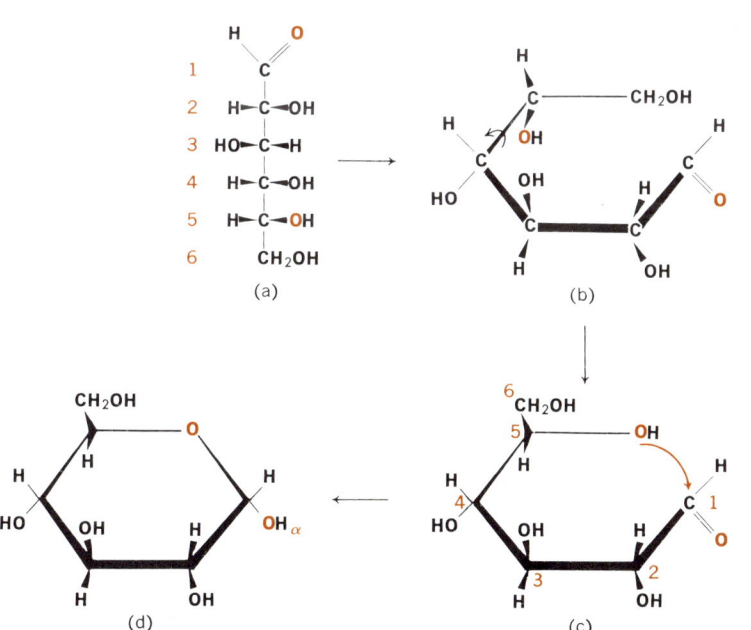

Fig. 29-5 Ring formation in D-glucose, shown in perspective for α-D-glucose.

bonds in the hemiacetal ring should be about the same length. The ring skeleton should look very much like that of cyclohexane, except that one of the six positions is occupied by an oxygen rather than by a carbon.

Figure 29–5 shows in a more realistic manner how ring closure can allow formation of a hemiacetal group in D-glucose. First of all, Fig. 29–5(a) attempts to show a little more perspective in the bonds of the aldose formula of D-glucose. That is, when we are looking at the molecule from the front, the H and OH groups at the sides are closest to us and their bonds are receding back toward the carbon skeleton. (This matter of perspective is discussed in greater detail in Section 21–7.) In Fig. 29–5(b) we have merely turned the aldose structure over on its side and, *most important*, have shown the C—C bonds with their normal 109° angles. Note that doing this allows the carbon skeleton to curl back upon itself. Now in conformation (b), when rotation of the bond between the fourth and fifth carbons occurs, the OH group at position five will come up as seen in (c). In this location, the OH is at just the right distance for bonding to the C=O group. If addition of the OH takes place while the C=O is oriented as it is in (c), the α hemiacetal ring (d) is produced. Formula (d) is a much more accurate representation of α-D-glucose than is the formula in Fig. 29–4. Henceforth, we will represent α-D-glucose by the more satisfactory formula (d) in most of our discussions.

It is entirely possible, of course, for each of the C—C bonds in a glucose molecule to rotate, so that the carbon skeleton can assume many shapes other than (c). But the whole point is that frequently the skeleton *will* get into conformation (c). As soon as it does, the hydroxyl at C-5 can add across the carbonyl group, forming the hemiacetal (d) and, incidentally, closing a ring.

A few other comments concerning Fig. 29–5 should be made. First of all, notice that (c) is still an open-chain, aldose structure. Second, a very important point: for any glycose, a hemiacetal structure such as (d) is very stable. Most of the molecules will exist in this form rather than in the open aldose form (c). This situation is in contrast, of course, to simple aldehydes, for which the hemiacetal structure is *less* stable than the corresponding aldehyde; see Section 29–8. Next, observe that the hydroxyl groups which are on the *right* in the structures of Fig. 29–4 are *down* in structure (d) of Fig. 29–5. Finally, although D-glucose has been used throughout this discussion on ring formation via the hemiacetal reaction, other glycoses show the same behavior.

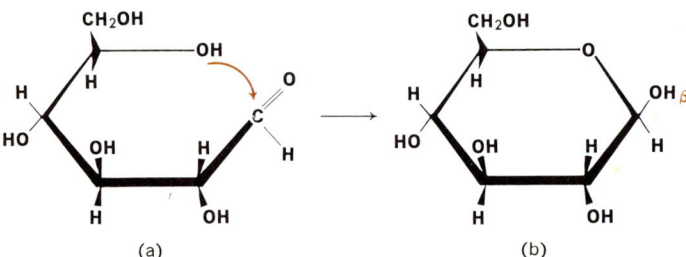

Fig. 29–6 The structure of β-D-glucose, an alternative ring form of D-glucose (compare Fig. 29–5d); (a) shows the mode of formation and (b) shows the final structure of the β-form.

Figure 29–5 shows a D-glucose molecule in conformation (c). If the reaction occurs at the instant which catches it in this form, then an alpha hemiacetal (d) results. However, rotation of the bond between C-1 and C-2 could throw the aldehyde group of glucose into the conformation represented by (a) in Fig. 29–6. If the glucose molecule reacts at the moment it is in conformation (a), the beta hemiacetal ring results, as in Fig. 29–6(b).

It was mentioned that a hemiacetal structure for a glycose is actually more favorable than is the aldose or ketose—so much so, in fact, that crystals of D-glucose, for example, consist *only* of the hemiacetal form. Glucose can be crystallized out of a solution and will form crystals built up from α molecules only. Or by doing the crystallization at a different temperature and from a different solvent, one can obtain a solid made up entirely of D-glucose in the β form. Only these two kinds of solid D-glucose are known. It has never been possible to get any of the aldose form out of solution.

When glucose dissolves in certain solvents, the molecules in the crystals will pass into solution with exactly the same molecular structure. Thus, crystals of α-D-glucose will dissolve in pyridine to give a solution containing only α-D-glucose molecules. Similarly, from β crystals one can obtain a pyridine solution of β-D-glucose. A cold solution of either one—let's use β for illustration—will react with an excess of benzoyl chloride so that each of the five hydroxyl groups of the hemiacetal structure is converted to a benzoate ester group. This final ester product will still be exclusively in the β form.

However, when glucose is dissolved in *water*, or in solutions containing acid, all three forms of glucose will be present. Recall that the reaction to form a hemiacetal is easily reversible, so that there can be an equilibrium between hemiacetal molecules and aldehyde molecules; see Eq. (29–14). Thus, when α-D-glucose crystals dissolve in water, initially we find only molecules of the α hemiacetal form in the solution. But in many molecules the ring will soon break open to give the aldose form:

$$\alpha\text{-D-glucose} \rightleftarrows \text{aldose form of D-glucose} \rightleftarrows \beta\text{-D-glucose} \quad (29\text{-}25)$$

hemiacetal form ⇌ aldose form of D-glucose ⇌ hemiacetal form

Some of the aldose molecules now present in solution will close the ring again, thus going back to the α hemiacetal form. But we have seen that when a ring is closed on the aldose form, some of the molecules will go over to the β hemiacetal form, as well as to α. And the β molecules will also establish the reverse reaction, with some of them reopening to give aldose molecules. Eventually both sets of reversible reactions shown in Eq. (29–25) will reach equilibrium, so that some of all three forms of D-glucose will exist in a water solution. By the same token, if we started with crystals of β-D-glycose, they would dissolve in water and eventually give an equilibrium mixture containing the three forms in the same proportions as those that resulted when we started with α-D-glucose. It is interesting to note that in the equilibrium mixture, less than 1% of the molecules exist in the aldose form; all the rest are in a hemiacetal form, either α or β.

Since all three forms of glucose can exist in an aqueous solution, any one of the three can react, depending on conditions. Thus, if we add a reagent which customarily reacts with a carbonyl group, then the aldose form of glucose present in the equilibrium mixture will react. For instance, if we apply the Tollen's reaction, shown in Eq. (29-7), to glucose we would have this:

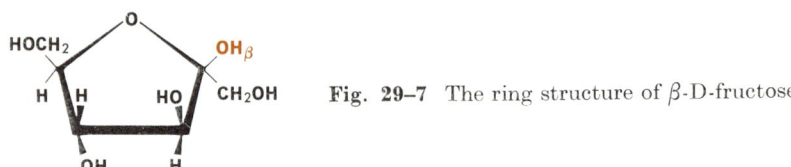

(29-26)

Since it is the aldehyde group of glucose which is oxidized by this reaction mixture, only those molecules in the aldose form react. But we must keep in mind that the reactions of Eq. (29-25) are reversible. Therefore, as fast as aldose molecules are removed from solution by reaction (29-26), more of the α and β molecules will change over to aldose form in an effort to reestablish the equilibrium represented by Eq. (29-25). However, since the newly formed aldose molecules will be promptly consumed by reaction (29-26), fairly soon *all* the glucose molecules from the solution will be oxidized.

Fig. 29-7 The ring structure of β-D-fructose.

A ketose sugar is just as likely as an aldose to form a ring structure, by addition of one of its hydroxyl groups to its carbonyl group. The ring structure of D-fructose, the most common ketose, is shown in Fig. 29-7, with the hydroxyl group of the hemiketal structure (the one derived from the original keto group) marked in color. To form the type of ring shown in Fig. 29-7, the OH group from position 5 of fructose has added to its carbonyl group, which is at position 2. This addition results in a five-membered ring. This is the form in which D-fructose exists when combined with other compounds in nature. In Fig. 29-7, the *beta* form of the five-membered D-fructose ring is shown because this is the form invariably found for *combined* fructose in natural products.

It would appear that fructose could, instead, form a six-membered ring if it utilized its OH group from position 6. This does indeed happen. If one crystallizes *free* fructose from an aqueous solution such as fruit juice, the molecules in the resultant solid have six-membered hemiketal rings, in either alpha or beta forms. However, in *solutions* of free fructose (as in honey or fruit juice) there would be an equilibrium mixture of α, β, and open-chain ketose forms.

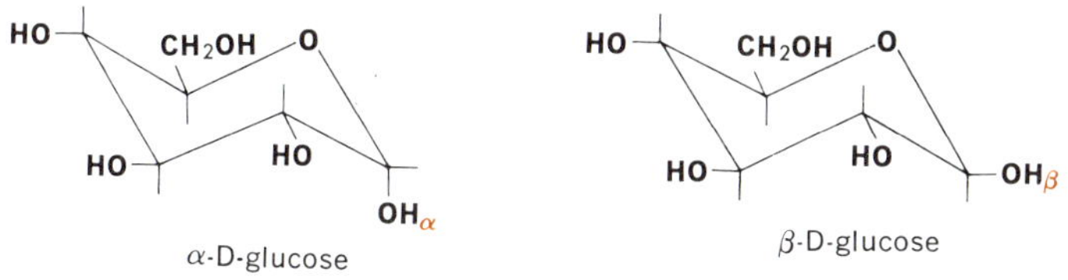

[Fig. 29-8 structures: α-D-glucose and β-D-glucose chair conformations]

Fig. 29-8 The actual conformation of a D-glucose molecule, shown for both α and β forms. (Hydrogens are not shown.)

C. Ring Conformations

One more refinement in our understanding of the ring structure of glycoses is illustrated by D-glucose in Fig. 29-8.

So far, we have represented the six-membered ring of glucose as being *flat*, for example in Fig. 29-5(d) and 29-6(b). Actually, as we found in Section 21-10, a six-membered ring will be *puckered* in order to accommodate the 109° angles of carbon–carbon single bonds. Therefore, the conformation shown in Fig. 29-8 more accurately portrays the true shape of a glucose molecule. For the chemist studying subtle differences in the reactivity of various groups in the glucose molecule, it may be necessary to think in terms of the conformations in Fig. 29-8. However, for most discussions such extreme refinement is not necessary, and we will continue to use the less complex planar formulas, such as Fig. 29-5(d).

D. Glycosides

A glycose will react with an alcohol, in the presence of an acid catalyst, to form a compound called a *glycoside*. This process is illustrated in Eq. (29-27) by the reaction of glucose and methanol.

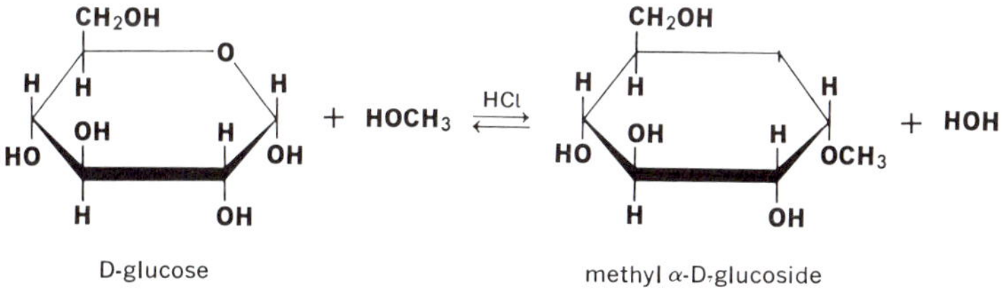

$$\text{D-glucose} + \text{HOCH}_3 \underset{}{\overset{\text{HCl}}{\rightleftharpoons}} \text{methyl } \alpha\text{-D-glucoside} + \text{HOH} \quad (29\text{-}27)$$

The glucoside product shown in Eq. (29-27) is really just an acetal; it has been produced by the reaction of one molecule of an alcohol with a hemiacetal, which in this case is α-D-glucose; compare Eq. (29-15). *An acetal compound produced from a glycose is called a* **glycoside**. This is a special term for this type of acetal; recall that in the case of amino acids we had a special name, peptide, for that particular kind of amide structure. *Glycoside* is a general name for a glycose acetal. The specific product shown in Eq. (29-27) is a *glucoside*.

Since glucose alone could already exist as a hemiacetal, it could become an acetal (i.e., a glucoside) by reacting with only one other molecule of an alcohol compound. Glucose and other glycoses can similarly react to form glycoside structures with a great variety of other compounds that have hydroxyl groups. These glycoside bonds constitute the principal way in which glycose molecules can link to other organic structures, including other glycose molecules. Many of these compounds are extremely important in nature. Examples will be discussed in the next three sections.

E. Disaccharides

Two molecules of a glycose (a monosaccharide) can combine, by splitting out a molecule of water, to form a disaccharide. The two glycose units become linked together by means of a glycoside group. For example, two molecules of α-D-glucose can unite to form maltose, a disaccharide.

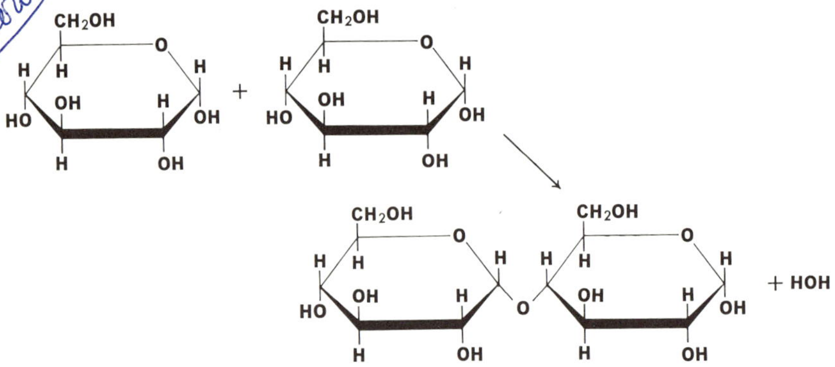

(29-28)

We see in this equation that one molecule of glucose, on the left, is acting as a hemiacetal; while the other molecule on the right is acting as an alcohol. Loss of water between the hemiacetal and the alcohol functional groups produces a glucoside link which holds together the two glucose units. Since the glucose unit on the right used its OH group at position 4 to form the glucoside bond, this link is described as an $\alpha 1 \rightarrow 4$ link. Because of this $\alpha 1 \rightarrow 4$ link, the right-hand glucose unit in maltose still has its hemiacetal structure and in solution could open and close to give the aldose form, as well as α and β. However, the left-hand unit of glucose in maltose is *held* by the glucoside link, which is an acetal. Therefore, the left ring is fixed and will *not* open, even when dissolved in solution. Since the glucoside link between the two glucose units is fixed, it will also remain, as it was formed, specifically as an *alpha* bond. Maltose is involved in the metabolism of many organisms, particularly as an intermediate hydrolysis product of starch, for instance during intestinal digestion.

Another important disaccharide is lactose, the sugar present in the milk produced by female mammals. Lactose, shown in Fig. 29-9, results from combining a β-D-galactose molecule with the number 4 hydroxyl group of D-glucose. Therefore, this link is called a $\beta 1 \rightarrow 4$ link. Due to the β configuration of the galactose unit, the bond between the glycoside oxygen and the 4-position of the glucose unit is shown in a bent form. A more realistic (but less convenient) way to show lactose would be to turn over the whole D-glucose unit, so that its bond from position 4 was aimed upward to meet the β glycoside oxygen as it should.

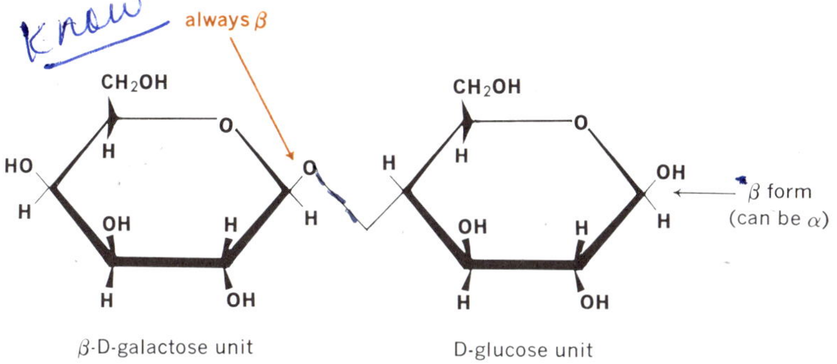

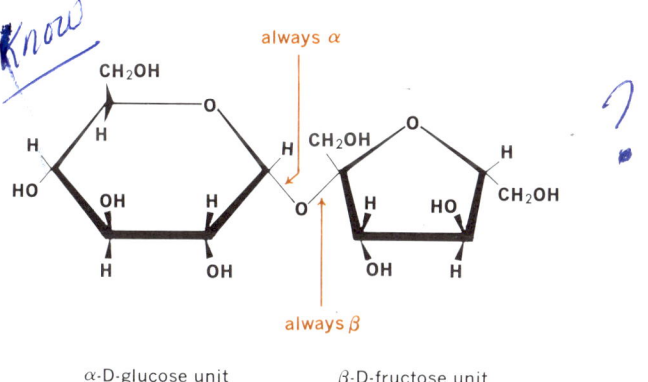

Fig. 29-9 The structure of lactose (β form).

Fig. 29-10 The structure of sucrose.

The other disaccharide we will show is sucrose, the most abundant one on earth due to the fact that we raise cane and sugar beets to produce millions of tons of sucrose annually. The structure of sucrose is given in Fig. 29-10. The D-fructose unit has been flipped over end for end compared to the one shown in Fig. 29-7. This orientation is necessary so that the β OH group (derived from the keto group originally present at position 2 in fructose) will be able to meet the α OH from D-glucose to form the glycoside link in sucrose. There is another aspect of this linkage which makes sucrose different from the other disaccharides we have discussed. Notice that to form the glycoside link, D-glucose used its hemiacetal hydroxyl and fructose used its hemiketal hydroxyl. Thus, the glycoside link "freezes" the ring structure of *both* rings; neither one can open to yield a free carbonyl group, even in solution.

F. Polysaccharides

Living organisms build quite a variety of polysaccharides, each with a specific function in the organism. These large molecules (some even gigantic) are often complex. Polysaccharide compounds are found in cartilage and internal organs, in the exoskeletons of insects and crustaceans, and in plant gums, to mention a few sources. We will discuss the two produced in the largest quantities, starch and cellulose. Interestingly, both of these are constructed from molecules of D-glucose, which is itself produced from carbon dioxide and water by the process of photosynthesis in green plants.

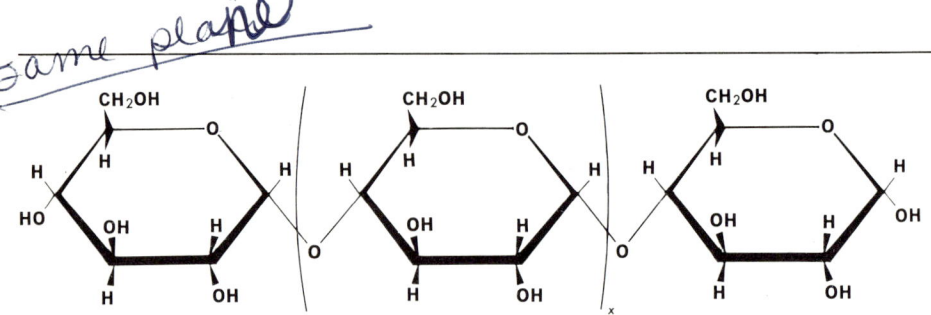

Fig. 29–11 The structure of starch. (Side chains attached at position 6 are not shown.) Note that all links are α.

Starch is a food-reserve compound stored by many plants. A starch molecule, portrayed in Fig. 29–11, is built up as follows. We saw in Section E that two α-D-glucose molecules can join α1 → 4, by loss of water, to form maltose. The right-hand glucose unit of maltose still has its α1 hydroxyl group (the hemiacetal OH) free. Therefore, it could react with the number 4 hydroxyl of yet a third α-D-glucose molecule, which in turn could link to a fourth, and so on until a great many glucose units have been linked together via glucoside bonds. For each glucoside link formed, one molecule of water is split out. This process reminds one of the fact that amino acids can produce proteins by linking to each other via a characteristic bond, a peptide, while splitting out water.

In Fig. 29–11 the *x* beside the brackets indicates a repetition of many glucose units; *x* represents a very large number. It is difficult to determine the exact size of a very large molecule such as starch, but the size seems to vary from one plant to another. Apparently *x*, the number of repeating units in starch, is commonly anywhere between about 1000 and 6000.

There is one other important aspect of the structure of starch which we have not attempted to show in Fig. 29–11. Rather than being one very long chain of connected

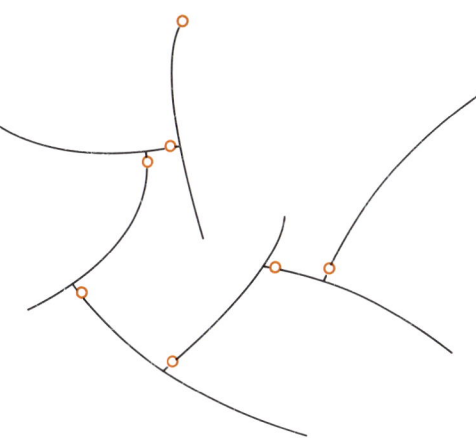

Fig. 29-12 The branched structure of a starch molecule. Each segment is attached to another by an α1→6 glycoside link, shown in color.

glucose units, starch actually consists of many short chains attached to each other in a branching fashion. Each segment of chain is about 25 glucose units long. Attached to the side of a segment, at some random glucose unit, is the "head" end of another segment, from which branches off yet another segment, and so on. The branched nature of starch chains is represented schematically in Fig. 29-12. One chain segment is attached to another by an α1 → 6 glucoside link. Such a bond results when the hemiacetal hydroxyl at the "head" end of one segment bonds to the number 6 hydroxyl of a glucose unit in another segment.

In an economic manner, a plant can also use some of the glucose available from photosynthesis to build **cellulose,** which serves an entirely different function than does starch and is even more abundant. Cellulose is the main structural material of most plant tissues, providing them with form and durability. Of all the carbon atoms in the vegetation of the earth, approximately half exist in cellulose. The structure of cellulose is shown in Fig. 29-13. Once again, it is glucoside bonds which link one D-glucose molecule to another; as in starch, the link is formed by splitting out water between a hemiacetal hydroxyl and a number 4 hydroxyl. The only way the glucoside link of cellulose differs from that of starch is in the configuration. In cellulose, the link is *beta*. Because of the beta link, the bond at position 4 appears to be distorted when we try to represent a

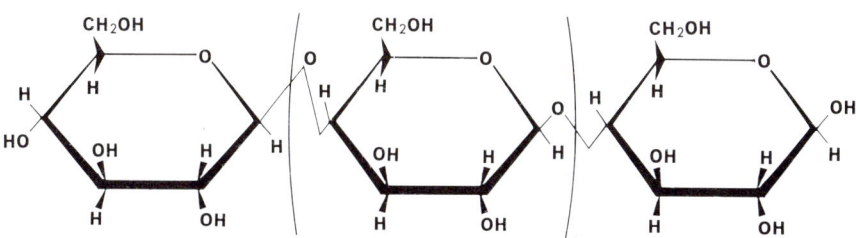

Fig. 29-13 The structure of a cellulose molecule. Note that all links are β1 → 4.

cellulose chain as in Fig. 29-13 (but in the real molecule, it is not distorted). It would be more accurate to show alternate glucose units in the chain turned over. To summarize, we can say that cellulose is composed of D-glucose units held by $\beta 1 \rightarrow 4$ links.

There is a second way in which cellulose differs from starch. Cellulose has a completely linear chain, with no branches. This structural feature is the basis of the strength and toughness of a cellulose fiber, such as cotton.

The number of glucose units in a cellulose chain, represented by x in Fig. 29-13, is roughly similar to the number in a starch molecule. Although x appears to be quite large in some samples of cellulose, it most often seems to be in the range of 1500 to 3000.

G. Other Natural Glycosides

In the previous two sections, all the glycoside linkages we examined were in disaccharides and polysaccharides. Living organisms produce numerous other glycosides, a few of which are discussed in this section.

Vanillin, whose structure is shown in Table 29-2, is the flavor agent found in natural vanilla extract. The identical compound can also be prepared synthetically. Vanillin is conjugated with glucose when deposited in the vanilla bean. A β-D-glucoside unit replaces the hydrogen from the hydroxyl group of vanillin, giving the glucoside structure shown in Fig. 29-14(a). When the vanilla beans are cured, an enzyme causes hydrolysis of the glucoside link to release free vanillin. Extraction of the beans with ethyl alcohol then yields a solution containing dissolved vanillin.

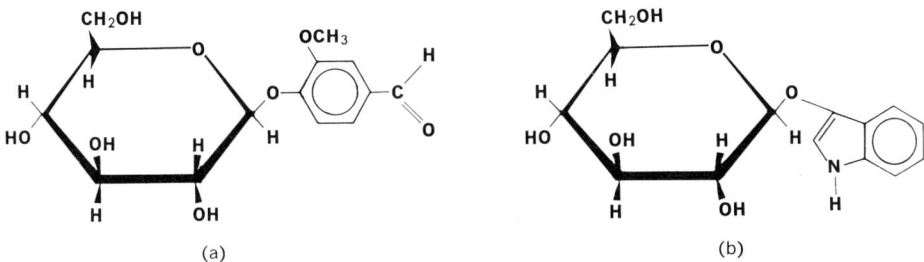

Fig. 29-14 Two naturally occurring glycosides. (a) The β-D-glucoside of vanillin. (b) Indican, the β-D-glucoside of 3-hydroxyindole.

The compound 3-hydroxyindole is a dye precursor found in various species of indigo plants. It occurs in the leaves as a glycoside, coupled via its hydroxyl group to a β-D-glucose. The resultant glucoside, shown in Fig. 29-14(b), is called **indican**. At harvest, the indican can be hydrolyzed either by the enzymes present in the leaves or by very dilute acid solutions, to release the 3-hydroxyindole. Exposure of the solution to air causes oxidation of the 3-hydroxyindole to indigo, a lovely, bright blue dye. Known since ancient times, the dye originated in India.

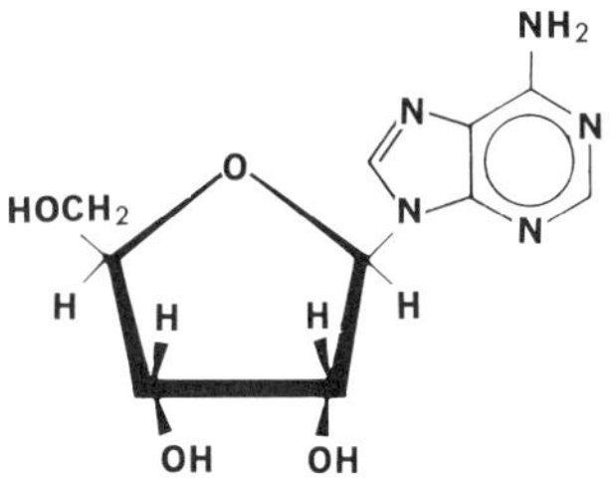

Fig. 29-15 The structure of adenosine.

A glycoside of vast importance is **adenosine,** shown in Fig. 29–15. It consists of a molecule of adenine held to the glycose D-ribose by a glycoside link. (Such a glycoside is sometimes called a nitrogen-glycoside, since the linkage is to a nitrogen atom rather than to an oxygen, as in an ordinary glycoside.) Note that this D-riboside link has a β configuration, and that ribose is a five-carbon glycose.

When a triphosphate group is attached to the $HOCH_2$— group of adenosine, the compound is ATP, or **adenosine triphosphate.** Molecules of ATP can provide the energy required for the completion of a great number of essential biochemical reactions. Eventually, of course, the ATP molecules, and the energy they contain, must be replenished. One way of accomplishing this is by using the energy released during the oxidation of fats or carbohydrates. The production and consumption of ATP is intimately involved in so many of the chemical processes of a living cell that it may be thought of as energy coinage which is earned, saved, and at the proper time exchanged for other valuable compounds.

Equally important are three or four other β-D-ribosides which differ from adenosine only in the structure of their heterocyclic rings. Members of this small group of natural compounds, consisting of β-D-ribosides holding certain nitrogen heterocycles, are called nucleosides. Many thousand nucleosides linked to each other by phosphate groups constitute a substance called RNA, **ribonucleic acid.** There are many different RNA molecules, the character of each depending on the sequence in which the various nucleosides are assembled. RNA molecules direct the synthesis of the fantastically complex protein molecules in living organisms. That is, one particular RNA dictates the exact structure of a certain protein molecule by controlling which amino acids should be added, one by one, to the newly forming protein. This control depends upon the sequence of nucleosides (i.e., the order in which they are arranged) along the chain of an RNA molecule. The sequence constitutes a code of information for building a particular protein molecule.

The masters in this process of controlling the formation of living cells are molecules of DNA, **deoxyribonucleic acid.** These are built up, as in RNA, from nucleosides linked by phosphate groups. However, in DNA the glycose utilized is 2-deoxy-D-ribose. *Deoxy* indicates that the hydroxyl group at position 2 is missing; there is a hydrogen in its place. DNA molecules are able to perform two truly vital functions which make them literally the keys to the physical existence of an organism. First of all, DNA molecules direct the synthesis of the various RNA molecules. The DNA molecules possess the master codes of information for protein synthesis. The DNA passes along portions of the code to the various RNA's as they are being formed. Each RNA in turn uses its portion of information to direct the synthesis of a protein. Second, each DNA molecule is able to *replicate*, or copy itself, so that the vital code can be passed along from one generation to another, and within one living creature, from one cell to another cell.

The first vital task of DNA molecules is to dictate the formation of RNA. The sequence in which nucleoside units occur up and down the length of the very long DNA chain amounts to a kind of code of information. Some fairly long segments of a DNA chain will be the pattern for the formation of an RNA molecule. As nucleosides are being linked together to form the RNA, only the correct ones will match their counterparts along the DNA chain. The result is that the RNA structure has copied part of the "message" from a DNA, and in turn uses this information to direct the construction of a particular protein. And thus arises the delicately balanced, interdependent cluster of substances we recognize as a living cell. Proteins are found in the cell membrane, the cytoplasm, and the enzymes. Also, some of the hormones are proteins. The enzymes, hormones, and other proteins in one way or another supervise the formation of all other substances in the cell and the occurrence of all the necessary chemical reactions which provide energy, and so forth.

The second vital task of DNA is to replicate; that is, to copy itself. DNA molecules are found in all the regions of a chromosome designated as genes, and indeed they *are* the genetic material. As we have seen, DNA carries all the essential information which determines that a particular organism shall become a rose, a frog, or a prince. Therefore, when a cell divides, the DNA must be able to duplicate itself, so that a complete set of DNA will be available for each of the nuclei in the two new cells. DNA also plays the central role at the moment of conception, when sperm meets egg. It is this event which transfers inherited traits from each parent to the offspring. What this process really amounts to is a donation of DNA, with half of it, that is half of the genes, coming from each parent. This process, of course, also requires that the parent's DNA be able to replicate itself, so that egg and sperm each have a set of the parent's DNA.

The copying of a DNA molecule is quite similar to the formation of an RNA from a DNA pattern. An original DNA molecule serves as a template alongside which the pattern of a new DNA is built by a matching of complementary nucleoside units in the new molecule with those down the length of the original chain. The two processes of DNA replication and DNA-directed synthesis of RNA are discussed in greater detail in Chapters 36 and 39.

H. Hydrolysis of Glycoside Bonds

The reaction of glucose, in its hemiacetal form, with an alcohol to produce a glucoside plus water, Eq. (29–27), is catalyzed by acid and is reversible. The reaction in the reverse direction, represented by Eq. (29–29), is easily accomplished.

methyl α-D-glucoside + HOH $\underset{}{\overset{HCl}{\rightleftarrows}}$ α-D-glucose + HOCH$_3$ (29-29)

Equation (29–29) is one more example of a hydrolysis reaction. To promote reaction in this direction we would shift the equilibrium by including a large amount of water in the mixture. Acid catalyst must also be included. By this procedure, methyl α-D-glucoside is converted back to D-glucose. The product initially obtained from the hydrolysis reaction is α-D-glucose, as shown in Eq. (29–29). However, we must keep in mind that as soon as this α hemiacetal form of D-glucose is released into solution, the ring will begin to open and close, providing a mixture of α, β, and aldose forms. The interconversion among the three forms will be rapid, due to the acid catalyst present.

The methyl α-D-glucoside just discussed is a laboratory product. However, natural glycosides undergo hydrolysis just as readily, provided a catalyst is present; two examples were briefly mentioned in Section G. Indican is a glycoside and vanillin exists as a glycoside in the plant. In the laboratory, either of these could be hydrolyzed in warm water containing some acid, such as hydrochloric, to function as catalyst. Strong acid could not be tolerated by plant tissues, but they have enzymes which can catalyze the hydrolysis reaction.

The relative ease with which a glycoside link can be hydrolyzed is very important to living organisms. As we have seen, organisms often build molecules held together by glycoside links. But then at some other time it may be necessary for the organism to reverse the process, that is, to use water to split the glycoside bond and recover the original components. Hydrolysis of a disaccharide or polysaccharide to provide a glycose (monosaccharide) is an excellent example.

Starch is food-storage material for a plant. The glucose produced during periods of rapid photosynthesis can be converted to starch and saved for future need, for example in a seed. Later, when the seed falls into moist soil and swells with water, one of its enzymes can promote hydrolysis of the starch back to glucose. This glucose can be metabolized to provide the energy needed by the sprouting seed for all its other vital chemical reactions until it can develop a green leaf and begin its own photosynthesis. The hydrolysis of starch is expressed by Eq. (29–30). As before, x represents some large number. A catalyst is necessary; in the plant it would be a very specific enzyme which promotes the hydrolysis of only α-glucoside linkages. (In Eq. (29–30) the hydrogens attached to carbons have not been shown.)

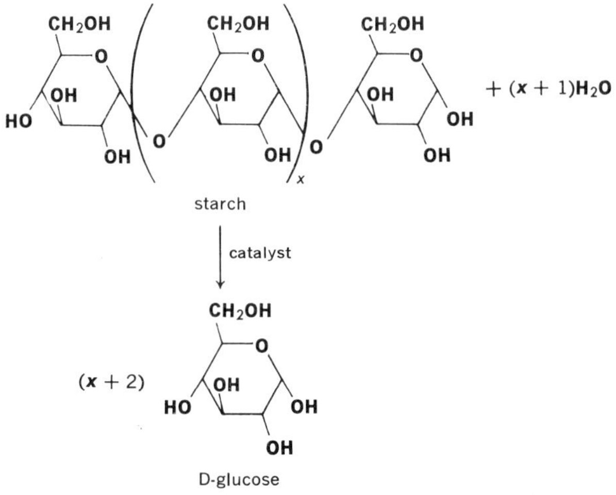

(29-30)

Alternatively, if an animal eats the seeds (corn, wheat, nuts, and so on), enzymes in his digestive tract can hydrolyze the starch, providing glucose for *his* metabolism. Here again we observe that the process of digestion requires just one type of chemical reaction, hydrolysis. The same is true for digestion of fats and of proteins. Subsequent stages of metabolism, of course, require numerous other reactions, some of them quite complex.

An enzyme is a protein; it is so assembled that it has a highly specific shape. This shape matches the shape of the compound (often called a *substrate*) with which it will be reacting. Enzyme and substrate must fit closely together if the enzyme is to catalyze the reaction. Thus, the enzyme which promotes digestion of starch in the human intestine will catalyze hydrolysis of only α-D-glucoside links. It cannot influence the β linkages found in cellulose. Therefore, humans cannot digest cellulose and make use of the glucose it contains. It is a curious fact that even cattle, deer, and termites are unable to produce an enzyme which hydrolyzes the β linkages of cellulose. However, their digestive tracts are hospitable to the growth of certain microorganisms which *do* possess such an enzyme and therefore perform the task of hydrolysis for the cow and the termite. In the laboratory, an acid catalyst can be used to promote hydrolysis of either starch or cellulose.

IMPORTANT TERMS

Acetal	Dehydrogenation	Hemiacetal	Monosaccharide
Aldehyde	Disaccharide	Hydrogenation	Oxidation
Aldose	Glycose	Ketone	Polysaccharide
Carbohydrate	Glycoside	Ketose	Reduction
Cyanohydrin			

WORK EXERCISES

1. Name each structure:

    a) CH_3CH_2CH
            ||
            O

    b) $CH_3CH_2CCH_3$
              ||
              O

    c)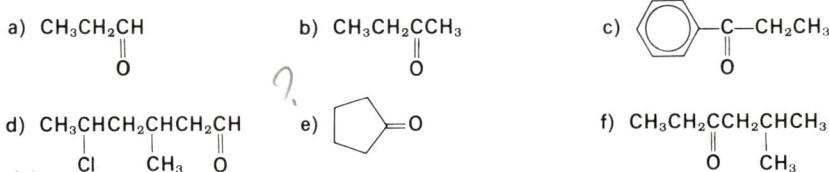

    d) $CH_3CHCH_2CHCH_2CH$
         | | ||
         Cl CH_3 O

    e) (cyclopentanone structure) =O

    f) $CH_3CH_2CCH_2CHCH_3$
              || |
              O CH_3

2. Write an equation to show how each of the following could be synthesized from the appropriate alcohol.

    a) 2-pentanone   b) formaldehyde   c) propanal   d) cyclohexanone

3. Show the structure that would result when each of these compounds is treated with hydrogen and nickel catalyst. [Consult Table 29–2 for structures.]

    a) 3-pentanone   b) vanillin   c) citral   d) testosterone

4. Show how butanone (methyl ethyl ketone) could be synthesized from an alkene. [More than one step is required.]

5. Benzaldehyde, C_6H_5CHO, is a liquid. After it has been exposed to air for some time, a white solid appears at the mouth of the bottle. As time goes on, more and more of the solid settles to the bottom of the bottle. The solid is not soluble in water, but does dissolve readily in dilute sodium hydroxide solution. Suggest a probable structure for the white solid and a reason for its formation.

6. Write an equation for:
    a) the formation of a hemiacetal from an aldehyde.
    b) the conversion of a hemiketal to a ketal.
    c) the reductive amination of 3-hexanone.
    d) the conversion of butanal to a cyanohydrin.
    e) the hydrolysis of an acetal.

7. For what purpose might one wish to prepare the 2,4-dinitrophenylhydrazone of butanone?
    Write an equation for the reaction.

8. Lactic acid (Table 26–1) can be synthesized in three steps from ethanol.
    a) First, the ethanol is converted to ethanal (acetaldehyde). Write an equation.
    b) Now, show how the acetaldehyde can be converted to lactic acid in two more steps.

9. Write an equation for:
    a) a biological example of reductive amination.
    b) the reaction expected when indican is placed in warm water containing some hydrochloric acid.
    c) the hydrolysis of cellulose.

10. Show the structure of the product resulting when β-D-glucose is treated with benzoyl chloride in pyridine.

11. Distinguish the terms glucose and glycose.

12. Compare starch and cellulose in terms of:
    a) the similarities and differences in their chemical structures.
    b) their functions in the plant.

13. Explain the reasons for these experimental facts:
    a) Methyl-α-D-glucoside does *not* react with Tollen's reagent.
    b) After methyl-α-D-glucoside has been warmed briefly in a dilute solution of hydrochloric acid in water, the resultant solution *does* react with Tollen's reagent.

14. Write equations for the conversion of D-erythrose to a mixture of D-arabinose and D-ribose. (See Section 29–7C)

15. List all the types of structural units (both carbon skeletons and functional groups) that you can find in vanillin (Table 29–2).

SUGGESTED READING

DePuy, C. H., and K. L. Rinehart, Jr., *Introduction to Organic Chemistry*, Chapter 9, John Wiley, New York, 1967.

Guild, W., Jr., "Theory of Sweet Taste," *J. Chem. Educ.* **49**, 171 (1972).

Hart, H., and R. D. Schuetz, *Organic Chemistry*, 4th ed., Chapter 9, Houghton Mifflin, Boston, 1972.

POLYMERS 30

30–1 POLYMERS, NATURAL AND SYNTHETIC

As we have seen, the concept of structure in covalent compounds led to a vigorous development of organic chemistry from 1860 onward. Chemists became skilled at synthesizing compounds because, as a rule, they could reliably predict the outcome of reactions. They also became adept at proving the structures of the molecules, both natural and synthetic, with which they worked. Some of these molecules were highly complex and moderately large. However, it was not until the 1920's that chemists began to realize that the covalent structures so characteristic of organic substances could lead to the formation of *extremely* large molecules called **polymers** (meaning *many parts*). A single polymer molecule contains not thirty or one hundred or two hundred atoms, but *thousands* of atoms linked together covalently.

Most of the very stuff of life is polymeric material. Aside from water, a large amount of an animal body consists of proteins, which are polymers. The proteins make up not only the soft tissues but also such structures as hair, feathers, cartilage, horns, fingernails, and turtle shells. Starch, the chief food storage material of plants, is a polymer. Cellulose, the structural material of many plants, is a polymer. In wood the cellulose fibers are further strengthened by being cemented together by lignin, another polymer. Plant cells, like animal cells, contain protein. Nucleic acids, which play one of the most critical roles in living material, are also polymers. The nucleic acids carry the genetic information from one generation to the next and dictate the construction of the protein polymers which carry on many of the other functions of life.

In addition to consuming plants and animals for food, man has for ages adapted many of the natural polymers for other useful purposes. Some notable examples are wood, cotton, linen, hemp, wool, and silk. Rubber has found widespread application only within the past hundred years. Not until 1930 were chemists able to make completely new *synthetic* polymers. The resultant plastics, elastomers (synthetic rubbers), adhesives, fibers, and coatings (paints and enamels) have profoundly affected our technology and standard of living.

In this chapter we will discuss a number of the common synthetic polymers. Several of the natural polymers have already been discussed: starch and cellulose in Section 29–11F, the proteins in Section 28–5, and the nucleic acids in Section 29–11G. The biochemical functions of these natural polymers will be considered in greater detail later in the book.

30-2 ADDITION POLYMERS

Under the influence of the proper catalyst, one molecule of a substituted alkene, such as vinyl chloride, Eq. (30–1), can use its "extra" electrons to form a covalent bond to a second molecule of vinyl chloride, which can in turn link to a third molecule, and this to a fourth, and so on, to form an immensely long chain of such units. The gigantic molecule thus produced is a polymer. In Eq. (30–1) the n represents a very large number, from several hundred up to a few thousand.

$$n\ \underset{\underset{\text{Cl}}{|}}{\text{CH}}=\text{CH}_2 \xrightarrow{\text{peroxide}} \left(\underset{\underset{\text{Cl}}{|}}{\text{CH}}-\text{CH}_2\right)_n \tag{30-1}$$

vinyl chloride poly(vinyl chloride)
a **monomer** a **polymer**

The number n is called the degree of polymerization. Each of the small molecules from which the polymer is built is called a monomer (meaning *one part*). A structural formula for the polymer is written by drawing one portion of the structure in parentheses. The n beside the parentheses indicates that this portion of structure, called the repeating unit, is repeated n times. Note that there is a definite structural relation between the repeating unit and the monomer from which it is derived. For one particular polymer molecule the value of n may differ from that of another molecule in the same sample. It is not possible to determine n accurately, but experiments can tell us that for a certain polymer, n has a value of 1000–1400, for example.

Poly(vinyl chloride) is one example of an addition polymer, one of the two fundamental types of polymers. An **addition polymer** is so named because it is formed by an addition reaction, similar in some respects to other addition reactions of alkenes (Section 23–4). The structural characteristic of an addition polymer is that its backbone is a chain of carbon atoms built up during the polymerization reaction. One addition polymer differs from another in the nature of the groups—alkyl, aryl, or functional groups—which may be attached at alternate carbons along the chain.

> The manner in which a polymerization reaction occurs is now reasonably well understood. The formation of poly(vinyl chloride) and many other polymers can be initiated by a peroxide catalyst. Figure 30–1 summarizes the stages in such a polymerization. First, the peroxide decomposes to form radicals (Section 22–7). A radical, $R\cdot$, is highly reactive; it can become more stable if it finds one more electron to pair with its one odd electron. In the presence of an alkene, the radical can make use of one of the two electrons in the extra bond of the alkene. This, however, leaves one odd electron at the other end of the former alkene bond. Consequently, the newly-formed particle is a radical also. As soon as it collides with another alkene molecule it binds one electron, once again leaving one unpaired. Thus, the stage is set for an indefinite number of repeated reactions, whereby the growing polymer chain adds to itself one monomer molecule after another.
>
> It is apparent from Fig. 30–1 that at each stage of the polymerization a reactive site (in this example, a radical) is produced at the end of the growing chain. In fact, it might appear that the polymer chain would stop growing only when all the monomers had been used. If so, it would contain not a thousand or two but billions and billions of repeating units. The reason the polymer chain stops growing is that occasionally the odd electron at the end of the growing polymer chain will encounter another radical, $R\cdot$, with which it can form a covalent bond *without* generating a new reactive site. Because only a trace of peroxide is added to the alkene monomer, only a few

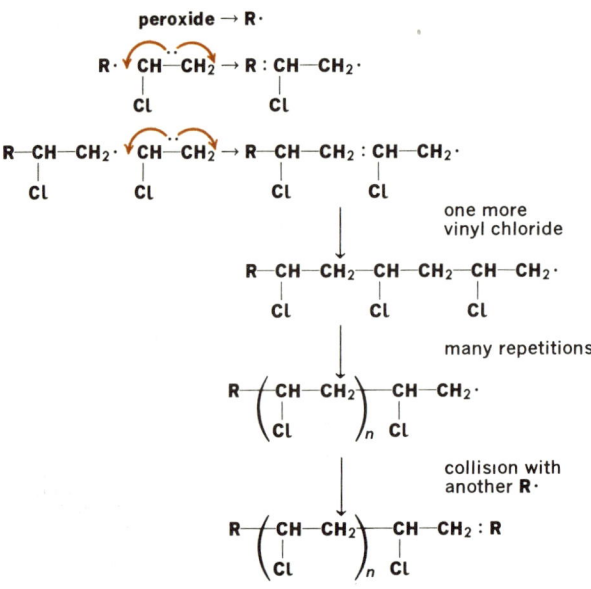

Fig. 30–1 Mechanism of a typical addition polymerization catalyzed by peroxide; vinyl chloride converted to poly(vinyl chloride).

radicals are present at any given time. Therefore the growing polymer radical may typically collide with, and add, about a thousand alkene molecules before a collision with another radical causes termination. Still other radicals, released by the decomposing peroxide, simultaneously initiate the growth of numerous other polymer chains. In general, the greater the number of radicals supplied to the mixture, the faster will be the rate of polymerization, because more polymer chains will be growing at any given moment. The greater concentration of radicals will also cause the formation of shorter polymer chains (that is, n will be smaller); chain growth will terminate sooner because there will be a greater chance that the reactive site will encounter another radical.

Polymerization of an alkene can also be initiated by either a positive ion (Eq. 30–2) or a negative ion (Eq. 30–3), rather than by a radical.

$$R^+ \; CH\text{—}\underset{\underset{CH_3}{|}}{\overset{\overset{CH_3}{|}}{C}} \; \longrightarrow \; R:CH_2\text{—}\underset{\underset{CH_3}{|}}{\overset{\overset{CH_3}{|}}{C^+}} \; \longrightarrow \; \text{etc.} \tag{30-2}$$

$$R:\; CH\text{—}CH_2 \; \longrightarrow \; R:CH\text{—}CH_2:^- \; \longrightarrow \; \text{etc.} \tag{30-3}$$
$$\underset{CH_3}{|} \phantom{CH\text{—}CH_2} \phantom{\longrightarrow\; R:CH\text{—}}\underset{CH_3}{|}$$

However, in every case the attack of the reactive particle on an alkene generates a particle having the same kind of reactive site. The new particle then tries to stabilize itself by attacking another alkene molecule, but in the process only generates a new reactive site, and so on.

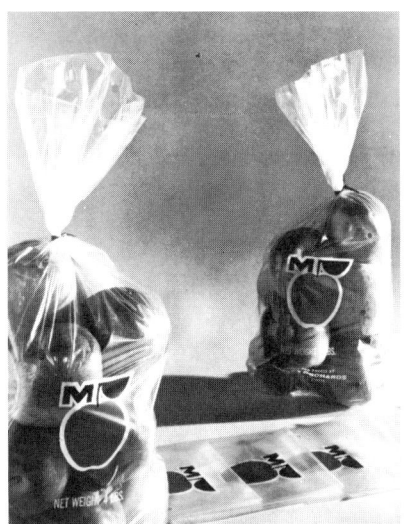
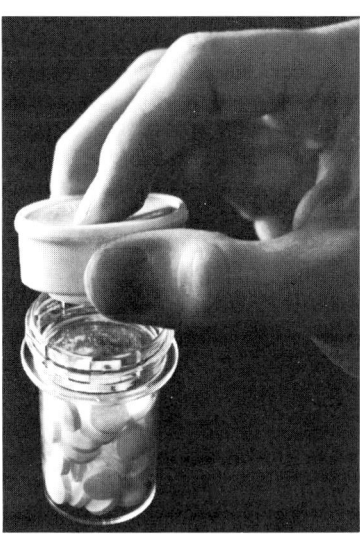

Fig. 30-2 Clear, flexible film of polyethylene, an addition polymer, is widely used for food packaging. [Courtesy of Eastman Chemical Products, Inc.]

Fig. 30-3 A molded bottle cap of polypropylene, an addition polymer. [Courtesy of Eastman Chemical Products, Inc.]

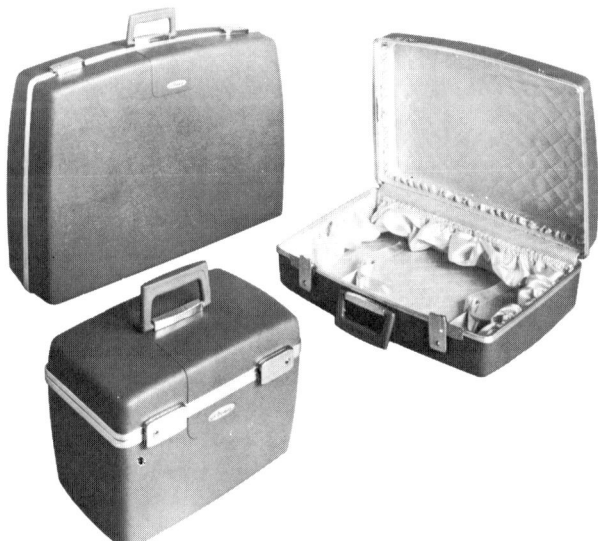

Fig. 30-4 Sturdy, attractive luggage can be fashioned from tough polypropylene. [Courtesy of Eastman Chemical Products, Inc.]

Fig. 30-5 Teflon, the familiar "non-stick" lining for cooking utensils, is also an effective electrical insulator. In this instrument used to guide jet aircraft, the printed circuit (dark lines in center and upper section) is embedded within a sheet of Teflon 0.01 in. thick. [Courtesy of du Pont de Nemours.]

It is possible to synthesize a great variety of polymers, each having different properties, by starting with different alkene monomers. For example, polystyrene is a stiff, relatively brittle plastic used to mold toys, model trains, knobs for radios, and inexpensive picnic spoons and cups. In contrast, polypropylene, although rigid, is extremely tough, impact-resistant, tear-resistant, and in thin sections is quite flexible. A contoured chair having the seat and back all in one piece can be fashioned from polypropylene. A complete box with hinge and lid can be molded in one operation from polypropylene. At the hinge the polypropylene is simply molded paper-thin so that it will be flexible. The hinge is so tough that it can be bent back and forth a great many times (probably more than a million) without breaking. Despite the flexibility of the hinge, the box and lid, being thicker, are rigid.

Several other common varieties of addition polymers are described in Table 30–1.

30–3 DIENE POLYMERS

A **diene** is an organic compound having *two* carbon–carbon double bonds somewhere in its structure. Of particular value in polymerization reactions are compounds such as 1,3-butadiene,

$CH_2=CH-CH=CH_2$

in which the two double bonds are separated from each other by one single bond. A polymer prepared from such a diene is just another example of an addition polymer.

Neoprene (Eq. 30–4) was the first diene polymer successfully commercialized and the first synthetic rubber developed in the United States. Neoprene has been manu-

TABLE 30-1 Some common addition polymers

Structure	Chemical Name	Common Name or Brand Name	Uses
$\left(\begin{array}{c}\text{CH}-\text{CH}_2\\ \mid \\ \text{Cl}\end{array}\right)_n$	poly(vinyl chloride)	PVC, "vinyl"	floor tile; pipes; wire covering; clear film for food packaging
$\left(\begin{array}{c}\text{CH}-\text{CH}_2\\ \mid \\ \text{C}_6\text{H}_5\end{array}\right)_n$	polystyrene	"styrene"	toys and models; household articles; instrument panels and knobs; styrofoam insulating material; hot drink cups, picnic jugs, etc.
$\left(\begin{array}{c}\text{CH}-\text{CH}_2\\ \mid \\ \text{CH}_3\end{array}\right)_n$	polypropylene		pipes; valves; packaging film; sterilizable bottles for babies and hospitals; utility boxes; fish nets; waterproof carpeting
$+\text{CH}_2-\text{CH}_2\text{)}_{\overline{n}}$	polyethylene		squeeze bottles; flexible cups and ice cube trays; clear, soft film for food bags and clothes packages; weather balloons; moisture barriers in buildings
$+\text{CF}_2-\text{CF}_2\text{)}_{\overline{n}}$	polytetrafluoroethylene	Teflon	linings for cooking pans; electrical insulation; gaskets, valves, and machine parts; heat and chemical-resistant pipes and sheets
$\left(\begin{array}{c}\text{CH}-\text{CH}_2\\ \mid \\ \text{CN}\end{array}\right)_n$	polyacrylonitrile	Acrilan, Orlon	fabrics and knitted goods
$\left(\begin{array}{c}\text{CH}_3\\ \mid \\ \text{C}-\text{CH}_2\\ \mid \\ \text{COOCH}_3\end{array}\right)_n$	poly(methyl methacrylate)	"acrylic," Lucite, Plexiglass	airplane windows; models; light fixtures; costume jewelry; dental fillings and false teeth; tooth brush and hair brush handles; reflectors; automobile tail-lights

factured since the late 1930's for use in shoe soles, coverings for electric wires, and for rubber hoses, gaskets, and fittings which must withstand attack by oil or gasoline. Neoprene is superior to natural rubber in its toughness and its resistance to weathering and to oil. This polymer would make excellent tires, but is too expensive for this purpose.

$$n\text{CH}_2\!=\!\underset{\underset{\text{Cl}}{|}}{\text{C}}\!-\!\text{CH}\!=\!\text{CH}_2 \xrightarrow{\text{peroxide}} \left(\!\text{CH}_2\!-\!\underset{\underset{\text{Cl}}{|}}{\text{C}}\!=\!\text{CH}\!-\!\text{CH}_2\!\right)_{\!n} \tag{30-4}$$

chloroprene
2-chloro-1,3-butadiene

polychloroprene
neoprene

It is important to note that when chloroprene or a similar diene polymerizes (Eq. 30–4), each monomer is linked into the polymer chain by the opposite ends of the original diene unit, and that a double bond remains in each repeating unit of the polymer.

The manner in which the diene becomes bonded at its opposite ends is shown in Eq. (30–5). There is a shift of electrons on all four carbons of the diene skeleton.

$$\text{R}\cdot\ \text{CH}_2\!-\!\underset{\underset{\text{Cl}}{|}}{\text{C}}\!-\!\text{CH}\!-\!\text{CH}_2 \quad \rightarrow \quad \text{R}:\text{CH}_2\!-\!\underset{\underset{\text{Cl}}{|}}{\text{C}}\!-\!\text{CH}\!-\!\text{CH}_2\cdot \quad \rightarrow \quad \text{etc.} \tag{30-5}$$

Consequently, the new reactive site (an odd electron) appears at the end of the chain. However, this process is just a variation of the fundamental addition reaction (compare with Fig. 30–1).

Natural rubber is a polymer of isoprene (Eq. 30-6):

$$n\text{CH}_2\!=\!\underset{\underset{\text{CH}_3}{|}}{\text{C}}\!-\!\text{CH}\!=\!\text{CH}_2 \xrightarrow{\text{catalyst}} \left(\!\text{CH}_2\!-\!\underset{\underset{\text{CH}_3}{|}}{\text{C}}\!=\!\text{CH}\!-\!\text{CH}_2\!\right)_{\!n} \tag{30-6}$$

isoprene
2-methyl-1,3-butadiene

polyisoprene
natural rubber

A number of plants, especially *Hevea* trees, can produce rubber, using enzymes to catalyze the synthesis. For years all attempts to form rubber in the laboratory, by polymerizing isoprene, failed. Then during the early 1950's, Karl Ziegler in Germany and Guilio Natta in Italy conducted extensive research on new types of catalysts which regulate reactions in a special way. Because of the fundamental principles they established during these and subsequent studies, Ziegler and Natta shared the Nobel Prize for chemistry in 1963. By applying their principles, chemists at the Firestone Rubber Company and at Goodrich-Gulf developed methods which would, in the laboratory or factory, polymerize isoprene into a material virtually identical to natural rubber. The Firestone product has the brand name Coral Rubber; the Goodrich-Gulf material is called Ameripol. These rubbers are now manufactured in large quantities for use in tires or in any of the other applications for natural rubber.

The synthetic polyisoprene described above is truly a synthetic *rubber*, in that it is a manufactured material which actually duplicates the substance produced by rubber trees. The other materials commonly called "synthetic rubber" are really different compounds which have elastic, rubbery properties. It is less ambiguous to call them *synthetic elastomers*.

Isoprene can be manufactured readily from petroleum. A method for polymerizing chloroprene was developed in the 1930's. Why, then, did it take until the mid-1950's before a catalyst could be found which would produce a polyisoprene that was a true duplicate of natural rubber? This is another case in which the *shape* of the molecule is a critical factor. Note that the repeating unit of natural rubber (Eq. 30–6), still contains one double bond. The portions of carbon skeleton attached to the double bond could be in either a *cis* or a *trans* arrangement. In natural rubber, all the double bonds throughout the polymer chain have the *cis* configuration. This can be duplicated synthetically only with a special catalyst which maintains close control over the growing polymer chain, forcing it to assume the *cis* configuration at each step.

A number of the catalysts used to polymerize chloroprene or the alkenes *will* convert isoprene to a polymer. However, they produce a polymer chain having both *cis* and *trans* bonding randomly distributed. The polymer is therefore not equivalent to natural rubber and its properties are not especially useful. Recently still other catalysts have been discovered which will produce all-*trans* polyisoprene. This material is very stiff, in contrast to the highly elastic quality of all-*cis* polyisoprene, be it natural or synthetic.

Polybutadiene can be prepared from its monomer (Eq. 30–7) by the same catalysts used for polymerization of isoprene.

$$n\text{CH}_2\!=\!\text{CH}\!-\!\text{CH}\!=\!\text{CH}_2 \xrightarrow{\text{catalyst}} \left(\text{CH}_2\!-\!\text{CH}\!=\!\text{CH}\!-\!\text{CH}_2\right)_n \qquad (30\text{–}7)$$
butadiene   polybutadiene

Polybutadiene is more elastic and has a better bounce than natural rubber. It is the component of the toy "superballs" which bounce so high.

30–4 COPOLYMERS

When a mixture of *different* monomers is polymerized the product is a **copolymer.** By varying the kinds of monomers used and the percentage of each in the mixture, a chemist can often create a polymer having some property, such as toughness or heat stability,

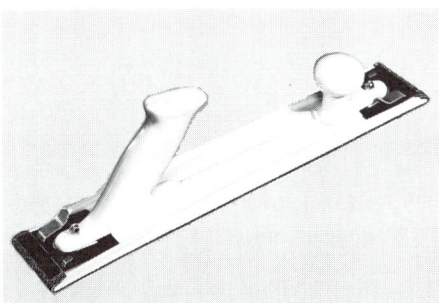

Fig. 30–6 Body and handle of a tool molded from acrylonitrile-butadiene-styrene (ABS), a copolymer. [Courtesy of the Marbon Chemical Division, Borg-Warner Corporation.]

Fig. 30–7 Since ABS plastic can be chrome-plated, it is used for decorative parts in cars and homes. Shown here is a lavatory faucet. [Courtesy of the Marbon Chemical Division, Borg-Warner Corporation.]

required for a certain application. Table 30–2 lists some of the copolymers presently being manufactured. These are further examples of addition polymers. Figures 30–6 and 30–7 illustrate how these materials can be utilized.

TABLE 30–2 Examples of (addition) copolymers

Monomer Components	Polymer Name	Uses
vinyl chloride, vinyl acetate	"vinyl," vinylite	phonograph records; shower curtains; rain wear
vinyl chloride, vinylidene chloride	Saran	"Saran wrap" for food; fibers for auto seat covers; pipes
vinyl chloride, acrylonitrile	Dynel, Vinyon	fibers for clothing
ethylene, propylene	ethylene-propylene rubber, EPR	tires
styrene, butadiene	styrene-butadiene rubber, SBR	tires
isobutylene, isoprene	butyl rubber	inner tubes
acrylonitrile, butadiene, styrene	ABS	crash helmets; women's spike heels; luggage; pipes; parts and cases for batteries, telephones, and other instruments; knobs and handles

30-5 CONDENSATION POLYMERS

The second fundamental type of polymer is a condensation polymer, formed by a **condensation reaction.** A small molecule such as water or methanol is split out as each unit is added to the polymer chain. Structurally, condensation polymers are distinguished by the fact that there are almost always functional groups *within* the polymer backbone. (In contrast, the backbone of an addition polymer consists of carbon atoms—see Section 30–2).

Nylon is a good illustration of a condensation polymer. It is formed by the reaction of two monomers, a diamino compound and a dicarboxylic acid (Eq. 30–8):

$$NH_2(CH_2)_6NH_2 + HOC(CH_2)_4COH \longrightarrow \cdots NH(CH_2)_6NHC(CH_2)_4CNH(CH_2)_6NHC(CH_2)_4C \cdots$$
$$\overset{\|}{O}\ \overset{\|}{O} \qquad \overset{\|}{O}\ \overset{\|}{O}\ \overset{\|}{O}\ \overset{\|}{O}$$

1,6-diaminohexane   1,6-hexanedioic acid   nylon
                     (adipic acid)

(with H₂O H₂O H₂O split out)

(30-8)

The amino groups react with the carboxylic acid groups in the usual way (Eq. 28–24), forming amide groups. Note that the amide functional groups become part of the polymer backbone, and that for each amide link formed one molecule of water is split out.

It is apparent from Eq. (30–8) that for a condensation reaction of this type to lead to a polymer, each participating monomer must have *two* functional groups. Thus, after one end of the diamine has reacted with an acid molecule, the amino group at the opposite end is available for reaction with another acid molecule. Since the acid molecule also has another functional group at the opposite end, it can react with yet another diamine molecule. Consequently, innumerable repetitions of the reaction are possible. It would be good practice for the reader to rewrite Eq. (30–8) step by step, to see how the polymer chain grows.

$$\left(NH(CH_2)_6NHC(CH_2)_4C \right)_n$$
$$\qquad\qquad\ \overset{\|}{O}\qquad\overset{\|}{O}$$

Fig. 30-8 The structural formula of nylon (a polyamide).

In Eq. (30–8) a moderately long section of the polymer chain was written out in order to show the nature of the reaction. Figure 30–8 is a structural formula for the nylon polymer written in the usual more abbreviated form. Since two different monomers go into the structure of nylon, a repeating unit in the polymer formula must show each of them. In general, nylon is called a **polyamide** because amide functional groups are the links holding the polymer together.

Nylon makes strong, long-wearing fibers particularly suitable for hosiery, sweaters, and other articles of clothing. It was the first completely synthetic fiber marketed,* and

---

* The manufacture of rayon began at an earlier date. Rayon is natural cellulose which has been *modified* by chemical reactions. The process depends on the fact that the polymer backbone of rayon was present originally in the cellulose. The structural changes are made only in the groups attached to the backbone.

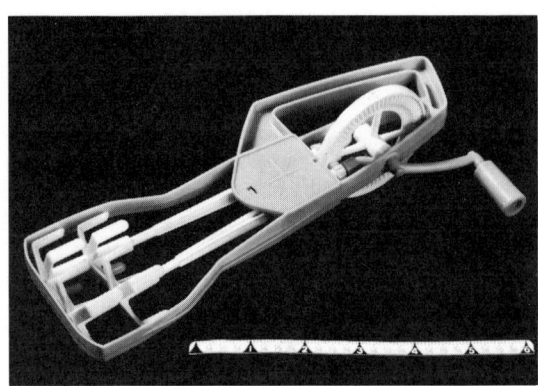

Fig. 30-9 Articles molded of nylon: left, machine gears; right, gears and blades of a beater. The frame and handle of the beater are of ABS. [Courtesy of du Pont de Nemours.]

to a very large extent has replaced silk which it resembles in many of its properties. Nylon is superior to silk in most respects, and it is cheaper to produce. It is not entirely coincidental that silk and nylon have similar properties. Silk is a protein and all proteins are polyamides (Section 28-5). Nylon can be molded into solid articles (see Fig. 30-9). It is quite strong and has a slippery surface; gears made of nylon need no lubrication. It is also frequently used for glides on drawers and zippers on clothing, for combs and cups, and for knobs and handles of various sorts. In the United States the production of nylon now exceeds the astounding total of one billion pounds annually.

The most common type of nylon has the structure shown in Fig. 30-8. Other nylon polymers having somewhat different properties can be made by using a diamine or a dicarboxylic acid having a different carbon skeleton.

The common natural polymers we have previously considered—starch, cellulose, and proteins—are all *condensation* polymers. For instance, let us review the pertinent features of a protein molecule first described in Section 28-5. A protein (the polymer) is built up from amino acids (the monomers) by means of a condensation reaction; a small molecule, water, is split out as each amide link (peptide bond) is formed. Furthermore, this protein polymer is characterized by functional groups (the amide links) throughout the backbone of the polymer. We see also that the monomer unit, an amino acid, possesses two functional groups, an amino group and an acid group. It is the existence of *two* functional groups within one monomer molecule which sets up the possibility of repeated reaction as each monomer unit is condensed onto the growing polymer skeleton. Analysis of cellulose structure reveals that it too is a polymer formed by a condensation reaction of its monomer units (glucose), with simultaneous loss of a water molecule as each new link is created.

Another type of condensation polymer is a **polyester;** as the name implies, the functional groups repeated in this polymer chain are esters. The most familiar polyester is poly(ethylene terephthalate), fibers of which are known as Dacron or Terylene. The polymer is formed (Eq. 30–9) by the reaction of ethylene glycol with the methyl ester of terephthalic acid. In the process, a molecule of methanol is split out for each new ester group formed; the net result is that teraphthalic acid (a *di*carboxylic acid) is transformed from one type of ester to another. A polymer results because ethylene glycol, the alcohol forming the new ester groups, is also a *di*functional molecule.

$$n\text{HOCH}_2\text{CH}_2\text{OH} + n\text{CH}_3\text{OC}\!\!-\!\!\bigcirc\!\!-\!\!\underset{\underset{\text{O}}{\|}}{\text{COCH}_3} \longrightarrow \left(\underset{\underset{\text{O}}{\|}}{\text{OCH}_2\text{CH}_2\text{OC}}\!\!-\!\!\bigcirc\!\!-\!\!\underset{\underset{\text{O}}{\|}}{\text{C}}\right)_n + 2n\text{CH}_3\text{OH} \quad (30\text{–}9)$$

ethylene glycol   methyl terephthalate   poly(ethylene terephthalate) (Dacron, Mylar)

Polyester fibers, such as Dacron, are used extensively in fabrics for clothing and other goods. The material can also be extruded into clear, tough films having excellent tear resistance and dimensional stability. One application is in Cronar motion-picture film. Another is the Mylar film and sheets used for recording tapes and other purposes.

30–6 THERMOPLASTIC versus THERMOSET POLYMERS

In the previous sections, polymers were classified on the basis of chemical structure as either addition or condensation types. It is also convenient to classify polymers into two different groups on the basis of their behavior when exposed to heat, which profoundly affects their uses. **Thermoplastic** polymers become plastic (that is, moldable) when heated. They can be repeatedly heated until soft and remolded. In contrast, a **thermoset** polymer hardens or sets when heated. It can be molded only once; thereafter it is rigid and heat will not soften it.

Polyethylene is a good example of a thermoplastic polymer. It can be heated and molded over and over again. A Melmac dish is a thermoset polymer. It remains rigid even when placed directly in a flame, but would soon char and decompose at such a high temperature.

All the polymers, both addition and condensation, described in the previous sections are *thermoplastic*. However, it is also possible to make a *thermoset* polymer utilizing either the addition or the condensation type of linkage. The essential difference is that a thermoset polymer is cross-linked, whereas a thermoplastic polymer is not. The phenomenon of cross-linking is described in the next section.

30–7 CROSS-LINKED POLYMERS

We discovered in Section 30–5 that a condensation reaction can lead to a polymer only if each monomer component has *two* functional groups. Now let us see what can happen if at least one of the components has *three* functional groups on each molecule. One common mixture of this type consists of phthalic acid and glycerol. When the mixture is heated, the acid and the glycerol react with each other to form ester bonds and release water (Fig. 30–10). Long polymer molecules result. Although this product is similar in many ways to the polyester molecule shown in Eq. (30–9), it differs in one essential feature. Each glycerol molecule has three hydroxyl groups, only two of which are needed to form the ester links holding the polymer backbone together. The third hydroxyl group on each glycerol unit is available to react with phthalic acid, and frequently does so.

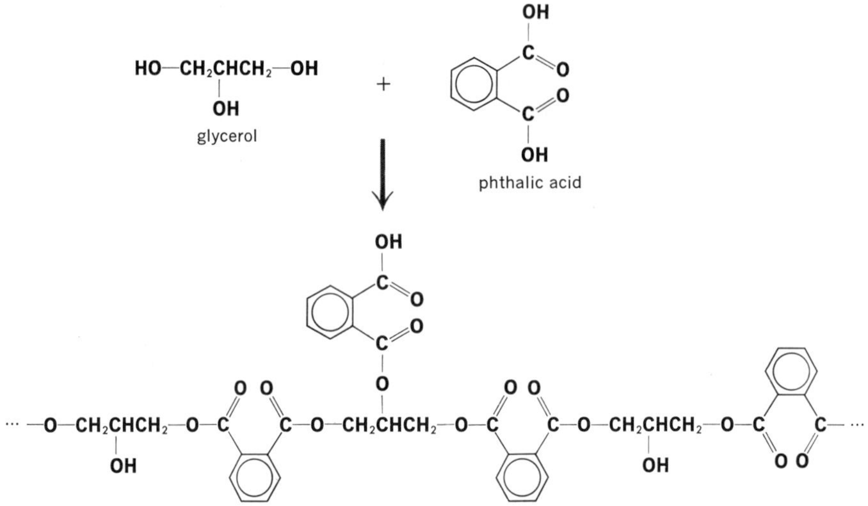

Fig. 30–10 The initial stage of reaction between glycerol and phthalic acid.

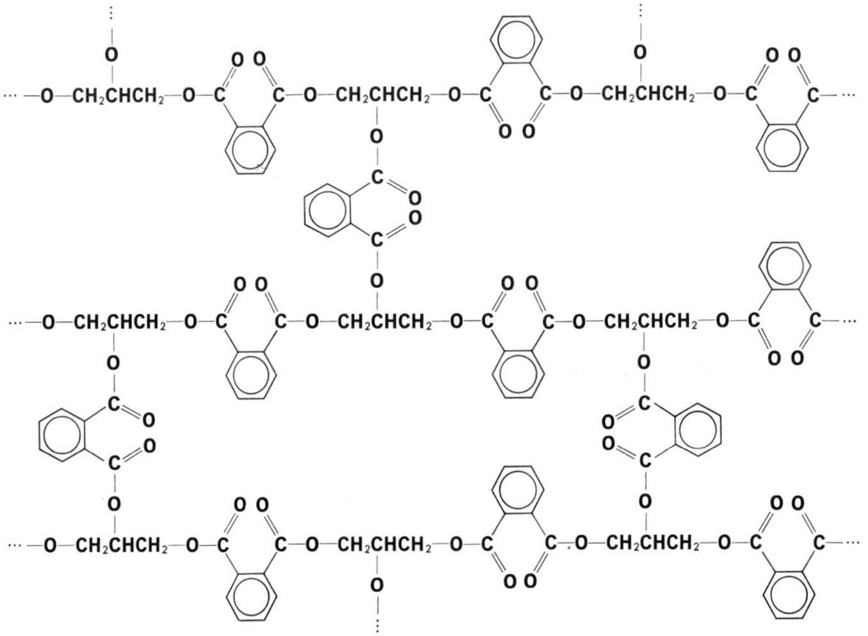

Fig. 30–11 The structure of glyptal, a thermoset polymer. The final stage of reaction between glycerol and phthalic acid produces a highly cross-linked, rigid material.

Figure 30-10 represents the initial stage of reaction between phthalic acid and glycerol, and shows two polymer chains. This stage can be achieved by proper control of reaction conditions; for example, moderate heat can be maintained for a limited time. We note however, that in each polymer chain carboxyl and hydroxyl groups are still available. If the mixture is heated further, these remaining functional groups react to form additional ester bonds, linking one polymer chain to another. The result, shown in Fig. 30-11, is a vast network in which each polymer chain is covalently bonded to many others. Indeed, since each chain is bonded to others, and these in turn to others, the whole mass of material could be considered as one giant molecule. After the final stage of reaction the polymer is said to be highly **cross-linked.** Physically, it has become rigid.

The initial polymer prepared from glycerol and phthalic acid has very few cross links (Fig. 30-10). It is a slightly viscous liquid which can be poured into a mold or treated in some other way. After one thorough heating the material becomes a rigid, cross-linked polymer which can never again be melted or remolded. We see now that cross-linking creates a thermoset polymer. In contrast, a thermoplastic polymer is said to have a linear structure, because each molecule is just one long chain.

Glyptal is the common name for the glycerol-phthalic acid polyester (Figs. 30-10 and 30-11). Vast quantities of the material are used for baked enamels on automobiles and household appliances, and much smaller amounts for molded articles. For enameling, the first-stage glyptal polymer is mixed with pigments and sprayed onto a metal surface. Exposure to a bank of infrared heat lamps then causes the cross-linking leading to the thermoset polymer.

A variety of well-known thermoset polymers are listed in Table 30-3. All have one feature in common: they are cross-linked polymers.

TABLE 30-3 Some common thermoset (cross-linked) polymers

Monomer Components	Polymer Name	Uses
glycerol, phthalic acid	glyptal	baked enamels; molded articles
phenol, formaldehyde	phenol-formaldehyde resin, phenolic, Bakelite, Formica, Micarta	table tops; wall panels; resin bond for plywood; radio and appliance cabinets; washing machine agitators; handles; dials; drawer pulls
urea, formaldehyde	urea-formaldehyde resin	translucent light panels in homes and cars; decorative wall panels; bottle caps; adhesives; enamels; housings for radios and appliances
melamine, formaldehyde	melamine-formaldehyde resin, Melmac	dishes; panel boards; buttons; cases for hearing aids

30-8 THE SOCIAL IMPORTANCE OF SYNTHETIC POLYMERS

One of the distinctly human traits is man's ability to construct useful and beautiful objects. Centuries ago he started using the materials he could find in nature, such as wood, silver, gold, clay, leather, horn, and plant fibers. Gradually he developed more complex processes for obtaining useful materials. He discovered new metals that could be isolated by heating certain rocks (ores), and found that a mixture of sand and alkali subjected to very high temperatures yielded glass, an *inorganic* polymer. Only within the last few decades has man learned to synthesize organic polymers, thereby providing himself with a host of new structural materials.

With synthetic polymers, man has an immense new field for creative expression, both artistic and utilitarian. The synthetic polymers have properties different from those of any natural material previously employed; in many cases they are profoundly different. Moreover, man understands that the polymerization reaction is a very general process. By the proper choice of monomers one can create a new material having, within reason, almost any property desired.

This flood of new materials has had a tremendous effect upon our culture within a short period of time. As recently as 1940 the only synthetic polymer often encountered by the average citizen was Bakelite. Nylon hosiery, neoprene rubber soles, and Plexiglas were relatively new. We now take for granted a wide variety of polymers used in fabrics, toys, gadgets, machinery, construction materials, and art objects—paintings, jewelry, dishes, and sculptures.

30–9 CONSERVATION OF ORGANIC RAW MATERIALS

We must bear in mind that nearly all the valuable synthetic organic compounds—polymers, medicines, dyes, etc.—are derived from two natural resources: coal and petroleum. At present the greatest quantities of both these resources are simply burned up to provide heat. Coal and petroleum are becoming so important as sources of material that we cannot afford much longer to waste them in combustion. Within a few generations man must develop other sources of energy, before these critical sources of carbon compounds are completely exhausted.

To reduce the consumption of carbon compounds, some plastics and other organic compounds can be recycled. Others cannot, and this fact poses a problem of disposal. The nonrecycled materials could probably be burned to provide some of the heat energy required by a complex civilization; this solution would be better than burning virgin petroleum. The compounds would have had at least one other use, perhaps many others, before being burned. Of course, any furnace for burning the disposed materials would have to be designed to provide complete combustion and the trapping of any noxious by-products, to avoid an unacceptable amount of air pollution.

IMPORTANT TERMS

Addition polymer	Diene	Polymer
Condensation polymer	Monomer	Repeating unit
Copolymer	Natural rubber	Synthetic rubber
Cross-linked	Polyamide	Thermoplastic
Degree of polymerization	Polyester	Thermoset

WORK EXERCISES

1. Give the name and structure of the monomer from which each of the polymers in Table 30–1 can be prepared.

2. Using any one of the examples from Table 30–1, write an equation for the formation of the polymer showing the catalyst, the degree of polymerization, and the repeating unit.
3. In Section 30–3 it was stated that all the double bonds in natural rubber (polyisoprene) have the *cis* configuration. Draw a portion of a rubber molecule in the correct *cis* arrangement.
4. Write a fairly long section of structure for styrene-butadiene rubber, keeping in mind that there will not necessarily be a regular alternation of monomer components as the polymer chain develops.
5. Write out, step-by-step, a fairly long segment of Dacron structure (Eq. 30–9), so that you see how the polymer chain grows. What small molecule is eliminated as each new covalent bond forms? What type of polymer is Dacron?

SUGGESTED READING

Asimov, I., *The New Intelligent Man's Guide to Science*, Basic Books, New York, 1965, pp. 462–486. A highly readable account of polymers, plastics, fibers, and rubbers.

Asimov, I., *A Short History of Chemistry*, Anchor Books, Doubleday, New York, 1965, pp. 182–188. Discussion of polymers.

Ayres, E., "The Fuel Situation," *Scientific American*, Oct. 1956. Discusses the rapidity with which our fossil fuels are being consumed.

Crick, F. H. C., "Nucleic Acids," *Scientific American*, Sept. 1955. This entire issue of *Scientific American* is devoted to polymers, both natural and synthetic. Other articles in this edition are "Giant Molecules," by H. F. Marks; "How Giant Molecules Are Made," by G. Natta; and "Proteins," by P. Doty.

DePuy, C. H., and K. L. Rinehart, Jr., *Introduction to Organic Chemistry*, Chapter 4, John Wiley, New York, 1967.

Fisher, H. L., "New Horizons in Elastic Polymers," *J. Chem. Educ.* **37**, 369 (1960). A comprehensive but readable survey of the field, with many references.

Fisher, H. L., "Rubber," *Scientific American*, Nov. 1956. Discussion of synthetic rubbers and a brief history of natural rubber.

Frazer, A. H., "High-Temperature Plastics," *Scientific American*, July 1969.

Jellinek, H. H. G., "Outlook for Polymer Science in This Decade," *J. Chem. Educ.* **49**, 148 (1972).

Lessing, L. P., "Coal," *Scientific American*, July 1955. Coal is more important as a raw material than as a fuel.

Levinthal, C., "Molecular Model-Building by Computer," *Scientific American*, June 1966 (Offprint #1043). The activity of biologically important giant molecules is profoundly influenced by the shapes into which they can twist. The possibilities can be studied with a computer.

Mark, H. F., "The Nature of Polymeric Materials," *Scientific American*, Sept. 1967. An excellent discussion of the properties of both natural and synthetic polymers and the possibilities of tailoring these molecules for new uses.

Price, C. C., "The Geometry of Giant Molecules," *J. Chem. Educ.* **36**, 160 (1959).

STEREOISOMERS 31

31–1 INTRODUCTION

This chapter deals with the fascinating isomers, called *stereoisomers*, which can exist because of particular spatial arrangements within the molecules. Since we have already encountered a few examples of one type, the *cis-trans* isomers, we will begin by investigating additional examples of *cis-trans* isomers. Then it will be possible to define some important terms and describe a very different type of isomer within the large category of stereoisomers.

31–2 *CIS-TRANS* ISOMERS: ADDITIONAL EXAMPLES

We discovered, in Section 23–4, that there could be both a *cis* and a *trans* isomer of an alkene such as 2-butene. The two isomers can exist because the rotation of groups about the double bond is restricted. A similar situation arises in substituted cycloalkanes, whose carbon skeletons fall approximately in a plane, with attached groups held either above or below the plane of the ring. For example, *cis*-1,3-dichlorocyclopentane and *trans*-1,3-dichlorocyclopentane (Fig. 31–1) are two different compounds. The two chlorine atoms can both be below the plane of the ring, or one can be above the plane and one below. In this case, it is the ring itself which acts as a more or less rigid structure, preventing part of the skeleton in one isomer from rotating to produce the other isomer. Although each

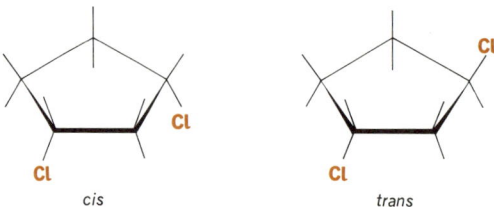

Fig. 31–1 Two different molecules of 1,3-dichlorocyclopentane.

bond in the skeleton of the compound is a single bond, the carbon atoms are linked in a ring. Therefore, one carbon cannot rotate relative to the others without breaking a bond and tearing the ring open. So the situation is the same as it is with the *cis-trans* isomers of alkenes: a *cis* compound can be changed to its *trans* isomer only by way of a chemical reaction involving bond-breaking and bond-formation.

31–3 STEREOISOMERS: DIFFERENCES IN CONFIGURATION

For any given pair of *cis-trans* isomers (for example, Fig. 31–1), the *cis* compound has the same *structure* as the *trans* compound. *The* **structure** *of a molecule refers to which atom is bonded to which other atom, and to the type of bonds involved*. A *trans* isomer differs from a *cis* isomer only in the arrangement in space of its atoms. We say it is their configurations which differ. **Configuration** *refers to spatial arrangement*. Compounds differing in this fashion constitute a particular class of isomers: **stereoisomers** *have the same structure, but different configurations*. The *cis-trans* isomers represent just one type of stereoisomers, whose differences in configuration are due to some rigid structure in the molecule (a double bond or a ring). The other type of stereoisomers, to be described below, are called *optical isomers;* their configurations depend upon a different factor.

31–4 THE STRANGE ISOMERS OF LACTIC ACID

Lactic acid occurs in a great variety of natural materials, including sour milk and sore muscles. It has a sour taste much like that of vinegar and other organic acids. Analysis of lactic acid reveals that its molecular formula is $C_3H_6O_3$; two of the oxygens are in a carboxyl group and the third is in a hydroxyl group. Furthermore, there is abundant evidence to prove that the hydroxyl group is on the carbon adjacent to the carboxyl group. Therefore, we are required to write the structural formula as in Fig. 31–2.

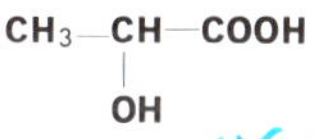

CH$_3$—CH—COOH
　　　|
　　　OH

Fig. 31–2 The structure of lactic acid.

For many compounds this would appear to be a satisfactory formula. Not so in this case. Lactic acid seems to have two different isomers, both of which, however, must have the same *structure* (Fig. 31–2). Although there are many sources for the different kinds of lactic acid, one kind can be obtained conveniently from certain bacteria, another kind from muscle tissue. The properties of the two are listed in Table 31–1.

In *nearly* all respects the two kinds of lactic acid are identical. Are we justified in suggesting that they are different compounds? In one unusual property, their effect on the rotation of polarized light, the two kinds of lactic acid are curiously similar and yet different. Each affects the polarized light to the same extent, rotating it 3.8°. However, the results are exactly opposite. In one case it is a positive value (rotation to the right), while in the other case it is a negative value (rotation to the left). Indeed, it is remarkable that lactic acid has *any* effect on polarized light; many organic compounds do not. For this

TABLE 31-1 Properties of the two isomeric lactic acids

Property	Isomer A	Isomer B
source	muscle	bacteria
melting point	26°C	26°C
rotation of polarized light	+3.8° (to the right)	−3.8° (to the left)
other physical properties	A and B behave identically	
chemical properties	reactions correspond to the structural formula (Fig. 31-2); A and B behave identically	
assigned name	(+)-lactic acid	(−)-lactic acid

reason we have never before bothered to mention this particular physical property, as we have properties such as boiling point, density, or solubility. The nature of polarized light will be discussed in the next section. For now, we will say that because the distinctive feature of the lactic acids involves light they were called **optical isomers.*** We shall soon see that numerous other organic compounds exhibit the same behavior.

In order to have names to designate each of the optical isomers of lactic acid, chemists have called them simply (+)-lactic acid and (−)-lactic, referring to their effect on light.

The idea that the two lactic acids really are different compounds is further supported by their behavior upon melting. Once again they are curiously similar yet different. Each type of lactic acid, when pure, melts (or freezes) at 26°C. However, if we mix a small amount of (+)-lactic acid with (−)-lactic acid, it melts at a temperature *lower* than 26°C. This suggests that the (+)-isomer is behaving as a foreign substance toward the (−)-isomer; that is, the molecules of the (+)-isomer must be different than those of the (−)-isomer. Note that, whereas pure water freezes at 0°C, if we dissolve in it a different compound, such as salt or sugar, the water freezes at a lower temperature.

Although the two varieties of lactic acid are exactly the same in most properties, there are these few differences which suggest that they are not the same compound. At least we will assume they are different compounds until we find out more about them. Even more important, we should try to discover what is causing this behavior. But before we do, we should learn a little more about polarized light.

31-5 POLARIZED LIGHT AND OPTICAL ACTIVITY

Many of the properties of light may be described by regarding it as wave motion (Fig. 31-3). That is, as a ray of light moves along in one direction, there are vibrations of an electromagnetic field in a direction perpendicular to the direction of travel. In the diagram only one wave is shown; an end view of its plane of vibration would appear as a line. In ordinary light the vibrations may be in *any* direction perpendicular to the line of travel. Therefore, we must imagine that the planes of vibration of other waves could be tilted at various angles, so that in the end view there are many lines.

---

* Optics is a branch of physics concerned with light. The term was derived from the Greek word *optikos*, relating to the eye or vision.

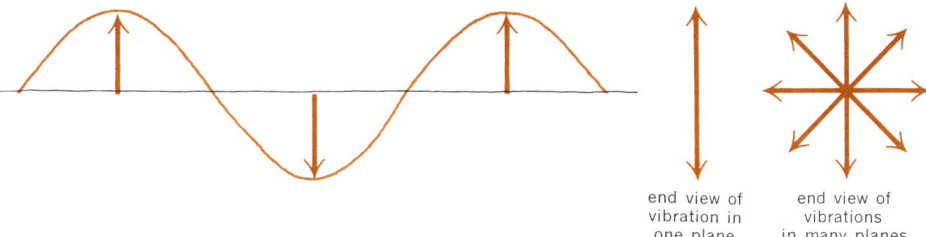

Fig. 31-3 Ordinary light as wave motion.

When ordinary light, vibrating in many different planes, passes through certain materials, the only light which gets through is confined to vibrating entirely in one plane. Light having only one plane of vibration is called plane-polarized light, or more briefly, **polarized light**. One material which has polarizing ability is calcite, a clear, crystalline form of calcium carbonate. The arrangement of atoms in the calcite crystal is such that light passing through it is forced to align its vibrations into one plane.

A **polarimeter** is an instrument designed to measure the effect of substances on polarized light. It is a long cylinder holding various bits of apparatus; from the outside it usually looks like a telescope. Figure 31-4 shows the functioning parts of a polarimeter, in diagram form. The front end of the cylinder holds a calcite crystal, called a *polarizer*. This converts the ordinary light entering the instrument to polarized light. The light then passes through a tube containing the sample being studied. If the sample is a compound such as lactic acid, it can rotate the plane of vibration of the polarized light. The light will still be vibrating entirely in one plane, but the plane will be tilted so that it lies at a different angle than it did originally. At the back end of the instrument is another calcite crystal, the same as the first, but called an *analyzer*. Before the sample is placed in the instrument, the analyzer is carefully lined up with the polarizer, so that the polarized light vibrating in a vertical plane can pass on through the analyzer. However, if polarized light which passes through the sample is no longer aligned in a vertical direction, it is blocked by the analyzer. In order to allow the light to pass, we must rotate the analyzer until it is tilted at the same

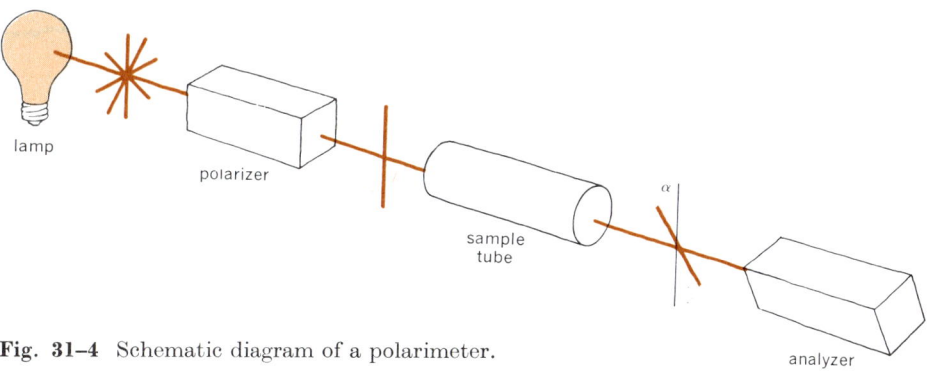

Fig. 31-4 Schematic diagram of a polarimeter.

angle as the polarized light from the sample. The analyzer is attached to a round dial marked off in degrees, like a protractor. Therefore we can read on the dial the number of degrees through which we need to turn the analyzer to get it properly aligned with the polarized light.

The ability to rotate polarized light is a property of individual molecules. Therefore the amount of rotation observed in the polarimeter will be greater if the polarized light encounters more solute molecules by passing through a more concentrated solution or through a longer tube. Meaningful rotation values can be obtained only if we compare one with another under a standard set of conditions. The value used, called the **specific rotation** and symbolized by $[\alpha]$, is the *amount of rotation given by a pure liquid or a solution at a concentration of one gram per milliliter, in a sample tube one decimeter long*. Very often the measurement will be observed using a tube of a different size or a solution of different concentration. In this case $[\alpha]$ can easily be calculated by the following equation:

$$[\alpha] = \frac{\alpha}{lc},$$

where $[\alpha]$ = the specific rotation,
α = the observed rotation in degrees,
l = the tube length in decimeters (dm).
c = the concentration in g/ml.

Thus if a 2-dm tube is used for a certain sample, α (the degrees of rotation actually observed on the polarimeter dial) would be twice as great as it would have been for the same sample in a 1-dm tube. But when one substitutes 2 for l in the equation, α is divided and brought back to the standard conditions, giving $[\alpha]$.

We say that lactic acid is *optically active* because it affects the polarized light. **Optical activity** *is the ability to rotate the plane of vibration of polarized light*. Acetic acid, ethanol, acetone, and many other organic compounds do not rotate polarized light at all; these we say are optically inactive. If the rotation caused by an optically active compound is to the right (clockwise), the compound is said to be *dextrorotatory* (L. *dexter*, right). When writing the numerical value, we give it a positive sign, thus: +3.8°. Conversely, a *levorotatory* substance (Fr. *lévo*, L. *laevus*, left) rotates the light to the left (counterclockwise) and its numerical value is given a negative sign.

One characteristic of an optically active compound is that somewhere, somehow, there can always exist a companion which behaves exactly like it in all respects but one: the companion rotates polarized light the same number of degrees but in the *opposite* direction. Thus we think of optical isomers as existing in pairs which somehow are opposites, like a right hand and a left hand. Even though we may first encounter only the right hand, sooner or later we will locate the left hand.

Quite a variety of organic compounds can display optical activity; a few examples are listed in Table 31–2. Surprisingly, many other organic compounds are optically inactive, despite the fact that they are similar to some of the active ones.

31–6 TETRAHEDRAL CARBON TO THE RESCUE

So far all we have done is to describe the observed properties of optical isomers. The big question is: *Why* are they that way? If they exist as pairs of different substances, yet have the same structure, how can we write two different formulas? Or, more properly, can the

TABLE 31-2 Comparison of optically active and inactive organic compounds

Optically Active	Optically Inactive

CH₃—CH—COOH
       |
      OH
2-hydroxypropanoic acid (lactic acid)
	CH₃—CH₂—COOH
	
	propanoic acid

CH₃—CH₂—CH—CH₃
             |
           OH
2-butanol
	CH₃—CH₂—CH₂—CH₂
	                   |
	                  OH
	1-butanol

CH₃—CH₂—CH₂—CH—CH₃
                   |
                  OH
2-pentanol
	CH₃—CH₂—CH—CH₂—CH₃
	              |
	            OH
	3-pentanol

CH₃—CH—COOH
       |
      NH₂
2-aminopropanoic acid
	CH₂—CH₂—COOH
	 |
	NH₂
	3-aminopropanoic acid

CH₃—CH—CH₂—COOH
       |
      Cl
3-chlorobutanoic acid
	CH₃—CH—CH₂—COOH
	       |
	      CH₃
	3-methylbutanoic acid

atoms somehow be assembled in two ways which will make most properties the same yet have opposite effects on light? Furthermore, why are other organic compounds, apparently rather similar, *not* optically active?

Louis Pasteur was aware of this problem as long ago as 1860; he had been studying optically active salts of tartaric acid, obtained from the sediment in wine kegs. Despite his genius, there was not sufficient chemical knowledge at that time to make it possible for him to see an answer to the puzzle. He did surmise, however, that an optically active compound must have its atoms arranged in some kind of nonsymmetric pattern, and that there would be a companion pattern in which the nonsymmetric arrangement was laid out in the opposite order. Examples of such nonsymmetric objects are a right hand and a left hand or a clockwise and a counterclockwise helix (a coil). These pairs have the same structure; they differ only in the arrangement of their parts in space.

The answer to the problem was proposed by two young men independently and almost simultaneously in 1874. One was a Dutch chemist, Jacobus van't Hoff, age twenty-two; the other was a Frenchman, Joseph le Bel, age twenty-seven. They perceived that *if* a carbon atom holds four different groups and *if* its bonds have a tetrahedral structure (Section 21–6), then the arrangement of atoms around the carbon will automatically be nonsymmetric. And precisely because it is nonsymmetric, one other (and *only* one other) arrangement is possible, which is nonsymmetric in the opposite sense. The three-dimensional models pictured in Fig. 31–5 illustrate this principle.

The two possible spatial arrangements for lactic acid (Fig. 31–5) are related to each other in the way that an object is related to its mirror image. That is, if we hold one of the models up to a mirror, the image we see in the mirror looks like the *other* model; the image

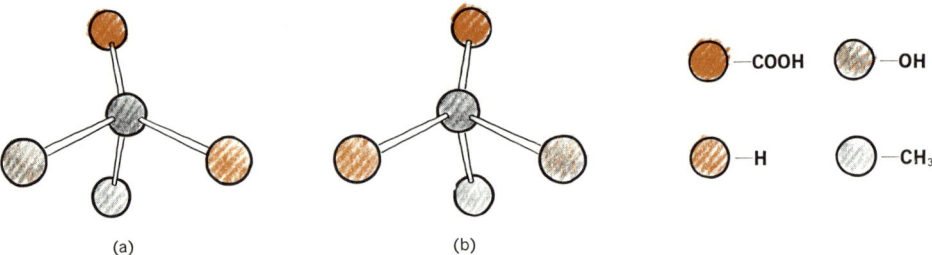

Fig. 31-5 Two nonsymmetric arrangements are possible when tetrahedral carbon holds four different groups. If lactic acid is an example, the groups would be: —COOH, —OH, —H, —CH$_3$.

is the opposite or reflected arrangement. Equally important, these two arrangements (configurations) are *different*. Regardless of how we turn one model in space it will never have the same configuration as its mirror-image companion. (Of course, we must not take the model apart and rebuild it.) A pair of hands illustrates this concept nicely. A right hand has a nonsymmetric arrangement in space. A left hand has the same *structure:* it has fingers of the same length attached to the hand by the same kind of bones at the same angles. The left hand differs only in its spatial arrangement, which is an opposite non-symmetric pattern. If we hold a right hand up to a mirror, the image we see there is a left hand! Furthermore, the hands are nonidentical. We cannot make a right hand become a left hand by merely turning it upside down or backward. To return now to the two different molecules of lactic acid, a pair of isomers related in this way are called enantiomers. **Enantiomers** *are compounds which are nonidentical mirror images*.

The fact that four different groups are important to make enantiomers possible is further emphasized if we examine a molecule of propanoic acid, CH$_3$CH$_2$COOH. This compound is the same as lactic acid except that it lacks a hydroxyl and therefore has *two* hydrogens bonded to the central carbon. We can make a model of the propanoic acid molecule. Like any object, this model has a mirror image (Fig. 31-6). However, in this case the object and its mirror image are *identical*. That is, there can be only one kind of object having this structure. This uniqueness is due to the fact that both the object and its mirror image are symmetric; this symmetry, in turn, depends on the presence of two groups of the same kind (in this case, two hydrogens) attached to the central carbon atom.

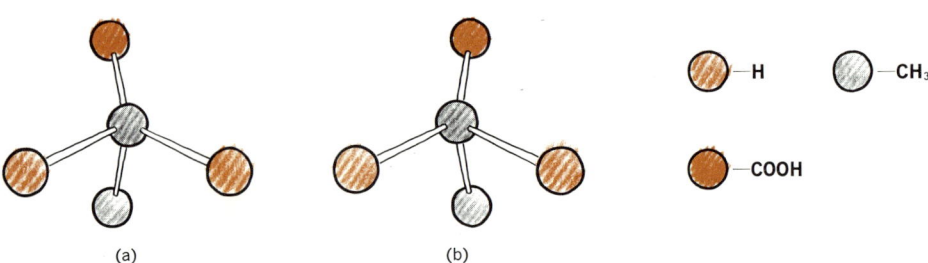

Fig. 31-6 A molecule of propanoic acid and its mirror image, which is identical.

Soon after van't Hoff grasped the idea that a tetrahedral carbon could possibly account for optical isomers, such as the lactic acids, he checked his hypothesis with many of the organic compounds that had been described in the chemical journals of his time. (For a few selected illustrations, see Table 31–2.) He found that compounds having one carbon holding four different groups were indeed optically active. However, a few of the compounds then listed as being optically active did *not* have four different groups at one carbon. Van't Hoff was confident enough to predict that these samples were contaminated with other compounds which were optically active. Because of the keen interest in the new concept, various chemists reexamined these "exceptions" and before long proved that van't Hoff's predictions were true. There is now no doubt that the explanation of optical isomerism offered by van't Hoff and le Bel is correct. It is supported by experimental evidence from thousands of compounds.

We say then that optical isomers have the same structure but different configurations. Their differences in configuration depend upon nonsymmetric patterns within the molecules. Since optical isomers differ only in configuration, they fall within the class of stereoisomers (Section 31–3).

31-7 FORMULAS FOR OPTICAL ISOMERS

Once again we are confronted by the problem of developing a written formula which will adequately represent a three-dimensional molecule. In fact, for optical isomers the problem is especially acute because their very nature depends entirely upon the subtle differences in spatial arrangement. Certain methods for writing formulas of optical isomers, shown below, have proved to be quite satisfactory and are widely used by chemists.

A perspective formula (Fig. 31–7) is easy to draw and effectively portrays the three-dimensional arrangement of groups in a molecule. It is evident that this formula is a close approximation to a picture of a three-dimensional model; compare Figs. 31–5 and 31–7. In a perspective formula the tapered bond lines imply that the two groups at the side are out in front of the central carbon atom, closer to the viewer. Also, it is understood that the other two groups, attached by broken lines, are behind the central carbon.

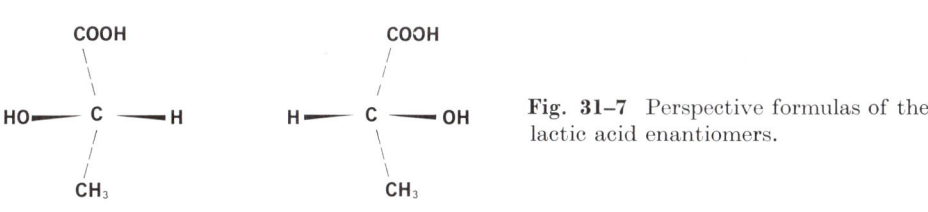

Fig. 31-7 Perspective formulas of the lactic acid enantiomers.

```
    COOH           COOH
    |              |
HO—C—H         H—C—OH
    |              |
    CH₃            CH₃
    (a)            (b)
```

Fig. 31-8 Valence bond formulas of the lactic acid enantiomers, corresponding to the perspective formulas shown in Fig. 31–7.

It is even possible to write a satisfactory valence bond formula for an optical isomer (Fig. 31–8), *provided* we understand exactly what it represents and how to use it properly. By arbitrary convention chemists have agreed that in a valence bond formula the groups written at the sides of the central carbon project forward toward the viewer, whereas the other two groups extend backward behind the central carbon. In other words, the lactic acid formulas in Fig. 31–8 represent the same configurations as those shown in Figs. 31–7 and 31–5.

A perspective formula is drawn in such a way that we can easily see the three-dimensional arrangement of groups. But when we look at a valence bond formula we must *imagine* the spatial arrangement; it is very important that we visualize the arrangement correctly, following the rule agreed upon. For example, one might be tempted to argue that configuration (b) in Fig. 31–8 is really identical to that in (a); it just looks different because of the way it was written. After all, if (b) were just lifted off the paper and flipped over it would look the same as (a). The error in this argument results from ignoring the rule that one must assume the groups written at the sides are projecting *forward*. If (b) were flipped over, the groups at the side would then project to the *rear* and (b) would *not* coincide with (a). To say it another way, although a valence bond formula *appears* to be flat, it represents a molecule which actually has three dimensions. When we inspect the models (Fig. 31–5) or the perspective formulas (Fig. 31–7) of the two lactic acid configurations we realize that (b) is really different from (a).

> Whenever valence bond formulas are used it is essential to remember that they represent three-dimensional objects. It is important that we all follow the same rule as to *how* we visualize the spatial arrangement of groups. All the rule really says is that whenever we write a valence bond formula for a configuration we will always view it from the same direction. The real molecules, or models of them, can of course be moved anywhere in space. However, when we wish to compare one with another and write their formulas, it is just a matter of convenience to view them always from the same direction.

Stop here

31–8 RACEMATES

In Section 31–4 we found that lactic acid could be obtained from a variety of natural sources. Lactic acid can also be made synthetically (Eq. 31–1) by catalytic hydrogenation of pyruvic acid. The product thus obtained is called *racemic* lactic acid.

$$CH_3-\underset{\underset{O}{\|}}{C}-COOH + H_2 \xrightarrow{Ni} CH_3-\underset{\underset{OH}{|}}{CH}-COOH \qquad (31\text{--}1)$$

pyruvic acid   racemic lactic acid

Is this yet a third type of lactic acid, differing from either the $(+)$-lactic acid or the $(-)$-lactic acid described in Table 31–1? Yes and no. The chemical properties of *racemic* lactic acid are identical to those of $(+)$- or $(-)$-lactic acid. However, it has a different solubility, its melting point is 18°C, and it is optically inactive ($[\alpha] = 0$).

Chemists have discovered that *racemic* lactic acid is composed of exactly 50% $(+)$-lactic acid and 50% $(-)$-lactic acid. Such a mixture, called a racemate, can be formed by any pair of optical isomers. *A **racemate** is composed of exactly equal amounts of a pair of enantiomers.* A racemate is optically inactive because there are just as many

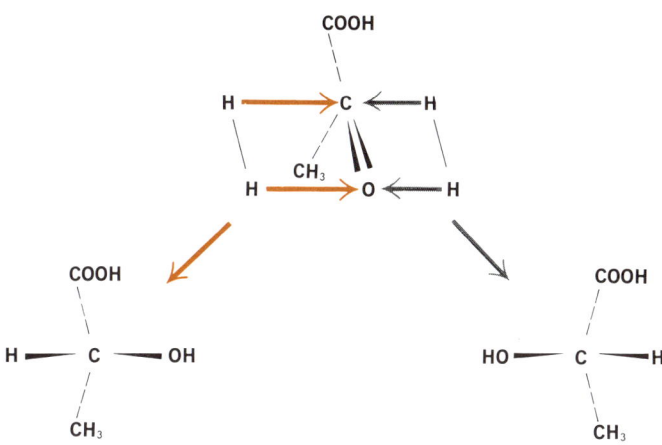

Fig. 31–9 A hydrogen molecule can add to either side of a pyruvic acid molecule, yielding either enantiomer of lactic acid with equal probability. The product thus obtained is a racemate.

molecules rotating the polarized light to the left as there are to the right; hence the net rotation observed in a polarimeter is zero.

When lactic acid is *synthesized* in the laboratory from pyruvic acid, *why* is a racemate produced? In contrast, some living organisms can build exclusively (+)-lactic acid and others, the (−) enantiomer. The laboratory reaction can be explained by considering the spatial arrangement of the molecules involved (Fig. 31–9). The starting material, pyruvic acid, is a symmetric molecule and hence is optically inactive. The molecule is flat; that is, the three groups attached to the central carbon all lie in one plane. In the diagram this plane is perpendicular to the page; the oxygen is nearest the viewer and the methyl and carboxyl groups are to the rear. Now when a hydrogen molecule adds to the double bond, it has an exactly equal chance to approach the flat pyruvic acid molecule from either side. Thus of the billions of molecules reacting, sheer chance will dictate that half of them will produce one enantiomer of lactic acid and the other half will produce the opposite enantiomer. Consequently, the product *as obtained in the reaction flask* is optically inactive; it is a racemate. This will always be the case, regardless of the type of reaction used. Whenever optically inactive materials are converted by chemical synthesis to new compounds, the resultant products are still optically inactive. Even though nonsymmetric configurations are created in the new molecules, the mixture as a whole will be a racemate.

If the two enantiomers constituting the racemate are separated from each other, then each will be optically active when alone. However, such a separation can *not* be accomplished by ordinary methods such as distillation or crystallization. Since the enantiomers have the same chemical and physical properties, both will show identical behavior during such treatment. A racemate can be separated into its component enantiomers only by the use of some other compound, which is already optically active, as a reagent or catalyst to influence the process.* In other words, we can generate new optically active material

---

* The exact details of such a process are described in more comprehensive texts on organic chemistry.

only if we use material which is already optically active! This is similar to the chicken-and-egg dilemma and the answer to it is similar: this is part of the mystery of life. The optically active material necessary to start the process must be obtained from a living organism.

To summarize, enantiomers behave differently from each other in the presence of another compound which is nonsymmetric (optically active). This is the second way in which enantiomers differ from each other. Their effect on polarized light is the other property in which they differ (Section 31–4).

We have just seen that an optically active reagent can be used to *separate* a racemate into two enantiomers, each of which will then be optically active. New optically active material can also be produced if an optically active catalyst or reagent is present at the time that new, nonsymmetric molecules are being formed. This optically active agent, being nonsymmetric, favors the formation of one enantiomer in preference to the other. A laboratory synthesis can be controlled by an optically active substance; the great majority of biochemical syntheses occur in this manner. We will discuss a biochemical example in the next section.

31–9 OPTICAL ISOMERS IN LIVING ORGANISMS

Muscle tissue can convert pyruvic acid to lactic acid. This reaction, Eq. (31–2),

$$CH_3-\underset{\underset{O}{\|}}{C}-COOH + 2H \xrightarrow[\text{(muscle tissue)}]{\text{enzyme}} HO-\underset{\underset{CH_3}{|}}{\overset{\underset{|}{COOH}}{C}}-H \qquad (31-2)$$

pyruvic acid (+)-lactic acid

is a biosynthesis. The crucial difference between it and the chemical synthesis (Eq. 31–1) is that the biosynthesis produces only *one* enantiomer, (+)-lactic acid. In Eq. (31–2) the two hydrogen atoms added to the double bond come not from H_2 gas, but from an organic compound in the tissue. This difference, however, happens to be unimportant in determining which optical isomers will be produced. The important factor in Eq. (31–2) is the catalyst, an enzyme.

The enzyme catalyzes the reaction because it binds the pyruvic acid to itself and weakens the double bond, thereby facilitating the addition of hydrogen. The enzyme is a nonsymmetric molecule. (Incidentally, this of course makes it optically active.) Consequently, the enzyme acts as a pattern; a pyruvic acid molecule can fit onto its surface aligned in only one way (Fig. 31–10). Hydrogen atoms can then be added to the pyruvic acid from just *one* direction, thus yielding only one enantiomer of lactic acid.

As we have seen, certain other organisms may produce the opposite enantiomer of lactic acid. This just means that they have a different enzyme to promote the biosynthesis of lactic acid. Still different organisms may form both enantiomers, i.e., an optically inactive racemate. The significant point is that living organisms *can* produce single enantiomers (optically active) by means of their enzymes which are optically active molecules. In fact, in those biochemical reactions which are most crucial in sustaining life, organisms nearly always produce specifically one optical isomer. We found in Section 28–5 that amino acids are the essential units from which all proteins are built.

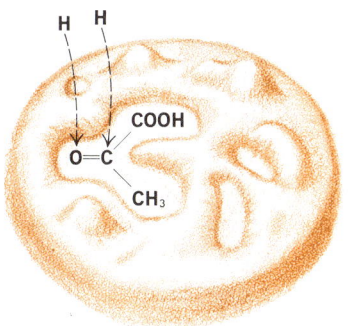

Fig. 31–10 An enzyme holding a pyruvic acid molecule, only one side of which is available for reaction. (The enzyme, represented schematically in color, is a complex organic molecule which would be much larger, in proportion, than suggested here.)

Every amino acid molecule used for this purpose must have the configuration shown below; the opposite enantiomer will not suffice.

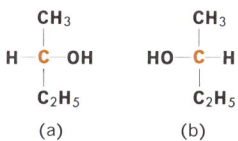

31–10 MORE COMPLEX OPTICAL ISOMERS

A carbon atom holding four different groups is often called an **asymmetric carbon,** meaning literally, *not symmetric*.* For instance, in Fig. 31–11 the C shown in color designates the asymmetric carbon in a molecule of 2-butanol. Because of the asymmetric carbon, 2-butanol can assume two different configurations. In a written valence bond formula, when the carbon skeleton is stretched out vertically, only two arrangements for the H and OH groups are possible: the OH can be either to the right or to the left.

```
    CH3            CH3
     |              |
H—C—OH          HO—C—H
     |              |
    C2H5           C2H5
    (a)            (b)
```

Fig. 31–11 The two enantiomers of 2-butanol.

Now suppose we add to the 2-butanol molecules one more asymmetric carbon, making 2,3-pentanediol (Fig. 31–12). Starting with one of the 2-butanol configurations (Fig. 31–11a) one could add to it, at the upper end, another asymmetric carbon in either of two ways, with the OH to the right or to the left (Fig. 31–12a,b). Similarly, a new asymmetric carbon can be added in two ways to the other configuration of 2-butanol

---

* The term *asymmetric carbon* is convenient, although strictly speaking the carbon itself cannot be asymmetric. It is the carbon plus its four groups which creates a nonsymmetric arrangement.

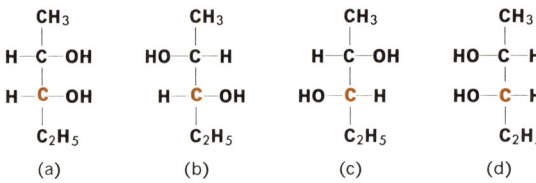

Fig. 31-12 The four configurations of 2,3-pentanediol.

(Fig. 31–11b) giving 31–12c,d. In this fashion, four different configurations of 2,3-pentanediol are possible.

In Fig. 31–12, compounds (a), (b), (c), and (d) are all optical isomers of each other. Furthermore, (a) and (d) are enantiomers; they are mirror images but are not identical. Compounds (b) and (c) make up another pair of enantiomers. A fifty-fifty mixtue of (b) and (c) would constitute a racemate. A fifty-fifty mixture of (a) and (d) would yield another racemate having solubility, melting point, etc., different from the racemate of (b) and (c).

A further examination of the configurations in Fig. 31–12 reveals some additional facets of optical isomerism. Since (b) and (c) are enantiomers, they have identical melting points, boiling points, solubilities, etc. Even their rotation of polarized light is the same in amount, although of opposite sign. But what is the relation between (b) and (a)? They are **diastereomers**; *they are optical isomers but are not mirror images*. Compounds (b) and (a) have *different* melting points, boiling points, solubilities, etc. Each is optically active, but (b) rotates polarized light to a very different extent than does (a). Since (a) and (b) both have OH functional groups, they will have the same type of chemical properties. However, the *rate* at which they react may be different. To summarize, each configuration in Fig. 31–12 has one companion which is its enantiomer, its mirror image. But either of the other two configurations will be diastereomers of it. Thus, (a) and (b) are diastereomers of each other; so also are (a) and (c), (b) and (d), or (c) and (d).

Many organic compounds, both natural and synthetic, have *several* asymmetric carbon atoms. How many optical isomers can we expect in such cases? We know that if there is 1 asymmetric carbon, 2 configurations are possible. Then if there are 2 asymmetric carbons, each of the first two possibilities is multiplied by 2 (Fig. 31–12). Hence, if there are three different asymmetric carbons, there would be $2 \times 2 \times 2$, or 2^3, or 8 possible configurations. In general, the number of different configurations will be 2^n, where n is the number of different asymmetric carbons. (This is often called *van't Hoff's rule*, for he was the first one to state the principle.) This simple mathematical relation leads to some astounding results. Glucose, a sugar occurring in humans and in many other organisms, was first discussed in Section 29–11. Careful inspection of the formula of D-glucose, shown in Fig. 29–3, reveals that it has four asymmetric carbons. This means there are 2^4 or 16 compounds having the same structure but different configurations. One of these is glucose; another is its enantiomer. The other 14 compounds in this family of optical isomers are all diastereomers of glucose.

The important compound cholesterol has molecular formula $C_{27}H_{46}O$. Of these 27 carbon atoms, only 8 are asymmetric. However, $2^8 = 256$! The enzymes controlling the formation of cholesterol are so specific that only the one compound, out of 256 possible isomers, is formed by living organisms!

The number of different configurations predicted by the 2^n rule is the *maximum* number possible. When all the asymmetric carbons in a molecule are of different type, the maximum will be realized. Occasionally two of the asymmetric carbons will be similar, in the sense that the four groups bound to one carbon are the same as the four groups bound to another carbon. In this case not so many combinations will be possible, and the total number of configurations will be less than 2^n.

31–11 SUMMARY OF TYPES OF ISOMERS

Isomers are different compounds having the same molecular formula.
a) **Structural isomers** have different structures.

$CH_3-CH_2-CH=CH_2$ and $CH_3-CH=CH-CH_3$

CH_3 O CH_3 and CH_3 CH_2 OH

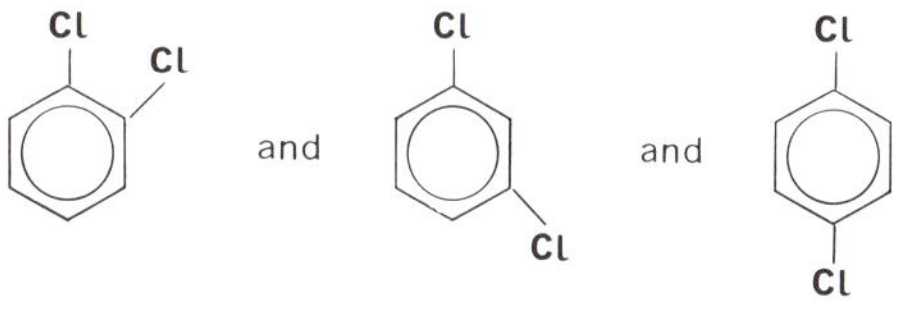

[three dichlorobenzene isomers: ortho, meta, para]

b) **Stereoisomers** have different configurations, but the same structure.
   1. *Cis-trans* isomers have different configurations because of rigid structures in the molecules. (These are also called *geometric isomers*.)

   [cis and trans 1,2-dichloroethylene; cis and trans dichlorocyclobutane]

   2. **Optical isomers** have different configurations because of nonsymmetric patterns in the molecules.

   Enantiomers are nonidentical mirror images.

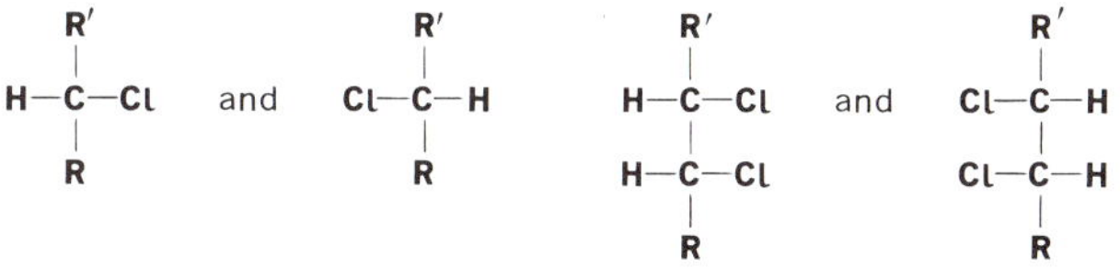

   [enantiomer pairs with R, R', H, Cl groups]

   A **racemate** is composed of exactly equal amounts of a pair of enantiomers.

478   Stereoisomers

Diastereomers are optical isomers which are not mirror images of each other.

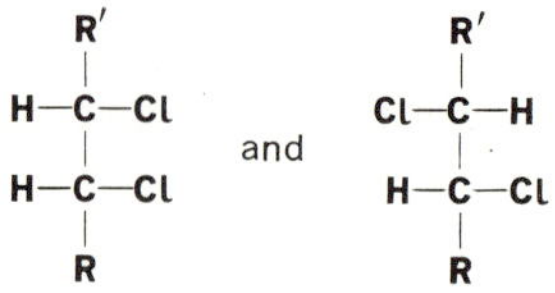

```
    R'                R'
    |                 |
H—C—Cl            Cl—C—H
    |      and        |
H—C—Cl            H—C—Cl
    |                 |
    R                 R
```

31-12 CONCLUSION: THE IMPORTANCE OF STRUCTURE AND CONFIGURATION

From our first encounter with organic substances, in Chapter 21, to these final discussions of organic chemistry, the importance of isomers has been apparent. Covalent bonds give organic compounds specific structures and configurations, which in turn establish their chemical and physical properties. In fact, we can go one step further. Subtle differences in the shape and structure of organic molecules are of prime importance to life itself. The function of organic compounds in living cells falls within the field of *biochemistry*, our next broad topic for study.

IMPORTANT TERMS

Configuration	Optical activity	Polarized light
Enantiomers	Polarimeter	Tetrahedral carbon atom

[See also the terms in Section 31-11.]

WORK EXERCISES

1. Which of these structures could exist in an optically active form?

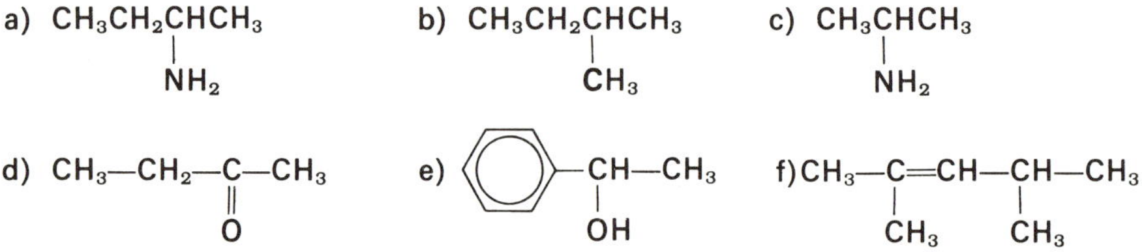

   a) $CH_3CH_2CHCH_3$
         |
         NH_2

   b) $CH_3CH_2CHCH_3$
         |
         CH_3

   c) CH_3CHCH_3
         |
         NH_2

   d) $CH_3—CH_2—\underset{\underset{O}{\|}}{C}—CH_3$

   e) $C_6H_5—CH—CH_3$
              |
              OH

   f) $CH_3—C=CH—CH—CH_3$
            | |
            CH_3 CH_3

2. The biologically important sugar ribose has the structure shown below. How many optical isomers (different configurations) are possible for this structure?

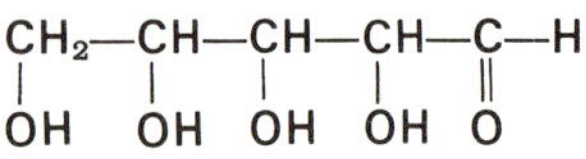

   $CH_2—CH—CH—CH—\underset{\underset{O}{\|}}{C}—H$
   | | | |
   OH   OH   OH   OH

3. For each pair of molecules below, tell what type of isomerism is involved, *if any*.

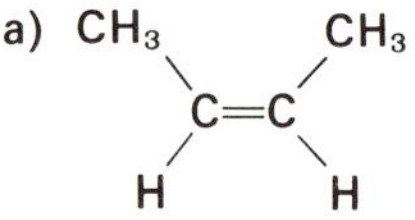

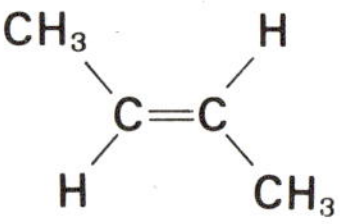

   a) $\underset{H}{\overset{CH_3}{\diagdown}}C=C\underset{H}{\overset{CH_3}{\diagup}}$ $\underset{H}{\overset{CH_3}{\diagdown}}C=C\underset{CH_3}{\overset{H}{\diagup}}$

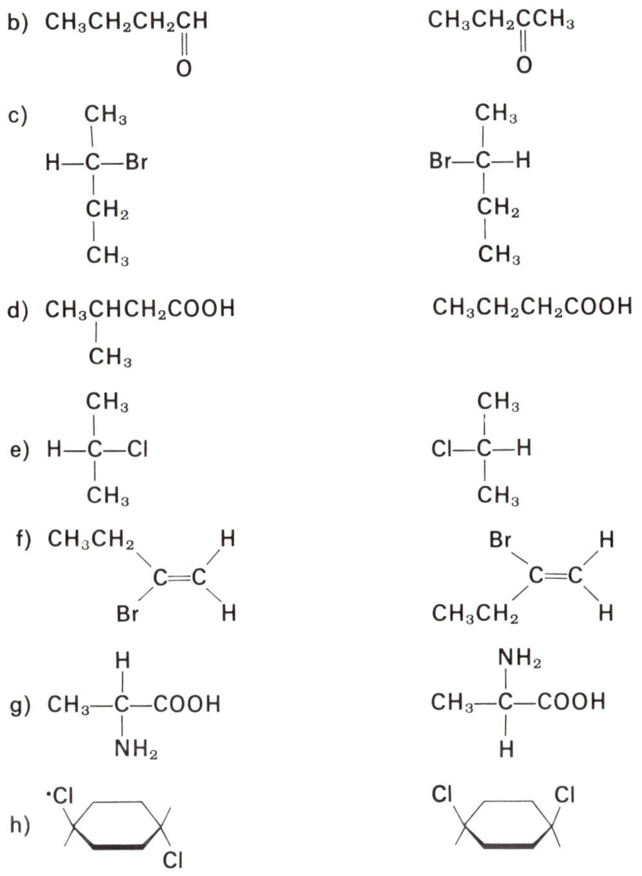

b) $CH_3CH_2CH_2\underset{\underset{O}{\|}}{C}H$ $CH_3CH_2\underset{\underset{O}{\|}}{C}CH_3$

c) $\underset{\underset{CH_3}{|}}{\overset{\overset{CH_3}{|}}{H-C-Br}}\overset{}{\underset{}{|}}CH_2$ $\underset{\underset{CH_3}{|}}{\overset{\overset{CH_3}{|}}{Br-C-H}}\overset{}{\underset{}{|}}CH_2$

d) $CH_3\underset{\underset{CH_3}{|}}{C}HCH_2COOH$ $CH_3CH_2CH_2COOH$

e) $\underset{\underset{CH_3}{|}}{\overset{\overset{CH_3}{|}}{H-C-Cl}}$ $\underset{\underset{CH_3}{|}}{\overset{\overset{CH_3}{|}}{Cl-C-H}}$

f) $\underset{Br}{\overset{CH_3CH_2}{\diagdown}}C=C\underset{H}{\overset{H}{\diagup}}$ $\underset{CH_3CH_2}{\overset{Br}{\diagdown}}C=C\underset{H}{\overset{H}{\diagup}}$

g) $CH_3-\underset{\underset{NH_2}{|}}{\overset{\overset{H}{|}}{C}}-COOH$ $CH_3-\underset{\underset{H}{|}}{\overset{\overset{NH_2}{|}}{C}}-COOH$

h) [cyclohexane with Cl substituents]   [cyclohexane with Cl substituents]

4. Review Problem 3, Chapter 21. In the light of your present knowledge of stereoisomers, what is the *total* number of isomers possible for each molecular formula?

SUGGESTED READING

Asimov, I., *A Short History of Chemistry*, Anchor Books, Doubleday, New York, 1965, pp. 116–124. Concerns optical isomerism and tetrahedral carbon.

DePuy, C. H., and K. L. Rinehart, Jr., *Introduction to Organic Chemistry*, John Wiley, New York 1967, pp. 117–124.

Dubos, R., *Pasteur and Modern Science*, Chapters 2 and 3, Anchor Books, Doubleday, New York, 1960. Describes Pasteur's exciting work with nonsymmetric crystals of tartaric acid salts and his astounding discovery that the racemic material could be separated into two kinds of crystals which were optically active but of opposite rotation.

Eglinton, G., J. R. Maxwell, and C. T. Pillinger, "The Carbon Chemistry of the Moon," *Scientific American*, Oct. 1972. The simple organic compounds found on the moon were not produced by living organisms, but they do provide additional clues to the origin of life.

Elias, W. E., "The Natural Origin of Optically Active Compounds," *J. chem. Educ.*, **49**, 448 (1972).

Hart, H., and R. D. Schuetz, *Organic Chemistry*, 4th ed., Chapter 15, Houghton Mifflin, Boston, 1972.

Lawless, J. G., C. E. Folsome, and K. A. Kvenvolden, "Organic Matter in Meteorites," *Scientific American*, June 1972.

Noller, C. R., *Textbook of Organic Chemistry*, 3rd ed., Chapter 17, W. B. Saunders, Philadelphia, 1966.

Roberts, J. D., "Organic Chemical Reactions," *Scientific American*, Nov. 1957 (Offprint #85). This article is very good at this stage of your study. It will provide both some new insights and a review of some of the important ways in which organic molecules react.

BIOCHEMISTRY III

This virus, a T$_2$ E. coli bacteriophage, consists almost entirely of DNA. Magnification 97,000x. Photo courtesy of A. K. Kleinschmidt.

Read (handwritten)

INTRODUCTION TO BIOCHEMISTRY

32

32-1 DEVELOPMENT OF BIOCHEMISTRY

It is difficult to set a date for the beginning of the study of biochemistry. Discoveries of facts, proposals of classifications, and the development of the theories we now recognize as comprising the body of knowledge of modern biochemistry were made by many persons. Fundamental discoveries were made by persons we would now call chemists, physiologists, physicians, and many others who would be difficult to classify. For a long time all these people were held back by the doctrine that special vital forces were needed for biological reactions, and that these reactions could not be understood by men. However, some men either from bravery, foolhardiness, or perhaps because they did not realize that what they were doing was supposed to be impossible began to study biochemical questions. Even before the Renaissance many studies were made of the composition of various parts of plants and animals, of the chemistry of the soil, fertilizers, and the requirements for plant growth.

Theophrastus Bombastus von Hohenheim, better known as Paracelsus, who lived from 1493 to 1541, brought together much of the old alchemical mystique and added new ideas of his own. He believed, among other things, that food contained both poisonous and wholesome parts and that digestion separated the two parts. His theory was that when the digesting agent, the *Archaeus*, was not working properly, the poisonous elements were not separated and eliminated, so the person became ill. Medicines were given to restore the *Archaeus* to its normal state. Some of the medicines used were poisonous salts (mercury II chloride and lead acetate); others were extracts from plants. Among the latter were extracts made by distillation, sometimes a distillate from fermented material which contained ethanol. The euphoria produced by ethanol no doubt led to widespread acceptance of the remedies of Paracelsus. Historically more important was the fact that the ideas of Paracelsus and his followers were interesting and stimulated much experimentation. Probably many of their patients died not so much from their illnesses as from the "cures" prescribed. However, the ways of approaching and looking at the chemistry of living things proposed by Paracelsus led to better theories.

Another early scientist who was interested in the individual reactions taking place in living organisms was the Frenchman, Louis Pasteur. He helped to develop the idea that fermentation of grapes (we know now that it was actually the sugar in the grapes) to produce alcohol really involved a series of chemical reactions, and that these reactions occurred in living yeast cells. He further realized that the products formed in these reactions could be changed if different types of cells (either different yeasts or bacteria) were present. A further step was taken when the Buchner brothers showed in 1896 that fermentation did not require whole cells, as Pasteur had believed, but that cell-free yeast juices which contained a "ferment" could bring about certain reactions. It was later realized that the "ferment" was not a single substance but was really a mixture of many biological catalysts which came to be called enzymes. When these catalysts were extracted and purified, and as highly purified forms of the reactants became available to scientists, the road was opened for the investigation of the reactions that take place in living organisms.

As investigation led to greater understandings, the early biochemists were nagged by a fear that the reactions observed when parts of plants or animals were placed in a glass container (*in vitro*) might not be the same as those that actually occurred in a living plant or animal (*in vivo*). Another difficulty some of the clearer thinkers recognized was that the substances isolated and purified from natural products might have been modified by the processes they used and thus might no longer be exactly the same as they were in the starting material. Whether or not they are producing "isolation artifacts" and the "*in vitro-in vivo*" problem still plague modern biochemists.

The past few decades have seen an extremely rapid increase in biochemical knowledge, and predictions for the future make the older science fiction seem mild. Why is our knowledge of biochemistry growing so rapidly? There are many answers. One is that the United States government has been and is supporting biochemical research. Another

Fig. 32-1 Methods of separation and analysis of mixtures by chromatography and electrophoresis.

(a) Paper chromatography. In paper chromatography a mixture of substances is applied to a spot on a piece of filter paper. The paper is then placed in a container with a solvent mixture which either rises on the paper by capillary action or is allowed to flow down the paper from a solvent reservoir. If this procedure does not separate all components (as with C and D above) a second solvent mixture can be used in a direction at right angles to that of the first solvent mixture. This gives a two-dimensional chromatogram. Thin-layer chromatography is similar to one-dimensional paper chromatography except that a plate of glass coated with a thin layer of absorbent material is used.

(b) Column chromatography. In column chromatography a mixture is applied to the top of a glass column filled with a powdered adsorbent substance, and a solvent mixture is allowed to flow through the column. This produces a separation of substances in the mixture. If test tubes are moved under the column at given intervals by the use of a turntable, the different components can be separated.

(c) Gas (gas-liquid) chromatography. In gas chromatography a mixture of substances is injected into the instrument. A current of helium or other inert gas carries the mixture through a small tube packed with an adsorbent material. The tube is heated to help in separation of the components of the mixture. As the components leave the instrument they are detected by some type of device and a curve is plotted. From such curves both the number of components and the relative amount of each may be seen. Some gas chromatographs have arrangements for trapping the individual components as they leave the instrument. These trapped materials can then be analyzed by other methods.

(d) Paper electrophoresis. In paper electrophoresis the mixture is applied to a filter paper or other similar material which is then placed in an electrophoresis cell having an anode at one end and a cathode at the other. Electricity is allowed to flow, and the components of the mixture migrate in the electric field. In the illustration, substances A and B are anions (they migrate toward the anode) and D and E are cations. C has not migrated.

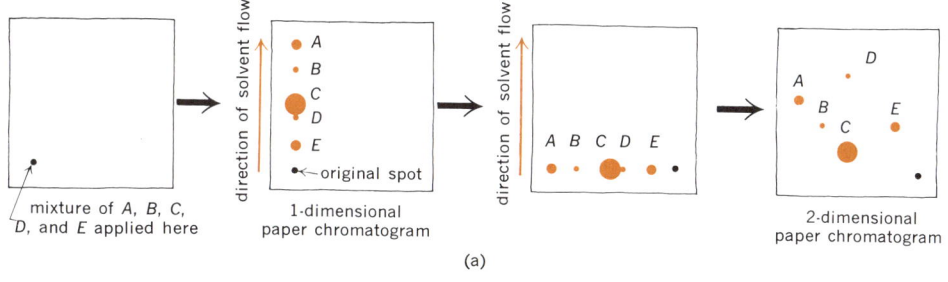

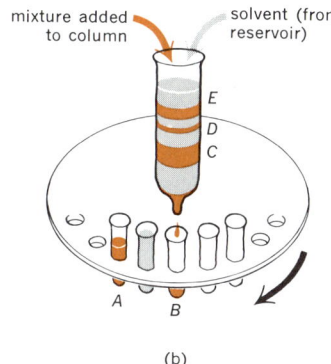

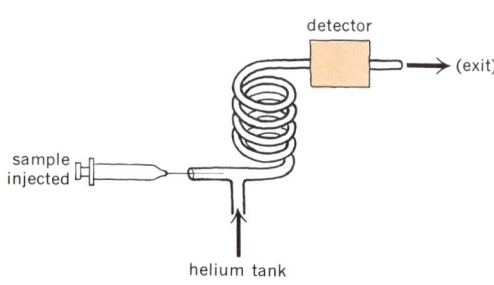
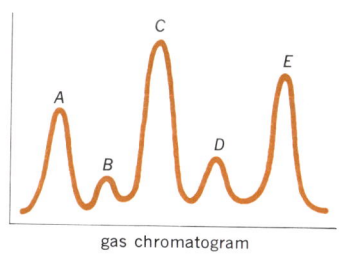

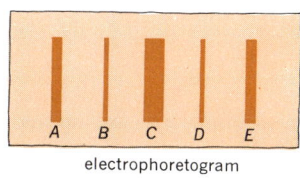

reason for the rapid growth is the discovery and development of new techniques, particularly those for separating and analyzing biochemical compounds. Researchers are purifying and separating substances of importance to biochemistry by using the techniques of **chromatography**. Either solid material packed in columns, paper, or the process of gas-liquid chromatography (popularly called "gas" chromatography) has been especially useful. Chromatographic techniques have made it possible to separate individual amino acids from complex mixtures—a task that previously had been almost impossible. Some of these analytical techniques are shown in Fig. 32–1.

Separation of components of cells by centrifugation is also an important technique (see Section 32–4). If we use an ultracentrifuge—one that produces forces hundreds of thousands times gravity—and use centrifuge tubes containing solutions that are more dense at one part than another, it is possible to separate large molecules from one another. These separations have led to knowledge of the composition and structure of many of the compounds most important to life, including RNA and DNA.

Analytical methods which involve the absorption of infrared and ultraviolet radiation, as well as other techniques, have made it possible to analyze a sample consisting of only a few milligrams of material. The availability of radioactive isotopes, particularly the radioisotope of carbon, carbon-14 (C-14), has led to amazing progress in the study of biochemical reactions. Now specific molecules can be labeled with C-14 and their "fate" in an animal or plant shown by determining which excretion products are labeled. Thus, finding labeled carbon dioxide in the air expired by a rat given glucose which contained carbon-14 is evidence that glucose is converted, *in vivo*, to carbon dioxide by the rat (Fig. 32–2). By administering very small quantities of radioisotopes to humans, medical

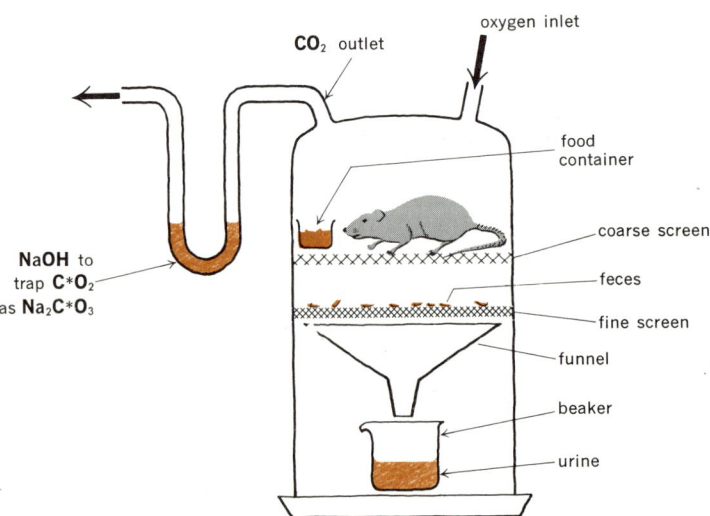

Fig. 32–2 Diagram of an apparatus used to measure metabolism of radioactive substances in an animal. An animal that has been fed or injected with a radioactive substance is placed in the apparatus shown above. Gases he exhales as well as feces and urine can be collected and analyzed. (The asterisk is used to indicate radioactive atoms.)

researchers can sometimes diagnose, and in other instances (usually with larger doses) treat, various diseases.

Many of the early, and some of our modern discoveries in biochemistry have resulted from accidental discoveries in a research laboratory. Discovery of things not searched for, which is called serendipity, has been responsible for countless biochemical developments. For instance, the important discovery by the Buchners that cell extracts could carry out the reactions of fermentation was made not as the result of an experiment planned for this purpose, but came about as the result of their attempts to produce a "yeast juice" that they hoped would have medicinal value. When Eduard Buchner added a solution of sugar to the yeast juice with the hope that it would act as a preservative, he observed that fermentation was occurring, and concluded that units smaller than the whole cells (which Louis Pasteur insisted were necessary) could carry out fermentation. The cell is so complex that studies of how it works are only now being done. However, the realization that smaller units—the "ferments" that we now call enzymes—could serve as catalysts was a significant biochemical advance. Discoveries such as this require a trained, perceptive scientist, accurate observations of unexpected results, and the correct interpretation of these observations. It is interesting to speculate as to when—or whether—antibiotics would have been developed if Alexander Fleming had merely discarded his culture of staphylococcus that had been "contaminated" by a penicillin-producing mold.

Although such accidental discoveries are important to the progress of science, many discoveries and most of the development of new products have been made as the result of well-planned experiments. While researchers are always looking for the unexpected event in their routine work, this prospect does not provide the only motive for biochemical research. Biochemists are usually driven by at least two other motives in their research. For some, the desire to know how living organisms function—just for the pleasure of knowing—is the major motive. Others are more concerned with practical results—curing or preventing diseases. To most researchers, of course, both motives are important.

Although it is interesting to know how biochemistry developed and to realize some of the difficulties it faces, the major question, "What is biochemistry?" has not yet been discussed.

The simplest definition of biochemistry is that it is the chemistry of living plants and animals. Chemistry, as we have learned, is concerned with the composition, structure, and reactions of various types of matter. Thus biochemistry is the study of the composition and structure of those materials found in living organisms and the substances derived from them. Further, it is concerned with the *reactions* which occur in living organisms.

32-2 BIOCHEMICAL COMPOSITION

Life exists in many varieties from the unicellular algae through the simpler animals and plants to the giant redwood trees and whales. In spite of their obvious differences, there is a remarkable similarity in the compounds which are found in these varied types of organisms. All living matter, as well as all inanimate matter, is composed of the approximately one hundred naturally occurring chemical elements. About half of these elements are found in measurable amounts in the human body and in most other organisms that have been studied. Of these fifty elements, less than half have some known function in the human body. Those elements for which no function is known may be present simply because they became incorporated in the organism passively—just because they happened to be in the environment. The other possibility is that further research will discover their function.

Only eleven chemical elements make up any appreciable percentage of the tissues of plants and animals. These elements are carbon, hydrogen, oxygen, nitrogen, sulfur, phosphorus, sodium, potassium, calcium, magnesium, and chlorine. The first three of these elements, carbon, hydrogen, and oxygen, are present in virtually all the compounds found in living plants or animals. Nitrogen is also a major component of biochemical compounds; sulfur and phosphorus are present in many instances. Sodium, potassium, calcium, and magnesium are often referred to as "mineral elements," and are usually found either as ions or in relatively simple, "inorganic" compounds, although they may be loosely bound to a protein or other biochemical substance. Table 32–1 shows the composition of the human body.

TABLE 32–1   Chemical elements found in the human body*

oxygen	65 %	potassium	0.35%
carbon	18.5	sulfur	0.25
hydrogen	9.5	chlorine	0.20
nitrogen	3.3	sodium	0.15
calcium	1.5	magnesium	0.05
phosphorus	1.0		

* Data from Best and Taylor, *The Living Body*, 4th ed., New York, Holt, Rinehart, and Winston, 1958.

Although one must begin by discussing the elements found in biochemical substances, these elements are present primarily in the form of compounds. A real understanding of biochemistry, therefore, must involve a description of compounds. If one were to attempt to answer the question, "What compounds are found in living organisms?", the list of such compounds would be extremely long.

When scientists are confronted with such a long list, they usually look for similarities and differences and, on the basis of these, they develop a system of classification. While no simple system of classification can include all the compounds found in living matter, biochemists have developed a system which includes the large majority of these compounds, with the usual qualifications that some compounds may belong to two or more classifications, and that there be a classification of "miscellaneous."

The most abundant compound in all living organisms is water. The percentage of water varies from well over 90% in some of the simpler aquatic animals to about 50% in a severely dehydrated camel. Various parts of any plant or animal will contain different amounts of water. Muscle tissue is about 70% water, human blood is 80%, and even the proverbially dry bony tissue is about 40% water. The importance and functions of this simple but remarkable compound were discussed in Chapter 12.

If we remove the water from any biochemical substance, we find that the dehydrated material contains three major types of compounds. The relative amounts of these three types of compounds may vary widely, but all plants and animals contain at least small amounts of each and together these three account for almost all the "dry weight" of a plant or animal.

One of the early attempts to describe these three types of compounds was made in 1827 when William Prout classified food substances as either saccharine, albuminous, or oily. With increasing knowledge came the realization that not only foods, but all animal and plant tissues contained these three types of compounds. We now use the terms carbohydrates, proteins, and lipids to describe these compounds which were originally differentiated so long ago. A fourth type of compound—the nucleic acids—is present in much lesser quantities, but is a vital component of living organisms.

Table 32–2 gives the approximate composition of several substances of interest to biochemists.

TABLE 32–2 Approximate composition of materials from plants and animals

	Water	Carbo-hydrate	Protein	Lipid	Nucleic* Acid	Ash†
mammalian muscle (lean)	72–80%	1%	17–21%	5%	0.3%	1%
brain	79	1	8–10	10	0.2	—
hen's egg	74	1	13	11	—	1
cow's milk	87	5	3	4	—	0.7
rat liver	69	4	16	5	0.3	1.4
fish, cod	83	0.4	16	0.4	—	1.2
bananas	75	23	1.2	0.2	—	0.8
rice	12	80	7.6	0.3	—	0.4
peanuts (roasted)	2.6	24	27	44	—	2.7
honey	20	80	0.3	0	—	0.2
potatoes, white	78	19	2	0.1	—	1.0
potatoes, sweet	68	28	2	0.7	—	1.1
spinach	93	3.2	2.3	0.3	—	1.5
beer (4% alcohol)	90	4.4	0.6	0	—	0.2
soybeans, dry	7.5	35	35	18	—	4.7

* Figures for nucleic acid content are, in general, not available for foods and other animal products. It is interesting to note that some viruses, the bacteriophages, may contain from 12 to 61% DNA, and that tobacco mosaic virus has from 5 to 6% RNA.
† The value given as ash indicates the amount of material remaining when all organic matter has been oxidized. It is an indication of the amount of mineral elements present.

Since the carbohydrates, proteins, lipids, and nucleic acids are complex compounds, although made of simple building blocks, it is important to know their structures. Most of the naturally occurring representatives of these compounds, with the exception of the lipids, may be considered to be polymers. The two-dimensional structure of many of these substances has been studied for some time; however, the three-dimensional structure is now known for several of these important polymers and is a subject of research at the present time. Separate chapters are devoted to the description of the composition and structure of the carbohydrates, the proteins, the lipids, and the nucleic acids.

In addition to substances in these four groups, there are other compounds of major importance in biochemistry. Although these compounds make up less than 1% of most organisms, they may be extremely important for the continued life of the organism. Some of these groups of compounds are the vitamins, the porphyrin-containing compounds such as hemoglobin and chlorophyll, various plant pigments, and the medically important substances, the alkaloids.

32-3 BIOCHEMICAL REACTIONS

When we understand the composition and structure of the basic biochemicals, we are ready to consider their reactions. The similarities of reactions in various types of organisms are much more striking than are their differences. The term used by biochemists to describe the reactions which take place in living organisms is **metabolism**. Metabolism is subdivided into a "building" aspect called **anabolism** (anabolic reactions), and a process of degradation or "tearing-down" of compounds called **catabolism** (catabolic reactions).

Since the basic reactants in metabolic reactions are organic chemical compounds, biochemical reactions are much like organic reactions. There are, however, some important differences. If organic reactions are to proceed at an appreciable rate, it is often necessary that the reactants be heated or that acidic or basic catalysts be added to the reaction mixture. Since high temperatures and acidic or basic conditions almost always result in the death of any living thing, another milder method is utilized to bring about biochemical reactions.

Almost all biochemical reactions, both catabolic and anabolic, are catalyzed by enzymes. An enzyme is itself a protein, and may be defined as a catalyst of biological origin or alternately as a catalytically active protein. A further discussion of enzymes and of the energy relationships of biochemical reactions is given in Chapter 37, "Biochemical Reactions."

32-4 THE CELL AND ITS PARTS

The various compounds just described are found as parts of structures in living organisms. Although there is extracellular material, the cell is the basic unit of all living things. Cells may be combined to make tissues, groups of tissues to make organs, and organs to make organisms. While most aspects of these organizations are the subject matter of physiology, biochemists are concerned with the basic structures of cells.

The modern idea of a cell is that it is composed of many small parts called *organelles*. We can see these in an electron micrograph such as that shown in Fig. 32–3. In this and other similar photographs taken through an electron microscope, we see not only the nucleus which was seen in the ordinary light microscope but other units we call mitochondria, microsomes, and lysosomes. Figure 32–4 shows an idealized diagram of our modern concept of a cell.

We can separate these subcellular organelles by centrifugation. If we remove the liver from an animal, cut it, and then grind it with sand or disrupt the tissue in a blender or homogenizer, we get a thick semisolid much like a milkshake in consistency. This mixture is called a *homogenate*, and when it is spun in a centrifuge, we can separate the different units or organelles which make up the cells. This separation is possible because the subcellular units are of different sizes. Figure 32–5 shows the way such particles would be distributed after a short period of centrifugation. By centrifuging at definite speeds and pouring off the unsedimented fraction (the supernatant fraction), we can obtain relatively pure preparations of cell nuclei, the mitochondria and microsomes. The clear supernatant liquid from which all organelles have been removed is a rich source of enzymes—as are all the other fractions.

The **mitochondrion**, Fig. 32–6, even though extremely small, is a complex unit, the site of many of the metabolic reactions that release and store the energy from foods. It has been postulated that the enzymes which catalyze a series of reactions are arranged in

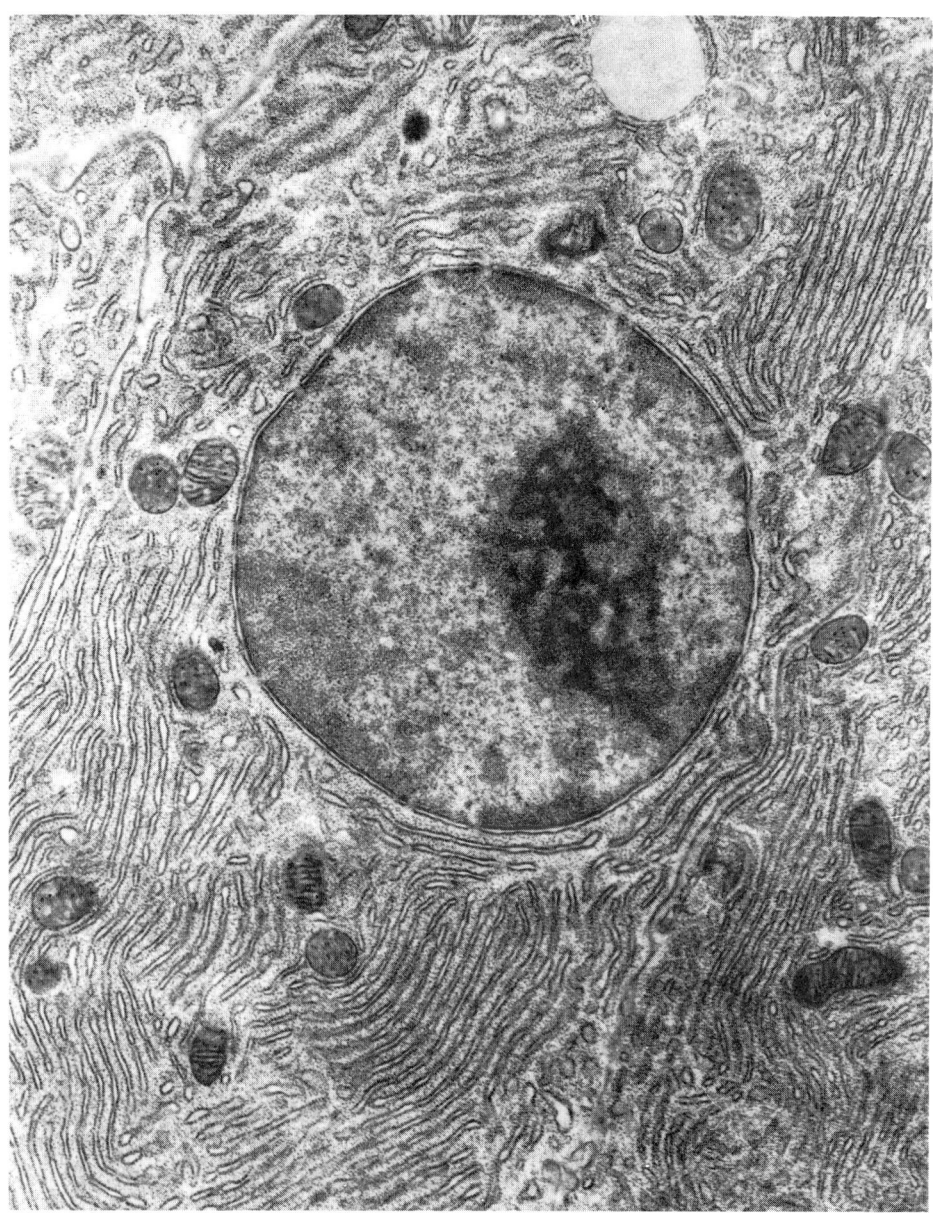

Fig. 32–3 An electron microscope photograph of the nucleus and part of the cytoplasm of a cell from the pancreas of a bat (magnified 18,000 times). The double-layered nuclear membrane, ribosomes lining channels that probably serve for secretion, and many mitochondria are clearly seen. [From John W. Kimball, *Biology*, Addison-Wesley, Reading, Mass., 1965. Photograph courtesy of Dr. Don W. Fawcett.]

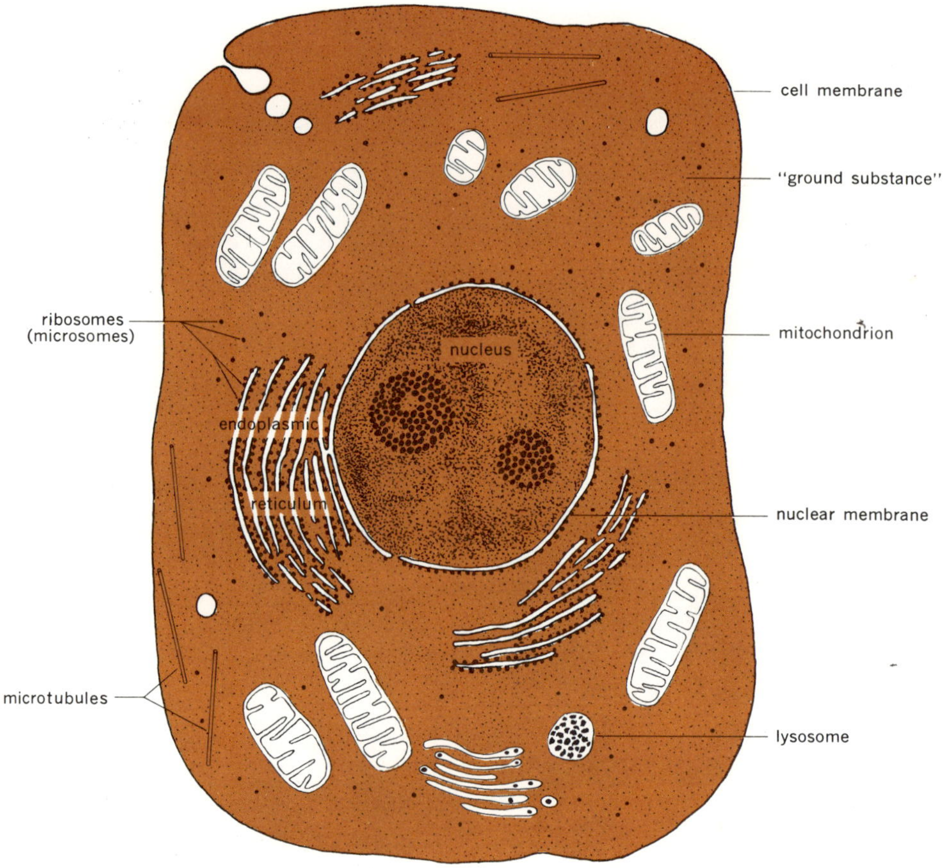

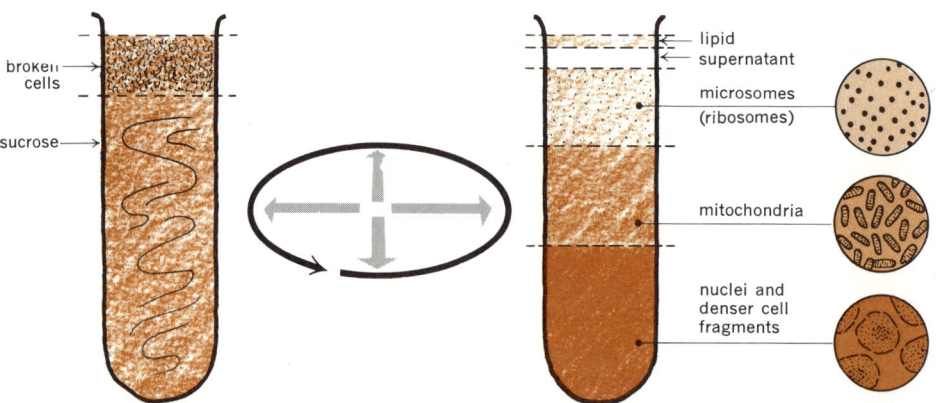

Fig. 32–4 Drawing of a typical animal cell, showing subcellular units (organelles) believed to be present, according to studies of electron micrographs. Not all these organelles are observed in every cell.

Fig. 32–5 Diagram showing separation of subcellular fractions by centrifugation. [Adapted from an illustration of a centrifuge by the Fisher Scientific Company.]

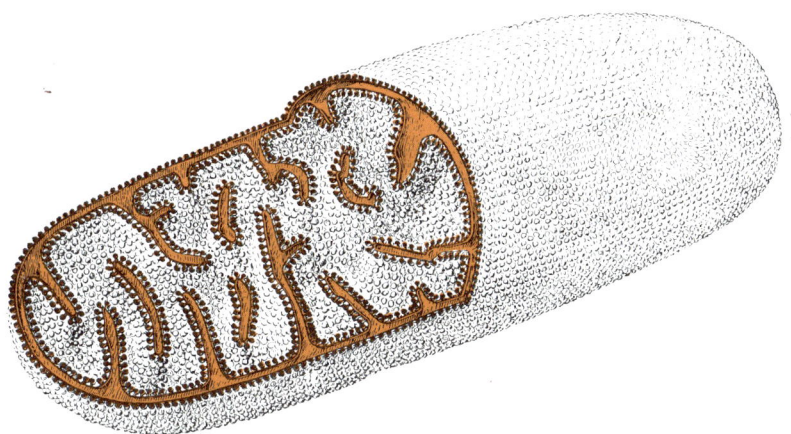

Fig. 32-6 Cross section drawing of a typical mitochondrion. This represents an idealized view based on electron micrographs. There are many small particles attached to the inner membrane. [From "The Mitochondrion" by David E. Green. Copyright © 1964 by Scientific American, Inc. All rights reserved.]

sequence on the cristae (inner membranes) of mitochondria. Thus a molecule may be metabolized by passing from one enzyme to another on a mitochondrial assembly or, perhaps more correctly, a disassembly line.

The name **microsomes** was given to a preparation isolated by centrifuging a cell homogenate at a high speed after the nuclei and mitochondria had been removed. The microsome fraction contains the ribosomes and some of the threads of material that tie these small spherical bodies together. The *ribosomes*, so named because they are rich in ribonucleic acids (RNA), are important in the synthesis of proteins by the cell.

By other techniques we can isolate **lysosomes,** which contain enzymes that catalyze the decomposition of many important compounds in the cell. It is believed that these "suicide bags" are ruptured after the death of a cell and are responsible for the process of self-destruction—*autolysis*—of dead cells. Articles further discussing these fascinating subcellular organelles are listed at the end of this chapter. In the following chapters, the involvement of these organelles in specific reactions or series of reactions will be given.

32-5 SUMMARY

In this chapter we have traced briefly the beginnings of biochemistry and have indicated some of the answers to the question, "What is biochemistry?" Another possible answer to this question might be, "Biochemistry is the things a biochemist does." While this is less satisfactory as a formal definition of biochemistry, perhaps it is the best one. For biochemistry is still a growing, developing science. What it will be a few years from now depends on what the biochemists and their colleagues who call themselves biophysicists, molecular biologists, or cytochemists find interesting enough to experiment with and to speculate about. Like the living things with which it deals, biochemistry is changing. If we would really understand it, we must know not only what it seems to be today, but what it has been and the direction in which it seems to be going.

Biochemistry is a fascinating science because it is the chemistry of living things—plants and animals. Most of us are primarily concerned about one particular kind of animal—the human. There are, of course, many ways of answering the question, "What is a human?" The biochemical answer presents a picture in which some areas are well-defined and often highly complex while others are still hidden by the mists of ignorance. Nevertheless, the whole picture makes a beautiful kind of sense. In the next several chapters we will try to show you this biochemical way of looking at life. It is not a view that can be really understood after a brief, superficial glimpse. Some concentration on details is demanded if you would eventually see at least a bit of the grand design of life as seen by the biochemist.

IMPORTANT TERMS

Anabolism	Fermentation	Metabolism
Biochemistry	In vitro	Microsome
Catabolism	In vivo	Mitochondrion
Chromatography	Isolation artifacts	Ribosome
Enzyme	Lysosome	

SUGGESTED READING

Biochemistry Texts. The reader may wish further details on many of the items discussed throughout the section on biochemistry. The following list represents a selection of basic biochemistry texts containing different degrees of detail.

Cheldelin, V. H., and R. W. Newburgh, *The Chemistry of Some Life Processes*, Reinhold, New York, 1964. A short, paperback guide to biochemistry.

Conn, E. E., and P. K. Stumpf, *Outlines of Biochemistry*, 3rd ed., John Wiley, New York, 1972. A brief textbook of biochemistry filled with formulas and equations.

Harrow, B., and A. Mazur, *Textbook of Biochemistry*, 10th ed., W. B. Saunders, Philadelphia, 1971. One of the standards among the shorter biochemistry texts. The present edition is up to date and well written.

Mahler, H. R., and E. H. Cordes, *Biological Chemistry*, Harper and Row, New York, 2nd ed., 1971. A comprehensive biochemistry text. A good, detailed basic reference.

White, A., P. Handler, and E. L. Smith, *Principles of Biochemistry*, 4th ed., McGraw-Hill, New York, 1968. A good, extensive biochemistry text written primarily for medical students.

Other Reading. The following references are both more general and, in some instances, more specific.

Asimov, I., *The Chemicals of Life*, Signet Science Library Book, New American Library, New York, 1962. A highly readable, elementary discussion of biochemistry by Mr. Asimov, a biochemist with many other interests.

Asimov, I., *A Short History of Biology*, Natural History Press, Garden City, New York, 1964. A well-written introduction to many biochemical topics.

Baker, J. J. W., and G. E. Allen, *The Study of Biology*, Addison-Wesley, Reading, Mass., 2nd ed., 1971. A modern presentation of biology for those who want additional background for their study of chemistry.

Baldwin, E., *An Introduction to Comparative Biochemistry*, 4th ed., Cambridge University Press, London, 1964. The fourth edition of a classic in the area of comparative biochemistry—the study of differences in biochemistry found in different kinds of animals.

Borek, E., *The Atoms Within Us*, Columbia University Press, New York, 1962. This amazing book contains an interesting discussion of biochemistry and biochemical principles without using a single structural formula.

Brachet, J., "The Living Cell," *Scientific American*, Sept. 1961 (Offprint #90). The key article in an issue devoted to the topic of cells.

Crowe, J. H., and A. F. Cooper, "Cryptobiosis," *Scientific American*, Dec. 1971. The ability of some small animals to survive in the absence of water, oxygen, and heat calls for a reevaluation of the differences between life and death. This article discusses the state of cryptobiosis, or "hidden life."

deDuve, G., "The Lysosome," *Scientific American*, May 1963 (Offprint #156). A discussion of a cell organelle which sometimes digests the cell itself.

Green, D., "The Mitochondrion," *Scientific American*, Jan. 1964 (Offprint #175). A discussion of the molecular architecture and function of this important cellular component.

Jevons, R., *The Biochemical Approach to Life*, Basic Books, New York, 1964. A well-written discussion of biochemistry, its relation to other sciences, and its significance.

Kamen, M. D., "Tracers," *Scientific American*, Feb. 1949 (Offprint #100). An introduction to the technique of using isotopes to trace biochemical processes.

Kennedy, D., *The Living Cell*, W. H. Freeman, San Francisco, 1965. The book includes twenty-four articles reprinted from the *Scientific American*, nine of them from the Sept., 1961, issue devoted to the cell.

Margulis, L., "Symbiosis and Evolution," *Scientific American*, Aug. 1971. Were mitochondria and chloroplasts once independent organisms? This fascinating theory is discussed in this well-illustrated article.

Morrison, J. H., *Functional Organelles*, Reinhold, New York, 1966. A readable but detailed discussion of mitochondria, chromosomes, and other subcellular organelles.

Scientific American Book, *The Physics and Chemistry of Life*, Simon and Schuster, New York, 1955. Although this book is a bit outdated, it includes many basic articles from the *Scientific American*.

Siekevitz, P., "Powerhouse of the Cell," *Scientific American*, July 1957 (Offprint #36). An older discussion of the mitochondrion.

Stein, W. H., and S. Moore, "Chromatography," *Scientific American*, Mar. 1951 (Offprint #81). A good introduction to technique that is so important in biochemistry today.

Zamecnik, P. C., "The Microsome," *Scientific American*, Mar. 1958 (Offprint #52). Somewhat outdated, but a basic discussion of this subcellular organelle.

CHEMISTRY OF THE CARBOHYDRATES

33

33–1 INTRODUCTION

When we heat a sugar in a test tube we observe several successive changes. The sugar melts and then begins to darken and finally, after prolonged heating, there is a black residue, which seems to be carbon, in the tube. Water condenses on the upper walls of the tube during the heating. From this experiment we might deduce that sugar can be decomposed into carbon and water, so that we could classify sugar as a hydrate of carbon or a carbohydrate. If we found a variety of compounds which behave similarly when heated, we could classify these compounds as carbohydrates. Chemical analysis of sugar and similar compounds would show that sugar and the other carbohydrates are composed only of carbon, hydrogen, and oxygen, and that their formulas can be expressed as $C_x \cdot (H_2O)_y$. In many cases x would be equal to y ($C_6H_{12}O_6$) while in others ($C_{12}H_{22}O_{11}$) accurate analysis would be required to show that x and y were not really equal.

Relatively early in the development of biochemistry as the result of experiments like those listed above, certain compounds were classified as carbohydrates. Further observations have shown, however, that it is inaccurate to describe those compounds we know as carbohydrates as "hydrates of carbon." Not only do the hydrogen and oxygen atoms of carbohydrates fail to exhibit the properties of water of hydration, but the way these compounds react leads to the conclusion that there are alcohol and either aldehyde or ketone functional groups in the carbohydrate molecule. The modern definition for the classification known as **carbohydrates** is that the compounds are polyhydroxy aldehydes or ketones. This definition is usually expanded to include polymers and other compounds derived from, or closely related to, the polyhydroxy aldehydes and ketones.

A nutritionist often describes the carbohydrates as the sugars and starches. This description is incomplete, however, since it does not include the many polysaccharides other than starch. Sugars, which biochemists usually define as sweet-tasting carbohydrates, are the simplest carbohydrates. Their names usually contain the suffix -*ose*. The starches, which may be defined as the complex, granular or powdery carbohydrates found in seeds, bulbs, and tubers of plants, represent only one group of complex carbohydrates—the polysaccharides—which includes glycogen, cellulose, the pectins, and many other compounds. Carbohydrates are usually classified as follows: (1) monosaccharides, (2) oligosaccharides, (3) polysaccharides, and (4) derived or related compounds.

33-2 MONOSACCHARIDES

The monosaccharides may be described as those carbohydrates that cannot be further degraded by hydrolysis. They contain a single chain of carbon atoms. Compounds containing three carbon atoms, the *trioses*, are the simplest monosaccharides.

The general names for the monosaccharides are formed from a prefix indicating the number of carbon atoms and the suffix *-ose* which we have already mentioned. Thus we have trioses, tetroses, pentoses, hexoses, and heptoses containing 3, 4, 5, 6, and 7 carbon atoms, respectively. Those monosaccharides which contain an aldehyde group are **aldoses,** those with a ketone group, **ketoses.** The indication that a sugar contains a ketone group may also be made by including the letters *-ul-* in the name of the compound. Thus we may speak of *aldopentoses*, *ketoheptoses*, and *heptuloses*, the latter two terms being synonymous.

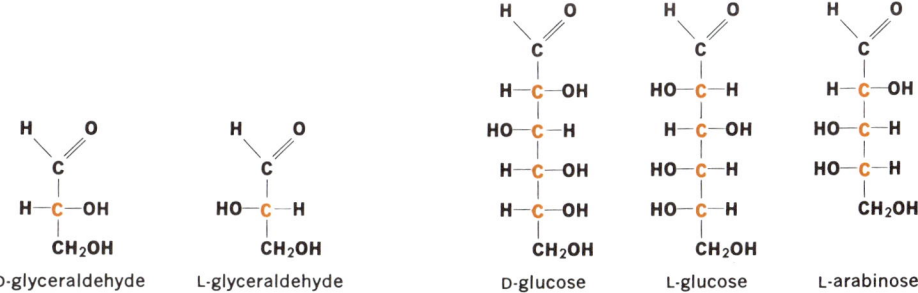

The two trioses are glyceraldehyde, which may exist in either of two optically active forms, and dihydroxyacetone. Glyceraldehyde is especially important because most of the common sugars may be considered related to it by addition of carbon atoms to lengthen the three-carbon chain, although biosynthesis of sugars does not follow this course. Sugars are designated as D- or L-sugars, depending upon whether the asymmetric carbon atoms farthest from the aldehyde or ketone group is similar to the D-form or the L-form of glyceraldehyde. In the structures shown below, asymmetric carbons are shown in color. (See Sections 31–5, 31–7.)

Although D-glyceraldehyde rotates plane-polarized light to the right (i.e., it is *dextrorotatory*) and L-glyceraldehyde rotates plane-polarized light to the left (hence it is *levorotatory*), compounds which are related to these parent compounds may have different rotations, since they usually contain other asymmetric carbon atoms. In the previous examples, glucose has four asymmetric carbon atoms and arabinose has three. The form of the ketohexose, *fructose*, which is related to D-glyceraldehyde and is thus D-fructose, rotates polarized light to the left. To designate this rotation of polarized light, we use the plus (+) sign for dextrorotatory compounds and the minus (−) sign for those compounds that are levorotatory. Thus fructose may be classified as D(−) fructose, the D indicating that it is related to D-glyceraldehyde, and the (−) indicating that it rotates plane-polarized light to the left.

The most abundant monosaccharides are the hexoses and pentoses. Since these compounds may contain several asymmetric carbon atoms, many isomers are possible. In general, these isomers have been given common or trivial names which often reflect the source or a property of the sugar. Structures of the more common monosaccharides are shown below. We indicate specific carbon atoms in the chain by numbers, the carbonyl group being number 1. In the case of fructose and other ketoses, we assign the number 1 to the carbon at the end of the chain nearest the carbonyl group.

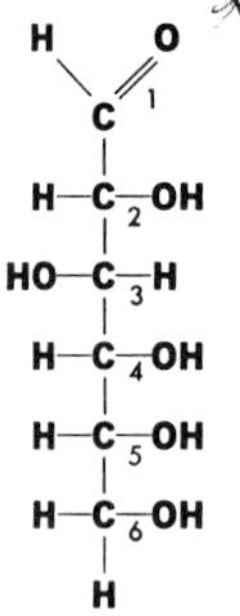

(hexoses)

D-glucose (dextrose)   D-fructose (levulose)   D-galactose 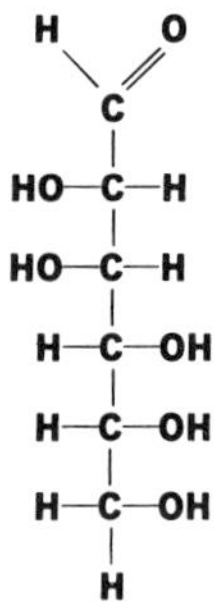   D-mannose

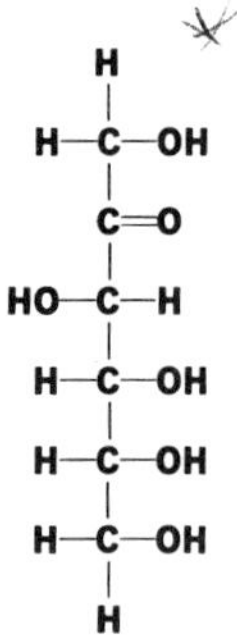

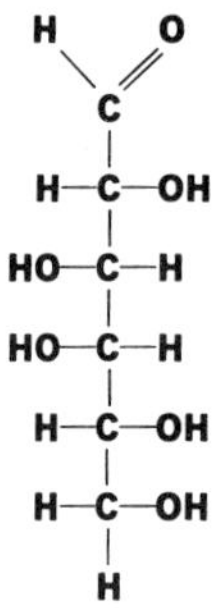

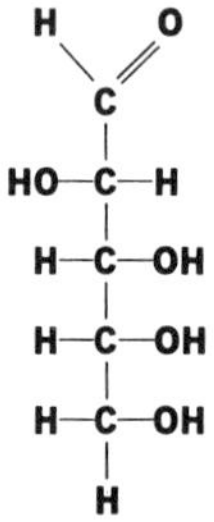

pentose

D-arabinose 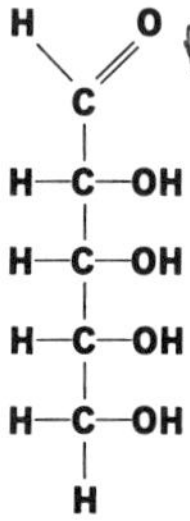   D-ribose 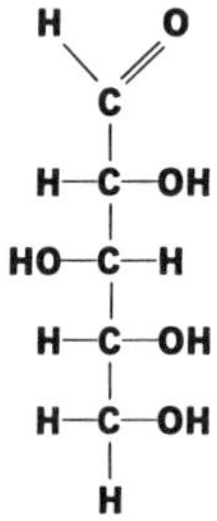   D-xylose

The name *glucose* which is derived from a Greek word for sweet wine, reflects the fact that this sugar was originally isolated from grapes. The alternative name, *dextrose*, resulted from the observation that the common form of this sugar rotates polarized light to the right. The name *fructose* is derived from the Latin word for fruit, *fructus;* and the other common name for this sugar, *levulose*, indicates its levorotatory property. *Galactose* (a component of the disaccharide lactose) is found in milk and owes its name to the Greek prefix *galact-*, meaning milk. *Lactose* is derived from the Latin prefix *lact-*, meaning milk. The name *mannose* reflects the fact that its related hexahydric alcohol, *mannitol*, was found in the dried juice of the manna tree. *Arabinose* is produced from gum arabic, an exudate from certain types of acacia trees. Apparently the name *ribose* was coined by selection and rearrangement of the letters in arabinose, which is appropriate, since the two are isomers. *Xylose* is found in woody parts of plants; its name is derived from the Greek word, *xylon*, meaning wood. The name *sucrose* is derived from the French word for sugar, *sucre*, and *maltose* is named for its source, malt, which is formed in germinating cereals. The word *saccharide* is derived from words for sugar that are similar in Arabic, medieval Latin, and Greek.

The formulas given so far for the pentoses and hexoses are called *open-chain formulas*. They were proposed by the great German chemist, Emil Fischer. Further study of the properties of these sugars, especially properties of sugars in solution, led to the conclusion that the carbonyl group of the sugars reacts with one of the alcohol groups. The reaction involves the shift of one hydrogen atom from an alcohol group to the carbonyl group, with the formation of a bond between the oxygen of the alcohol and the carbonyl carbon to give a ring structure. When it forms this ring structure the carbonyl carbon becomes asymmetric, and two isomers are possible. Rather than giving a totally new name to each form, we call these isomers the *alpha* (α) and *beta* (β) forms. These structures can be represented in either one or the other of the following ways.

*

β-D-glucose

α-D-glucose

or

β-D-glucose

α-D-glucose

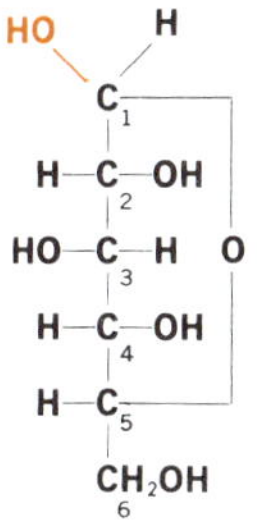

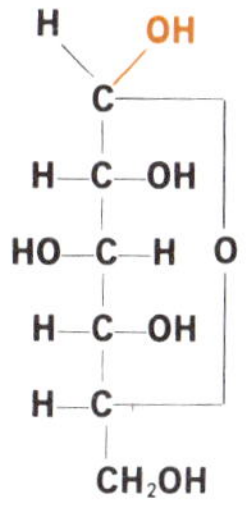

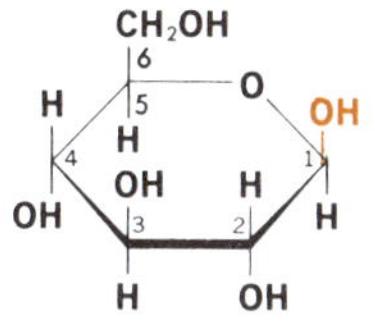

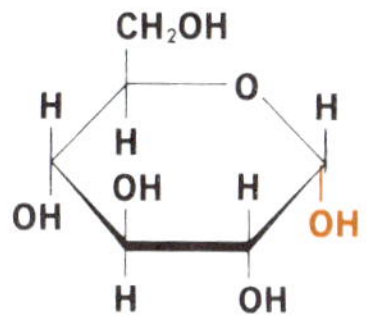

The two representations on the right are called *Haworth structures*. The most common forms of the monosaccharides (using the Haworth formulas) are

α-D-glucose

β-D-fructose

β-D-galactose

β-D-ribose

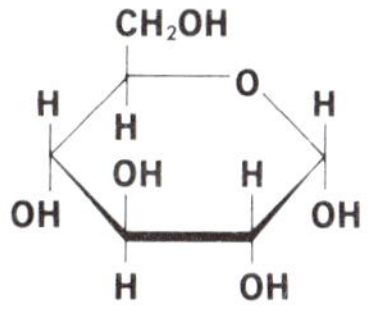

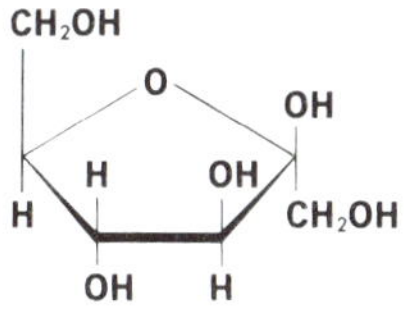

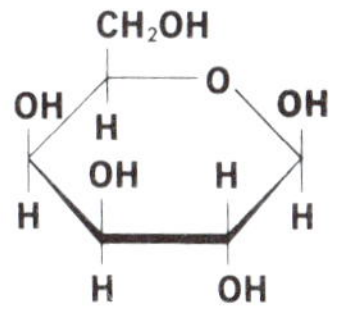

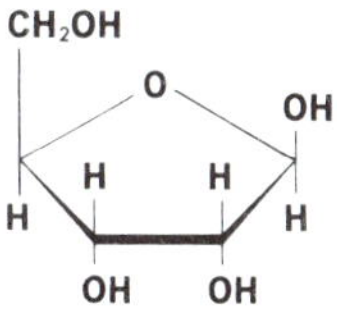

We have chosen, for reasons of simplicity, to use Fischer and Haworth formulas for carbohydrates. There is another method for writing such formulas in conformational structures. This method takes into account the fact that the structures of the carbohydrate ring are not flat but that two of the atoms are either above or below the other four. The formulas for some common sugars using these formulas are given below.

β-D-glucose

β-D-galactose

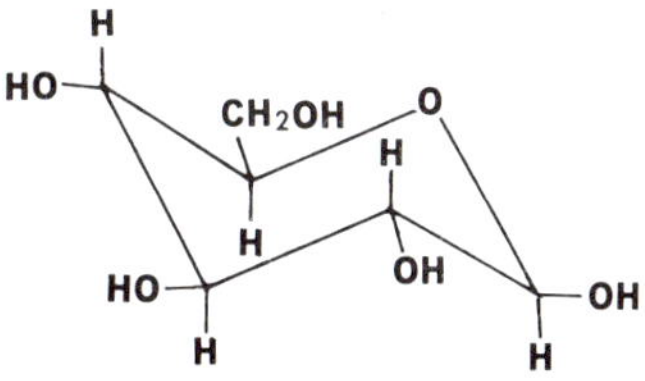

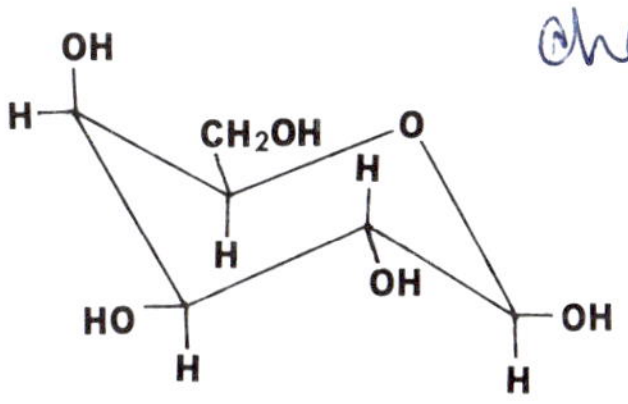

chair form – more favorable

as opposed to boat form which is less stable.

The monosaccharides as such are not found in any appreciable quantity in nature. A monosaccharide is usually found combined with other monosaccharides in the form of polymers called *polysaccharides*, such as starch or cellulose. In some cases, monosaccharides are combined with compounds other than carbohydrates, as ribose is in ribonucleic acid (RNA).

Many fruits, berries, and plant juices do, however, contain some free monosaccharides. Glucose was first found in grapes, which may contain as much as 27% glucose. However, most fruits contain much less glucose, and animals contain very little. The proportion of glucose in human blood, for example, is approximately 100 mg/100 ml. Even though the concentration is low, it must be maintained very near this level; persons with levels below normal go into *hypoglycemic* (low blood sugar) shock, and *hyperglycemia* (excess sugar in the blood) can result in coma.

Fructose, like glucose, is found in small quantities in many fruits. Honey contains relatively large amounts of both glucose and fructose. The other monosaccharides are generally found as components of polysaccharides.

Monosaccharides may be isolated from their natural sources by various physical methods. They are more commonly prepared by the hydrolysis of polysaccharides. Commercial preparation of glucose, for instance, is accomplished by the hydrolysis of starch. Although most of the monosaccharides have been synthesized in small amounts for theoretical purposes, such as the need to prove a certain structure or to establish the relationship of a given monosaccharide to other compounds, chemical synthesis is not commercially profitable.

Monosaccharides are colorless, crystalline solids. Crystallization is often difficult because of the presence of small amounts of impurities in the original solution. Since sugars are so similar to one another in chemical and physical properties, it is hard to separate them, and a small amount of fructose present as an impurity in a solution of glucose prevents formation of glucose crystals.

Solutions of sugars are viscous and have a sweet taste. While most of the common monosaccharides belong to the D-category, the actual (observed) rotation may be either to the right or to the left. Measuring the degree of rotation of plane-polarized light by a solution is a technique that may be used to find the amount of a sugar in the solution, provided only one sugar (or only one optically active compound) is present.

The chemical properties of monosaccharides are those of the aldehydes and the alcohols. Heat, catalysts, and strong acids, bases, or other reagents are usually required if we wish to observe these chemical properties in the laboratory.

Oxidation of Monosaccharides

Monosaccharides can be oxidized by hot alkaline solutions of certain metallic ions. In the process, the metallic ions are reduced. Since the sugars are the reducing agents, they are often referred to as **reducing sugars**. Such reactions are used to identify monosaccharides. Fehling's and Benedict's tests, for example, use an alkaline solution of copper(II). In both cases the test is positive if it results in a yellow or red compound of copper(I).

glucose + Cu^{2+} → oxidized glucose + Cu^+
(blue) (yellow-red)

These reactions can be adapted to test for the presence of glucose either in the urine or blood. A qualitative test is used on the urine of persons suspected to have diabetes.

Quantitative tests (ones that tell how much of a substance is present) based on the reducing properties of sugars are routinely used to determine the concentration of glucose in blood. Tollens' test uses a solution containing silver ions (Ag^+). A test is considered positive if a silver mirror forms on the wall of the test tube. In fact, before aluminum was used as the reflective substance for mirrors, a reaction similar to Tollens' test was used in the manufacture of mirrors.

Oxidations can convert glucose to various oxidization products such as those shown below.

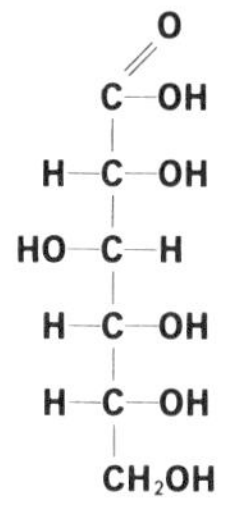

gluconic acid (Fischer formula) or (lactone form of gluconic acid) 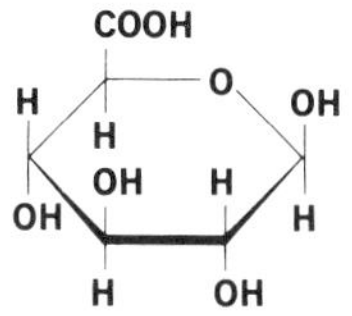 glucuronic acid β-form 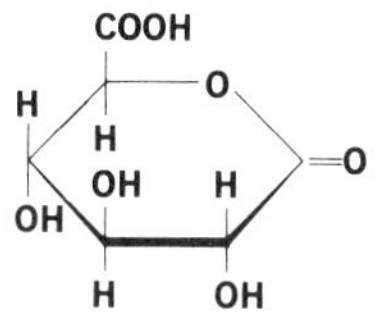 glucaric (saccharic) acid (lactone form)

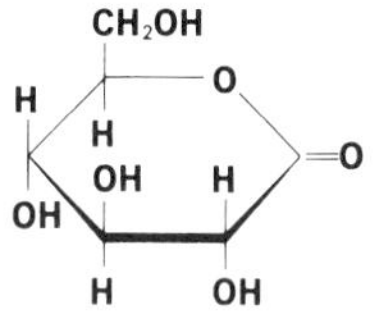

Oxidation processes which are catalyzed by enzymes yield products in which specific alcohol or aldehyde groups are oxidized. Although the examples above represent oxidation products of glucose, similar compounds result from the oxidation of other monosaccharides. Complete oxidation of monosaccharides yields CO_2 and H_2O.

Condensation of Monosaccharides

Two monosaccharide molecules can react to form a disaccharide.

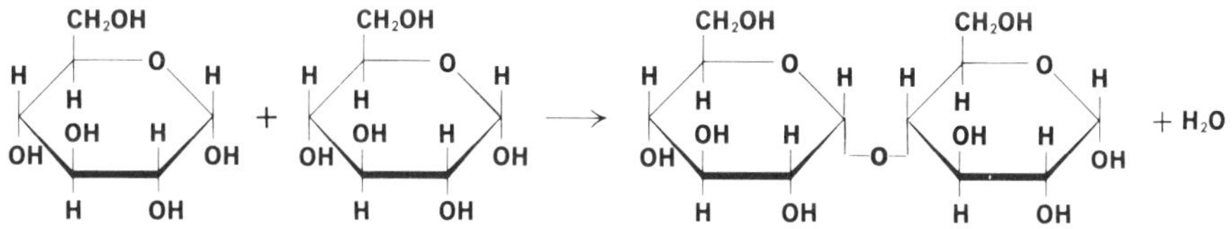

α-D-glucose + α-D-glucose → α-maltose + H_2O

The most common bonding in such sugars is from carbon number 1 of one unit to the carbon number 4 of the other, as shown above. However, 1-6, 1-3, and other linkages are possible. If the carbon number 1 is in the alpha form (as illustrated), the linkage between the units is an **alpha linkage**. **Beta linkages** are formed when the beta form of the monosaccharide reacts. The product of the reaction of two molecules of β-glucose is cellobiose.

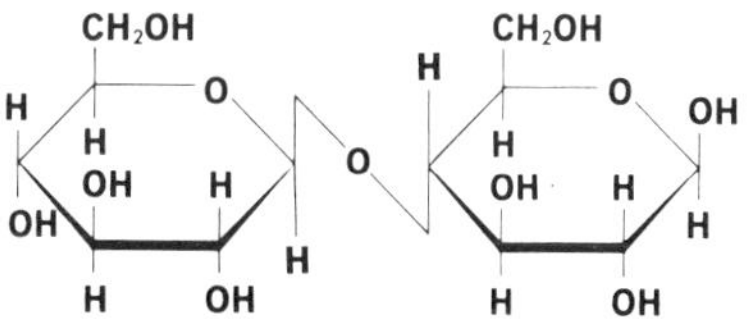

β-cellobiose

If this condensation process continues, trisaccharides, tetrasaccharides, and eventually, polysaccharides, may be formed.

Formation of Phosphate Esters

A structure of great biochemical importance is formed when either an alcohol or aldehyde group of a sugar is converted to a phosphate ester.

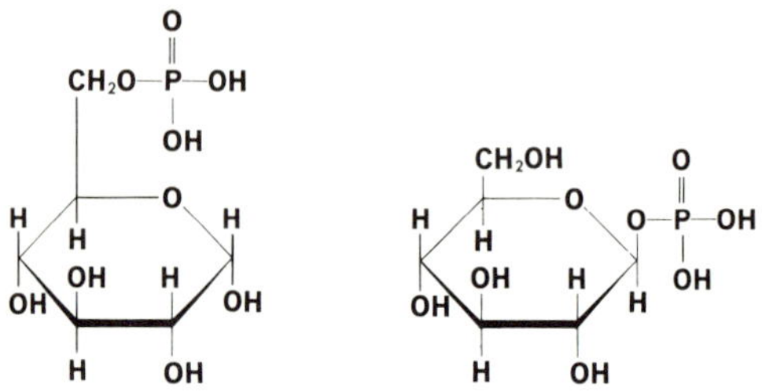

glucose-6-phosphate

glucose-1-phosphate

For simplicity, the phosphate group is often indicated by the symbol ⓟ as shown below.

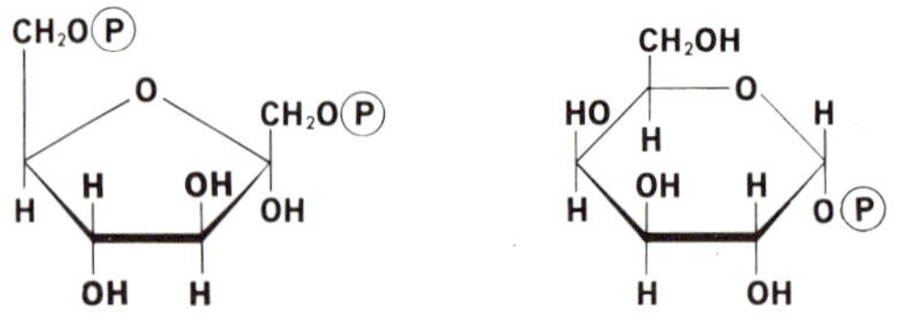

fructose-1,6-diphosphate

galactose-1-phosphate

In living organisms the source of the phosphate group in most such reactions is adenosine triphosphate (ATP) rather than phosphoric acid. In such reactions the ATP is converted to ADP (adenosine diphosphate).

glucose + ATP → glucose-6-phosphate + ADP

(See Chapter 36 for the formula for ATP.) The phosphorylation of glucose with ATP is probably the only important reaction of glucose in the human body.

33–3 COMPOUNDS RELATED TO MONOSACCHARIDES

Some compounds which are similar to carbohydrates do not fit the definition since they lack certain groups or contain others. Since these compounds are similar in structure to the monosaccharides and may occur as units in polymeric carbohydrates, they are usually included in a discussion of carbohydrates.

The compound 2-deoxyribose is found as a constituent of the deoxyribonucleic acids (DNA), which are important in the transmission of inherited characteristics. L-rhamnose and L-fucose are found in plants and animals, particularly in the cell walls of microorganisms. They are usually present as units of a polysaccharide.

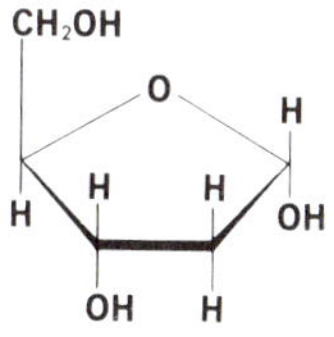

[Structures: 2-deoxy ribose, L-rhamnose, L-fucose, α-glucosamine]

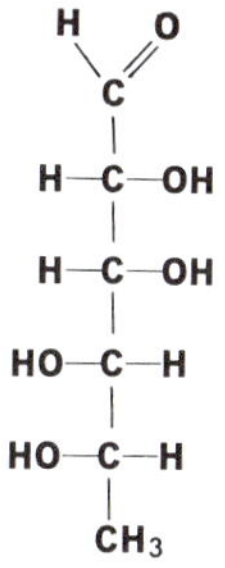

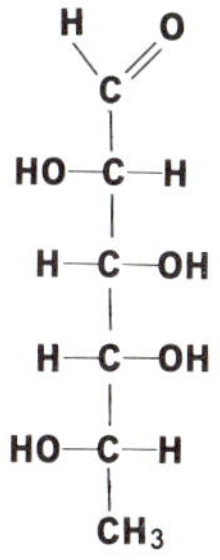

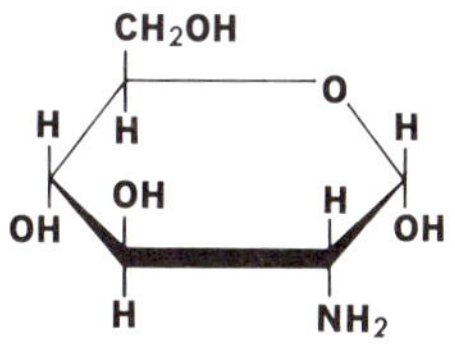

Glucosamine is a compound having an amino group in place of the alcohol group on carbon 2 of glucose. It is an important component of the polymer that makes up the tough outer covering (exoskeleton) of insects. Those compounds which can be regarded as oxidized monosaccharides, such as gluconic and galacturonic acids, are components of polymers which will be discussed in greater detail in the section on polysaccharides.

33-4 OLIGOSACCHARIDES — *series of dimers which hydrolyse & give monosacchar.*

Oligosaccharides are those carbohydrates which, when hydrolyzed, give two or more monosaccharide units. The prefix *oligo-* means few, and the upper limit of this classification is indefinite, but most authorities indicate that not more than ten or twelve monosaccharide units are found in oligosaccharides. While the number of possible oligosaccharides is great, only a few have been found in plants and animals. The most common oligosaccharides are the disaccharides.

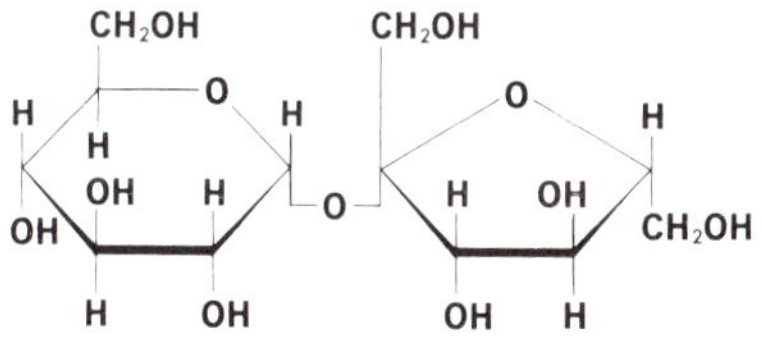

[Structures: sucrose, β-lactose, β-maltose]

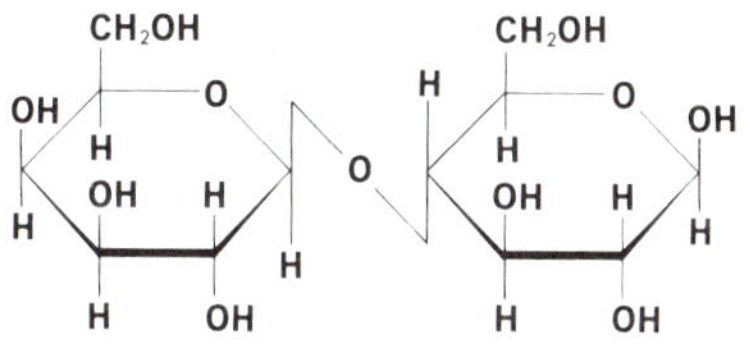

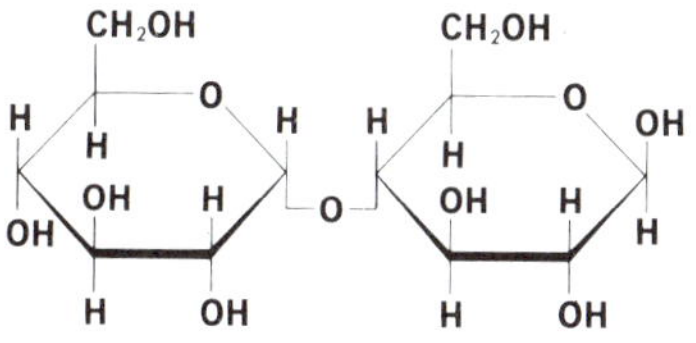

Sucrose is especially abundant in sugar beets, and in sugar and sorghum canes. It is also found in most common fruits and vegetables. Lactose is found primarily in milk. The milk from cows and goats is about 5% lactose; human milk contains from 6 to 7% of this sugar. The other disaccharides are not found in appreciable amounts in natural products.

While tri- and tetra-saccharides are apparently present in many plants, they have not been studied extensively, and the higher oligosaccharides are apparently quite rare.

The disaccharides are obtained commercially by isolation from natural products. Sucrose is commercially important and has the distinction of being one of the few chemically pure substances that can be obtained in a grocery store. Maltose is present, along with other substances, in corn syrup which is made by the hydrolysis of corn starch. Maltose is also found in germinating seeds where enzymes have produced it from the polysaccharide, starch.

Biosynthesis of the disaccharides involves the synthesis of monosaccharides and finally the coupling of two monosaccharides (or their phosphate esters) to give the disaccharides. Only small amounts of disaccharides have been made this way in the laboratory; as with monosaccharides, the synthesis is not commercially feasible.

TABLE 33-1 Solubility and relative sweetness of sugars

	Solubility (g/100 ml of H_2O)	Sweetness
sucrose	179	100
glucose	83	74
galactose	10	32
★ fructose	very soluble	173 *sweeter than sucrose.*
xylose	117	40
lactose	17	16
maltose	108	33
saccharin*	—	55,000

* Saccharin, which is not a sugar, is included for comparison.

Both the chemical and physical properties of the disaccharides are similar to those of the monosaccharides. Table 33-1 shows the solubility and relative sweetness of common sugars. The measurement of sweetness involves the use of human tasters who determine the dilution that can be made before a solution no longer tastes sweet. Because they tend to form supersaturated solutions, the exact solubility of sugars is difficult to measure. It is interesting to note that there is a correlation between solubility and sweetness.

Disaccharides can be hydrolyzed to give monosaccharides. This process is easily catalyzed by either strong acid or enzymes. The hydrolysis of sucrose is especially interesting. Since the aldehyde group of glucose and the ketone group of fructose are combined to make the bond between these two molecules in sucrose, there is no reducing group present, and sucrose does not reduce Fehling's or Benedict's reagents. After hydrolysis, however, molecules of glucose and fructose are present and the hydrolyzed solution is now "reducing." Since a mixture of fructose and glucose is actually sweeter than one of sucrose alone, hydrolysis of a solution of sucrose increases its sweetness.

Sucrose has an optical rotation of $+66.5°$; glucose has a rotation of $+52.7°$; and fructose, $-93°$. Consequently, when one molecule of sucrose is hydrolyzed to yield one each of glucose and fructose, the net rotation of the mixture is to the left. This change in rotation has been called inversion, and hydrolysis of sucrose is sometimes called **inversion of sucrose.** An equimolar mixture of glucose and fructose is called *invert sugar;* the enzyme which catalyzes this hydrolysis is often referred to as *invertase* instead of by its systematic name, *sucrase*.

Food preparation or cooking that involves exposing sucrose solutions to high temperatures under acid conditions produces hydrolysis. While this may make the food concerned a bit sweeter, it is more likely that the real purpose of the cooking is to produce a mixture of sugars. Such a mixture is less likely to crystallize (as when a jelly "sugars") than a pure solution.

Sucrose is commercially the most important of the disaccharides. It is an important item of world trade because of its desirability as a sweetening agent. In spite of the use of synthetic noncaloric sweeteners, each person in the United States consumes an average of about 170 pounds of sucrose per year. Although other countries consume less sucrose, more than 35 million tons are produced in the world each year. Approximately one-third of this sugar comes from beets and two-thirds from cane.

Lactose is found in the milk of mammals. It can be purified and has some commercial uses but has never been used as widely as a sweetener because of its low solubility and because it is not nearly so sweet as sucrose.

33-5 POLYSACCHARIDES

The polysaccharides, which contain many monosaccharide units linked together, are the most abundant carbohydrates. It is difficult to determine experimentally exactly how many such units there are in a polysaccharide chain. Molecular weights ranging from several thousand to values in the millions have been reported for various polysaccharides. It is generally assumed that the higher molecular weights represent the true molecular weight of the polysaccharide as it occurs in the plant, and that the lower weights are caused by degradation of the polymers during purification. However, we cannot totally disregard the possibility that some polymerization occurs during the purification process. Probably the best conclusion is that the number of units may vary somewhat without affecting the overall properties of the polysaccharide. There is a possibility that the molecular weight of a polysaccharide is characteristic of the species of plant or animal in which it is found.

One difference in structure which is of great biological significance is the one between the alpha- and beta-linked polysaccharides. Those polysaccharides containing the alpha-linked monosaccharide units are much more digestible than β-linked polymers. Animals use α-linked starch as a major source of food, but only a few animals can digest the β-linked cellulose. Actually the cellulose-utilizing animals, such as cattle and sheep, and even the lowly termite, are able to use this β-linked polymer only because of microorganisms in their digestive tract which actually perform the hydrolysis of cellulose to glucose for them.

Compare the structures of starch and cellulose, shown on page 506.

The polysaccharides make up between 60% and 90% of the dry weight of plants, where they serve both as structural materials and as reserve food supplies. They are found in much smaller amounts, usually less than 1% of the total weight, in animals. Typical polysaccharides are starch, which may be obtained from the seeds and tubers of a variety of plants; cellulose, the structural material of plants; and glycogen, which is found in small amounts in animals.

The polysaccharides are noncrystalline, white solids which are only slightly soluble in water. In contrast to the mono- and oligosaccharides, the polysaccharides are not sweet-

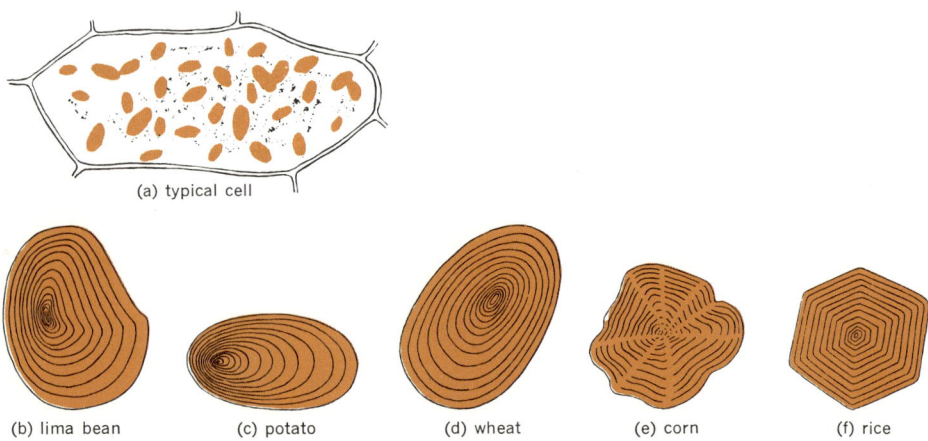

starch (α-linked)

cellulose (β-linked)

tasting. Nor are the polysaccharides very reactive chemically, both because of their insolubility and the fact that their most active groups (the aldehyde and ketone groups) are tied up in the linkage between monosaccharide units. Strong acids or enzymes are required to catalyze the hydrolysis of polysaccharides to give monosaccharide units.

Polysaccharides composed of pentose units are called **pentosans.** They are found in wood, straw, and leaves of plants. The gums from some trees are the richest sources of relatively pure pentosans. Pentosans are not very useful commercially since most animals cannot digest them; however, chemical degradation can give furfural which is used commercially for the manufacture of nylon and in the refining of petroleum.

The most important and abundant polysaccharides are polymers of the hexoses and are called **hexosans.** Glucose is the only monomeric unit of most of the abundant and important polysaccharides.

Fig. 33–1 Characteristic starch grains. Starch grains are scattered throughout the cytoplasm of many plant cells. (a) In the sketch of a typical cell the starch grains are indicated as colored bodies. Drawings of enlarged starch grains of typical plants are shown in (b) through (f). [Adapted from *Principles of Plant Physiology*, by James Bonner and Arthur W. Galston. W. H. Freeman and Company. Copyright © 1952.]

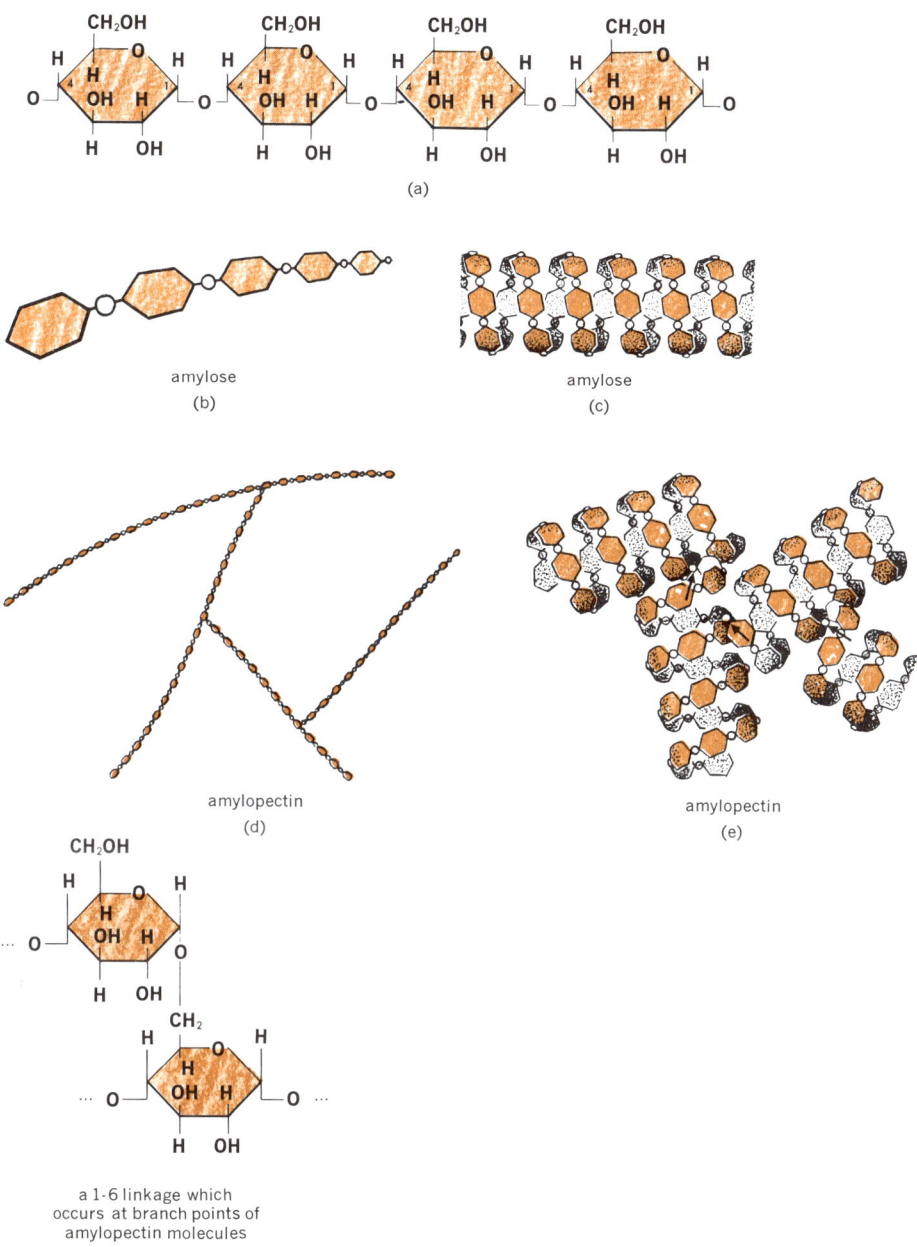

Fig. 33-2 The structure of the amylose and amylopectin components of starch. The structure of amylose can be represented as (a) or as simplified in (b). The three-dimensional structure of amylose is given in (c). A simplified representation of a part of a molecule of amylopectin is shown in (d), and a three-dimensional diagram of amylopectin is represented in (e), (f) shows the 1-6 linkage which occurs at branch points of amylopectin molecules. [Adapted from *Principles of Plant Physiology*, by James Bonner and Arthur W. Galston. W. H. Freeman and Company. Copyright © 1952.]

Starch

Various kinds of starch are found in the cereal grains and in many kinds of tubers. The starch from each type of plant has a characteristic form of granule. Experienced persons can tell the source of the starch by examining a sample microscopically (see Fig. 33–1). Chemically, all forms of starch are quite similar; all are composed totally of glucose. Most sources of starch yield two fractions, amylose and amylopectin (Fig. 33–2). These two forms are very similar and separation is difficult. While both kinds of starch contain alpha 1-4 links between monosaccharide units, the amylopectin molecules contain many more glucose units, are highly branched (1-6 linkages), and are somewhat more soluble in water.

The starches in the form of rice, potatoes, and wheat or cereal products supply about 70% of the world's food. In many cultures starch makes up much more than 70% of the diet, since most highly starchy foods are relatively inexpensive to buy or relatively easy to grow.

Glycogen

Glycogen is found in the liver and muscles of animals, and it is often called "animal starch." Glycogen is similar to the amylopectin fraction of starch in that it is made totally of glucose and is highly branched. Since animals do not synthesize starch, glycogen represents the only form in which they can store excess glucose. The total amount of glycogen in an animal is small; about 1% of the muscle and between 1.5 and 4% of the liver is glycogen. However, this small amount is extremely important in maintaining the correct level of glucose in the blood and in supplying the muscles with a source of energy.

Cellulose

Cellulose is the most abundant organic compound. It has been estimated that 50% of all carbon atoms in the vegetation on the earth is tied up in cellulose molecules. Like starch and glycogen, cellulose is made exclusively of glucose units, but unlike them, the glucose units are β-linked. This β-linked polymer of glucose is extremely tough and indigestible, and for this reason is an effective structural material both within plants and, in turn, for man who uses many materials which contain cellulose. The woody part of plants contain large amounts of cellulose. Cotton is 98% cellulose. Cellulose-containing materials such as lumber, cotton, and rayon (which is a chemically-modified cellulose) are important commercially.

Chitin

The tough outer covering of many insects and of crustaceans like crabs and lobsters is made of another β-linked polymer. In this instance, however, the monomeric unit is not glucose but glucosamine, and the substance is called chitin. Complete hydrolysis of chitin yields both glucosamine and acetic acid, and apparently the monomeric structural unit is N-acetyl glucosamine.

33–6 HETEROPOLYSACCHARIDES

Those polysaccharides discussed so far have been made of a single kind of monosaccharide. There are other polysaccharides which usually contain two different monosaccharides, and occasionally more than two monomeric units. These are the heteropolysaccharides. In many cases the linkage seems to be from the number 1 carbon of one unit to the 2, 3, or 6 carbon of the other unit, instead of the more common 1-4 linkage found in starch, glycogen, and cellulose. These materials are now the subject of a great deal of study, because they seem to play important roles in the chemistry of living things. Among the more important heteropolysaccharides are the pectic acids, which are polymers of the uronic acids combined with other substances. They are responsible for the solidity of many fruits, and are used commercially in making jams and jellies. Another group comprises the mucopolysaccharides which are usually found associated with proteins. The principal products obtained following hydrolysis of these polysaccharides are the amino sugars and the "uronic" acids. Some examples of substances which contain mucopolysaccharides are heparin, a naturally occurring anticoagulant of blood; cartilage; and hyaluronic acid which has been called the "cement" or ground substance between cells.

33–7 SUMMARY

The carbohydrates are found primarily in polymeric form in the structural parts of plants. Although the basic monosaccharide unit contains the relatively active aldehyde or ketone group, these groups are involved in linking the monosaccharide units into the polymeric form and do not influence the chemical properties of the polysaccharides to any extent. The way in which monosaccharides are linked determines not only the shape of the polymer, but also influences the possibility of hydrolyses that are catalyzed by enzymes. The β-linked polysaccharides, particularly cellulose, are insoluble, tough polymers that are not attacked by digestive enzymes of animals.

The more active, relatively soluble monosaccharides function as intermediate compounds in the breakdown and synthesis of polysaccharides. The monosaccharides formed from the hydrolysis of polysaccharides are a major source of energy for the life of all animals. The concentration of glucose in the blood of animals is of major significance. Without the monosaccharide, glucose, life as we know it would not be possible.

IMPORTANT TERMS

Aldose	Hexose	Oligosaccharide	Sugar
Carbohydrate	Hyperglycemia	Pentose	
Cellulose	Hypoglycemia	Polysaccharide	
Glycogen	Ketose	Starch	

WORK EXERCISES

1. Define the following terms: ketotetrose, triose, aldopentose, octulose, deoxysugar.
2. Draw the structural formulas for (a) all possible aldopentoses (indicate asymmetric carbon atoms with asterisks); (b) L-glucose; (c) a trisaccharide; (d) two galactose molecules with a β-1-4 linkage between them; (e) a molecule of glucosamine linked β-1-3 to a molecule of β-glucuronic acid.
3. Explain what is meant by (a) D (−) fructose, (b) a reducing sugar, (c) a β-linkage between monosaccharides.

4. Using structural formulas write equations for
   a) the hydrolysis of sucrose,
   b) the complete oxidation of glucose, of gluconic acid, of glucaric (saccharic) acid.
5. How is a fruit jelly that has been boiled for ten minutes different from the uncooked mixture of ingredients?
6. What is the nutritional significance of the β-linkage between the units of a polysaccharide?
7. What reason might be deduced for the fact that starch does not taste sweet?
8. Benedict's test is used to see whether glucose is present in the urine of persons with *diabetes mellitus*. During the period of time a mother produces milk after the birth of a child, she may excrete lactose or galactose in the urine. Would urine containing either lactose or galactose give a positive Benedict's test?
9. Consult a general biochemistry text to find the following:
   a) the formulas for mucopolysaccharides not given in the previous chapter,
   b) the names and formulas of hexoses other than glucose, galactose, mannose, and fructose.

CHEMISTRY OF THE PROTEINS

34

34-1 INTRODUCTION

Many important parts of the body are largely made of protein. Muscle, blood, and the nonmineral part of bone are protein. The enzymes, those vital catalysts without which biological reactions would be too slow to support life, are also proteins.

The first definition of proteins was given in Gerardus Johannes Mulder's *General Physiological Chemistry*, published in 1844-1851:

"There is present in plants as well as animals a substance which ... performs an important function in both. It is one of the very complex substances, which under various circumstances may alter their composition, and serves ... for the regulation of chemical metabolism ... It is without doubt the most important of the known components of living matter, and it would appear that, without it, life would not be possible. This substance has been named protein."*

While Mulder did basic research on the compounds that he finally called **protein,** apparently the name (derived from the Greek word *proteios*, which means of the first rank or importance) was suggested to him in 1838 by the great Swedish chemist, J. J. Berzelius.

Later, more precise studies of the substance Mulder called protein indicated that it was not just one substance as he and others believed, but that it was composed of many substances, each differing slightly from the others. When faced with complexities of this sort, scientists can discard the classification, but they usually try to save it by separating and classifying the individual members of the group. By 1907 the British Physiological Society proposed a classification of proteins based on solubility. The new classification included

1) albumins: proteins soluble in water and salt solutions;
2) globulins: proteins slightly soluble in water, soluble in dilute salt solutions;
3) prolamines: proteins insoluble in water, but soluble in 70-80% ethanol;
4) glutelins: proteins insoluble in water and ethanol, but soluble in acid or base;
5) scleroproteins: proteins insoluble in aqueous solvents.

---
* Cited from Fruton and Simmonds, *General Biochemistry*, 2nd ed., New York, Wiley, 1958.

Although other classifications have been added to this list, it is still the most common method for classification of proteins. However, only the first two names, albumin and globulin are commonly used today.

While classification of proteins by solubility was relatively easy and provided the basis for a simple classification, it did not give real insight into the nature of proteins. Elemental analysis had shown that proteins contained nitrogen. This knowledge was useful in describing proteins but was not adequate for defining them, since many other biochemical substances also contained nitrogen. Modern biochemists know that the formula for lactoglobulin, the globulin obtained from milk, is approximately $C_{1864}H_{3012}N_{468}S_{21}O_{576}$. Since this is a representative protein, not really one of the larger ones, it is easy to see that writing the formula for a protein is both cumbersome and probably useless as a device for describing or understanding the nature of these compounds.

How, then, can we characterize the proteins? We saw that the hydrolysis of carbohydrates gives certain basic building blocks, the monosaccharides. What happens when we hydrolyze proteins? When hydrolysis is done under acid conditions we get a mixture of about twenty different amino acids. All proteins so treated give a similar mixture. Some proteins give other substances also, but they *all* give amino acids. Here, then, is the common characteristic of proteins—amino-acid building blocks. Our current definition of proteins is that *they are high-molecular weight substances which, upon hydrolysis, yield amino acids.* Any attempt to understand the nature of proteins, therefore, must begin with a discussion of the amino acids.

34–2 AMINO ACIDS

The amino acids, as their name implies, are organic compounds which contain both an acid and an amino group. There are hundreds of possible amino acids, but only twenty are found in appreciable quantities in proteins. These are all α-amino acids—that is, the amino group is on the α carbon—the carbon next to the carboxyl carbon. In the case of proline and hydroxyproline, the α amino group is in a ring structure; these are more correctly called imino acids. A list of the formulas and names of some of the common amino acids is given in Fig. 34–1.

> Except for those amino acids containing an additional acid group which are named as acids, the names of amino acids end in the suffix -ine.
>
> Cystine was first found in bladder stones in 1810. Its name is derived from the Greek *kystis*, meaning a pouch or bladder. Cystine is one of the least soluble amino acids and is still occasionally found in the form of stones in the urinary bladder.
>
> The simplest amino acid, glycine, was one of the first to be isolated by hydrolysis of proteins. Since it has a slightly sweet taste, it was named from the Greek word for sweet, *glykys*. When leucine was isolated as white, glistening plates, its name was chosen as a derivative of *leukos*, the Greek word for white. Isoleucine was named for its structural analogy to leucine.

Fig. 34–1 (pp. 513–517) The common amino acids. The names, symbols, R-groups, and a photograph of a Courtauld space-filling model are given for the common amino acids. The R-groups are indicated in color. [Adapted from Snell, Shulman, Spencer, and Moos, *Biophysical Principles of Structure and Function*, Addison-Wesley, 1965.]

Amino Acids

Name	Symbol	R-group	Model
Glycine	Gly	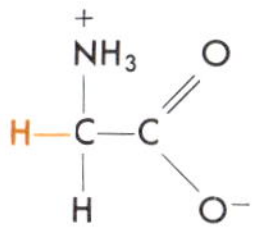 $\overset{+}{N}H_3$ / H—C—C(=O)O$^-$ / H	
Alanine	Ala	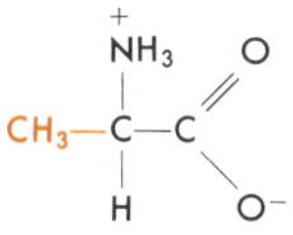 $\overset{+}{N}H_3$ / CH$_3$—C—C(=O)O$^-$ / H	
Valine	Val	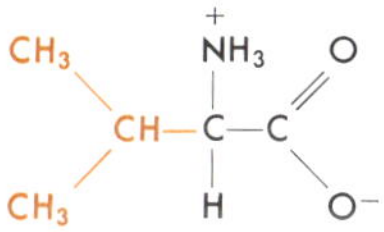 CH$_3$—CH(CH$_3$)—C($\overset{+}{N}H_3$)(H)—C(=O)O$^-$	
Leucine	Leu	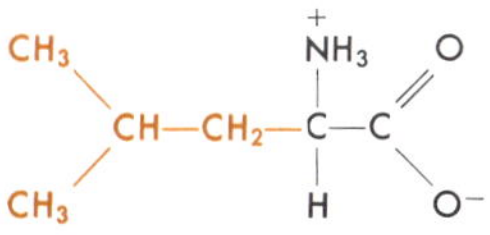 (CH$_3$)$_2$CH—CH$_2$—C($\overset{+}{N}H_3$)(H)—C(=O)O$^-$	
Isoleucine	Ileu or Ile	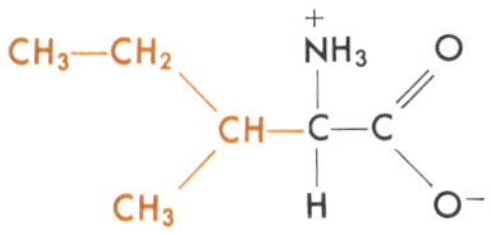 CH$_3$—CH$_2$—CH(CH$_3$)—C($\overset{+}{N}H_3$)(H)—C(=O)O$^-$	

Name	Symbol	R-group	Model
Serine	Ser	HO—CH₂—C(NH₃⁺)(H)—C(=O)O⁻	
Threonine	Thr	HO—CH(CH₃)—C(NH₃⁺)(H)—C(=O)O⁻	
Phenylalanine	Phe	C₆H₅—CH₂—C(NH₃⁺)(H)—C(=O)O⁻	
Tyrosine	Tyr	HO—C₆H₄—CH₂—C(NH₃⁺)(H)—C(=O)O⁻	
Tryptophan	Try	(indole)—CH₂—C(NH₃⁺)(H)—C(=O)O⁻	

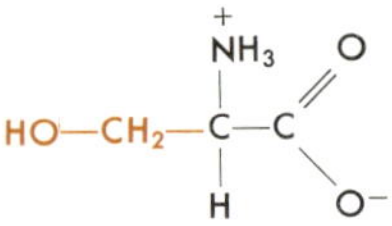

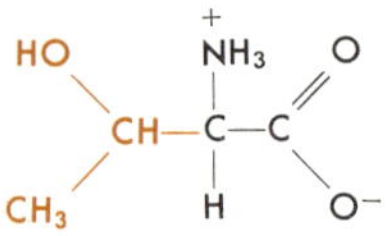

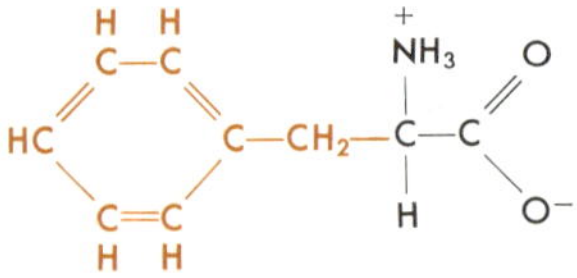

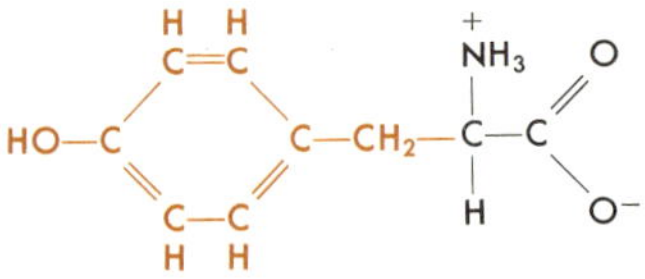

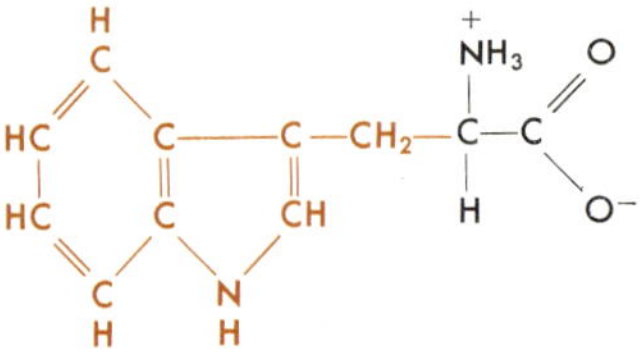

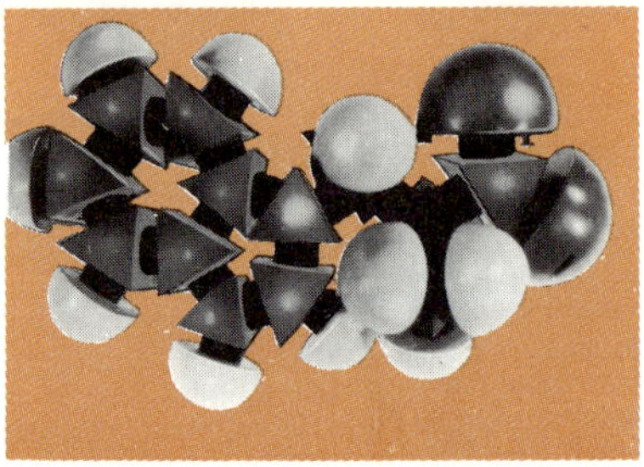

Amino Acids

Name	Symbol	R-group	Model
Cystine	$(CyS)_2$	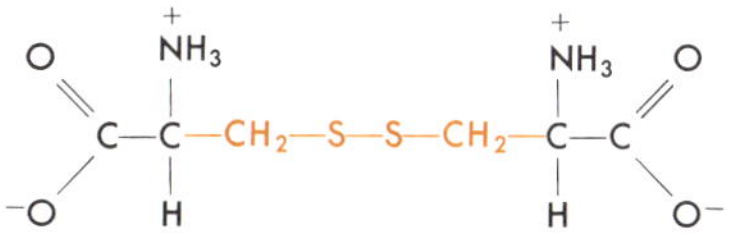 $\overset{O}{\underset{O^-}{\|}}C-\underset{H}{\overset{\overset{+}{NH_3}}{C}}-CH_2-S-S-CH_2-\underset{H}{\overset{\overset{+}{NH_3}}{C}}-\overset{O}{\underset{O^-}{\|}}C$	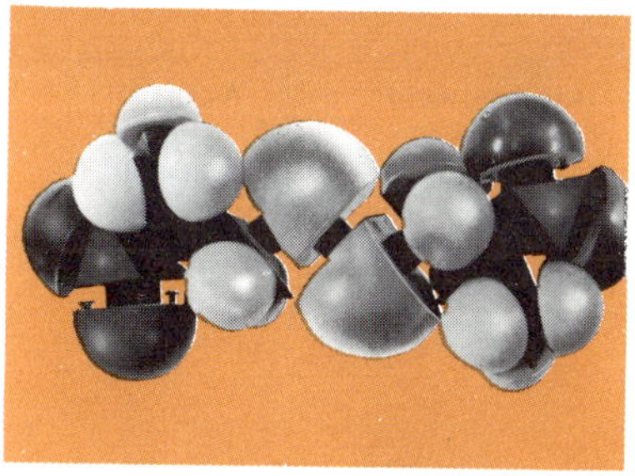
Cysteine	CySH	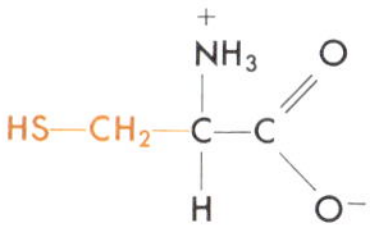 $HS-CH_2-\underset{H}{\overset{\overset{+}{NH_3}}{C}}-\overset{O}{\underset{O^-}{\|}}C$	
Methionine	Met	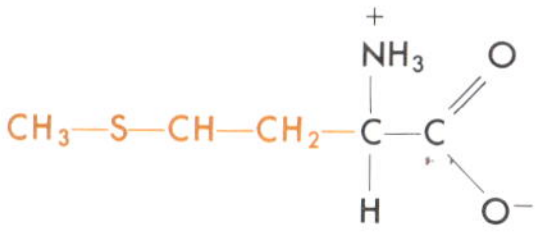 $CH_3-S-CH_2-CH_2-\underset{H}{\overset{\overset{+}{NH_3}}{C}}-\overset{O}{\underset{O^-}{\|}}C$	
Proline	Pro	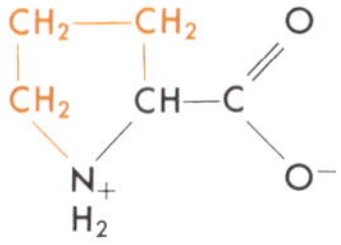 $\begin{array}{c} CH_2-CH_2 \\ \| \quad \quad \| \\ CH_2 \quad CH-C{\overset{O}{\underset{O^-}{}}} \\ \diagdown \diagup \\ \overset{+}{N} \\ H_2 \end{array}$	
Hydroxyproline	Hyp	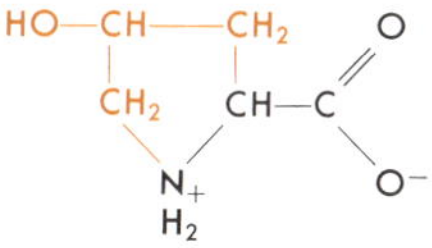 $\begin{array}{c} HO-CH-CH_2 \\ \| \quad \quad \| \\ CH_2 \quad CH-C{\overset{O}{\underset{O^-}{}}} \\ \diagdown \diagup \\ \overset{+}{N} \\ H_2 \end{array}$	

Name	Symbol	R-group	Model
Aspartic Acid	Asp	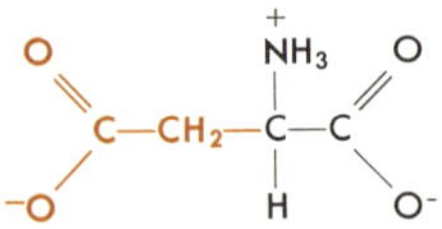 $\overset{O}{\underset{O^-}{\|}}C-CH_2-\overset{\overset{+}{NH_3}}{\underset{H}{C}}-\overset{O}{\underset{O^-}{\|}}C$	
Glutamic Acid	Glu	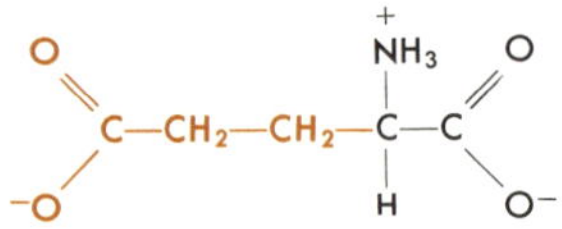 $\overset{O}{\underset{O^-}{\|}}C-CH_2-CH_2-\overset{\overset{+}{NH_3}}{\underset{H}{C}}-\overset{O}{\underset{O^-}{\|}}C$	
Asparagine	Asp-NH$_2$ or Asn	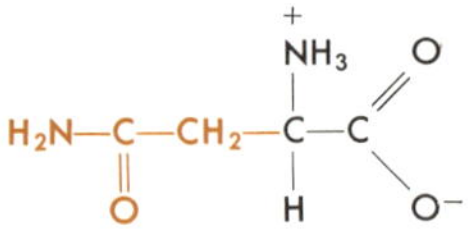 $H_2N-\overset{O}{\underset{\|}{C}}-CH_2-\overset{\overset{+}{NH_3}}{\underset{H}{C}}-\overset{O}{\underset{O^-}{\|}}C$	
Glutamine	Glu-NH$_2$ or Gln	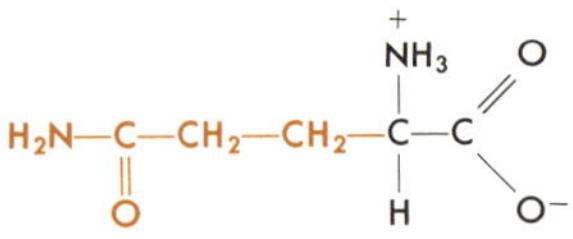 $H_2N-\overset{O}{\underset{\|}{C}}-CH_2-CH_2-\overset{\overset{+}{NH_3}}{\underset{H}{C}}-\overset{O}{\underset{O^-}{\|}}C$	
Histidine	His	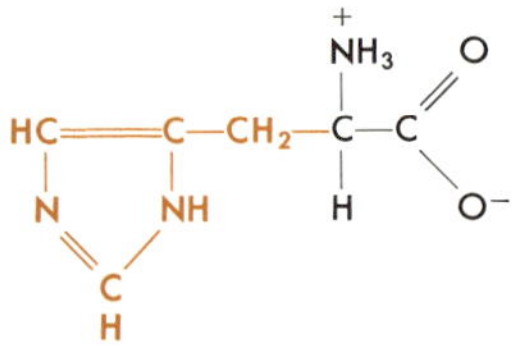 imidazole-$CH_2-\overset{\overset{+}{NH_3}}{\underset{H}{C}}-\overset{O}{\underset{O^-}{\|}}C$	

Name	Symbol	R-group	Model
Arginine	Arg	$\overset{+}{C}-NH-CH_2-CH_2-CH_2-\underset{H}{\overset{NH_3}{\overset{|}{C}}}-\overset{O}{\underset{O^-}{C}}$	
Lysine	Lys	$H_3\overset{+}{N}-CH_2-CH_2-CH_2-CH_2-\underset{H}{\overset{NH_3}{\overset{|}{C}}}-\overset{O}{\underset{O^-}{C}}$	

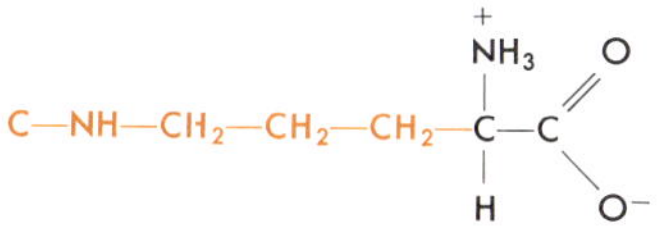

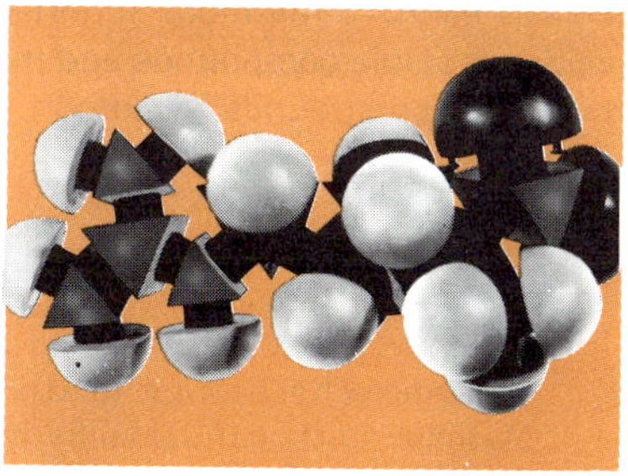

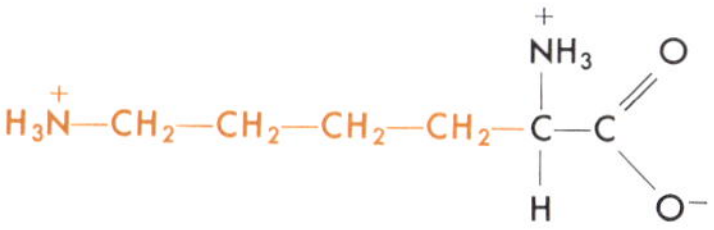

The silk protein sericin (from the Latin word for silk) yields large amounts of serine. Cheese (*tyros* in Greek) was the first source of tyrosine. The amide asparagine is found in asparagus, and its hydrolysis product is aspartic acid. Arginine was named because it forms silver (L., *argentum*) salts. Proline derives its name from its chemical similarity to pyrrolidine; valine is named for its similarity to valeric (pentanoic) acid; and glutamic acid for its similarity to the dicarboxylic acid, glutaric acid.

Although small quantities of amino acids are found in the blood after the digestion of a protein-containing meal, they are found primarily in the form of their polymers, the proteins. It is from the naturally occurring proteins that commercial quantities of amino acids are usually produced. Plants can synthesize amino acids from ammonia and carbon dioxide. Animals, however, have to use larger units for making amino acids, and usually take an amino group from one compound and transfer it to a keto-acid. Green plants can synthesize all the amino acids they require for the production of protein, but most animals can make only approximately half of the amino acids they require. Consequently, they must get the others from their diet to build the proteins of muscles, enzymes, and blood components.

An interesting, but commercially unfeasible, process for synthesizing amino acids was discovered by Miller and Urey, working at the University of Chicago. They were able to find very small quantities of amino acids after passing an electric spark through a mixture of methane, ammonia, water, and hydrogen. Other experimenters have proved that the gases used as starting materials can be varied somewhat and that very small amounts of amino acids are relatively easy to produce. These experiments are interesting because the gases used, often called the *primitive* gases, are the ones known to be present in the atmospheres of the other planets, and were the ones probably present on the primitive earth. These experiments indicate a way amino acids may have been synthesized on a lifeless earth. Even more important, these discoveries led to the realization that amino acids are stable compounds and that any reactions that supply large amounts of energy to hydrogen compounds of carbon, nitrogen, and oxygen are likely to synthesize amino acids.

By adding hydrocyanic acid (HCN) to mixtures of primitive gases, pyrimidine and purine bases like those of the nucleic acids can be produced. When phosphates are also present, ATP can be produced. Other researchers have produced protein-like substances by heating mixtures of amino acids with phosphate or other catalysts. Methane, carbon dioxide, hydrogen, and ammonia are prevalent in the atmospheres of the planets of our solar system and HCN has been detected in the tails of comets. Thus, we could predict that we should find most of the building blocks of life in materials from the surface of Mars and other planets. Samples of soil brought back from the moon by Apollo 12 and Apollo 14 may contain very small amounts (nanogram quantities) of glycine and even smaller quantities of alanine, serine, aspartic and glutamic acids.

Physical Properties of the Amino Acids

The amino acids are white, crystalline solids, and most are quite soluble in water. They have relatively high melting points for compounds of low molecular weight. These properties suggest that the amino acids resemble the inorganic salts and probably exist as polar molecules. Measurement of the properties of amino acids in solutions confirms the fact that they are indeed charged molecules and also that their charge may change from negative to positive or vice versa with changes in the acidity of the solution. This change in charge is accompanied by a change in chemical and physical properties. At pH values near neutral, most amino acids have both a negative and a positive charge, and thus are dipolar ions. In acid solutions they tend to have a proton attached to their negative group and are cations, while in basic solutions they are usually anions. The pH at which a protein will not migrate when placed in an electrical field is called the *isoeletric point*. (See Fig. 34–2.) The isoelectric point is the pH at which the negative charges on the protein molecule exactly balance its positive charges. This point varies for the different amino acids and provides a method of separating one amino acid from all others, or at least all others having different isoelectric points.

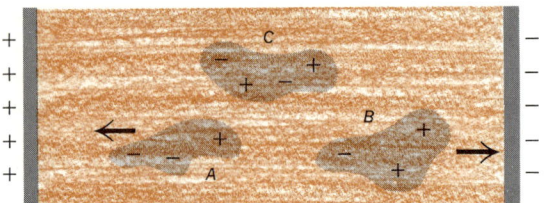

Fig. 34–2 Migration of charged particles in an electric field. This drawing shows how various amino acids or protein molecules can be separated by electrophoresis. Substance *A*, which has an excess of negative charges, migrates toward the anode; substance *B* has an excess of positive charges and thus is a cation. Substance *C*, although it has charged groups, has equal numbers of positive and negative charges and will not migrate when placed in an electrical field. Substance *C* is at its isoelectric point.

ionic forms of the amino acid alanine

$$\underset{\text{in an acid solution}}{\overset{\overset{O}{\underset{\|}{C-OH}}}{\underset{CH_3}{\overset{+}{H_3N}-C-H}}} \quad \underset{\text{in a neutral solution}}{\overset{\overset{O}{\underset{\|}{C-O^{(-)}}}}{\underset{CH_3}{\overset{+}{H_3N}-C-H}}} \quad \underset{\text{in a basic solution}}{\overset{\overset{O}{\underset{\|}{C-O^{(-)}}}}{\underset{CH_3}{H_2N-C-H}}}$$

Since amino acids take up hydrogen ions (protons) in acid solutions and release them in basic solutions, they resist changes in the pH of a solution and are good buffers. The alpha carbon of amino acids usually has four different groups attached to it. We could predict, therefore, that the amino acids are optically active. This is indeed true with most of the naturally occurring amino acids having the L-configuration.

Chemical Properties of the Amino Acids

Several biochemical reactions are common to all amino acids. Among these are formation of peptide bonds, transamination, oxidative deamination, and decarboxylation.

When the amino group of one amino acid reacts with the acid group of another amino acid, a compound called a **peptide** is formed. The group

$$-\overset{\overset{O}{\|}}{C}-NH-$$

is called the *peptide bond*. It is a substituted amide bond.

$$\underset{\text{alanine}}{\overset{\overset{O}{\underset{\|}{C-O^{(-)}}}}{\underset{CH_3}{\overset{+}{H_3N}-C-H}}} + \underset{\text{glycine}}{\overset{\overset{O}{\underset{\|}{C-O^{(-)}}}}{\underset{H}{\overset{+}{H_3N}-C-H}}} \longrightarrow \underset{\text{alanylglycine}}{\overset{\overset{O\;H\;\overset{O}{\overset{\|}{C}}-O^{(-)}}{\underset{|}{|\;|}}}{\underset{CH_3}{\overset{+}{H_3N}-C-H\;\;H}}} \quad \text{or} \quad \underset{\text{glycylalanine}}{\overset{\overset{O\;H\;\overset{O}{\overset{\|}{C}}-O^{(-)}}{\underset{|}{|\;|}}}{\underset{H}{\overset{+}{H_3N}-C-H\;\;CH_3}}}$$

$+ H_2O$

In addition to the two dipeptides given above, alanylalanine, glycylglycine, or tri-, tetra- or polypeptides may be produced whenever two amino acids react to produce peptides. The *chemical* synthesis of a specific peptide is difficult; it usually requires that the amino group of one amino acid and the carboxyl group of the other one be "blocked" so that only one reaction takes place. The *biological* synthesis of peptides is more specific and often gives only one product because of those remarkable catalysts—the enzymes. Although there is only one peptide bond in alanylglycine, the compound is called a **dipeptide**. Three amino acids, when combined, give a **tripeptide**. The combination of many amino acids gives first polypeptides, and finally the extremely high-molecular-weight compounds, the proteins.

The structure of peptides may be indicated by names, as was done for alanylglycine. A shorter method is routinely used, however, whereby the three-letter symbols for each amino acid, Fig. 34–1, are strung together. It is understood that the free amino group is at the left end and the free carboxyl group at the right. However, the symbols (NH_2) and (COOH) can be used for unambiguous presentation. Thus alanylglycine is shown as ala·gly or (H_2N) ala·gly (COOH).

The reaction of an amino acid with a keto acid in the presence of an enzyme called a transaminase gives a different amino acid and a different keto acid. The process is called **transamination** since the amino group is transferred from one carbon chain to another one.

$$\underset{\text{alanine}}{\overset{\text{COO}^{(-)}}{\underset{\text{CH}_3}{\overset{+}{\text{H}_3\text{N}}-\text{C}-\text{H}}}} + \underset{\substack{\alpha\text{-ketoglutaric}\\\text{acid}}}{\overset{\text{COO}^{(-)}}{\underset{\underset{\text{COO}^{(-)}}{|}}{\underset{(\text{CH}_2)_2}{\text{O}=\text{C}}}}} \rightleftharpoons \underset{\substack{\text{pyruvic}\\\text{acid}}}{\overset{\text{COO}^{(-)}}{\underset{\text{CH}_3}{\text{O}=\text{C}}}} + \underset{\substack{\text{glutamic}\\\text{acid}}}{\overset{\text{COO}^{(-)}}{\underset{\underset{\text{COO}^{(-)}}{|}}{\underset{(\text{CH}_2)_2}{\overset{+}{\text{H}_3\text{N}}-\text{C}-\text{H}}}}}$$

The reaction is easily reversible, and at equilibrium there are appreciable quantities of both reactants and products. This reaction provides a method for synthesis of amino acids when the corresponding keto acid is available. Although it is generally said that certain amino acids are essential, i.e., they cannot be synthesized in a given animal, it is really the keto acid or the basic carbon skeleton which is essential. The reaction given above shows that either glutamic acid or alanine can be synthesized when α-ketoglutaric acid or pyruvic acid are present along with a source of amino groups. Since α-ketoglutaric acid, pyruvic acid, and sources of amino groups are common in the bodies of animals, glutamic acid and alanine are synthesized by most animals and, therefore, are not essential amino acids.

Amino acids may also be converted to keto acids by the process of **oxidative deamination**. This is a two-step process which involves first a loss of two hydrogen atoms, followed by the addition of a molecule of water. (For simplicity, the formulas for the amino acids are written in a nonionic form.)

$$\underset{R}{\overset{H_2N}{\underset{|}{H-C-COOH}}} \xrightarrow{2H} \underset{R}{\underset{|}{H-N=C-COOH}} \xrightarrow{HOH} \underset{R}{\underset{|}{O=C-COOH}} + NH_3$$

Unlike transamination, this series of reactions is not readily reversible. The overall reaction releases appreciable amounts of energy from the amino acid, and the reverse reaction does not serve as an important source of amino acids in animals. A process essentially the reverse of this, **reductive amination**, is used by plants for synthesis of amino acids.

Amino acids may react to produce carbon dioxide and an amine by the process of **decarboxylation**.

$$\underset{R}{\overset{H_2N}{\underset{|}{H-C-\overset{O}{\overset{\|}{C}}-OH}}} \rightarrow CO_2 + \underset{R}{\overset{H_2N}{\underset{|}{H-C-H}}}$$

The extent to which such reactions occur in humans is probably slight, but histamine may be synthesized from histidine in this way. Histamine produces allergic reactions and promotes the motility of the stomach. The decarboxylation products of lysine and ornithine give the vile-smelling amines which have been given the expressive names of putrescine and cadaverine.

In addition to the general reactions which can occur with all amino acids, there are specific reactions. These are reactions of the side chain of the amino acid. Some illustrations are given below.

The hydrolysis of arginine to give urea and ornithine is a major source of urea in animals. Urea, which is found in the urine, is the major nitrogen-containing excretion product of the metabolism of amino acids. (See Chapter 39–5.)

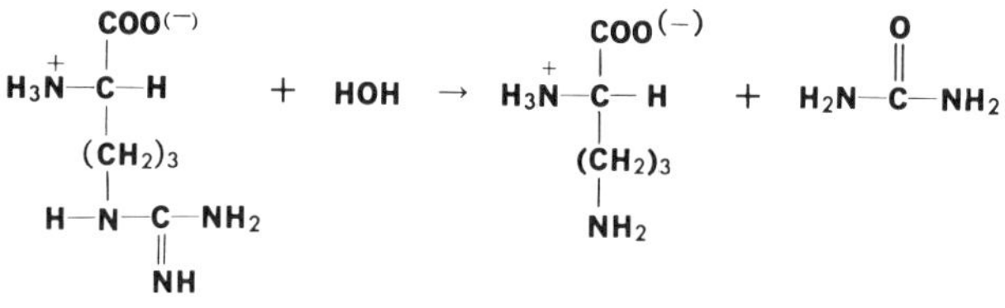

$$\underset{\text{arginine}}{\overset{\text{COO}^{(-)}}{\underset{\underset{\underset{\text{NH}}{\overset{\|}{\text{C}-\text{NH}_2}}}{\overset{|}{\text{H}-\text{N}}}}{\overset{|}{\underset{(\text{CH}_2)_3}{\overset{|}{\text{H}_3\overset{+}{\text{N}}-\text{C}-\text{H}}}}}}} + \text{HOH} \rightarrow \underset{\text{ornithine}}{\overset{\text{COO}^{(-)}}{\underset{\text{NH}_2}{\overset{|}{\underset{(\text{CH}_2)_3}{\overset{|}{\text{H}_3\overset{+}{\text{N}}-\text{C}-\text{H}}}}}}} + \underset{\text{urea}}{\overset{\overset{\text{O}}{\|}}{\text{H}_2\text{N}-\text{C}-\text{NH}_2}}$$

Two molecules of cysteine can be oxidized to form the disulfide cystine. This reaction is important in forming cross linkages between polypeptide strands in proteins. Mild reducing conditions can convert cystine to cysteine.

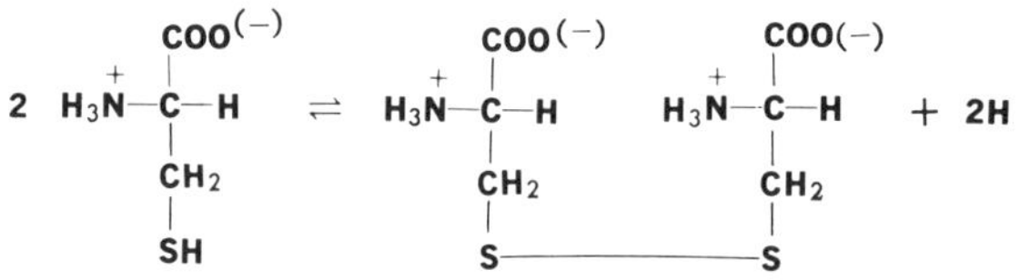

$$2\ \underset{\text{cysteine}}{\overset{\text{COO}^{(-)}}{\underset{\text{SH}}{\overset{|}{\underset{\text{CH}_2}{\overset{|}{\text{H}_3\overset{+}{\text{N}}-\text{C}-\text{H}}}}}}} \rightleftharpoons \underset{\text{cystine}}{\overset{\text{COO}^{(-)}\quad\quad\text{COO}^{(-)}}{\underset{\text{S}\quad\quad\quad\quad\text{S}}{\overset{|\quad\quad\quad\quad\quad|}{\underset{\text{CH}_2\quad\quad\quad\text{CH}_2}{\overset{|\quad\quad\quad\quad\quad|}{\text{H}_3\overset{+}{\text{N}}-\text{C}-\text{H}\quad\text{H}_3\overset{+}{\text{N}}-\text{C}-\text{H}}}}}}} + 2\text{H}$$

A combination of general and specific reactions occurs in the conversion of the amino acid phenylalanine to the hormone epinephrine which is also known as adrenalin. This conversion involves a series of reactions and illustrates how a normal dietary component is converted to a substance of vital importance to the continued life of the animal.

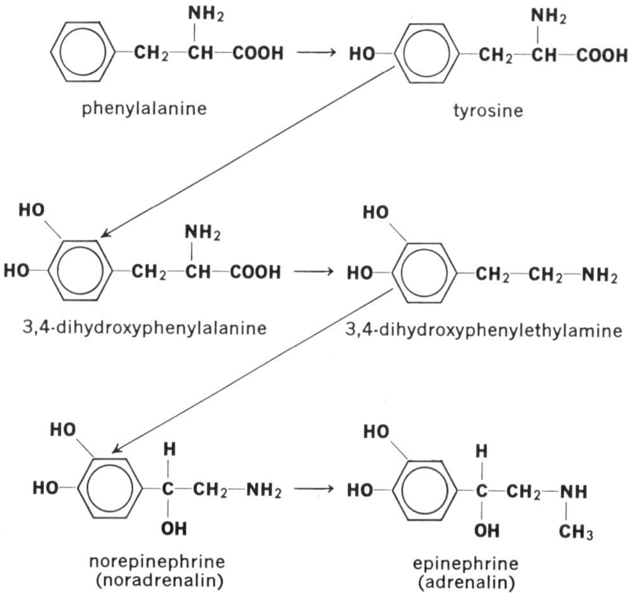

phenylalanine → tyrosine

3,4-dihydroxyphenylalanine → 3,4-dihydroxyphenylethylamine

norepinephrine (noradrenalin) → epinephrine (adrenalin)

Although amino acids can be converted to many relatively simple compounds which are of vital importance, their major function is to serve as building blocks for proteins.

34–3 PROTEINS

Compounds characterized as proteins are found throughout all living things. Muscles which are major structural material in all animals are proteins. While blood is about 80% water, the remaining 20% is largely protein. Almost all biochemical reactions require enzymes as catalysts if they are to proceed at a rate sufficient to maintain life. The enzymes may contain other compounds, but they are basically protein. The nucleic acids which form an important part of the hereditary material in the nucleus of the cell and which occur in the viruses are always found in close association with specific types of protein. Even the nonmineral matrix of the bone is protein. It has been estimated that probably 700 different proteins are known and that between 200 and 300 have been studied. Approximately 150 proteins have been obtained in crystalline form. These figures will, no doubt, soon be much too low as research in the area of protein chemistry continues.

While some proteins are extremely stable, such as those found in the hide of cattle which can be converted to a tough, durable leather, others are very delicate. Merely passing a current of air through a solution of some enzymes is sufficient to alter them so much

Fig. 34–3 A pentapeptide. The structure of this peptide is shown in the diagram, and models of two of its possible configurations are shown above the diagram. [Adapted from Snell, Shulman, Spencer, and Moos, *Biophysical Principles of Structure and Function*, Addison-Wesley, 1965.]

that they can no longer function as catalysts. The protein hemoglobin is remarkable for its ability to bind oxygen strongly enough so that it can be carried from the lungs to all parts of the body, and then to release the oxygen (and pick up carbon dioxide) in the tissues. The alteration of only 1 of the amino acids—in the more than 150 found in one of the polypeptide chains of hemoglobin—from a glutamic acid to a valine unit changes the properties of hemoglobin drastically. Persons having this altered hemoglobin (called sickle-cell hemoglobin) instead of normal hemoglobin suffer from sickle-cell anemia which is a fatal disease. This and other hereditary diseases are discussed in greater detail in Chapter 42.

The principal problem of protein chemistry is to explain in terms of structure or physical and chemical properties just how the proteins function.

The Structure of Proteins

The exact boundary between polypeptides and proteins is not precisely defined. However, it is generally accepted that 50 or more amino acids must be in the chain for a peptide to be called a protein. (See Fig. 34–3.)

Although the strongest bond between the amino acids in a protein is the peptide bond, experimental evidence indicates that there are other bonds holding the chains of amino acids to one another. This bonding is similar to the cross-linking which is found in many synthetic polymers. Hydrogen bonds and disulfide bonds have been shown to be responsible for forming these cross-links in protein molecules. Hydrogen bonds are found between a keto group of one chain and an amino group (and its hydrogen) of another chain, and between the side chains of some amino acids. Disulfide bonds are formed when a cysteine molecule in one chain reacts with the cysteine molecule of another chain.

Hydrogen bonds are weak and are easily ruptured. They are important, however, since there are so many of them in each protein molecule. Disulfide bonds are covalent bonds

hydrogen bonds between amino acid chains

formation of disulfide bonds between protein chains

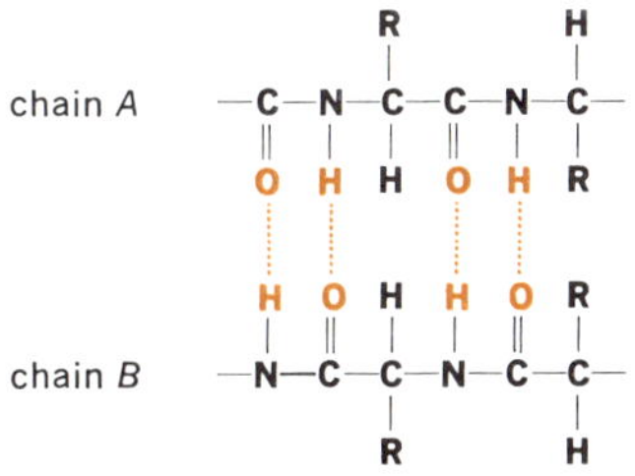

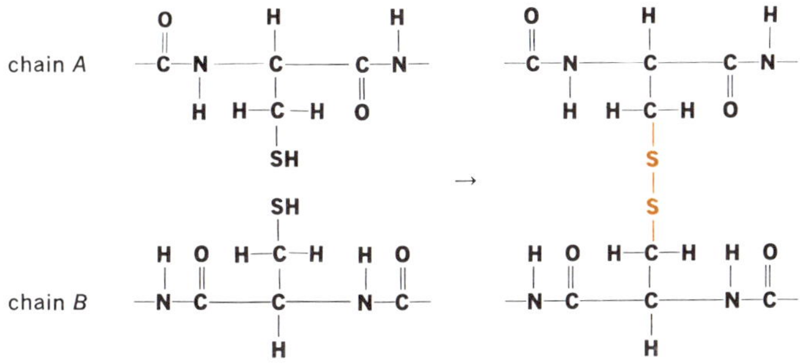

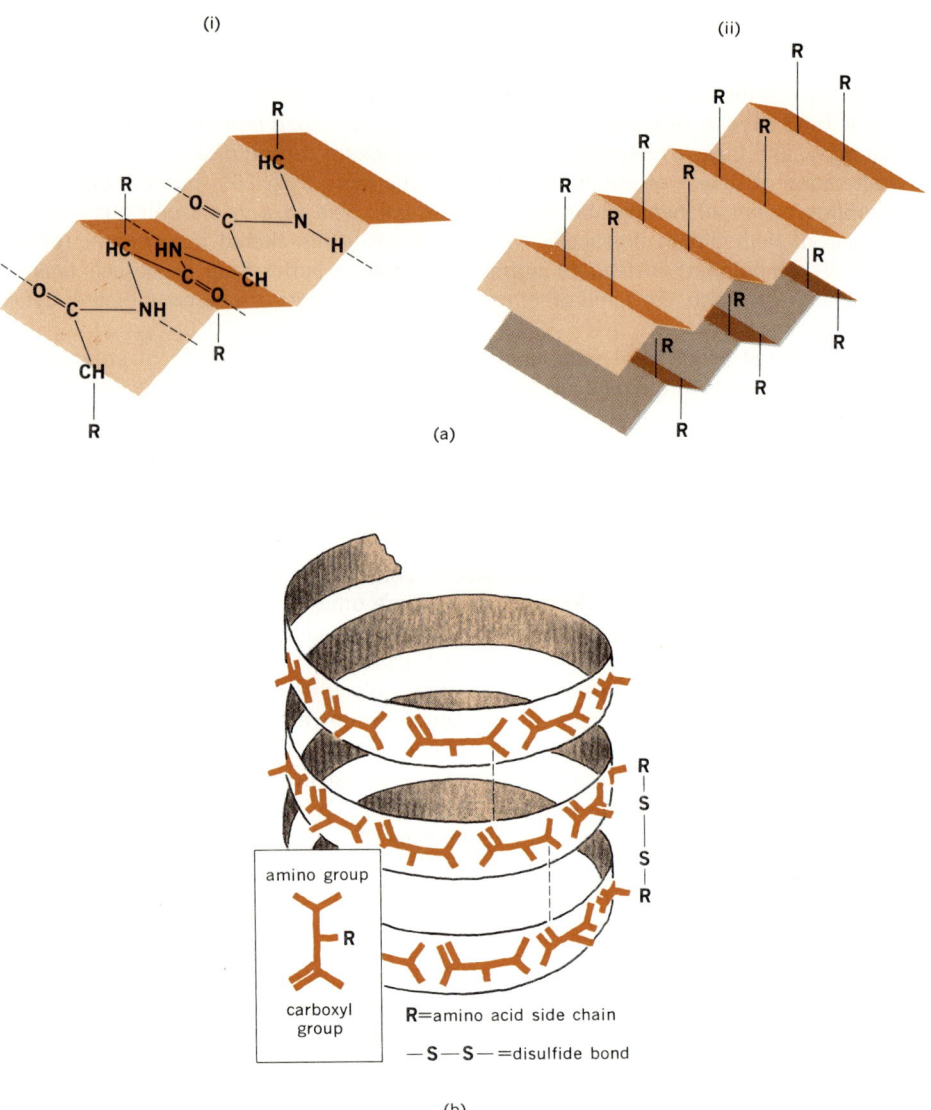

Fig. 34–4 Three-dimensional forms of polypeptides. Peptide chains in proteins are held together by secondary bonds, especially hydrogen bonds. Two forms of protein structures that are known to exist are the pleated sheet shown in (a) and the alpha-helix shown in (b). In (a), the pleated sheet form, the chains are linked together by the hydrogen bonds between the carbonyl groups of one chain and the amino groups of the other. Because of the natural bond angles, a structure such as this forms a pleated sheet. Several sheets may be held together as shown on the right. [Adapted from Setlow and Pollard, *Molecular Biophysics*, Addison-Wesley, 1962.] The alpha-helix shown in (b) represents another possible configuration of a polypeptide chain. The dotted lines between amino acid units represent hydrogen bonds. The covalent disulfide bonds are also shown. Other bonds are possible. X-ray diffraction studies show that many proteins contain the alpha-helix in parts of their structure. [Adapted from Baker and Allen, *Matter, Energy, and Life*, Addison-Wesley, 1965.]

34–3   Proteins   525

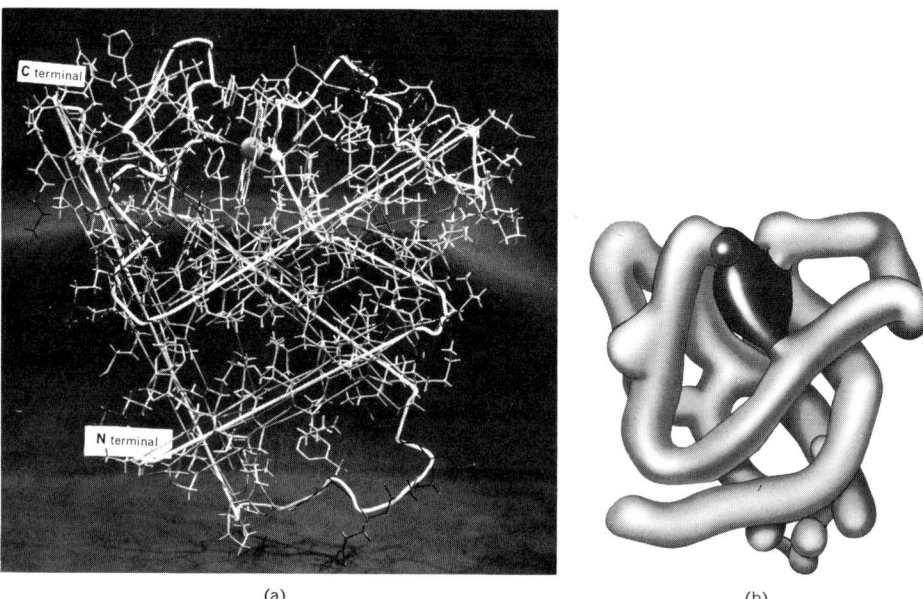

Fig. 34–5 The protein myoglobin: (a) photograph of a model of myoglobin from the sperm whale; (b) drawing of a molecule of myoglobin. In (b) the dark, disc-shaped portion is the heme group. Much of the structure of myoglobin is due to the attraction of non-polar side chains of amino acids within the molecule. [Photograph courtesy of Professor John C. Kendrew; drawing adapted from Epstein, *Elementary Biophysics*, Addison-Wesley, 1963.]

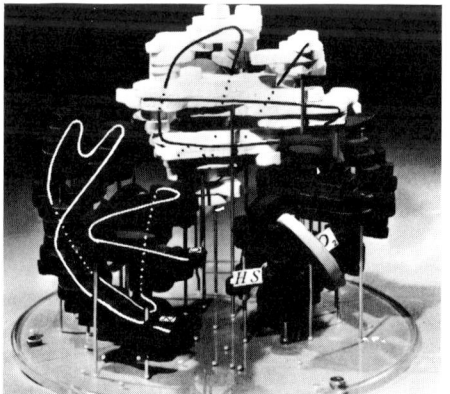 

Fig. 34–6 The protein hemoglobin. Each of the four subunits of hemoglobin contains one heme group. The a-chains are represented in white; the b-chains in black. The model on the left shows two black b-chains and one white a-chain. The basic peptide backbone is shown by the white and black lines. The model on the right shows the complete hemoglobin molecule containing two a-chains and two b-chains. Two heme groups are shown; the other two are on the opposite side of the molecule. In this model oxygen is shown attached to the heme groups. [From *Nature* **185**, 416 (1960). Photographs courtesy of Dr. M. F. Perutz.]

and are much stronger. There are probably other types of cross-linkages between polypeptide chains, such as secondary peptide bonds (between an amino acid which has an additional acid group and one which has an amino group in its side chain). Bonds which involve the linkage of the —OH groups of two serines by a phosphate group may also be present. Another way in which proteins maintain three-dimensional structure is by folding so that nonpolar side chains (like those of valine, leucine, and isoleucine) are close together. There is also a tendency of the polar side chains (glutamic acid, aspartic acid, and histidine) to be attracted to each other and to repel nonpolar side chains.

While cross-linkages can form between two separate chains, it is also possible for bonds, particularly hydrogen bonds, to form between some of the groups in a single chain. This type of linkage causes the chain to coil into a spiral or helix. X-ray diffraction studies indicate that many proteins exist in this helical form. Recent studies of the structure of the proteins hemoglobin and myoglobin indicate that part, but not all, of their structure is in the spiral form. Studies of insulin indicate that the helical structure is present there to only a small extent. (See Figs. 34–4, 34–5, and 34–6.)

Hydrogen bonds between polypeptide chains are relatively easy to break. Many procedures of protein chemistry probably involve changing the number and nature of hydrogen bonds between chains. The disulfide bonds are harder to break. Most of the processes for "permanent waving" of the hair involve first the rupture and then the reestablishment (in different relative positions) of disulfide bonds in the proteins of hair.

Protein Synthesis

Because of their great complexity and of the difficulties involved in synthesizing individual dipeptides and tripeptides, only a few proteins have been chemically synthesized. Insulin, which contains 51 amino acids, has been synthesized, however, as has ACTH which contains 39 amino acids, but these substances are near the polypeptide-protein border of classification. Relatively new techniques of synthesis such as that described by Merrifield in the Suggested Reading make more complex syntheses possible. The enzyme ribonuclease, which is a small protein, has already been synthesized, and we can expect others soon. Newer, less time-consuming, and less expensive techniques will be required before we can expect commercially feasible syntheses of proteins.

Proteins are, of course, synthesized by all living things. This process again is complex since it involves the linking of hundreds of specific units in a specific way. It can be calculated that there are approximately 1.35×10^{167} (135 followed by 165 zeros) possible arrangements of the amino acids found in the relatively small protein molecule of hemoglobin. This number is much greater than the number of atoms in the universe. The ability of a cell to synthesize just the right kind of proteins is truly remarkable.

The process of protein synthesis *in vivo* is under hereditary control. Although some of the details remain to be discovered, it now appears that the master plan for protein synthesis is found in the DNA (deoxyribonucleic acid) in the nucleus of the cell. This plan or code is transferred to the RNA (ribonucleic acid) which actually serves as the pattern for the extremely complex task of making just the right kind of protein. Protein synthesis is discussed in detail in Section 39–4. Several hereditary disorders are now known in which an individual is unable to synthesize one kind of protein. In some anemias there is a change in only one amino acid in the hemoglobin molecule. A similar "inborn error of metabolism" is responsible for the condition known as phenylketonuria which, if untreated, produces idiocy in persons with the defect. The deficiency in this case is in a specific enzyme required for the metabolism of the amino acid phenylalanine. (See Chapter 42 for a further discussion of these hereditary defects.)

Physical Properties of Proteins

From the large molecular weights of the proteins, which range from about twelve thousand to hundreds of thousands, we would expect these molecules to be insoluble in water. Many proteins are, in fact, insoluble in water, but many very important ones are quite water soluble, probably because of the attraction of water molecules to charged groups in the protein molecule. Although most amino and acid groups are tied up in the peptide bond, other groups of amino acids (their side chains) are often in a charged state. The additional acid groups in glutamic acid and aspartic acid, the amino groups in lysine and arginine, as well as other side chains can be charged. As with the primary amino and acid groups, the state of charge of the side chains depends upon the pH of the solution. When the net charges on a protein are at a minimum, i.e., when the negative charges are equal to the positive charges, the protein will not migrate when placed in an electrical field. The process of placing a protein solution on paper or some other relatively inert substance and allowing the proteins present to migrate toward the anode or cathode is called electrophoresis (see Fig. 34–2). This technique is used to separate individual proteins from mixtures. It is now finding wide application for studying the various proteins in blood serum. (See Fig. 34–7.)

The pH value at which a protein will not migrate when placed in an electrical field is called its **isoelectric point.** This is the point of minimal solubility for that protein. We can make use of such knowledge in separating it from other proteins, or in helping us decide which protein (or proteins) are present in a solution.

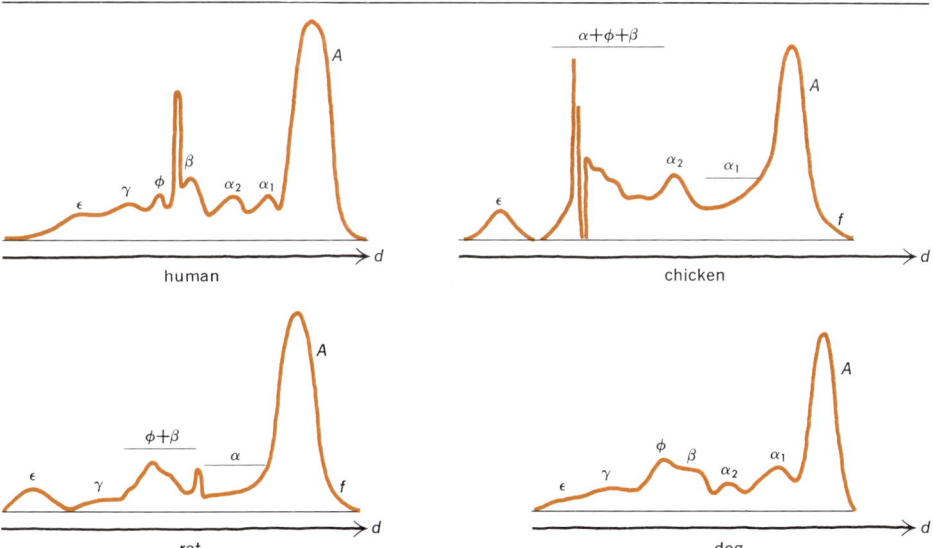

Fig. 34–7 Electrophoretic patterns of the blood plasma from the human, the rat, the chicken, and the dog. Albumin is identified by an A, fibrinogen by ϕ, and the various globulins by the letters α, β, and γ. The letter ϵ designates the so-called "salt boundary." [Adapted from an article by H. F. Deutsch, *J. biol. Chem.* **161**, 1, 1945.]

Chemical Properties of Proteins

The chemical reactions of the proteins are difficult to study. In most cases only a relatively small part of the huge protein molecule is actually changed or undergoes a reaction. It is now believed that only a small part of the molecule of an enzyme, called its active site, is involved when the enzyme catalyzes a reaction. The combination of the iron-containing protein hemoglobin with oxygen is believed to involve the iron atom and one of the histidine molecules in the protein chain. Structural studies of the enzyme ribonuclease show an opening in the molecule just the right size for molecules of the substrate upon which the enzyme acts. Biochemists are only beginning to understand the true nature of the chemical reactions of proteins. More knowledge in this fascinating area will bring us much closer to understanding the chemical nature of living processes.

Other, *in vitro*, chemical reactions of proteins are used for their identification and characterization. In some cases a colored product is produced, as in the xanthoproteic reaction in which proteins are treated with nitric acid, or in the biuret test which uses a copper salt in conjunction with a basic solution of the protein. Other tests involve the precipitation of the protein by heat or salts. A simple method for differentiation of the albumins from the globulins is based on the fact that the former stay in solution and the latter precipitate when in a solution that is half saturated with ammonium sulfate.

Most of these reactions used to identify proteins result in the gross alteration of the protein. The term **denaturation** has been used to describe these changes. While there is some disagreement as to the use of the term denaturation, it is best defined as *any change in any property of a protein that does not involve rupture of the basic peptide bonds*. Some of these changes can be reversed, as when a globulin that has been "salted out" of solution is redissolved. Others, such as the heat treatment of egg albumin, cannot be reversed, i.e., you can't "unboil" an egg. (See Fig. 34–8.) Some biochemists restrict the definition of denaturation to those processes which are irreversible. Denaturation, of course, causes a loss of biological activity.

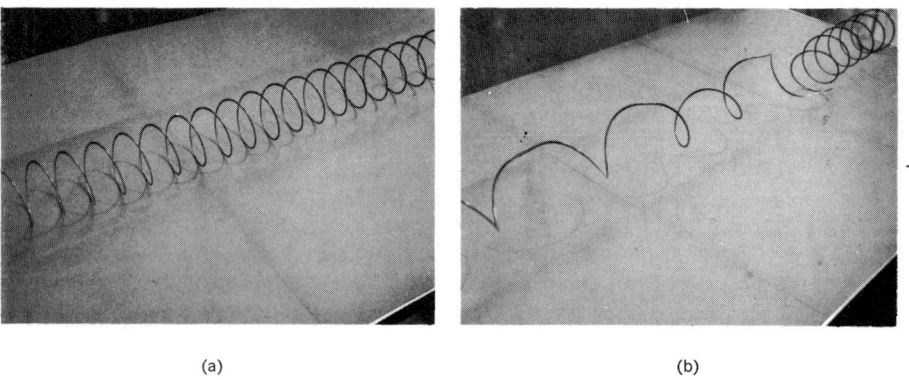

(a)   (b)

Fig. 34–8 Denaturation of proteins. (a) The coiled spring toy known as a "slinky" is used to represent the helical structure of a protein. When the elastic limit of the helix is exceeded, the shape is irreversibly altered as shown in (b). This is analogous to irreversible denaturation of a protein. [From Baker and Allen, *Matter, Energy, and Life*, Addison-Wesley, 1965.]

Purification of Proteins

The purification of proteins has played an important role both in the study of these vital compounds and in the uses of the purified substances. Before determining the composition of a protein, the chemist must be sure that it is pure, i.e., that he has only one specific protein. The study of the exact nature of catalysis by enzymes also requires pure substances. Since any foreign protein that is injected into an organism usually produces antibodies to that protein, it is essential that any material that is to be injected into a human for medical purposes be highly purified. For these reasons it is necessary that insulin, which is used to treat diabetes, and the polypeptide ACTH, which is sometimes used to treat local inflammations, as well as many other proteins be as pure as possible. Oral administration of proteins is ineffective since digestion breaks them down into individual amino acids.

Protein purification is based on a knowledge of the physical properties of the proteins. The proteins may be separated from lipids and carbohydrates on the basis of their solubility and of their charged state. If relatively crude materials are used, the material is first fractionated by spinning it in a centrifuge. Proteins have different solubilities. The differential solubility of the various proteins in salt solutions or in solutions containing organic solvents such as ethanol or acetone form the basis for many separations. Other purifications can be made by passing the material through a column of a chromatographic absorber which utilizes the differences in affinity of the various proteins for the adsorbent. Electrophoresis uses the difference in migration rate of proteins as influenced by their charged state which is, in turn, influenced by the pH of the solution. This method is useful in purifying relatively small amounts of proteins. (See Fig. 32–1.) All purification procedures, particularly those with organic solvents, are preferably done at low temperatures and with the use of other precautions to prevent irreversible denaturation of the protein. Since proteins are large molecules, they can be separated from salts and other small soluble molecules by dialysis. In this process a sac of semipermeable membrane retains the protein while the smaller particles escape into a surrounding liquid. Water may be removed from protein solutions under vacuum at low temperatures in a process called freeze drying. The dry powders so produced are much more stable in storage than are solutions of proteins. By using a variety of these techniques it is possible to separate the proteins of blood so that dried serum or gamma globulin become available for medical uses.

Importance of Proteins

Proteins play many vital roles in the living organism. Their importance in enzymatic catalysis, as oxygen carriers, and as structural materials have already been mentioned. While it is impossible in an elementary text to discuss all the important functions of proteins, a few more significant ones are given below.

Since animals synthesize their body proteins from amino acids, and especially since they can synthesize only half of these amino acids, it is extremely important that their diets include adequate amounts of the ones that are not made in their own bodies. The major source of these essential amino acids is the proteins of the diet. One of the greatest nutritional problems is that much of the world's population does not eat enough of the kind of protein that provides the critical amino acids. This topic is discussed further in Chapter 41, "Nutrition."

In addition to hemoglobin, the blood contains many other proteins. There are albumins which serve to maintain sufficient osmotic pressure to help water and other nutrients

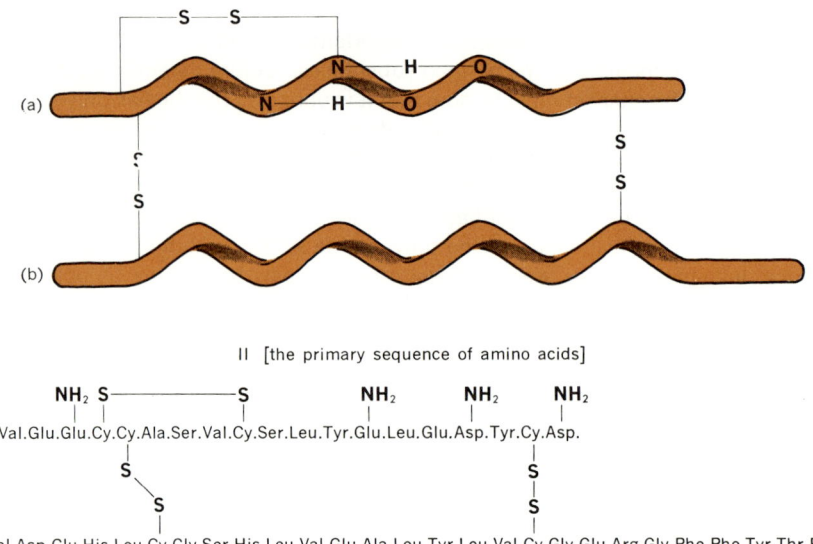

Fig. 34-9 Two representations of the structure of insulin. The lower diagram shows the sequence of amino acids in two protein chains called the a- and b-chains. These two chains are linked together with disulfide bonds. The upper representation is a drawing of the insulin molecule showing both the disulfide bridges and possible hydrogen bonds. The disulfide bridge in the a-chain links two parts of that chain and does not serve to bind two polypeptide chains together as do the other disulfide bonds. [Adapted from C. B. Anfinsen, *The Molecular Basis of Evolution*, 2nd. ed., New York, Wiley, 1963.]

pass freely through the walls of the capillaries. The particular mixture of proteins that has been called gamma globulin constitutes one of our major supplies of antibodies for resisting some diseases. The proteins fibrinogen and prothrombin are essential for the formation of the clump of red blood cells called a clot. Clotting is useful when it prevents excessive loss of blood, but clots can also form in the blood vessels of the heart or brain, causing a type of heart attack or a stroke.

Many of the hormones which serve as important regulators of the processes of the body are proteins. Most of the hormones of the master gland, the pituitary, are protein-containing. Two of them, vasopressin and oxytocin, are octapeptides; others are either glycoproteins (compounds containing both protein and carbohydrate) or pure proteins. The widely used hormone, insulin, contains 51 amino acids and may be classed as either a large polypeptide or a very small protein (see Fig. 34-9). The discovery and extraction of insulin and its use in the treatment of diabetes makes a dramatic chapter in the story of man's conquest over disease. Thousands of persons, whose days would have been limited, now live normal lives because large quantities of highly purified insulin are available. The use of the protein growth hormone has met with only limited success. Further study and purification of other hormones can be expected to lead to longer and better lives for some of those persons who today suffer from hormone insufficiency.

The importance of proteins in our lives may be traced from the time a sperm cell (made up of nucleic acids and proteins) unites with an ovum. In animals, the fertilized egg develops through stages of differentiation which are largely influenced by a succession of enzymes, to the birth of the individual. Birds' eggs, which provide nutrition for the embryo birds, are especially rich in proteins. The gradual growth of a person is due to the increase in bone and muscle under the control of the proteinaceous hormones and with the help of the enzymes. The production of new sperm and ova under the influence of other protein-containing hormones leads to the perpetuation of the species. The continued life and health of the individual depends on sufficient nutrition by proteins (along with other foods) and on his ability to manufacture sufficient enzymes to catalyze the essential reactions and sufficient antibodies to resist infections. The failure of any of these protein-influenced systems or, in some cases, the overactivity of some of these processes leads to the death of the individual.

While the importance of other classes of compounds cannot be forgotten, Mulder's statement that protein "is without doubt the most important of the known components of living matter, and it would appear that, without it, life would not be possible" is still true. Research studies of the nature and reactions of proteins are one of the most important ways we can achieve the fundamental objective of biochemistry—the explanation of life in chemical and physical terms.

IMPORTANT TERMS

Albumin	Electrophoresis	Hemoglobin	Polypeptide
Amino acid	Essential amino acid	Hormone	Protein
Deamination	Gamma globulin	Peptide	Salting out
Decarboxylation	Globulin	Peptide bond	Transamination
Denaturation			

WORK EXERCISES

1. List the common dicarboxylic amino acids.
2. List the common amino acids containing more than one amino group.
3. What functional groups other than the carboxyl and amino groups are found in the side chains of proteins?
4. List several natural products that are primarily or largely composed of protein.
5. Write all possible charged states of cystine, of glutamic acid, of lysine.
6. Write equations for each of the following reactions and name all the products of the reactions.
    a) the reaction of glycine and serine to form a dipeptide
    b) the transamination of alanine with alpha ketoglutaric acid
    c) the decarboxylation of histidine, of ornithine
    d) the complete hydrolysis of lysylphenylalanylasparagine
    e) the oxidative deamination of aspartic acid, of alanine
7. Write the structures for glycylcysteinylalanine and for threonylcysteinylproline, and show possible cross-linkages between these peptides.
8. Write the complete structure for the peptides indicated below:
    a) (H_2N)-tyr-ala-ser$(COOH)$
    b) (H_2N)-gly-gly-met-phe$(COOH)$

9. Why is insulin injected rather than taken orally in the treatment of diabetes?
10. A substance containing the elements C, H, O, N, and S, having a high molecular weight, was dissolved in water. Addition of sufficient ammonium sulfate to make the solution 50% saturated did not result in precipitation of the substance. What can you conclude about the substance? What test or tests would be necessary to confirm your identification?
11. From a general biochemistry text (see list at the end of Chapter 32) find the answer to the following:
    a) What is the structure of glutathione?
    b) List several dipeptides of biological importance.
    c) Using the simplest method possible, write the complete structure of ribonuclease and of insulin.

SUGGESTED READING

Altschul, A. M., *Proteins, Their Chemistry and Politics*, Basic Books, New York, 1965. As indicated in the title, this book discusses both the chemical characteristics of the proteins and their importance in world food problems.

Dayhoff, M. O., "Computer Analysis of Protein Evolution," *Scientific American*, July 1969 (Offprint #1148). Amino acid composition of cytochrome c can be used to trace relationships of various species; it also supports other indications of evolutionary theory.

Fraser, R. D. B., "Keratins," *Scientific American*, August 1969 (Offprint #1155). The type of protein known as keratin is found in skin, feathers, hair, and other external parts of many animals. This article describes methods used to establish the three-dimensional structure of this type of protein.

Merrifield, R. B., "The Automatic Synthesis of Proteins," *Scientific American*, March 1968 (Offprint #320). The term automatic is somewhat misleading. However, protein synthesis which involves anchoring peptide chains to polystyrene beads is a relatively simple process. The synthesis of insulin is described.

Patton, S., "Milk," *Scientific American*, July 1969 (Offprint #1147). The production and composition of a basic life-giving food are discussed at length in this well-written article.

Although the following four articles are slightly outdated, they present basic discussions of the important aspects of proteins.

Doty, P., "Proteins," *Scientific American*, Sept. 1957 (Offprint #7).

Fruton, J. S., "Proteins," *Scientific American*, June 1950 (Offprint #10).

Pauling, L., R. B. Corey, and R. Hayward, "The Structure of Protein Molecules," *Scientific American*, July 1954 (Offprint #31).

Stein, W. H., and S. Moore, "The Chemical Structure of Proteins," *Scientific American*, Feb. 1961 (Offprint #80).

In the following, the stories of the determination of the structure of the proteins, insulin, myoglobin, hemoglobin, and the polypeptide ACTH, are told by the workers who were active in this research. The illustrations in these articles are excellent.

Kendrew, J. C., "The Three-Dimensional Structure of a Protein Molecule," *Scientific American*, Dec. 1961 (Offprint #121).

Li, C. H., "The ACTH Molecule," *Scientific American*, July 1963 (Offprint #160).

Perutz, M. H., "The Hemoglobin Molecule," *Scientific American*, Nov. 1964 (Offprint #196).

Thompson, E. O. P., "The Insulin Molecule," *Scientific American*, May 1955 (Offprint #42).

Phillips, D. C., "The Three-Dimensional Structure of An Enzyme Molecule," *Scientific American*, Nov. 1966. This article not only discusses the structure of a protein as do those above, but has the added fascination of discussing the structure in terms of the way in which this enzyme works as a catalyst.

CHEMISTRY OF THE LIPIDS 35

35–1 INTRODUCTION

If we separate the water, carbohydrates, and proteins from a cell or from a total organism, the major component left is **lipid.** In practice the procedure used is just the opposite of that given above—the lipids are separated from all other biochemical materials by extracting them in ethyl ether, ethanol, chloroform, or similar solvents or mixtures of these solvents. This procedure is the basis for our definition of lipids as *biochemical substances that are soluble in nonpolar organic solvents*. The lipids are commonly known as fats; biochemists, however, reserve the name "fat" for one particular kind of lipid.

Since lipids are defined in terms of solubility, and since many types of compounds are soluble in the lipid solvents, we find the lipids include many heterogeneous compounds. They are not made of one particular kind of building block as are the proteins and carbohydrates, and almost no lipids are really polymers.

35–2 CLASSIFICATION OF LIPIDS

Since the lipids lack a really basic building block, the only way to study the lipids systematically is by a classification based on solubility and hydrolysis products. As you may expect, there are different systems of classifications. It is especially difficult to classify those lipids that yield two or more characteristic hydrolysis products. The following classification is widely accepted and is sufficient for a basic understanding of the lipids:

a) triglycerides   b) phospholipids (phosphatides)   c) waxes
d) steroids   e) terpenes   f) miscellaneous

The placement of a lipid in one of the categories above depends primarily on its hydrolysis products. Although there are many kinds of molecules in these hydrolysis products, as would be expected in a group of compounds defined only in terms of solubility, there is more similarity than one might expect. In fact, we find present in lipids relatively few of the thousands of possible organic compounds. This natural economy comes about because of the great similarity in synthetic pathways in different organisms. We must not forget that the composition of biochemical substances is the result of the synthetic processes in plants and animals.

TABLE 35-1 Common fatty acids

(handwritten annotation: Long side chain carboxylic acids)

SATURATED ACIDS

Name	Formula	Common Source
acetic acid	CH_3COOH	vinegar
butyric acid	C_3H_7COOH	butter
caproic acid	$C_5H_{11}COOH$	butter
caprylic acid	$C_7H_{15}COOH$	butter
capric acid	$C_9H_{19}COOH$	butter, coconut oil
lauric acid	$C_{11}H_{23}COOH$	coconut oil
myristic acid	$C_{13}H_{27}COOH$	coconut oil
*palmitic acid	$C_{15}H_{31}COOH$	animal and vegetable fats
*stearic acid	$C_{17}H_{35}COOH$	animal and vegetable fats
arachidic acid	$C_{19}H_{39}COOH$	peanut oil
lignoceric acid	$C_{23}H_{47}COOH$	brain and nervous tissue
cerotic acid	$C_{25}H_{51}COOH$	beeswax, wool fat

UNSATURATED ACIDS

Name	Formula	Number of Double Bonds	Common Source
*palmitoleic acid	$C_{15}H_{29}COOH$	1	animal and vegetable fats
*oleic acid	$C_{17}H_{33}COOH$	1	animal and vegetable fats and oils
linoleic acid	$C_{17}H_{31}COOH$	2	linseed oil, vegetable oils
linolenic acid	$C_{17}H_{29}COOH$	3	linseed oil
arachidonic	$C_{19}H_{31}COOH$	4	brain and nervous tissue

* Acids indicated with an asterisk are among the most widely distributed.

The most common hydrolysis products or building blocks of the lipids are the monocarboxylic organic acids which are commonly called the fatty acids. Of all possible fatty acids, we find that only those having an even number of carbon atoms occur to any appreciable extent. Although there is some variation as to the number of double bonds (degree of unsaturation) in these acids, branched chains and ring compounds are rare in the fatty acids. The commonly found fatty acids are given in Table 35–1. Some fatty acids are found loosely bound to albumins in the blood. The level of these "free fatty acids" (FFA level) is significant in diagnosis of cardiovascular disease (see Section 40–7).

In addition to the fatty acids, certain other components are found as hydrolysis products of lipids. The more common are glycerol, choline, and aminoethanol.

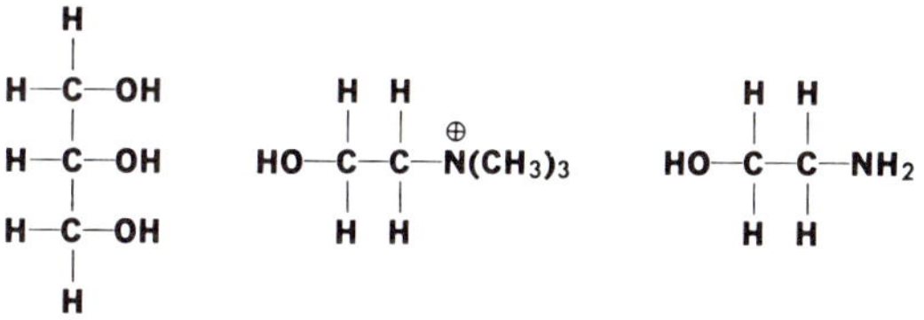

$$\begin{array}{c} H \\ | \\ H-C-OH \\ | \\ H-C-OH \\ | \\ H-C-OH \\ | \\ H \end{array} \qquad HO-\underset{\underset{H}{|}}{\overset{\overset{H}{|}}{C}}-\underset{\underset{H}{|}}{\overset{\overset{H}{|}}{C}}-\overset{\oplus}{N}(CH_3)_3 \qquad HO-\underset{\underset{H}{|}}{\overset{\overset{H}{|}}{C}}-\underset{\underset{H}{|}}{\overset{\overset{H}{|}}{C}}-NH_2$$

glycerol choline aminoethanol

Other lipids contain structural units which are related to isoprene.

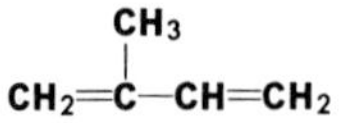

$$CH_2=\underset{\underset{CH_3}{|}}{C}-CH=CH_2 \quad \text{isoprene}$$

The isoprene units exist both in chain form and in ring structures, as in β-carotene.

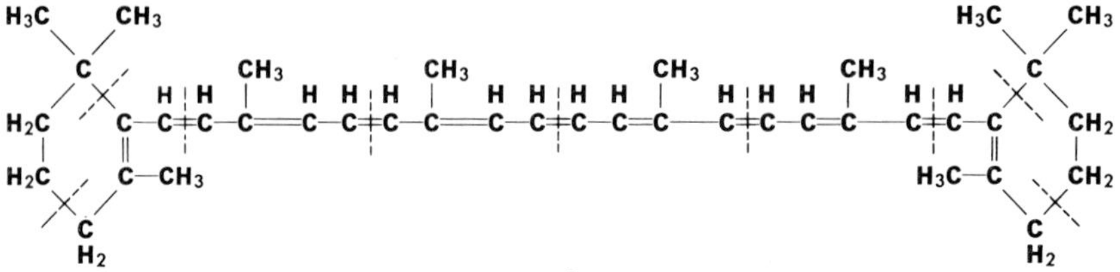

β-carotene

The dotted lines indicate isoprene units in this complex structure. Rubber is a polymer made of isoprene units (Eq. 30–6). Although it is more difficult to see, the complex structure known as the steroid nucleus contains several isoprene units. The biological synthesis of cholesterol involves the combination of several such units.

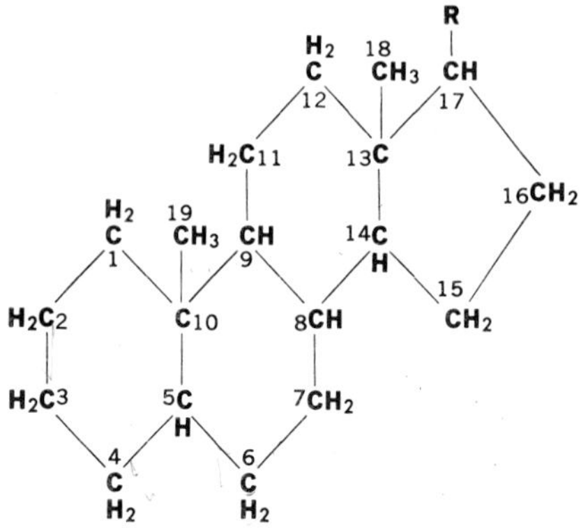

steroid nucleus

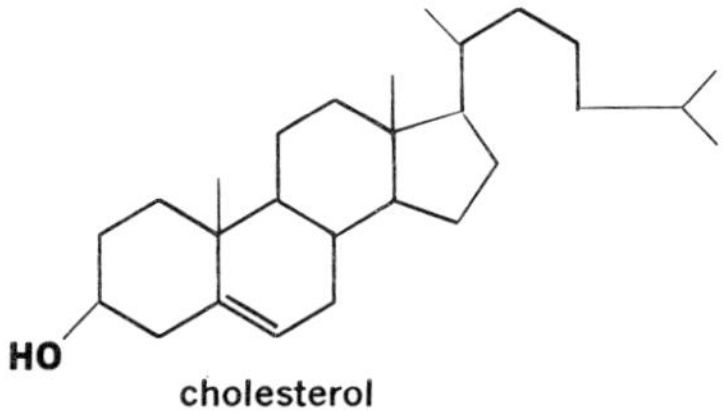

cholesterol

Recently a class of compounds known as the prostaglandins has been isolated and studied. They are variants of 20-carbon carboxylic acids which contain a five-membered ring, hydroxyl groups, and double bonds. Their solubility places them in the classification of the lipids. Although they were originally thought to come from the prostate gland and were named on this basis, they are now known to originate in the seminal vesicles of the male as well as in other parts of the body of both males and females. These substances affect blood pressure, contraction of the uterus, and other functions, apparently through their action on smooth muscle. They are effective in extremely small concentrations. For further discussion of these fascinating compounds, see the article by Pike listed at the end of this chapter.

35-3 TRIGLYCERIDES

The name triglyceride is used by biochemists to describe those esters of glycerol which contain three fatty acids. The triglycerides, which are commonly called the fats and oils, are the most abundant lipids. Animals accumulate fats when their food intake exceeds energy output. Although plants accumulate some fats and oils, particularly in their seeds, they usually store excess energy in the form of carbohydrates. The basic chemistry of the glycerides was discussed in Chapter 27. Here we will add details that are of special interest to biochemists.

The separation of most fats and oils from their natural sources usually involves only a physical process. Olive and cottonseed oils are separated by pressing, and butter by churning cream which contains the butterfat. Lard is separated from various fatty tissues of the swine by heating to melt the fat which is then filtered while hot to remove it from proteins and carbohydrates. In biochemical preparations, fats are usually extracted, along with the other lipids, by adding a mixture of organic solvents (usually ethanol and ethyl ether) to a natural product. Triglycerides are then separated from the other lipids by the use of other solvents, and finally by saponification. Since many lipids are not esters, they are not subject to alkaline hydrolysis (saponification). Consequently they may be separated from the soaps (which are water soluble) by extracting the unsaponified lipids in organic solvents. (See Section 27-6 for a discussion of saponification and soaps.)

Pure triglycerides are colorless, odorless, and tasteless. Characteristic tastes, odors, and colors of butter, olive oil, and cod liver oil are caused by compounds other than triglycerides. Many of the physical properties of the triglycerides depend upon length or degree of unsaturation of the carbon chain of the esterified fatty acids; see Section 27-3.

The major biochemical reaction of triglycerides is hydrolysis. Enzymes which catalyze the hydrolysis are commonly called *lipases*. Other reactions may occur at double bonds in the fatty-acid chain but these reactions, which include oxidation and addition of halogens, do not normally occur in living systems. These reactions are used primarily to identify and characterize lipids.

Butter and other triglycerides that have been left at warm temperatures for long periods of time tend to develop obnoxious odors and tastes. We say they have become **rancid.** The odors are due to free fatty acids (butyric acid in butter) and aldehydes. These free acids are formed as a result of two of the characteristic reactions of the triglycerides, oxidation of double bonds and hydrolysis of the ester bond. The hydrolysis which releases odorless glycerol in addition to the vile smelling acids is often catalyzed by enzymes present in bacteria. Oxidation of fatty acids occurs principally at the double bonds. Exposure to the oxygen of air at warm temperatures induces oxidation which results in the rupture of double bonds to form free aldehyde groups. These aldehydes have odors similar to those of the fatty acids and, in fact, are readily converted to fatty acids by oxidation. Oxidative processes are the primary causes of rancidity in most foods.

Triglycerides are of major importance because they represent the primary energy storage compounds in animals. The fact that they provide insulation both from heat loss and mechanical shock also makes them useful. An insulating layer of fatty tissues just below the skin helps the human body maintain its required temperature of 98.6°F. Many vital internal organs, such as the kidneys, are encased in a thick layer of fat which helps to protect the animal against mechanical shock. Subcutaneous fat also performs this function. However, persons with too much fat are less healthy and often die younger than average persons. High levels of triglycerides in blood are associated with heart attacks.

35-4 PHOSPHOLIPIDS

Know general structure

Many lipids contain the element phosphorus in addition to carbon, hydrogen, and oxygen. These phospholipids are found in many parts of living organisms, but are especially prevalent in the brain and nervous tissue. Liver and egg yolks also contain appreciable amounts of a particular type of phospholipid—lecithin. The outer membranes of most cells contain phospholipids.

Many phospholipids are considered derivatives of phosphatidic acid. The R-groups are those of relatively long-chain fatty acids.

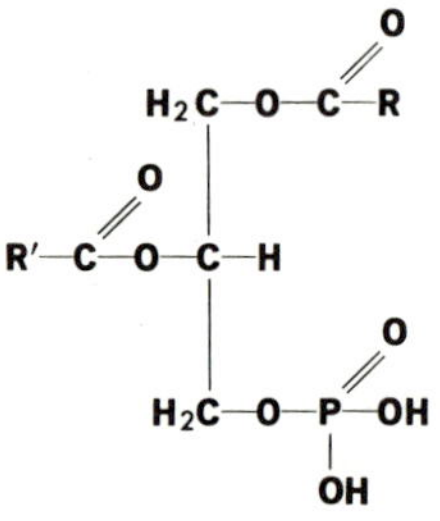

$$\begin{array}{l} H_2C-O-\overset{O}{\underset{\parallel}{C}}-R \\ R'-\overset{O}{\underset{\parallel}{C}}-O-C-H \\ H_2C-O-\overset{O}{\underset{\parallel}{P}}-OH \\ OH \end{array}$$

phosphatidic acid

Phosphoric acid
H_3PO_4
$$HO-\overset{O}{\underset{\parallel}{P}}-OH$$
OH

Phosphatidic acid does not occur as such in living animals. In derivatives of phosphatidic acid the phosphate group is generally esterified with an OH group of a nitrogen-containing alcohol. Substances most commonly esterified here are choline, ethanolamine, or the amino acid serine.

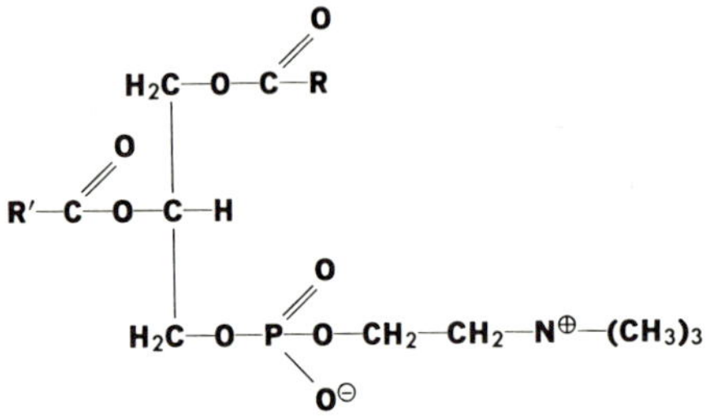

$$\begin{array}{l} H_2C-O-\overset{O}{\underset{\parallel}{C}}-R \\ R'-\overset{O}{\underset{\parallel}{C}}-O-C-H \\ H_2C-O-\overset{O}{\underset{\parallel}{P}}-O-CH_2-CH_2-N^{\oplus}-(CH_3)_3 \\ O^{\ominus} \end{array}$$

a lecithin

In fact, further classification of phospholipids, derivatives of phosphatidic acid, depends on the basic alcohol present. Lecithins contain choline; cephalins contain either ethanolamine or serine; and the plasmalogens contain an unsaturated long-chain ether and either choline or ethanolamine. Phospholipids which are not derivatives of phosphatitidic acid are also known.

The phospholipids, particularly the lecithins, are large molecules having both a nonpolar component and a highly polar one. Thus, we would expect them to be good detergents (see Section 27–7). Apparently the major biochemical function they perform is in associating the water-insoluble lipids and the water-soluble components of any living organism. It has been proposed that phospholipids help in the transport of lipids in the blood stream (which is basically water). As a matter of fact, tiny fat globules are present in the blood after digestion of a meal containing fat. It is interesting to speculate about whether the phospholipids are responsible for keeping these globules small enough so that they do not block the small blood vessels. Less vital uses of lecithins are as emulsifying agents to keep the fat, chocolate, and other components of many candies from separating. Egg yolks which are rich in lecithins are used to emulsify vinegar and salad oil into a mayonnaise or salad dressing.

The prevalence of phospholipids in brain and nervous tissue leads us to wonder about their significance in these important tissues. While it may be they are present only inadvertently, some biochemists are strongly influenced by the belief that "nature probably does nothing without a purpose," and they are trying to find the purpose or function of the phospholipids in these areas. At present, we can say only that their function is not yet fully understood.

35-5 WAXES   *not imp. as to names - know general structure*

The esterification of a fatty acid with a simple alcohol (i.e., one having only a single OH group rather than three as in glycerol) gives a wax. Examples were mentioned in Section 26-9E. Whale oil (which is not truly an oil, but a wax) contains cetyl palmitate, and beeswax contains myricyl palmitate.

$$CH_3(CH_2)_{14}-\overset{O}{\overset{\|}{C}}-O-(CH_2)_{15}-CH_3 \qquad CH_3(CH_2)_{14}-\overset{O}{\overset{\|}{C}}-O-(CH_2)_{29}-CH_3$$

cetyl palmitate   myricyl palmitate

35-6 STEROIDS

The term **sterol** means solid alcohol. Cholesterol was isolated from gallstones in 1775, and the prefix *chole* indicates its relation to the bile (*chole* is the Greek word for bile). We now know that cholesterol is present in many tissues of the body, especially in brain and nervous tissue. Cholesterol was only the first of a great variety of compounds having a steroid nucleus. Steroids having an —OH group are called *sterols*. (See Section 35-2 for the formula for cholesterol).

The steroid nucleus is generally a flat structure. If we consider the nucleus to be in the plane of the page of this book, substituent groups will be either above or below the ring. When we draw the structure, we use a solid line to indicate that the substituent is above or in front of the ring, and a dotted line to show it is below or to the back.

Since sterols are primarily hydrocarbons with a high molecular weight, they are soluble in the nonpolar organic solvents, and are classified as lipids. Modern biochemistry has found other relationships between the fatty acids and cholesterol that indicate the propriety of their being classified together. Acetate, in its active form (known as acetyl coenzyme A) is the starting material not only for the synthesis of fatty acids and cholesterol, but for many other lipids. The physiological function of cholesterol is not known, but there seems to be some correlation between the level of cholesterol in the blood and a tendency toward certain types of heart attacks. The presence of cholesterol in gallstones represents an instance of an unnatural concentration of a normal body constituent.

There are many steroids that have well-known specific functions. Testosterone, the male hormone, is responsible for secondary sexual development in the male, and estradiol and other related compounds perform a similar (although different) function in the female.

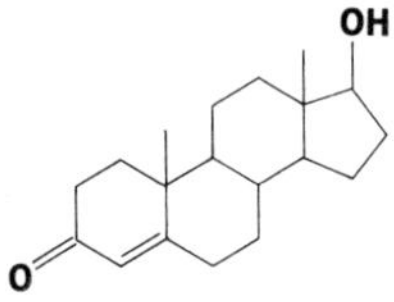

testosterone

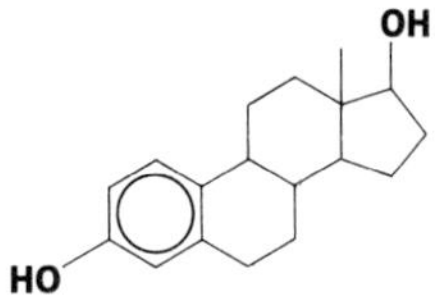

estradiol

The hormones of the adrenal gland, such as cortisol and aldosterone, as well as the hormone progesterone, are steroids. Progesterone helps to maintain pregnancy in the female, and is known to inhibit the release of ova. Cholesterol is the precursor of many of these sterols.

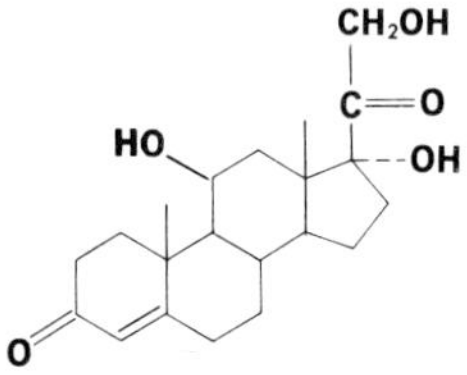

cortisol

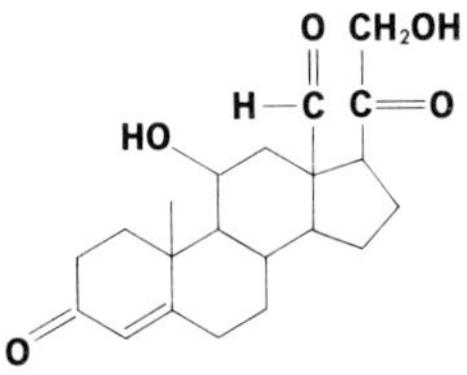

aldosterone

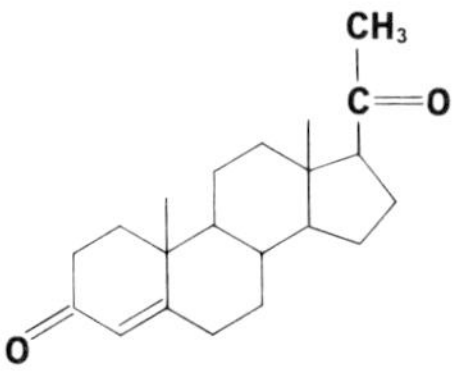

progesterone

A group of steroids which has recently received a great deal of attention comprises the oral contraceptives. Several compounds are used for this purpose, but they fall into two classes—the analogs of progesterone and those of the natural estrogens such as estradiol. Norethindrone and mestranol are widely used as oral contraceptives. Comparison of their formulas to those of progesterone and estradiol shows the obvious similarities.

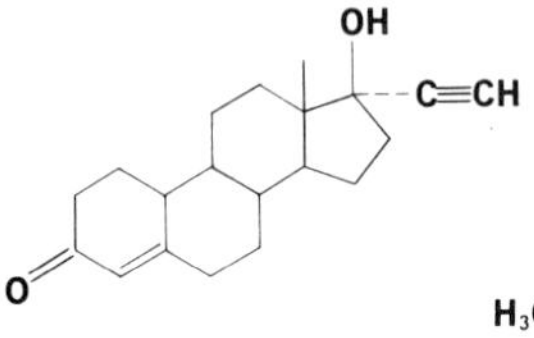

norethindrone

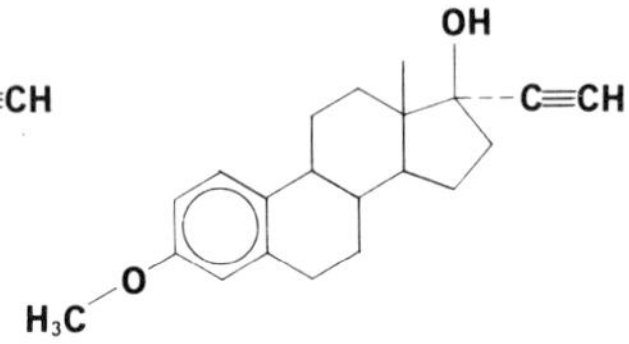

mestranol

When these two steroids are given either in a single pill or in separate pills at different times during the menstrual cycle they result in essentially normal menstruation with the exception that no ova (eggs) are released from the ovary. Similar results are produced by administering progesterone and estradiol; however, these hormones are relatively ineffective when given orally and must be injected for maximal effects. Thus we can see that a slight difference in chemical structure allows a steroid taken orally to function in a manner similar to one normally produced by the body. While estrogens are produced regularly following puberty, progesterone is normally produced only during pregnancy. The combination of the two steroids inhibits ovulation and prevents conception in a manner similar to that occurring during normal pregnancy. While the oral contraceptives are widely used, some bad side-effects have been reported. These drugs are currently available only upon prescription and should be taken under the supervision of trained medical personnel.

Vitamin D is not truly a steroid, since one of the rings of this compound has been ruptured, but it is closely related to the steroids. It is produced by the action of ultraviolet radiation on 7-dehydrocholesterol.

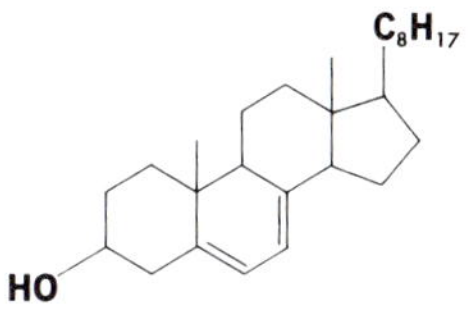

7-dehydrocholesterol

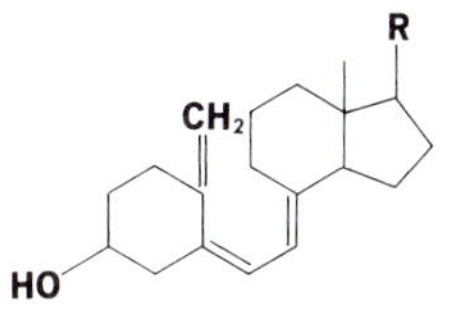

vitamin D

(If R is C_9H_{17}, the compound is vitamin D_2. If R is C_8H_{17}, the compound is vitamin D_3.)

The bile acids, which are necessary for the digestion of lipids, also contain a steroid nucleus. These acids are usually combined with an amino acid derivative when functioning biochemically.

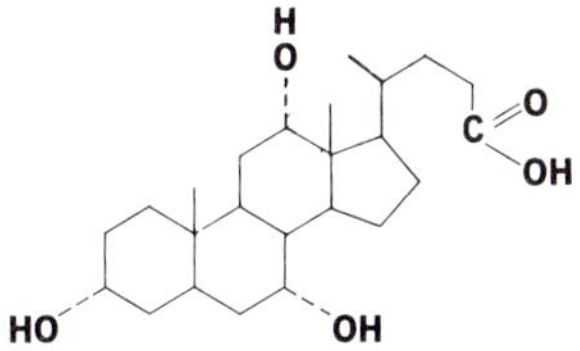

cholic acid

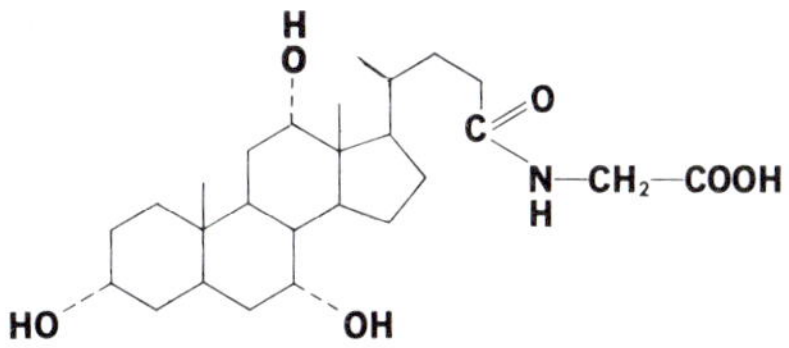

glycocholic acid (glycine derivative of cholic acid)

35–7 TERPENES

The substance we call turpentine can be obtained from the resin or gum of pine and fir trees. A great number of other compounds are related to turpentine in terms of both odor and chemical structure. As a group, these compounds are called terpenes, a name derived from an ancient spelling for turpentine.

The terpenes have a distinctive fragrance and can be isolated from natural sources by steam distillation. They are responsible for the odors of pine trees, citrus fruits, geraniums, and many other plants. Chemically, they can be considered as being derived from two isoprene units (which can, in turn, be derived from acetate); they contain 10 carbon atoms. Other similar compounds containing 15, 20, and 30 carbon atoms are called the sesquiterpenes, diterpenes, and triterpenes. Typical examples of terpenes are shown in the following structures:

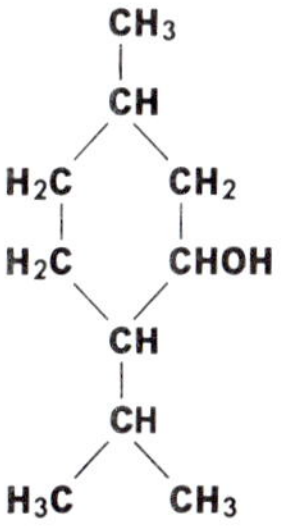

menthol

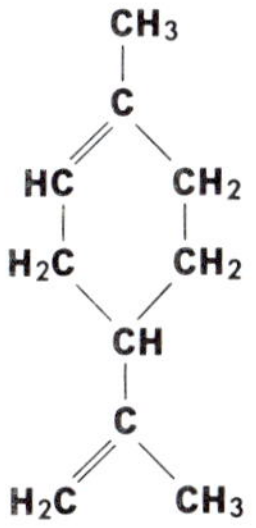

limonene
(from citrus oils)

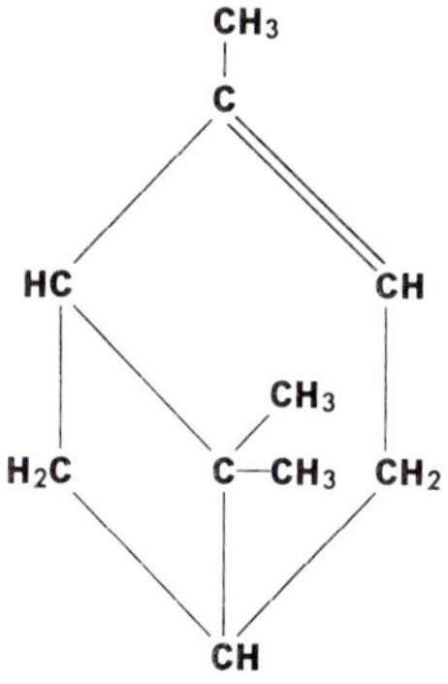

α-pinene

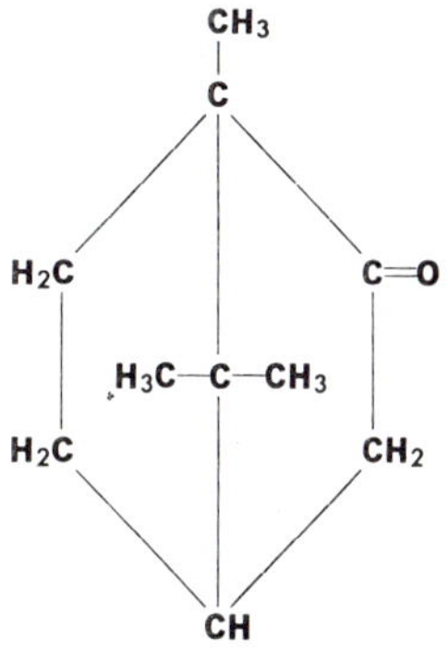

camphor

The carotenes, the orange pigment found in many green plants as well as in carrots, are related to the terpenes. Vitamin A also belongs to this classification although its molecular weight is sufficiently high to keep it from having the characteristic terpene odor.

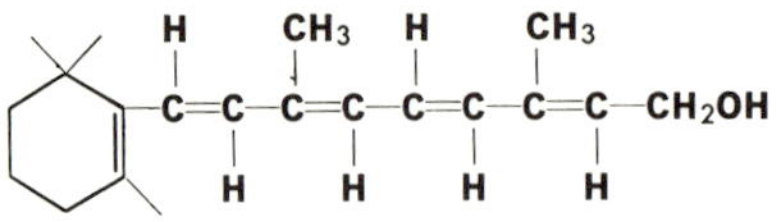

vitamin A (all *trans* form)

The odorous members of the terpenes apparently attract insects and thus aid the fertilization of plants. In photosynthesis the carotenes probably play some role in the absorption of light energy. Vitamin A has several functions in the human. It is present, for example, in rhodopsin (visual purple) and apparently plays an important role in vision (see Chapter 41).

35–8 SUMMARY

From the foregoing we can see that the classification lipid includes many different types of compounds. The triglycerides are probably of greatest importance, from a quantitative aspect, to most animals. These compounds provide energy when eaten, and serve as compounds in which energy can be stored for future use. The phospholipids are also widely distributed in the animal body. The role they play is not yet fully understood. Cholesterol is found in many places in animal tissues and is especially prevalent in the brain. Although found in only relatively small quantities, the steroid-derived vitamin D and the steroid hormones such as testosterone, estradiol, and cortisol are vital. When there is a deficiency or excessive amounts of them, the consequences are severe and may even be fatal.

Other lipids—particularly the terpenes—are of significance to plants, but only a few of them, such as vitamin A, are of biochemical importance to humans. However, the social and economic importance of these compounds, especially of rubber, is great.

IMPORTANT TERMS

Fat	Phospholipid	Triglyceride
Lipase	Saponification	Unsaturation
Lipid	Steroid	Wax
Oil	Terpene	

WORK EXERCISES

Before you do the following exercises, you may wish to review Chapter 27.
1. List the hydrolysis products of
   a) triglycerides b) phospholipids c) waxes
   d) steroids e) terpenes
2. Draw the structure of
   a) a triglyceride containing stearic, palmitic, and lauric acids. Is there more than one possible answer to this question?

b) a highly unsaturated triglyceride
   c) a fat
   d) a lecithin containing lignoceric and arachadonic acids
   e) cetyl stearate
3. Write equations for the
   a) hydrolysis of myristyloleyllaurin
   b) addition of bromine to palmitoleic acid
   c) saponification of a triglyceride
4. Indicate whether you think each of the following triglycerides would be a fat or an oil:

   a) tripalmatin   b) triolein   c) tributyrin

   [You may wish to check your predictions by consulting a handbook of chemistry.]
5. Would you expect fats or oils to have the greater tendency to become rancid?
6. On the basis of the two causes of rancidity, how might you protect triglycerides from developing this undesirable characteristic?
7. Cow's milk produced when cattle are grazing on green pastures is usually appreciably yellower than that produced when they are fed dried alfalfa or other dry food. Can you give a reason for this?
8. What information can you find on labels of various food products that are high in lipids, such as cooking oils, oleomargarines, and butter, about the source and content of lipids in these foods?
9. From a general biochemistry text,
   a) Find the formulas and significance of steroids other than those discussed in this chapter.
   b) List the formulas and explain the significance of other terpenes.

SUGGESTED READING

Dole, V. P., "Body Fat," *Scientific American*, Dec. 1959. A discussion of depot fat and obesity.

Fieser, L. F., "Steroids," *Scientific American*, Jan. 1955 (Offprint #8). A survey article written by an authority in the field of steroids.

Pike, J. E., "Prostaglandins," *Scientific American*, Nov. 1971 (Offprint #1235). A good discussion of these hormone-like lipids.

CHEMISTRY OF THE NUCLEIC ACIDS 36

36–1 INTRODUCTION

The carbohydrates, proteins, and lipids, along with the mineral elements and water, constitute about 99% of most living organisms. The remaining 1%, however, includes some of the most important biochemical compounds. Without these compounds, cells would be unable to exist for any appreciable length of time and could never reproduce.

Probably the most important of these minor constituents are the nucleic acids. These compounds were so named because they were originally found in the nuclei of cells. We now know that they are also found in the cytoplasm and that there is some difference both in location and function of the two basic kinds of nucleic acids. Deoxyribonucleic acid (DNA) is found almost totally in the nucleus, and ribonucleic acid (RNA) is found both in the nucleus and in the cytoplasm of cells. From the names it is obvious that the presence or absence of an oxygen in a ribose component of the nucleic acid molecule is a chemical difference between DNA and RNA. There are other differences as will be pointed out later. While the functions of RNA and, to some extent, of DNA still are not totally known, it seems that DNA plays a primary role in transmission of hereditary information from one cell to its daughter cells, and that the functions of RNA are secondary. The principal known function of RNA is its involvement in protein synthesis.

36–2 COMPOSITION AND STRUCTURE OF THE NUCLEIC ACIDS

The hydrolysis of a pure sample of a nucleic acid yields phosphoric acid, a pentose (either ribose or 2-deoxyribose), and a mixture of organic compounds that have the properties of bases. (See Figs. 36–1 and 36–2.) The five most common bases are adenine, guanine,

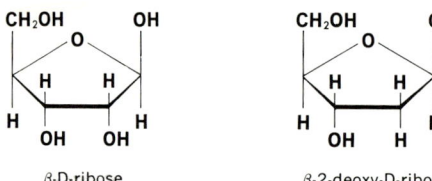

Fig. 36–1 Structures of the pentoses found in nucleic acids.

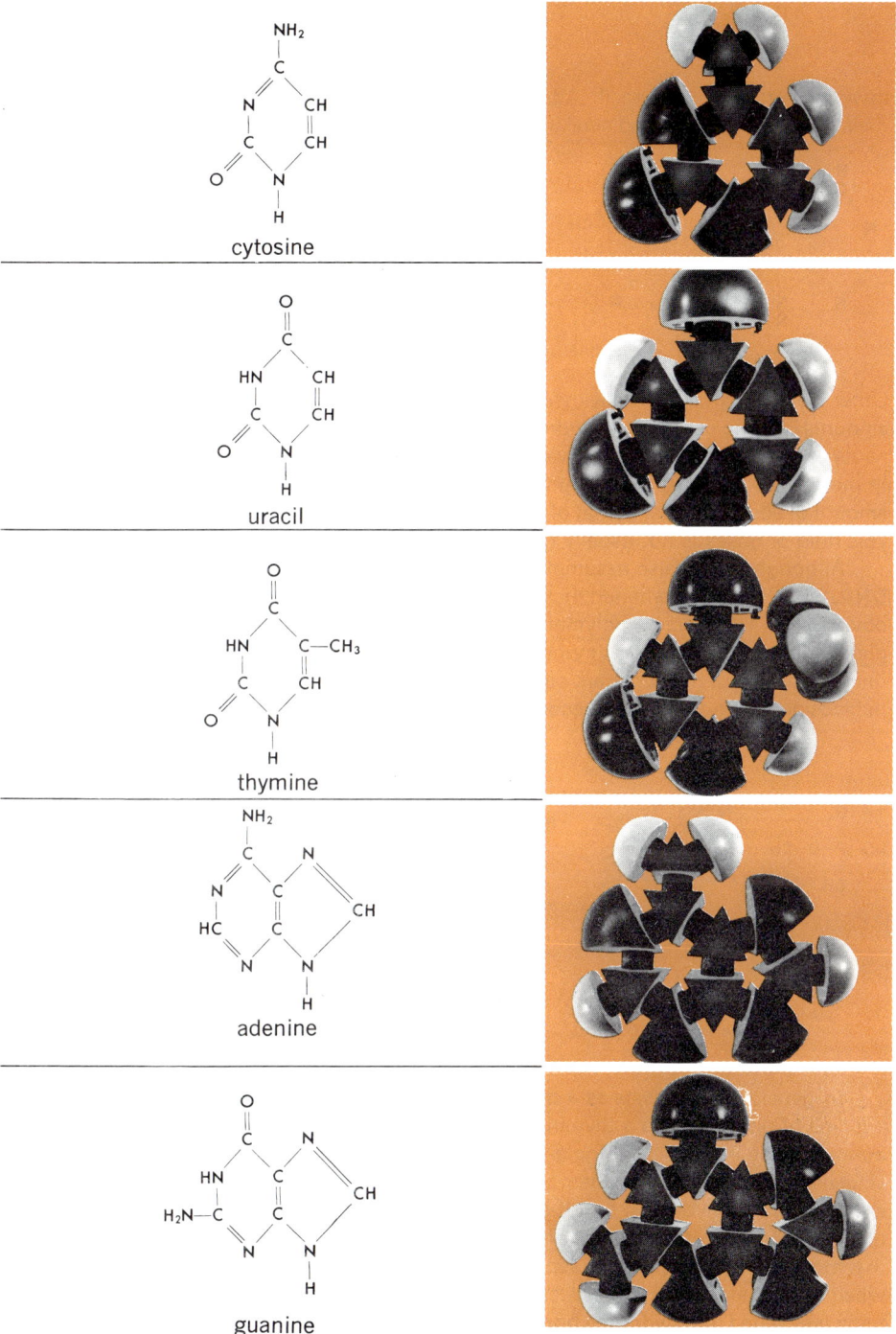

Fig. 36-2 Structures of pyrimidine and purine bases of nucleic acids. [Adapted from Snell, Shulman, Spencer, and Moos, *Biophysical Principles of Structure and Function*, Addison-Wesley, 1965.]

thymine, cytosine, and uracil. Adenine and guanine are called **purine bases** because of their relation to purine, and similarly thymine, cytosine, and uracil are **pyrimidine bases**.

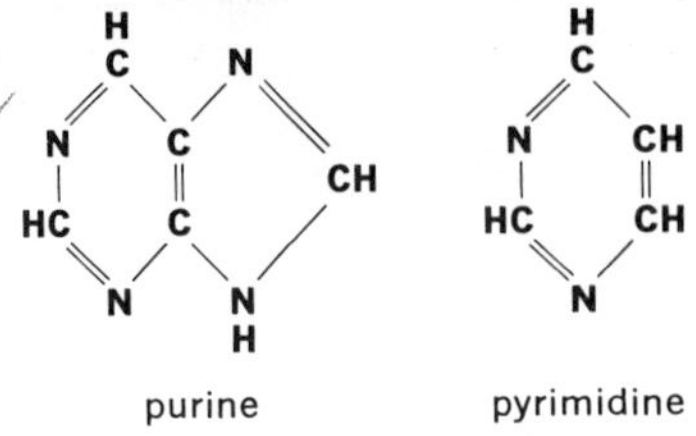

purine   pyrimidine

Other bases such as 5-methylcytosine and 5-hydroxymethylcytosine are found in small amounts in some nucleic acid hydrolysates. Although the amounts of the different components vary somewhat, there is always one phosphate group and one pentose for each base. In most kinds of DNA, the amount of purines is equal to the amount of pyrimidines, the amount of adenine equal to that of thymine, and the amount of guanine to that of cytosine. This ratio of bases is not usually found in RNA.

Although some false assumptions were originally made concerning the structure of DNA, when it was determined that the molecular weight of DNA was very high, it became obvious that DNA was a polymer of a basic unit containing a base, deoxyribose, and a phosphoric acid group. This monomeric unit is a **nucleotide**. The structures containing only the base and the sugar are called **nucleosides**. The structure of a nucleotide and a nucleoside are shown in the accompanying diagram. See also Table 36–1.

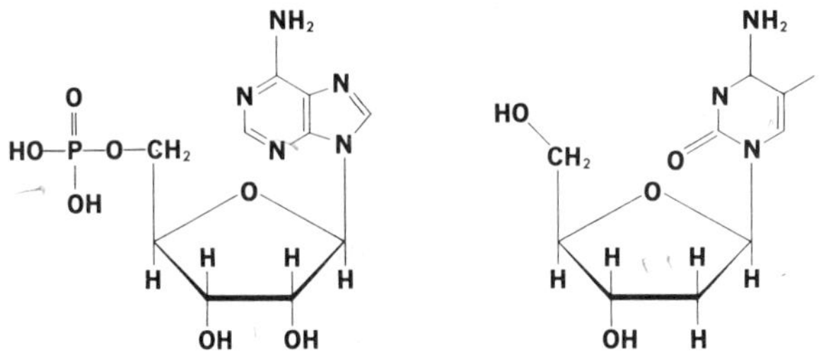

adenosine 5'-monophosphate
or adenylic acid
(a nucleotide)

deoxycytidine
(a nucleoside)

In general, DNA exists as a double-stranded polymer with hydrogen bonds holding adenine of one strand to a thymine in another strand, with cytosine and guanine similarly bonded. (See Fig. 36–3.) Single-stranded DNA is found in some viruses; however, such single-stranded forms now seem to be a minor exception to the rule that DNA is double-stranded. Three hydrogen bonds are possible between cytosine and guanine, while only two are present in the adenine-thymine pair. This finding correlates well with the fact that more energy is required to separate the two strands of DNA (a process called *melting*) when the guanine-cytosine content of the DNA is high. See Fig. 36–4. When DNA solutions are heated, the hydrogen bonds between strands are disrupted, and the strands separate as shown in (b). Slow recooling permits essentially normal double-stranded DNA. Rapid chilling produces some intrastrand bonds as well as some interstrand bonds.

The three-dimensional structure of the nucleic acids was established on the basis of X-ray diffraction studies. We know now that DNA is a double-stranded, spiral (helical)

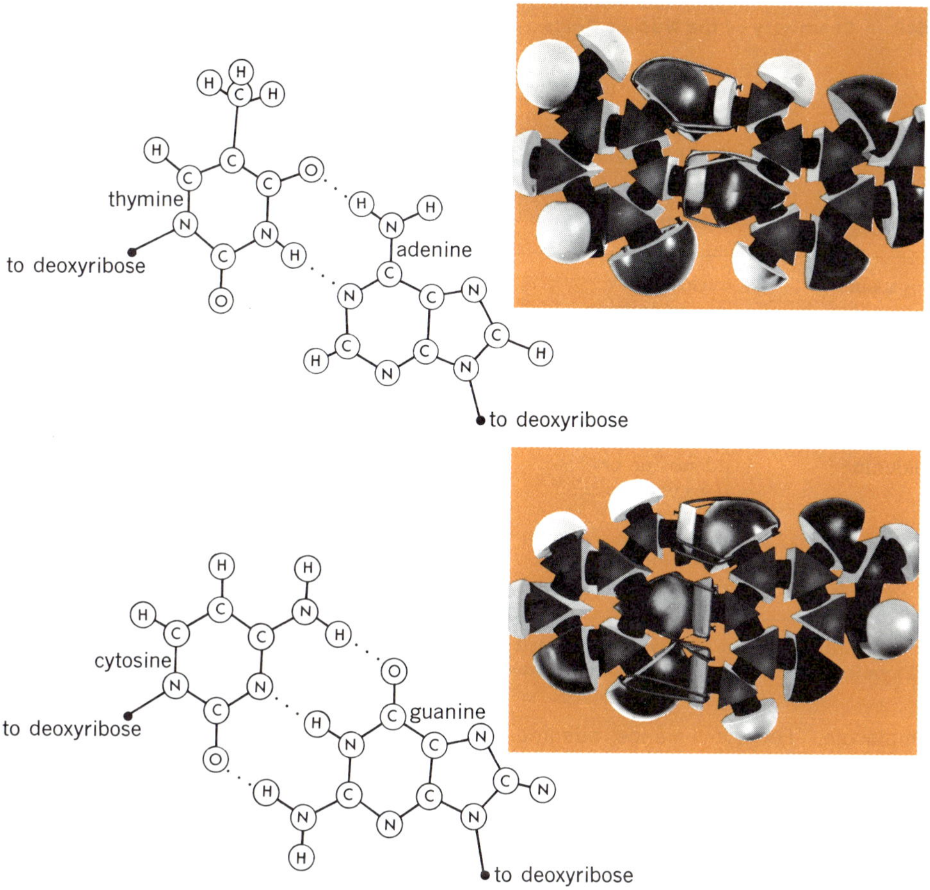

Fig. 36–3 Hydrogen bonds between the adenine:thymine and cytosine:guanine pairs of bases. Note that there are two bonds between adenine and thymine and three between guanine and cytosine. [Photographs from Snell, Shulman, Spencer, and Moos, *Biophysical Principles of Structure and Function*, Addison-Wesley, 1965. Drawings reproduced by permission of Academic Press from *Archs. Biochem. Biophys.* **65**, 164 (1956).]

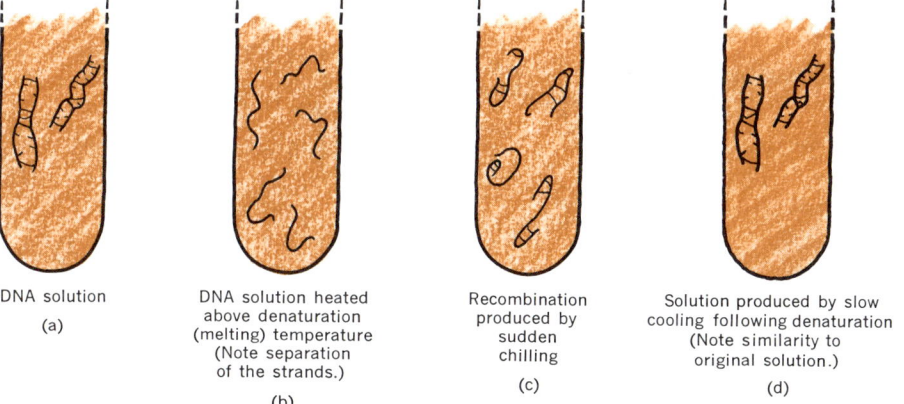

DNA solution
(a)

DNA solution heated above denaturation (melting) temperature (Note separation of the strands.)
(b)

Recombination produced by sudden chilling
(c)

Solution produced by slow cooling following denaturation (Note similarity to original solution.)
(d)

Fig. 36–4 A representation of the separation and recombination of DNA strands.

Chemistry of the Nucleic Acids

TABLE 36-1 Names of nucleotides and nucleosides

Base	Nucleoside	Nucleotide
cytosine	cytidine or deoxycytidine	cytidylic or deoxycytidylic acid
uracil	uridine or deoxyuridine	uridylic or deoxyuridylic acid
thymine	thymidine*	thymidylic acid*
adenine	adenosine or deoxyadenosine	adenylic or deoxyadenylic acid (adenosine phosphate)
guanine	guanosine or deoxyguanosine	guanylic or deoxyguanylic acid

* In nature thymine is always combined with deoxyribose, not ribose. Thymidine and thymidylic acid contain deoxyribose, not ribose.

molecule. For convenience, adenine, thymine, guanine, and cytosine are designated by the first letters of their names in the representation of the structure of DNA shown in Fig. 36-5. Other visualizations of the DNA structure are shown in Fig. 36-6.

In addition to the obvious difference in having ribose instead of deoxyribose, RNA contains uracil and does not contain thymine. Since its molecular weight is usually lower than that of DNA, and since there seems to be no uniform ratio of bases, we could postulate that RNA is only single-stranded. On the basis of research, it seems that most forms of RNA are single-stranded and have a helical structure in only part of their molecules. There seem to be at least three types of RNA that can be separated by essentially physical

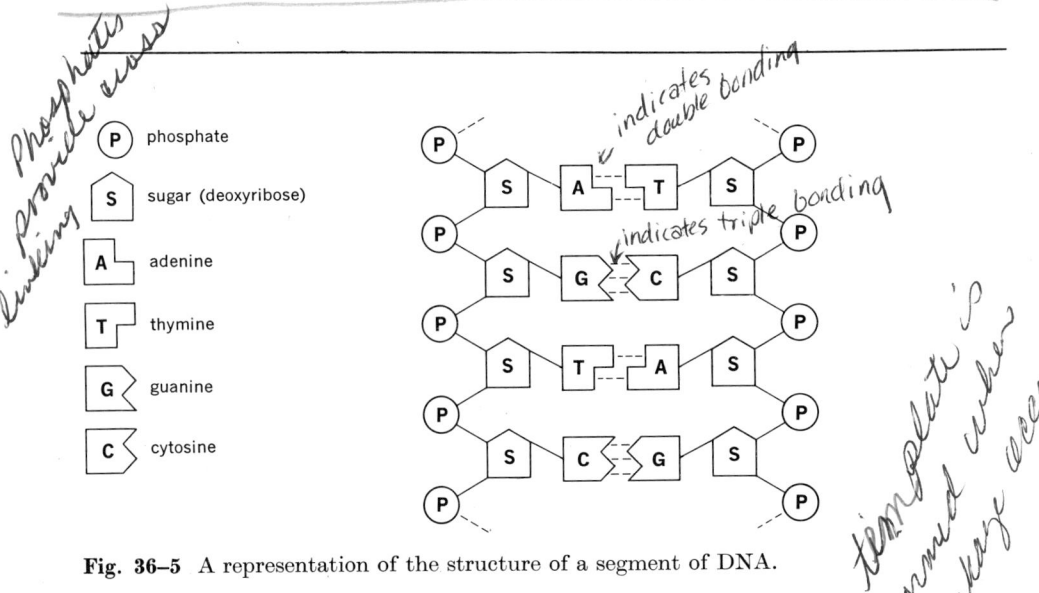

Fig. 36-5 A representation of the structure of a segment of DNA.

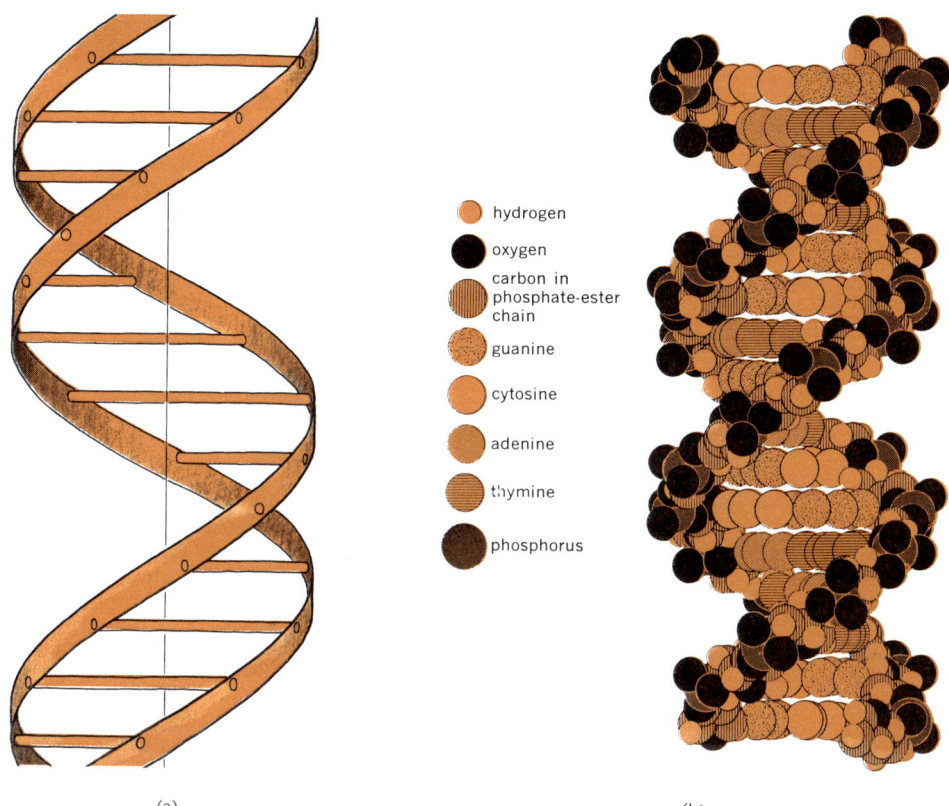

(a) (b)

Fig. 36-6 Schematic representations of the molecule of DNA: (a) a simplified model showing the two strands linked by ladderlike steps formed by the base pairs; (b) a more complex representation of the same structure. [(b) is reprinted by permission from *CA, A Cancer Journal for Clinicians* and from Dr. L. D. Hamilton; *CA, A Bulletin of Cancer Progress* **5**, 163 (1955) from material provided by Dr. W. E. Seeds and Dr. H. R. Wilson, Wheatstone Physics Laboratory, King's College, London.]

means: messenger RNA, ribosomal RNA, and soluble or transfer RNA. The names given these forms of RNA reflect the function they are believed to play in protein synthesis. The role of nucleic acids in protein synthesis is presented in Chapter 39.

36-3 OCCURRENCE OF NUCLEIC ACIDS

Nucleic acids are defined as those complex biochemical substances which yield, upon hydrolysis, purine and pyrimidine bases, ribose or deoxyribose, and phosphate. They are found in all living cells where they are usually present at a level of less than 1% of the total net weight of the cell. Yeast cells and some bacterial cells may contain up to 5% nucleic acid. The bacteriophages may contain as much as 60% DNA. It has been estimated that if all the DNA in an average adult human were uncoiled and laid out in a straight line it

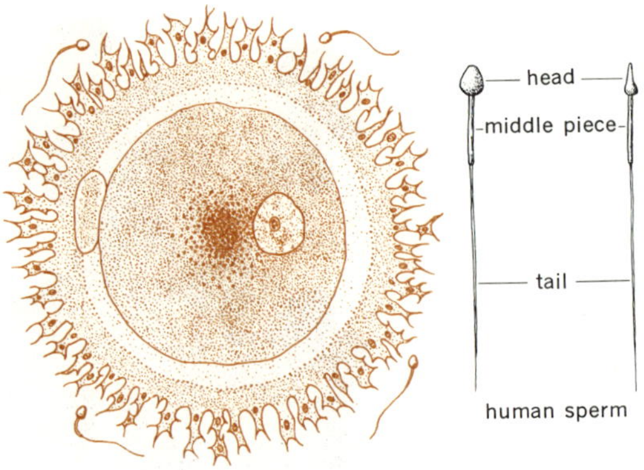

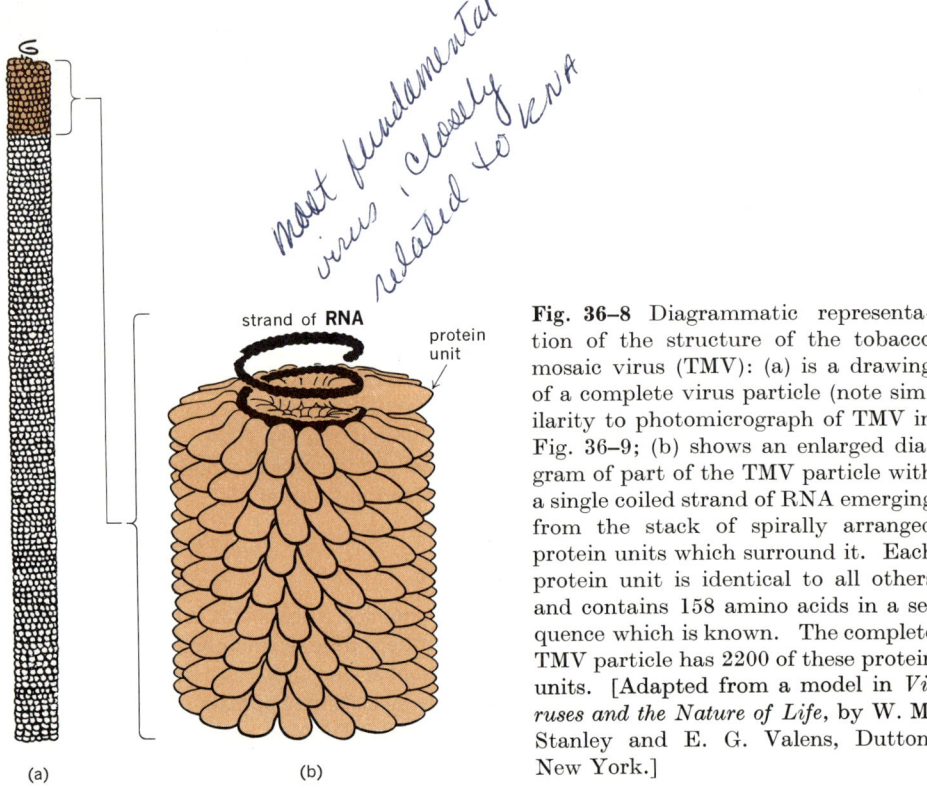

Fig. 36–7 Sperm cells. (a) shows a drawing of a human egg and sperm cells and (b) is reproduced from a microscopic slide of the sperm of a bull. [Adapted from Baker and Allen, *The Study of Biology*, Addison-Wesley, 1967.]

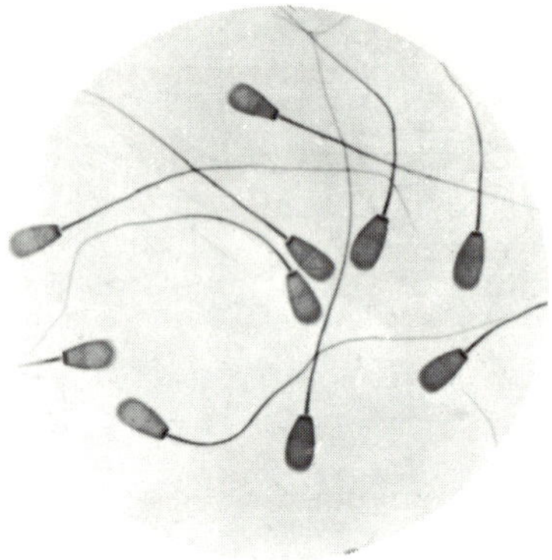

most fundamental virus closely related to RNA

Fig. 36–8 Diagrammatic representation of the structure of the tobacco mosaic virus (TMV): (a) is a drawing of a complete virus particle (note similarity to photomicrograph of TMV in Fig. 36–9; (b) shows an enlarged diagram of part of the TMV particle with a single coiled strand of RNA emerging from the stack of spirally arranged protein units which surround it. Each protein unit is identical to all others and contains 158 amino acids in a sequence which is known. The complete TMV particle has 2200 of these protein units. [Adapted from a model in *Viruses and the Nature of Life*, by W. M. Stanley and E. G. Valens, Dutton, New York.]

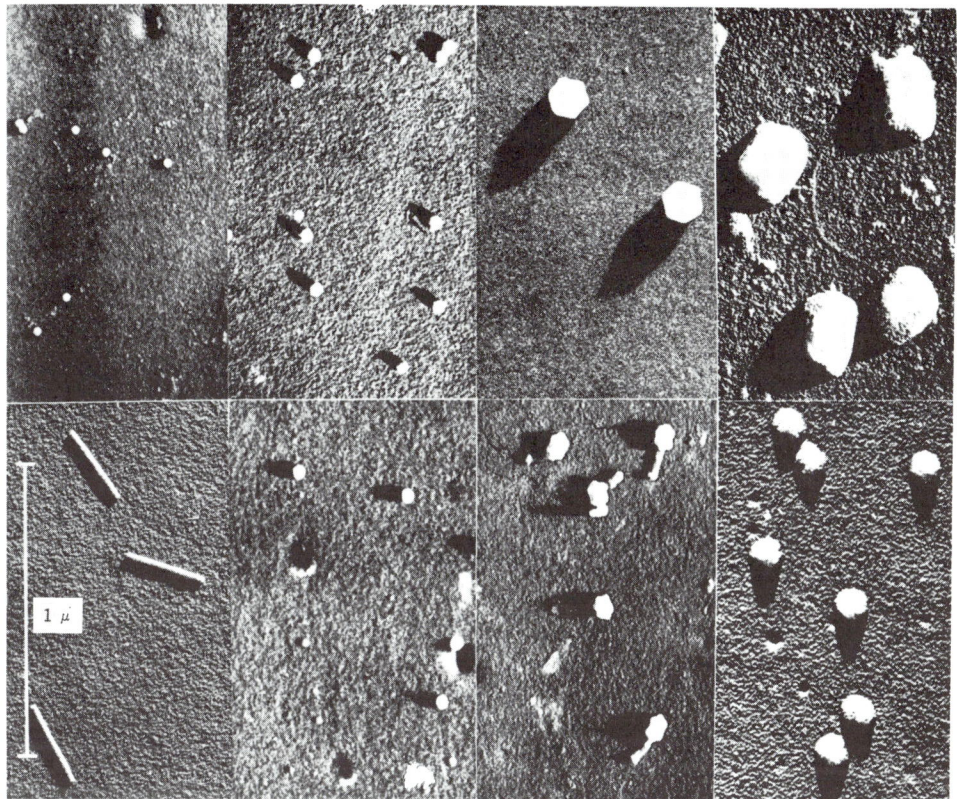

Fig. 36-9 Electron micrographs of eight typical viruses at the same magnification. Top (left to right): poliomyelitis, small bacteriophage, *Tipula irridescens* (an insect virus showing individual particles, not crystals), vaccinia (the virus that causes smallpox). Bottom: tobacco mosaic virus, rabbit papilloma, large bacteriophage, influenza virus. [Courtesy of the Virus Laboratory, University of California, Berkeley.]

would would reach five billion miles. This long string would, however, be only one or two angstroms wide.

In almost all instances the nucleic acids are found in combination with proteins in the form of nucleoproteins. Many of the functions of nucleic acids are performed only poorly, if at all, when the specific protein they ordinarily are associated with is missing. Sperm cells are composed of a DNA nucleoprotein. (See Fig. 36-7.) Viruses contain either DNA or RNA surrounded by a protein coating. (See Figs. 36-8 and 36-9.) In some cases it has been possible to separate the protein outer coat from the inner thread of the virus and then to recombine the two fractions into a virus which is able to infect living cells.

36-4 SYNTHESIS OF NUCLEIC ACIDS

The nucleic acids can be synthesized from simple materials by most organisms. Some bacteria require adenine or other bases in order to synthesize their nucleic acids, but most

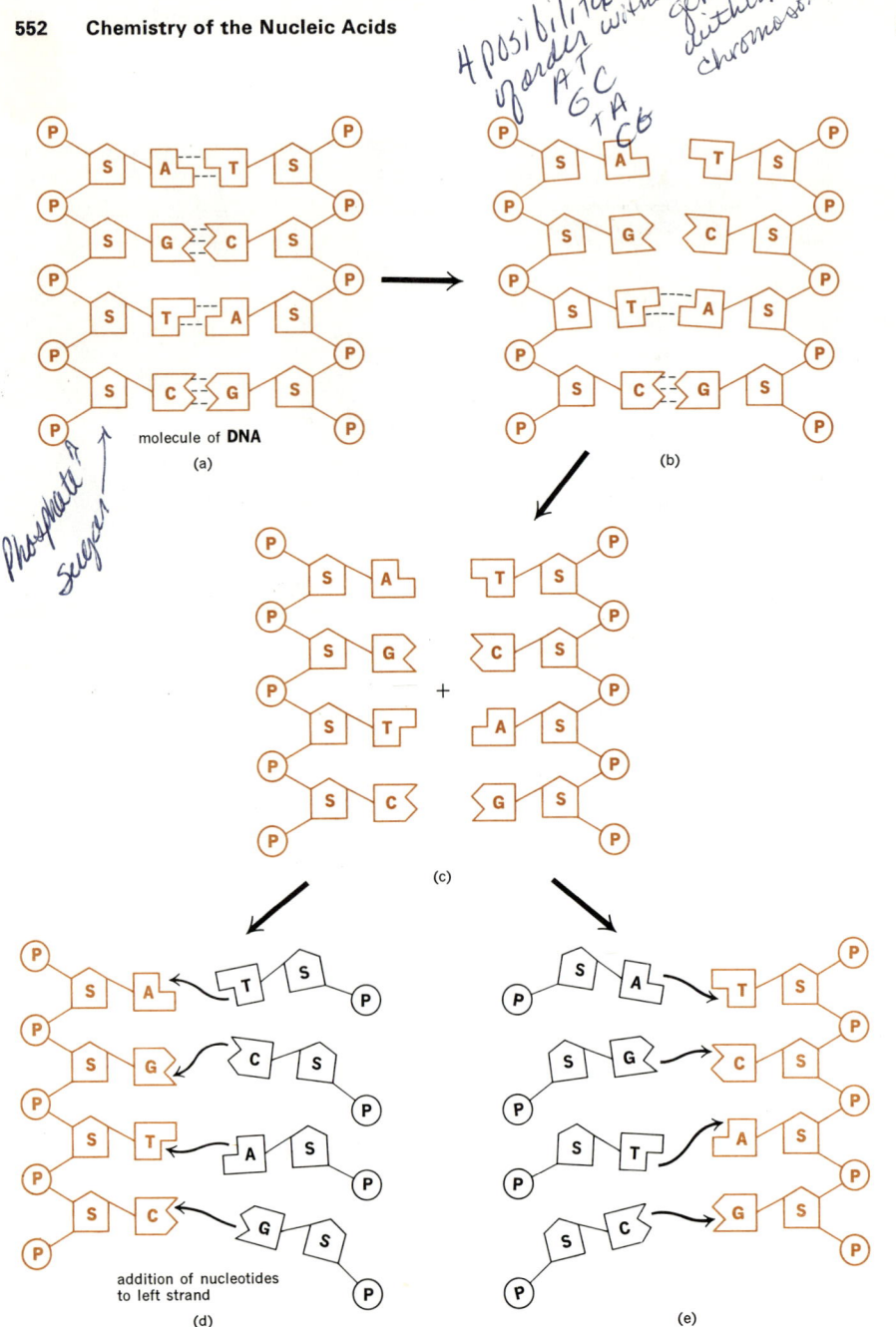

Fig. 36-10 The process of duplication (replication) of DNA. (a) shows part of the molecule of DNA (see Fig. 36-5 for identification of components). In (b) the two strands have begun to separate and in (c) they are completely apart. (d) shows how individual nucleotides join with the left strand of the DNA to form another double-stranded molecule, and (e) shows a similar process for the right strand of the original DNA. Since each half of the original DNA molecule has now produced a molecule identical to the original DNA molecule, duplication has occurred. This process occurs in the chromosomes of the cell nucleus prior to cell division.

organisms can synthesize the pentoses and bases, and combine them with phosphate from their diet to assemble the nucleic acids required. While each kind of organism makes its own specific kind of DNA, the RNA from different species seems to be quite similar. Some polymers that can function in place of natural RNA have been synthesized *in vitro*. The synthesis, however, requires an enzyme as a catalyst. Other enzymes can catalyze the synthesis of DNA when provided with the proper starting materials and a small amount of already-formed DNA as a primer. The synthesized DNA is similar in physical properties to the DNA of the primer, but until recently such synthetic DNA's had not been shown to have biological activity. Late in 1967, Kornberg and his co-workers reported the synthesis of a viral DNA that could infect microorganisms. Although it is probably incorrect to describe these experiments as the creation of life in a test tube, they do represent a major advance in the synthesis of biologically active molecules. Synthetic DNA may some day be used to change hereditary characteristics. The synthesis, or replication, of DNA is shown in Fig. 36–10.

36–5 PHYSICAL PROPERTIES OF THE NUCLEIC ACIDS

Samples of DNA have been found to have an extremely high molecular weight. Values in the millions have been found, and it is accurate to say that the nucleic acids are probably the largest molecules known. Although some lower values for molecular weights have also been reported, it is likely that the samples were torn apart in the process of separating them from other cell components. The alternate possibility, that the high molecular weight components were aggregations of smaller molecules present in the intact cell seems less likely, but cannot be ignored. It is probable that all the DNA in some bacterial and viral cells is represented by one single molecule. There are many forms of RNA, some of which have molecular weights as low as 75,000 while others have weights probably extending into the million range.

Because of their many polar groups and in spite of their high molecular weight, nucleic acids are water-soluble. The bases of the nucleic acids absorb ultraviolet radiation in the 250- to 290-millimicron range. This property is commonly used to identify and measure the amount of nucleic acids in a sample.

36–6 CHEMICAL PROPERTIES OF THE NUCLEIC ACIDS

The hydrolysis of nucleic acids is catalyzed either by acids, bases, or by enzymes. Enzymes which catalyze the hydrolysis of only RNA (RNA-ases) or DNA (DNA-ases) are widely used in studies of nucleic acids. The first products of hydrolysis of the nucleic acids are nucleotides, but these can be further hydrolyzed to give nucleosides, and finally to give the bases, pentose and phosphate. Each step in the hydrolysis requires a specific type of enzyme.

The amino-containing bases of the nucleic acids can react with nitrous acid with the removal of the amino group. It is interesting to correlate this with the finding that nitrous acid can produce mutations when it reacts with DNA.

36–7 THE IMPORTANCE OF NUCLEIC ACIDS

The role nucleic acids play in heredity and in protein synthesis is discussed in Chapters 39 and 42. A third important role played by nucleic acids is in the action of viruses. We tend

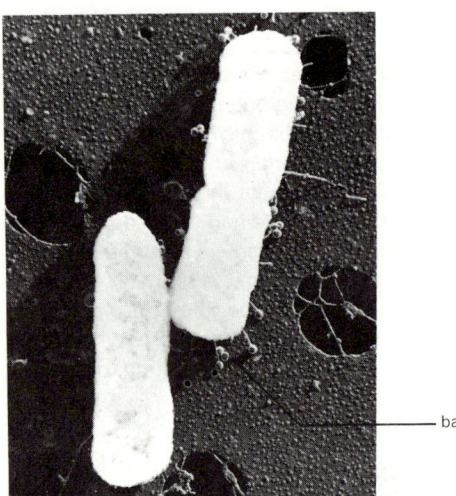

Fig. 36–11 A photomicrograph showing several bacteriophages attached to one of the two cells of the bacterium *E. Coli*. [Courtesy of Thomas F. Anderson, Francois Jacob, and Elie Wollman.]

extremely imp.

Fig. 36–12 The structure of ATP (adenosine triphosphate). [From Snell, Shulman, Spencer, and Moos, *Biophysical Principles of Structure and Function*, Addison-Wesley, 1965.]

sugar metab.
Insulin production

to think of viruses as causing diseases, which they do, but they are of great theoretical interest also. A virus represents a kind of bridge between living and nonliving materials. Many viruses have been isolated as crystallized nucleoproteins. In this form they can be stored in bottles on shelves for long periods of time. Yet, when added to living organisms or to cultures containing living cells, the virus can take over the function of the living cell so that it produces new virus particles instead of new cells. So far scientists have not been able to get viruses to reproduce even in a medium containing all known nutritional requirements of cells unless living cells are also present. They seem to function only when they assume control of the reactions of a cell.

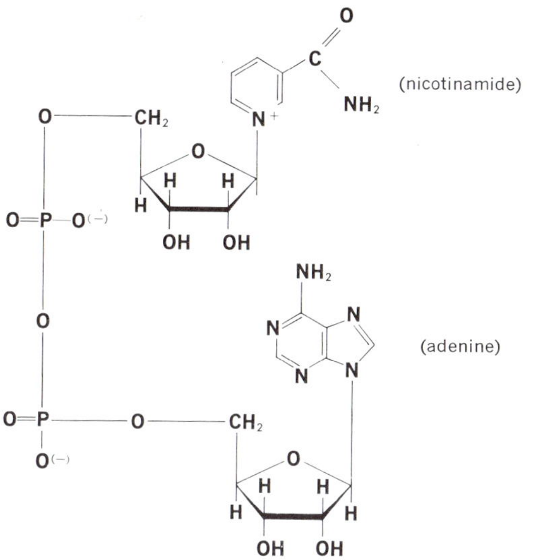

Fig. 36–13 The structure of NAD (nicotinamide adenine dinucleotide).

One type of virus, the bacteriophages, which are viruses that attack bacteria, has been studied extensively. These viruses consist of an inner core of nucleic acid surrounded by an outer coat of protein. They attach themselves to the wall of a bacterium and inject their nucleic acid into the bacterium. (See Fig. 36–11.) The injected nucleic acid then takes over the metabolism of the cell and so directs it that thousands of new virus particles, each capable of infecting other bacteria, are produced. In other instances the virus is apparently present in the bacterial cell in an inactive or latent form and the virus is, in fact, reproduced along with other cell components for several generations. When this infected cell is exposed to some kind of shock, such as ultraviolet irradiation or certain chemical agents the virus becomes active and directs the cell towards its own destruction by production of virus particles.

Some forms of cancer are caused by viruses. In fact, the Nobel prize-winning scientist, Dr. Wendell Stanley, stated many years ago that he believed all cancer was caused by viruses. Other scientists have speculated that many humans carry inactive viruses which can cause cancer when irradiated or otherwise stimulated. The development of new research techniques coupled with a better knowledge of the biochemistry of nucleic acids may lead to the understanding of cancer and to cures or controls for this disease.

Nucleotides have been discussed as the monomeric units from which the polymeric nucleic acids were made. Nucleotides which are present as such, not as polymers, also play an important role in biochemical reactions. Adenosine triphosphate (ATP), a nucleotide with two additional phosphate groups, serves as the energy source for hundreds of biochemical reactions (Fig. 36–12). When either of the two end phosphate groups is removed by hydrolysis, a large amount of energy is made available for driving energy-requiring reactions. Such diverse reactions as synthesis of proteins, muscular contraction, and the production of light by fireflies all require ATP as an energy source. Triphosphates of guanine and other bases serve similar roles.

The form of adenosine monophosphate in which the phosphate is bound to both the 3 and 5 carbons of the ribose is called cyclic AMP. This substance is an important regulator of metabolic reactions. It is involved in the action of many hormones.

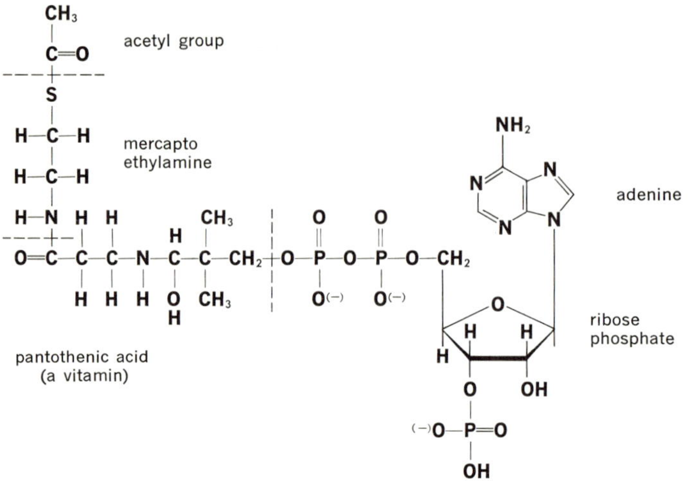

Fig. 36–14 The structure of coenzyme A (CoA). An acetyl group is shown attached to the CoA to give acetyl CoA.

The active form of many of the vitamins is a nucleotide. In some cases the vitamin itself is the base, combined with a ribose and phosphate unit; in others, adenine or some other base is required. (See Figs. 36–13 and 36–14.)

The reason why a nucleotide "handle" is so widely found in the structure of extremely important biochemical compounds is not known. While it may be only accidental or coincidental that the base-pentose-phosphate combination is so widespread, this apparent coincidence is the kind of observation that makes a biochemist want to track down the significance of this chemical structure.

IMPORTANT TERMS

Bacteriophage
Deoxyribonuclease (DNA-ase)
Dexoyribonucleic acid (DNA)
Nucleic acid
Nucleoprotein

Nucleoside
Nucleotide
Primer DNA
Purine base

Pyrimidine base
Ribonuclease (RNA-ase)
Ribonucleic acid (RNA)
Virus

WORK EXERCISES

1. Draw the formulas for the following:

   a) deoxycytidine   b) uridylic acid   c) thymidylic acid
   d) deoxyguanosine   e) adenosine diphosphate (ADP)

2. List the major hydrolysis products of each of the following:

   a) DNA   b) RNA   c) deoxyuridine   d) adenylic acid   e) guanine

3. Draw the structure for the trinucleotides specified by the symbols CGA and GGT. Locate possible hydrogen bonds between these nucleotides. What is the maximum number of hydrogen bonds possible between these two trinucleotides?
4. What would be the expected products if
    a) DNA-ase were added to a sample of nucleic acids purified from a specimen of liver?
    b) RNA-ase were added to a purified sample of DNA?
5. What are the major differences between DNA and RNA?

Consult a standard biochemistry text for answers to the following:

6. What bases other than guanine, adenine, cytosine, thymine, and uracil are found in the nucleic acids? What percentage of total bases is represented by these "other" bases?
7. What techniques are used to separate and analyze the components of the nucleic acids?

SUGGESTED READING

Britten, R. J., and D. A. Ross, "Repeated Segments of DNA," *Scientific American*, April 1970 (Offprint #1173). Apparently there are many copies of much of our DNA—the authors give the evidence and discuss the possible significance of this duplication.

Crick, F. H. C., "Nucleic Acids," *Scientific American*, Sept. 1957 (Offprint #54). A basic article by one of the pioneers in establishing the structure of nucleic acids.

Hoagland, M. B., "Nucleic Acids and Proteins," *Scientific American*, Dec. 1959 (Offprint #68). A discussion of the basic role of nucleic acids in protein synthesis. See the references at the end of Chapter 39 for further articles relating to this topic.

Holley, R. W., "The Nucleotide Sequence of a Nucleic Acid," *Scientific American*, Feb. 1966. A report of the first determination of structure of a molecule of RNA by the scientist who supervised the work.

Horne, R. W., "The Structure of Viruses," *Scientific American*, Jan. 1963. This article, which is illustrated with electron micrographs of virus particles, discusses the general structure of a variety of viruses.

Kornberg, A., "The Synthesis of DNA," *Scientific American*, October 1968 (Offprint #1124). Nobel Prize winner Kornberg describes his synthesis of DNA in this well-illustrated article.

Mirsky, A., "The Discovery of DNA," *Scientific American*, June 1968 (Offprint #1109). This account of Miescher's work with "nuclein" illustrates the processes of scientific discovery and shows how our ideas change.

Temin, H. M., "RNA-Directed DNA Synthesis," *Scientific American*, January 1972 (Offprint #1239). The transfer of information from RNA to DNA is the reverse of that usually observed—it may be significant in production of cancer by viruses.

BIOCHEMICAL REACTIONS 37

37–1 INTRODUCTION

With the exception of energy from radioisotopes, all energy used on earth comes from the sun. The energy released from petroleum and coal was trapped long ago; that from wood and from food we eat left the sun only a relatively short time ago. Devices for making direct use of the sun's energy through solar cells are now being developed.

For millions of years plants have used chlorophyll and other pigments to trap radiation from the sun and convert this energy into the potential energy of chemical compounds. This ancient process of energy conversion has only recently been scientifically investigated and there are still many unanswered questions concerning the nature of photosynthesis. As the name implies, **photosynthesis** involves synthesis of all the different compounds found in plants utilizing the energy of sunlight. The synthesis of glucose is often summarized as

$$6CO_2 + 6H_2O \rightarrow C_6H_{12}O_6 + 6O_2$$

The special biochemistry of plants and particularly of photosynthesis will be discussed in Chapter 43. In this chapter ways are discussed in which animals use the energy stored in food for the many processes that are required for life.

Although animals absorb some radiation from the sun, energy for the real work of the body is provided by the food the animal eats. The green plant is the original source of this useful energy; for flesh-eating organisms, the energy of the plant has been converted to energy stored in meat of the animal which is eaten. (See Fig. 37–1.) The processes by which an animal utilizes food have been categorized as ingestion, digestion, assimilation, followed by the elimination of unused material and waste products. Biochemists are concerned with digestion and especially with assimilation of digested materials. In some cases, they are interested in the excretions, particularly the urine, for evidence of the abnormal functioning of an animal.

37–2 DIGESTION

Following ingestion of food there is often some physical separation of the food particles. Whereas mammals and insects can chew their food, other animals must search for small morsels of food, since they have poorly developed mouth parts. In most animals that

Fig. 37–1 A biological food chain. Energy from the sun is trapped by a plant and converted into the potential energy of its leaves, seeds, and fruit. An animal, such as a cow, eats the plants and converts the energy stored in plants to that in its body or in the milk which it produces. A man may derive energy either by drinking the milk produced by the cow or by eating the flesh of the animal as steak or hamburger. At each step some of the energy is converted to heat or is lost to the surroundings by other means.

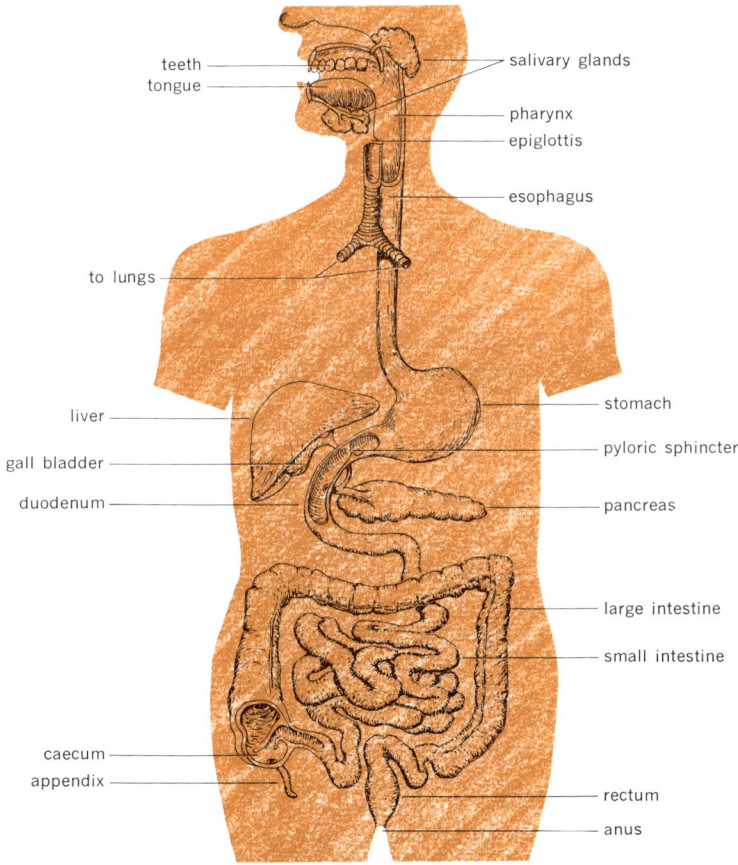

Fig. 37–2 The digestive system of the human. [From John W. Kimball, *Biology*, Addison-Wesley, 1965.]

are able to chew, chewing not only separates the food into small particles but also moistens and lubricates the food with saliva. In the human, as well as in many other animals, the saliva contains enzymes which catalyze the first steps in the process of digestion.

The **digestive system,** or **gastrointestinal tract,** is essentially a partially filled tube extending between the mouth and the anus. (See Fig. 37–2.) Food in the tract, which is called the alimentary canal, is not really "inside" the body. It enters and becomes part of the body only when it is absorbed through the intestinal wall. We can describe digestion and assimilation briefly by saying that as the food passes through this hollow tube, various secretions are added, some material enters into the body via either the blood or lymph, some waste products are added to the indigestible food residue, bacteria act upon the contents of the intestine, and finally the remaining material, the feces, is excreted.

Chemically, digestion is the hydrolysis of ingested food into smaller molecules. Polysaccharides are converted to monosaccharides, proteins to amino acids, and the triglycerides are broken down into fatty acids and glycerol. The function of digestion seems to be the production of molecules small enough to be absorbed through the intestinal wall.

The hydrolyses of digestion are catalyzed by enzymes which are secreted by a variety of organs and are added to the food at strategic points in its passage through the digestive tract. The systematic names for enzymes are formed by adding the suffix -ase to the name of the substance whose transformation is catalyzed by the enzyme. Thus a peptidase acts on a peptide, an amylase acts on starch (both amylose and amylopectin parts), and sucrase acts on sucrose.

It is true, in one respect, that "we are what we eat"—that is, we use the amino acids and monosaccharides and other components of our food to build our own bodies. However, human fat and human proteins are different from those of the milk, butter, and steak we eat. Since only simple units of the proteins, carbohydrates, and lipids are absorbed through the intestinal wall and since we are able to assemble the units only in our own way, a good analogy to digestion is that of a complex brick building being demolished—the mortar being dissolved, and the simplest units, the bricks, small pieces of wood, or steel being reused for construction of another building.

The digestive process in the human is complex, but it can be simply described to provide a background for the biochemistry of this important series of events. Detailed discussions of the digestion of specific biochemical compounds are given in Chapters 38, 39, and 40. As we have already noted, when food is chewed it is mixed with saliva which both lubricates the food and contains the enzyme (salivary amylase) which initiates digestion of starch. The food then enters the esophagus, which serves mainly as a connecting tube between the mouth and the stomach. It produces no digestive enzymes. When food reaches the stomach, action of salivary amylase is stopped by acidity which comes from the hydrochloric acid secreted by the stomach

Food normally stays in the stomach for a few hours while both hydrochloric acid and pepsin catalyze the hydrolysis of food, particularly of the proteins and carbohydrates. Pepsin, in contrast to salivary amylase, functions only in an acid solution and is well adapted to the pH of 1.0 to 2.5 which prevails in the stomach. Pepsin, which is found in birds, fish, and reptiles as well as in mammals, catalyzes the hydrolysis of proteins to yield a mixture of smaller polypeptides. Other digestive enzymes have been found in the stomach, but they are not effective at the pH of the stomach, and it has been speculated that perhaps these enzymes are holdovers from infancy when their secretion into a less acid stomach (characteristic of infants) may have been effective. Alternatively, they may become effective in the lower tract where the pH is neutral or slightly alkaline.

TABLE 37-1 Digestive enzymes

Site of Secretion	Enzyme	Site of Action	Substrate Acted Upon	Products of Action
mouth	salivary amylase	mouth	starches	maltose, dextrins
stomach	pepsin	stomach	proteins	polypeptides (peptone)
pancreas	pancreatic amylases	small intestine	starches	maltose
pancreas	trypsin	small intestine	proteins, polypeptides	small peptides
pancreas	chymotrypsin	small intestine	proteins, polypeptides	small peptides
pancreas	pancreatic lipase	small intestine	tryglycerides	fatty acids, glycerol
small intestine	disaccharidases	small intestine	disaccharides	monosaccharides
small intestine	nucleotidases	small intestine	nucleotides	nucleosides, phosphate

As the partially digested food enters the small intestine, it is met with a large number of catalysts for the digestive process. From the gall bladder come the nonenzymatic bile acids which aid in emulsification of lipids; from the pancreas comes a mixture of amylases, the peptidases trypsin and chymotrypsin, at least one lipase, cholesterol esterase which catalyzes the hydrolysis of esters of cholesterol, and perhaps other, yet to be discovered, enzymes. Finally, the intestine itself secretes several enzymes. Among these are various disaccharidases, peptidases, nucleosidases, and nucleotidases. In addition to the enzymes and bile acids, the various secretions contain sufficient alkaline materials to neutralize the acid of the stomach. It is believed that the secretions added to the food in the intestine are sufficient to do a complete job of digestion and in fact many persons digest food reasonably well following removal of large portions of their stomachs. However, they must eat more frequently and follow some dietary restrictions for adequate digestion. Digestive enzymes are listed in Table 37-1.

Digested foods pass through the intestinal walls into the blood or lymph. Lipids, in particular, are absorbed by the lymph. Undigested foods, along with bacteria which find the intestine a good home and breeding place, are passed into the large intestine or colon where water and other substances are absorbed. The contents of the large intestine are finally discharged as the feces.

The other major way the body rids itself of waste products is through the urine, but this represents a different type of excretion. Urine is produced by the kidneys, which may be thought of as a filter for the blood. In addition to water, urine contains waste products produced by the further reactions of materials absorbed through the intestinal wall. The urine is a much better indicator of the kinds of reactions going on in the body than are feces, since it contains only the end products of the vital reactions of the body, and does not contain undigestible substances which might be present merely by chance. Consequently urinalysis is much more important than analysis of feces in diagnosis of the state of health of a person.

37–3 FURTHER METABOLISM

The word **metabolism** is used to describe the sum total of all reactions that occur in a living organism. The study of metabolism emphasizes those reactions which follow the absorption of digested food; strictly speaking, however, digestion is also a part of metabolism. Those molecules which are absorbed into the body may either be built into more complex molecules in reactions described as **anabolic,** or degraded into smaller molecules in **catabolic** reactions. Anabolic reactions require energy and are characterized as **endergonic,** and catabolic reactions liberate energy and are **exergonic** reactions. Any chemical substance that is involved in metabolic reaction is a **metabolite.**

37–4 ENERGETICS

There are many ways of measuring the amount of energy released in or required for a chemical reaction. The most meaningful expression of biochemical energy is **free energy.** This term is defined as the energy available for doing useful work. Chemical reactions occur spontaneously when substances containing relatively high amounts of energy are converted into substances having less energy. Compounds with less energy are therefore less likely to react, i.e., they are more stable. Thus a spontaneous chemical reaction may be defined as a process in which matter proceeds to more stable states. Carbon dioxide contains less energy than the carbon and oxygen which react to produce it, and water contains less potential energy than its component elements, hydrogen and oxygen.

Since chemical substances tend to proceed to states in which their energy is released and they are more stable, the obvious question arises, "Why doesn't a potentially combustible substance, such as wood, start to burn spontaneously in air?" In this case, we know that the wood must first reach its kindling temperature. Speaking in more general terms, an **energy of activation** must be supplied before even a potentially spontaneous reaction will proceed. For some reactions, such as the burning of phosphorus in air, normal room temperature is sufficient to initiate the reaction.

A useful analogy to the energy of activation is provided by the water in a lake high on a mountain. Normally the water would run downhill until it reached sea level, but some barrier, perhaps a beaver dam or a landslide, prevents the water from running downhill and releasing the energy—energy which might be used to drive turbines and give rise to electrical energy. If we supply energy by using a pump, we can raise the water over the barrier and let it flow downhill. The pump is analogous to a device for providing the energy of activation. Ways in which energy can be stored, converted, and released are shown in Figs. 37–3, 37–4, and 37–5.

Although glucose does not burn at body temperature even when placed in pure oxygen, animals have systems for converting glucose, as well as other foods, into carbon dioxide and water. The oxidation of glucose which initially requires an energy of activation, takes place in many steps. At some of these steps, small amounts of energy are released. Trying to use the energy of a molecule of glucose in a single reaction would be similar to

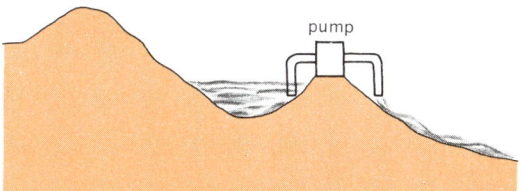

Fig. 37-3 Release of stored energy. The water in a mountain lake is analogous to chemical compounds which contain stored energy. Some of the energy can be released when the water runs downhill; however, the dam, which represents an energy barrier similar to the energy of activation in a chemical reaction, may prevent the water from flowing toward sea level. If a pump is installed to first lift the water over the dam, however, then release of energy can then take place spontaneously.

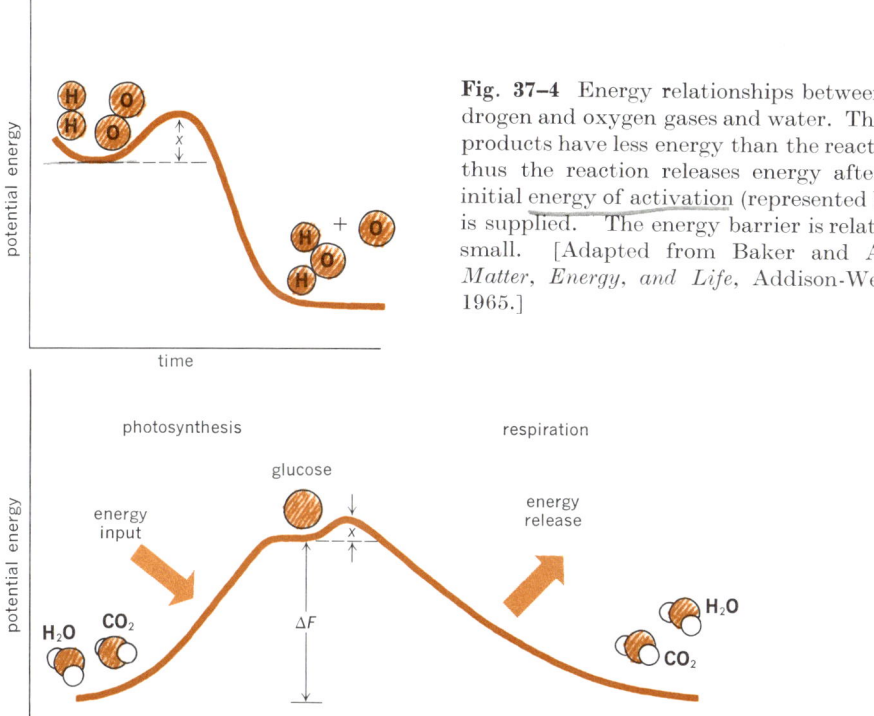

Fig. 37-4 Energy relationships between hydrogen and oxygen gases and water. The end products have less energy than the reactants; thus the reaction releases energy after the initial energy of activation (represented by x) is supplied. The energy barrier is relatively small. [Adapted from Baker and Allen, *Matter, Energy, and Life*, Addison-Wesley, 1965.]

Fig. 37-5 An energy diagram showing the conversion of water and carbon dioxide to glucose, followed by the oxidation of glucose. Energy from the sun is "stored" in the "uphill" process of photosynthesis. This energy is released in respiration. The difference in free energy between glucose and the two substances, water and carbon dioxide, is represented by ΔF. The value x represents the energy of activation necessary for the oxidation of glucose. This energy is supplied by ATP. [Adapted from Baker and Allen, *Matter, Energy, and Life*, Addison-Wesley, 1965.]

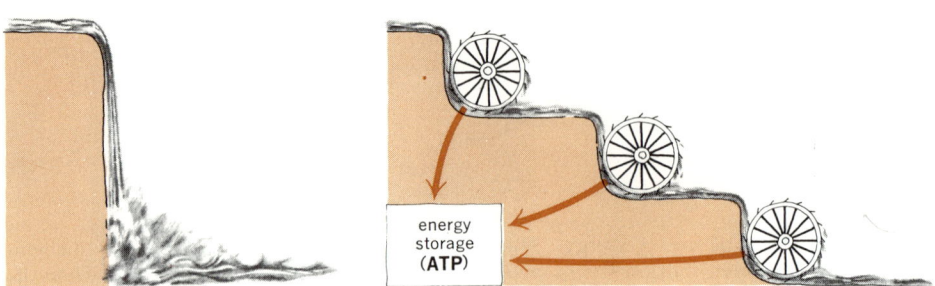

Fig. 37–6 Waterfall analogy to biochemical energetics. When water falls over a high waterfall it generates tremendous amounts of energy. A waterwheel at the bottom of such a fall would be smashed. If, instead of one precipitous waterfall, the stream is diverted into a series of smaller falls, each one having a water wheel, the energy which the water loses can be converted into useful energy. When biochemical reactions that yield large amounts of energy proceed in a series of reactions, each of which can synthesize ATP, the same principle is shown.

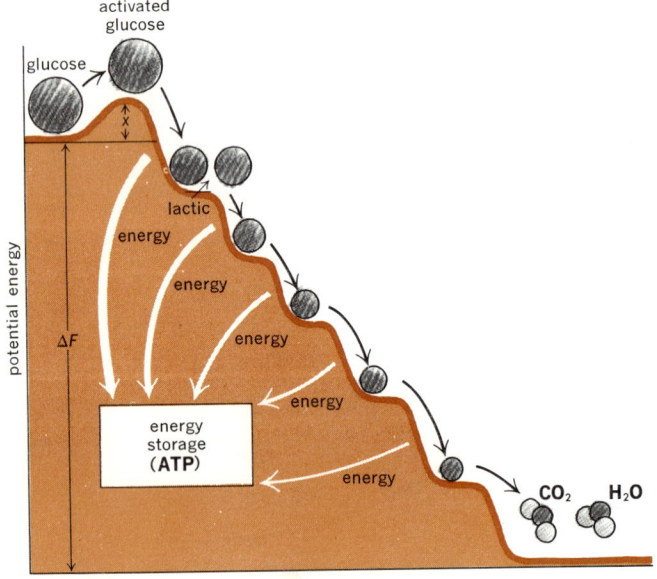

Fig. 37–7 The energetics of the oxidation of glucose. The diagram shows how glucose must first be activated (by ATP) before its energy is released. The release of this energy is shown as occurring in a series of steps, several of which generate ATP from ADP and phosphate. [Adapted from Baker and Allen, *The Study of Biology*, Addison-Wesley, 1967.]

the effect of placing a simple waterwheel at the base of Niagara Falls. Living organisms cannot utilize such large amounts of energy. Using the waterfall analogy, we can say that organisms divert the flow of water around a precipitous waterfall, and by using several small waterwheels at the bottom of small waterfalls, they are capable of trapping the energy in small, useful packets. (See Fig. 37–6.)

Some energy so released is used to drive the anabolic reactions of the organisms. Other energy is used to keep the heart beating, move food through the digestive tract, and move the animal from one place to another. The problem that living organisms have solved, that of release of large amounts of energy in small units, is similar to the problem still faced by nuclear scientists who are trying to make the tremendous energy from the fusion of hydrogen useful to drive generators or automobiles instead of functional only as a bomb blast.

Plants and animals are able to store and use the energy of metabolic reactions largely because of the compound adenosine triphosphate (ATP). This compound, which was discussed in Chapter 36, represents the principal "currency" of most living organisms. The energy of food is stored by the conversion of adenosine diphosphate to adenosine triphosphate, and the reverse reaction is used to drive energy-requiring reactions. (See Fig. 37–7.) The amount of free energy released in the conversion of ATP to ADP and phosphate is probably between 10,000 and 12,000 cal (10–12 kcal) per mole.*

ATP + HOH → ADP + phosphate + 8 to 12 kcal

37–5 ELECTRON TRANSPORT

The energy contained in foods is released primarily in the process of oxidation of the hydrogen atoms contained in the food. Hydrogen atoms, or, speaking more precisely, the electrons of these atoms are passed from one compound to another in the series of reactions known as electron transport. The enzymes and coenzymes necessary for electron transport are localized in the mitochondria of cells.

Electrons may take several different routes, and certain intermediates are not really known, but the scheme involving nicotinamide adenine dinucleotide (NAD), flavine adenine dinucleotide (FAD), and the cytochromes is probably the most important such system. The formulas for these compounds are given in Fig. 37–8. The substrate nicotinamide adenine dinucleotide phosphate (NADP), which functions much as NAD does, contains an additional phosphate on the ribose unit attached to nicotinamide. There are several cytochromes all having structures similar to cytochrome c (See Fig. 37–9).

In this system, two electrons along with one proton are passed from an energy-rich molecule to NAD. In the process NAD is reduced, and the donor molecule is oxidized. In turn, $NADH_2$ passes the electrons to FAD which oxidizes the $NADH_2$, reduces the FAD, and releases energy. Then $FADH_2$ passes the electrons to the cytochromes. The cytochromes are complex molecules which contain iron in the $3+$ state which is reversibly reduced to the $2+$ state by the electrons it receives from $FADH_2$. One of the cytochromes eventually passes electrons to an oxygen atom which also acquires two protons, and the end product of the series of reactions is water. In this series of reactions the combination

---

* The exact amount of energy released in a living organism by the conversion of ATP to ADP and phosphate is not accurately known. The reaction under certain specific conditions known as standard conditions, *in vitro*, yields less energy, a value nearer 8 kcal/mole.

Know what oxidoreductase reactions are!

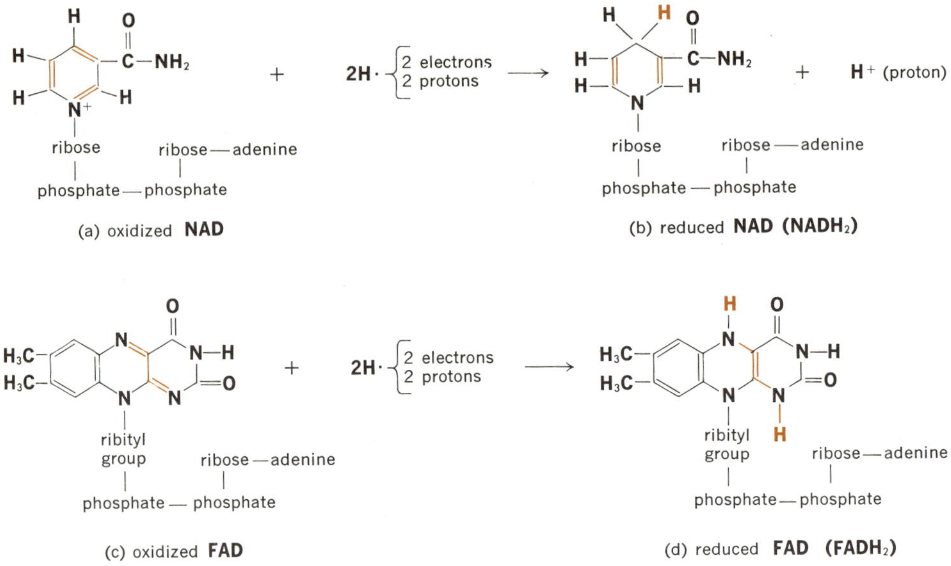

Fig. 37-8 Formulas of electron acceptors in the electron transport system: (a) shows oxidized NAD (see Fig. 36-13 for complete structure), (b) reduced NAD (NADH$_2$), (c) oxidized FAD, and (d) reduced FAD (FADH$_2$).

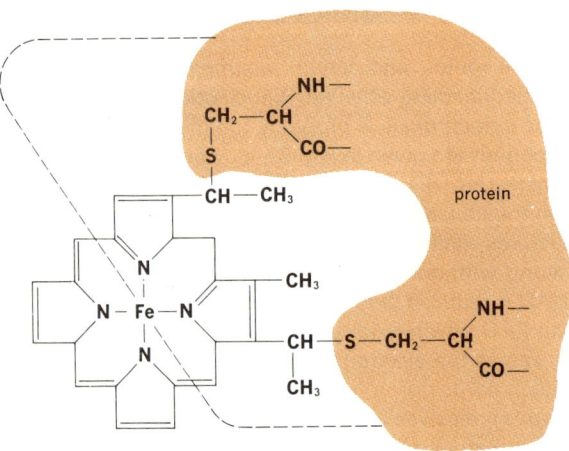

Fig. 37-9 The structure of cytochrome c. The heme group is bound to the protein molecule by sulfide linkages to the amino acid cysteine, which is part of the protein. [Adapted from Baker and Allen, *The Study of Biology*, Addison-Wesley, 1967.]

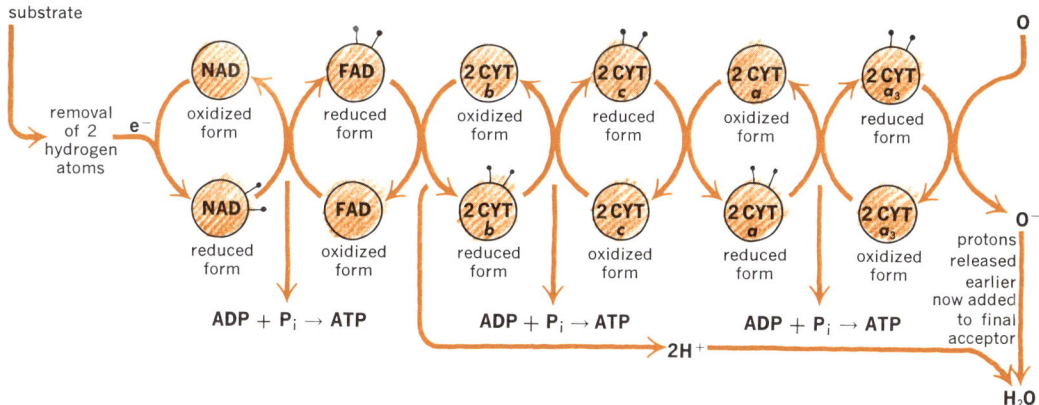

Fig. 37–10 A summary of electron transport. The diagram shows the removal of the two hydrogen atoms from an oxidizable substrate, the transport of the electrons, and the final acceptance of the electrons and two protons by an atom of oxygen to give water. The energy released is trapped in the synthesis of ATP from ADP and inorganic phosphate (P_i). [Adapted from Baker and Allen, *The Study of Biology*, Addison-Wesley, 1967.]

---

of hydrogen with oxygen to produce water, a process that releases large amounts of energy, has been broken into several steps, and the potentially destructive or unusable amounts of energy are made available in useful-size packets. There are more intermediate compounds than those given above and in Figs. 37–8 and 37–10. However, the exact role played by each is not so firmly established as is that of NAD, FAD, and the cytochromes. For this reason and for simplification, these other compounds have not been included in this presentation.

We can summarize the process of electron transport as follows:

1) The substrate loses two electrons and one proton to NAD, converting it to $NADH_2$, and releasing a hydrogen ion (proton) to the surroundings.
2) The $NADH_2$ gives two electrons and one proton to FAD. In addition, the FAD acquires a proton from its surroundings and becomes $FADH_2$.
3) Since each molecule of the cytochromes can accept only one electron, two molecules are required to accept the two electrons from $FADH_2$. The two protons lost in the transformation of $FADH_2$ to FAD are released into the surroundings.
4) An atom of oxygen acquires two electrons from two molecules of a cytochrome, oxidizing the cytochrome and reducing the oxygen.
5) The oxygen acquires two protons to form water.

This process is given diagrammatically in Fig. 37–10.

During the process of hydrogen transport, in a way still not fully understood, ATP is synthesized from ADP and phosphate. If NAD is the original electron acceptor there are usually three molecules of ATP synthesized for every two electron pairs (or two hydrogen atoms) or for every molecule of water formed. If FAD is the original electron acceptor, only two molecules of ATP are synthesized in the formation of water. Thus, by the process of electron transport, any chemical substance that can yield electrons (which are almost always released in combination with protons as hydrogen atoms) can result in the synthesis of ATP and the storage of energy. The synthesis of ATP, as well as electron transport, occurs primarily in the mitochondria of cells.

To illustrate, ethanol is oxidized to give acetic acid (probably in the form of acetyl coenzyme A) by two reactions, both of which involve NAD.

ethanol + NAD → NADH$_2$ + acetaldehyde

acetaldehyde + H$_2$O + NAD → NADH$_2$ + acetic acid

An alternative pathway involves the oxidation of acetaldehyde, with FAD as the hydrogen acceptor. Thus ethanol can yield a molecule of acetic acid with the resultant synthesis of either six molecules of ATP (if both reactions use NAD) or five molecules of ATP (if one reaction uses NAD and the other FAD).

The complete oxidation of acetic acid involves a more complex process and can yield a total of twelve molecules of ATP.

CH$_3$COOH + 2O$_2$ → 2CO$_2$ + 2H$_2$O

The energy stored as the result of the complete biological oxidation of ethanol can be compared with the amount of energy released when ethanol is oxidized *in vitro*, as follows. The *in vivo* oxidation of one molecule of acetate is known to yield 12 molecules of ATP, and the conversion of ethanol to acetate gives either 5 or 6 molecules of ATP depending upon the pathway taken. If we use an average total value of 17.5 moles of ATP synthesized as the result of the oxidation of one mole of ethanol, and multiply this by 12 (the energy stored when one ADP is converted to an ATP), we get a total of 210 which is the number of kilocalories of energy stored when one mole of ethanol is oxidized *in vivo*. If we oxidize one mole of ethanol in a bomb calorimeter we release 312 calories of energy. Theoretically this is the maximum amount of energy that is released by oxidizing one mole of ethanol. It is presumably the amount actually released in the oxidation by an animal. We can find the efficiency with which the animal traps the energy of oxidation of ethanol by dividing the energy he stores in 17.5 moles of ATP (210 kcal) by the amount of energy theoretically available (312 kcal); thus we divide 210 by 312 and get 67%. If we use 8 kcal as the energy stored per mole of ATP, the efficiency is 140/312 or 45%. Even this lower figure is appreciably better than the efficiency with which an internal combustion engine converts the energy of fuel to other forms of energy.

37–6 ENZYMES

In many respects biochemical reactions are similar to the chemical reactions occurring *in vitro*. They follow the laws of conservation of mass and energy and most require an energy of activation. The products formed are similar to those formed in organic reactions. Like many nonbiochemical reactions, they require catalysts in order to proceed at any appreciable rate. Although a few reactions that occur in living organisms proceed without catalysts, the greatest number require the biochemical catalysts, enzymes. An enzyme is usually defined in just these terms, as a catalyst of biological origin. All known enzymes contain protein. Some are composed only of protein; others require some cofactor other than protein in order to function. FAD and NAD are cofactors, usually called coenzymes, and function primarily in conjunction with a particular enzyme. (See Fig. 37–11.)

Enzymes are remarkable catalysts in that they can accelerate the rate of reactions without either an increase in temperature or a major change in pH. They differ from most

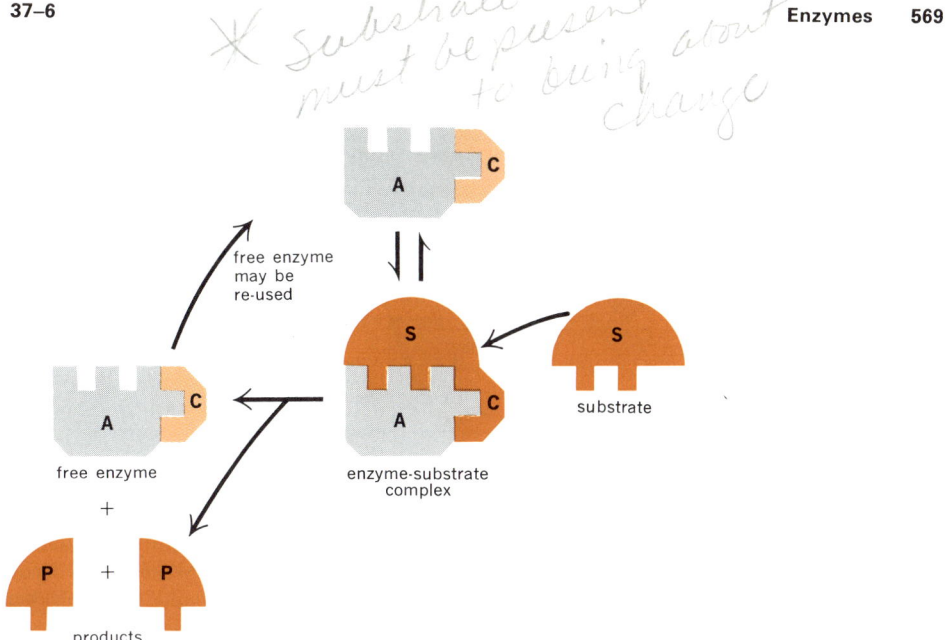

Fig. 37-11 A diagram showing the involvement of a coenzyme in an enzymatic reaction. The protein part of an enzyme (apoenzyme) is combined with its coenzyme, as shown upper center. The substrate molecule then attaches to the completed enzyme and is converted, first into an unstable enzyme-substrate complex, and then into the products of the reaction with the regeneration of the free enzyme. Not all enzymes require coenzymes. [Adapted from Baker and Allen, *Matter, Energy, and Life*, Addison-Wesley, 1965.]

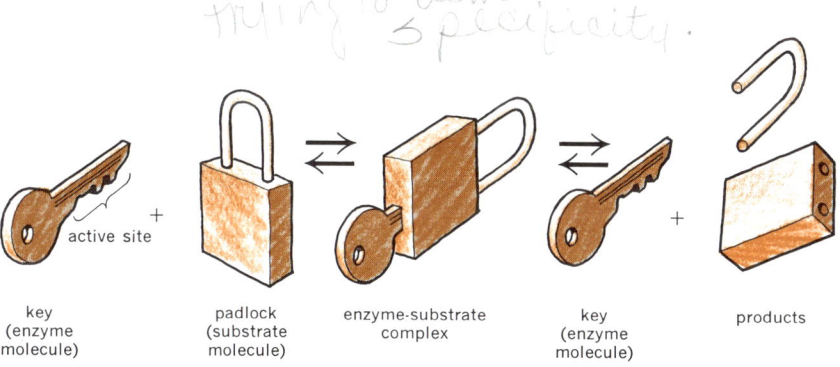

Fig. 37-12 The lock-and-key model for picturing the interaction of substrate and enzyme. [From Baker and Allen, *Matter, Energy, and Life*, Addison-Wesley, 1965.]

other catalysts since usually only one reaction is promoted and by-products are not formed. The exact way enzymes function is still the subject of active research. It is generally believed that the **substrate,** the substance reacting, is bound in some way to the surface of an enzyme. The binding site is apparently quite specific and can bind only one or only one kind of substrate molecule. This specificity has been compared to a lock and key. The enzyme (the key) functions only when a certain specific lock (substrate) or perhaps a certain type of lock is available. (See Fig. 37-12.)

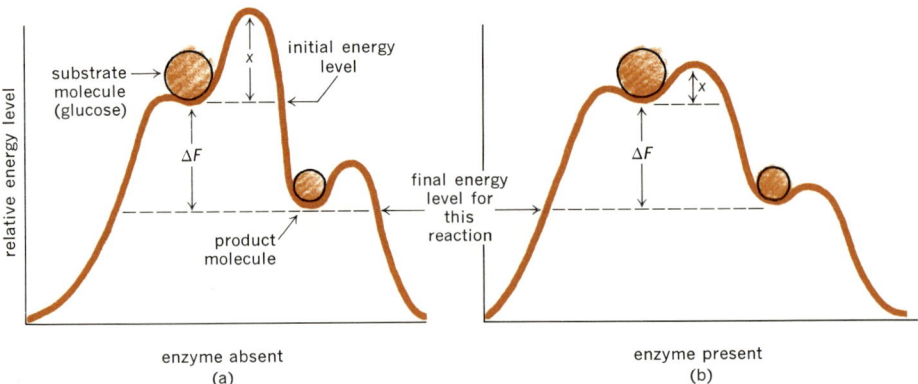

Fig. 37-13 The effect of an enzyme on the activation energy. The amount of activation energy is represented by x and the change in free energy by ΔF. The uncatalyzed reaction (on the left) requires a greater energy of activation. In reactions that proceed in a series of stepwise reactions, each reaction has its energy of activation. [Adapted from Baker and Allen, *Matter, Energy, and Life*, Addison-Wesley, 1965.]

One way of answering the question, "What is it that enzymes really do?" is to say that they influence the course of a reaction so that less energy of activation is required. They *do not*, however, provide this energy. Enzymes cannot make *impossible* reactions proceed; the energy driving the reaction is still the difference between reactants and products. Enzymes, in common with other catalysts, only hasten the rate of attainment of equilibrium in a chemical reaction. (See Fig. 37-13.)

Another way of answering the question of what enzymes do is to say that they hold reactants at sites on the enzyme surface so that reactions which would otherwise require random collisions can take place more often. It has also been postulated that when enzymes bind substrates, they actually stretch the molecule somewhat as the rack stretched the victims on this instrument of torture. This tension makes it easier for specific bonds to be broken. Perhaps this stretching exposes a specific bond (e.g., an ester bond) so that it is more easily attacked by another molecule (e.g., a water molecule) to bring about a reaction (e.g., the hydrolysis of an ester). (See Fig. 37-14.)

To discuss enzymes we must have methods for naming them. Originally enzymes were given trivial names, and we still use the names pepsin, catalase, and steapsin for enzymes that catalyze the hydrolysis of proteins, the decomposition of H_2O_2, and a mixture of lipid-hydrolyzing enzymes. A better nomenclature consists of adding the suffix *-ase* to the name of the substrate upon which an enzyme acts, and we still use the names sucrase, urease, and arginase for enzymes that catalyze reactions involving sucrose, urea, and arginine. Since one substrate may undergo any one of several possible reactions, however, we now find it better to include not only the name of the substrate but also the type of reaction catalyzed when naming the enzyme.

The most recent system for naming enzymes is based on the type of reaction they catalyze. The categories used are given in Table 37-2. The complete name of the enzyme in this system includes the name of the substrate (or the general type of compound to which it belongs), as well as the type of reaction catalyzed. Under this system urease belongs to class 3, the hydrolases, and is called *urea amidohydrolase*, which indicates that the sub-

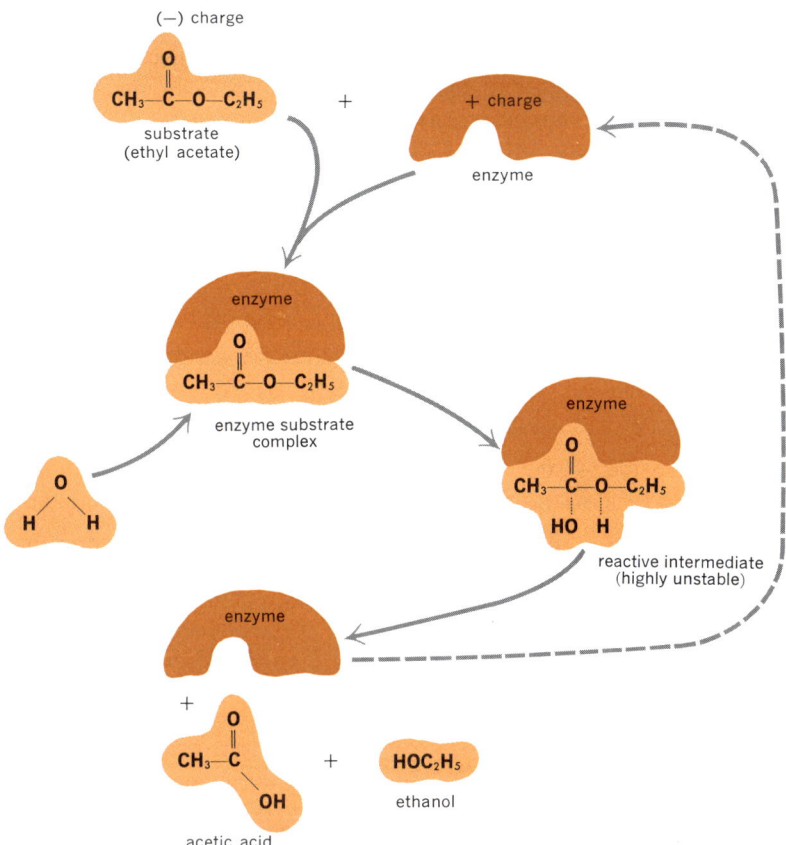

Fig. 37-14 A diagrammatic representation of an enzymatic reaction in which the enzyme makes the substrate more vulnerable to attack by the reacting molecule. In this reaction, the substrate (ethyl acetate) is bound to the enzyme both by complementary shapes and by opposite charges. This binding facilitates the hydrolysis of the ester bond by a molecule of water.

TABLE 37-2 Types of enzymes

Enzymes	Reaction Catalyzed
1. oxidoreductases	oxidation and reduction
2. transferases	transfer of a chemical grouping from one compound to another
3. hydrolases	hydrolytic reactions
4. lyases	nonhydrolytic addition or removal of groups
5. isomerases	conversion of a substance into one of its isomers
6. ligases	syntheses or, more specifically, the formation of various types of bonds

strate urea is hydrolyzed and that the reaction is the hydrolysis of an amide. Sucrase is named as a β-D-*fructofuranoside fructohydrolase*, that is, an enzyme that hydrolyzes a fructose group from a compound which contains D-fructose in a beta linkage and in the furanose form. The basic classifications are broken down into subgroups and we can refer to an enzyme by a series of numbers. Urease, for instance, is 3.5.1.5. For the purpose of this book, many aspects of the formal nomenclature are too specific, so we will use trivial names as a general rule, sometimes with explanations as to the exact reaction catalyzed.

37-7 FAILURE OF METABOLIC REACTIONS

Metabolism has been characterized as a series of conversions from one compound to another, each conversion requiring an enzyme. We can illustrate the process as follows:

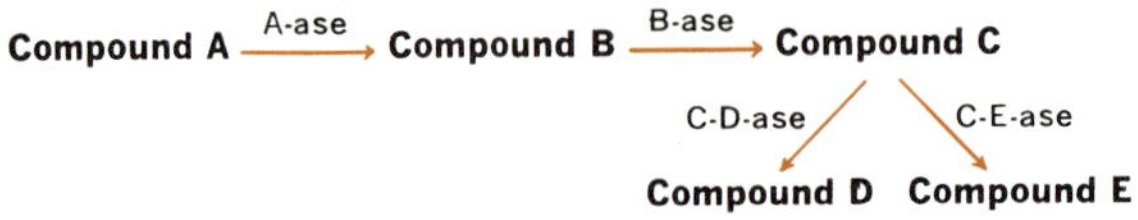

$$\text{Compound A} \xrightarrow{\text{A-ase}} \text{Compound B} \xrightarrow{\text{B-ase}} \text{Compound C} \begin{array}{c} \xrightarrow{\text{C-D-ase}} \text{Compound D} \\ \xrightarrow{\text{C-E-ase}} \text{Compound E} \end{array}$$

In this illustration, as long as sufficient compound **A** is available in the ingested food and as long as all enzymes are present and functioning normally, the reactions will proceed with the production of compounds **D** and **E**. (In reality, a better illustration would go as far as compounds **X, Y,** and **Z**.) If there is an inadequate amount of compound **A** in the diet, all other compounds would be deficient, unless perhaps **B** or **C** happened to be included in the diet. Such dietary deficiencies can be quickly corrected when they are detected.

More serious consequences result when an enzyme does not function. Failure of an enzyme may come about for several reasons: an organism may be unable to synthesize the enzyme either because of lack of components or because of some hereditary defect which prevents the synthesis. Normal enzymes may also fail to function when inhibitors are present. Many of the potent poisons such as cyanide and fluoride act as enzymatic inhibitors. Enzymes may also be inhibited by an excess of a normal metabolite other than their normal substrate.

In the illustration above, if the enzyme **B**-ase were absent or not functioning normally, we would find a deficiency of compounds **C, D,** and **E**. If these were vital compounds, such as the hormones adrenalin or thyroxine, the organism would function abnormally and might die. Another effect of the malfunction of **B**-ase would be that large amounts of **B** and perhaps of **A** (depending upon the reversibility of the conversion of **A** to **B**) would accumulate. This accumulation might result only in increased excretion of **B** which might go almost unnoticed unless **B** were colored or especially odorous. If, however, large amounts of **B** inhibited the reaction in another metabolic sequence (for instance, the conversion of compound **M** to compound **N**) here again the result might be extremely deleterious or even fatal. Several illustrations of such enzyme abnormalities are given in the chapters that follow. (See Sections 38-11, 42-4.)

37-8 HORMONES

A major problem faced by any organism is the control of the *rates* of the various reactions which are constantly taking place within the organism. While the enzymes are very important in this control, a further role is played by hormones. A **hormone** *may be defined as a product of an endocrine gland*. There are several such glands, such as the pituitary, the thyroid, parts of the pancreas, the adrenals, and parts of the ovaries or testes. (See Fig.

44-4.) The products of these glands are secreted directly into the blood stream and distributed to all parts of the body of an animal. Some of the glands listed above are also exocrine glands—they produce some secretions that flow through ducts to a definite site of action.

Hormones of higher animals are usually found to be either protein-containing or steroids. In many cases nonprotein matter is associated with the proteinaceous hormones. Insulin, which is the hormone of the pancreas, is a polypeptide which normally contains small amounts of the metal zinc. The hormones of the pituitary, such as the growth hormone, the thyroid-stimulating hormone, the hormones which influence the ovary and testes, as well as others from this source, are primarily proteins or polypeptides. The sex hormones produced by the ovary or testes as well as most of the hormones of the adrenal gland are steroid hormones. (See Section 35–6.) The hormones of the thyroid gland and of the medulla of the adrenal gland are derivatives of amino acids.

Although the physiological effects of the hormones, such as the increase of the basal metabolic rate by thyroxine or the effect of testosterone on the secondary sexual characteristics in the male are well known, the exact manner in which hormones exert their control on body processes is not known. Recent research indicates that hormones function by increasing the rate of synthesis of certain proteins, perhaps by first increasing the rate of synthesis of ribonucleic acid (RNA) which, in turn, influences the rate of protein synthesis. If the protein synthesized is an enzyme, the effects on metabolism are obvious. The nucleotide cyclic adenosine monophosphate is important in initiation of many hormone-controlled reactions. (See the article by Pastan in the Suggested Reading for details of how this amazing nucleotide works.)

A great deal of interrelation exists between the various hormones. One of the simplest examples is that between the thyroid-stimulating hormone of the pituitary and the thyroid hormone thyroxine and/or triidothyronine. The thyroid-stimulating hormone induces the thyroid gland to produce its hormone thyroxine. Thyroxine, in turn, influences the pituitary gland and inhibits the production of the stimulating hormone. This kind of process has been called "feedback." If the thyroid gland cannot produce sufficient thyroxine because of a lack of iodine in the diet, excessive amounts of thyroid-stimulating hormones are produced, causing an enlargement of the thyroid gland, which is called a goiter.

There are many other feedback relationships between various endocrine glands, some of them extremely complex. For example, the sexual cycles in females, the production of inflammation and wound healing, and the control of the level of glucose in the blood are all influenced by relationships of endocrine glands and their hormones.

37-9 SUMMARY OF METABOLISM

Metabolic reactions are remarkably similar in all plants and animals so far studied. Not only are the same types of molecules involved as reactants, products, and catalysts, but in all cases the reactions occur in a number of small, sequential steps. The presence of a number of steps in an overall reaction not only aids in trapping energy from foods, but also transforms anabolic reactions requiring large amounts of energy to a series of small reactions, each of which can be driven by available energy. The introduction of a number of small steps also makes possible more sensitive control of reactions.

Each metabolic reaction is catalyzed by an enzyme. Failure of an enzyme to function normally usually causes a deficiency of some substance or substances and an accumulation of other substances. These deficiencies or excesses may have serious or fatal consequences for living organisms.

Fig. 37-15 An illustration of dynamic equilibrium. Water flows from a faucet into a container that has an overflow outlet. Although the amount of water in the container remains constant, individual molecules are constantly being added from the faucet and removed from the overflow. If we added a dye to the water, we would see that it would gradually be washed away. Essentially, this is what we see when we give radioactive isotopes to a plant or animal and note their disappearance even though the plant or animal is neither growing nor shrinking in size.

Anabolic reactions require energy which is provided by catabolic reactions, particularly oxidations. The ATP helps transfer energy to energy-requiring (endergonic) from energy-releasing (exergonic) reactions. Any substance that can be oxidized by NAD or FAD usually brings about the synthesis of ATP from ADP and phosphate. A molecule of reduced NAD normally yields three molecules of ATP, while one of FAD yields only two molecules of ATP when complete oxidation has taken place.

Both anabolic and catabolic reactions are going on all the time in all living organisms. Although an animal may maintain a given weight for a long time, the molecules of which he is composed are constantly being changed. When animals are given isotopically labelled foods, we soon find these isotopes incorporated in their body tissues. They then begin to be eliminated slowly. We characterize this condition as being a dynamic equilibrium. It is not like a chemical equilibrium, because new molecules are constantly being added and eliminated. The situation is better illustrated (Fig. 37-15) if we picture a vessel of water that maintains a constant level even though water is constantly entering and leaving.

We often speak of a **metabolic half-life.** This is the time it takes to replace half of a given body component. Half of the protein in the liver of a rat is replaced in 7 days so we say that rat-liver protein has a metabolic half-life of 7 days.

The next several chapters are concerned with the metabolism of various types of compounds. We will discuss specific instances that led to the generalizations we have stated in this chapter. We will examine normal catabolic and anabolic reactions and trace the effects of blockages of such pathways or abnormal reactions, and the relation of these abnormalities to disease.

IMPORTANT TERMS

ATP	Chymotrypsin	FAD	Nucleosidase
Alimentary canal	Coenzyme	Free energy	Nucleotidase
Amylase	Cytochromes	Ingestion	Pepsin
Anabolic	Digestion	Inhibitor	Peptidase
Assimilation	Endergonic	Lymph	Photosynthesis
Bile	Energy of activation	Metabolite	Saliva
Catabolic	Enzyme	Metabolism	Substrate
Chlorophyll	Exergonic	NAD	Trypsin

WORK EXERCISES

1. Write equations for the digestion of
   a) sucrose   b) a triglyceride   c) ethyl acetate
2. List the enzymes necessary for the complete digestion of
   a) a protein   b) a triglyceride   c) a polysaccharide
3. What type of organic compound is being hydrolyzed in the digestion of (a) a protein, (b) a triglyceride, (c) a polysaccharide?
4. Substance A gives 200 kcal/mole when completely oxidized *in vitro*. Its metabolism to the same end products results in the synthesis of 15 moles of ATP from ADP and phosphate. What is the efficiency of conversion of the energy of compound A to the energy stored in ATP?
5. How do enzymes differ from other catalysts?
6. Name the enzyme catalyzing each of the following reactions:
   a) triglyceride + water $\rightarrow$ fatty acids + glycerol
   b) maltose + water $\rightarrow$ glucose
   c) RNA + water $\rightarrow$ organic bases + ribose + phosphate
   d) adenylic acid + water $\rightarrow$ adenosine + phosphate
   e) adenosine + water $\rightarrow$ adenine + ribose
7. Some persons secrete little or no hydrochloric acid in their stomachs. How would their digestion differ from that of a normal person? What treatment might be used for these persons?
8. Why must the growth hormone be administered by injection rather than by mouth in order to be effective?
9. How does an enzyme affect the equilibrium of a reaction?
10. Using electron dot formulas, show the actual number of electrons in the oxidized and reduced part of the riboflavin molecule that changes when FAD $\rightarrow$ $FADH_2$.

Consult a basic biochemistry text to find the answers to the following:

11. What are some urine tests that are of metabolic significance?
12. What are the digestive hormones, and how do they function?

SUGGESTED READING

Changeux, J. P., "The Control of Biochemical Reactions," *Scientific American*, Apr. 1965 (Offprint #1008). The author discusses the control of enzymes which in turn control other syntheses—a complex theory is involved, but is well explained in this article.

Chappell, G. S., *Through the Alimentary Canal with Gun and Camera*, Dover, New York, 1963. A humorous account of a totally imaginary, pseudoscientific trip.

Davenport, H. W., "Why the Stomach Does Not Digest Itself," *Scientific American*, January 1972 (Offprint #1240). This article discusses how the stomach "avoids digesting itself and what happens when the safety mechanisms fail."

Kleiber, M., *The Fire of Life—An Introduction to Animal Energetics*, John Wiley, New York, 1961. The title explains the book, but does not indicate that it is well written and interesting.

Lehninger, A. L., *Bioenergetics*. W. A. Benjamin, New York, 1965. A modern, readable (paperback) introduction to the processes of energy transfer in living systems.

Lehninger, A. L., "How Cells Transform Energy," *Scientific American*, Sept. 1961 (Offprint #91).

Lehninger, A. L., "Energy Transformation in the Cell," *Scientific American*, May, 1960 (Offprint #69).

Mayerson, H. S., "The Lymphatic System," *Scientific American*, June 1963 (Offprint #158). This well-illustrated article describes the function of this important circulatory system.

Pastan, I., "Cyclic AMP," *Scientific American*, Aug. 1972. Cyclic AMP appears to control the rates of many metabolic processes. This control is complex but is well described in this article. The discovery of cyclic AMP is discussed, as well as theories of its action.

Racker, E., "The Membrane of the Mitochondrion," *Scientific American*, Feb. 1968. The process of oxidative phosphorylation, the place where it occurs, and the inner membrane of the mitochondrion are discussed in this well-illustrated article.

Stumpf, P. K., "ATP," *Scientific American*, Apr. 1953 (Offprint #41). For further information about this important molecule we recommend this article.

METABOLISM OF CARBOHYDRATES

38

38-1 INTRODUCTION

Carbohydrates that are ingested by an animal are usually in the form of high-molecular-weight polymers, principally the starches and cellulose. The animal is then faced with two major problems: first, that of degrading the polymer into units small enough to be absorbed through the wall of the intestine, and second, that of releasing the energy of the carbohydrate and converting it into a useful form. Some of the degraded carbohydrate is reconverted to the polymer glycogen which is then stored in the liver and muscles; however, the amount of glycogen present at any time is relatively small. Most digested carbohydrates are oxidized to carbon dioxide and water with the release of energy or are converted to lipids and stored.

Plants metabolize carbohydrates quite differently. Instead of ingesting polymeric carbohydrates, green plants make their own carbohydrates through the process of photosynthesis. It is true that some of this carbohydrate may be oxidized to yield energy for energy-requiring reactions in the plant; however, the remainder of the saccharide units are polymerized to yield starches and cellulose. Plant biochemistry and photosynthesis are discussed in detail in Chapter 43. The following discussion is concerned primarily with the metabolism of carbohydrates in animals.

38-2 DIGESTION OF CARBOHYDRATES

The digestion of starch begins in the mouth where salivary amylase is added to food as it is being chewed. **Amylases** catalyze the hydrolysis of starch to give molecules of the disaccharide, maltose, and smaller starchlike molecules called dextrins. (See Fig. 38-1 for a representation of the action of an amylase.) The action of salivary amylase continues as food passes through the esophagus but stops when the food is exposed to the acidity of the stomach. Although the low pH of the stomach stops the action of salivary amylase, the hydrogen ions themselves may serve as catalysts for hydrolysis of bonds between saccharide units. These acid-catalyzed hydrolyses tend to act at random sites in the starch molecules and produce a mixture of mono, di-, tri-, and higher saccharides. Following most normal meals, only part of the starches are converted to the monosaccharide level by the time the food leaves the stomach and passes into the small intestine.

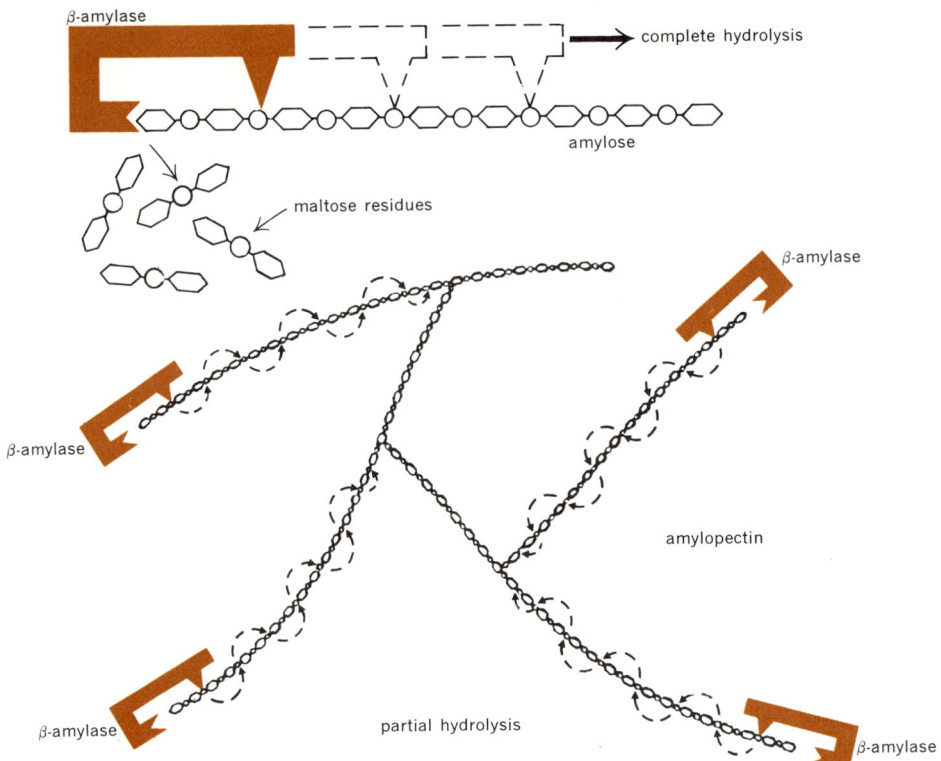

Fig. 38-1 The action of an amylase on starch. The enzyme β-amylase aids in the digestion of starch by catalyzing the hydrolysis of every second α 1-4 linkage in a starch molecule. This reaction yields maltose which must be further digested under the influence of maltase to yield glucose. Since β-amylase catalyzes the hydrolysis of only α 1-4 linkages, its action stops when a 1-6 branch is reached. A starch molecule in which degradation has stopped at the 1-6 branches is called a dextrin. [Adapted from *Principles of Plant Physiology*, by James Bonner and Arthur W. Galston, W. H. Freeman and Company. Copyright © 1952. Reprinted from an original article by W. Z. Hassid and R. M. Mc-Cready, *J. Am. chem. Soc.* **65**, 1943, p. 1159.]

In the small intestine, pancreatic amylases attack the remaining starch and dextrin molecules, and the disaccharidase, maltase, completes the digestion by converting maltose to glucose. Disaccharides such as sucrose are not digested by enzymes until they reach the small intestine, where the disaccharidases convert them to monosaccharides. Monosaccharides are not digested but are absorbed in the same form in which they were eaten.

The carbohydrate cellulose is not digested by higher animals. In most animals, cellulose is excreted as such; it makes up the so-called roughage of the diet and contributes to the bulk of the feces. Cattle, sheep, and similar animals, however, are able to use the cellulose of their diet. Those animals have several food pouches or stomachs; the first of them contains various microorganisms, some of which can convert cellulose to glucose.

Since this conversion occurs long before the food enters the intestine, the glucose produced by the hydrolysis of cellulose can be absorbed and utilized by ruminants.

> The name *ruminant* is derived from the name given to the first food pouch—the rumen. In addition to sheep and cattle, goats, antelopes, giraffes, deer, and camels have such a food pouch and are known as ruminants. These animals swallow their food without really chewing it well and later, while resting, regurgitate the food and chew it in a process described as chewing their cud. The process seems to mix saliva and microorganisms from the rumen with the food prior to its being carried to another food pouch, the abomasum, where it is acted upon by the acid gastric juice.

One other type of animal—the insect called the termite—can also make use of cellulose. This is, of course, why termites are so destructive. They eat their way through wooden structures using the cellulose of the wood for energy. Again it is not the termite itself, but a microorganism, a protozoan, living in the gut of the termite that contains the enzymes necessary to degrade cellulose into a useful form—glucose.

38-3 ABSORPTION AND FURTHER METABOLISM OF CARBOHYDRATES

After complete digestion has converted carbohydrates into monosaccharides, they are absorbed through the wall of the intestine into the blood stream. The portal vein carries the absorbed monosaccharides to the liver where they may be either synthesized into glycogen, oxidized to CO_2 and H_2O, or perhaps released as monosaccharides and allowed to circulate to other parts of the body. Although other organs are capable of metabolizing glucose, the liver is responsible for much of the carbohydrate metabolism. Under the influence of the pancreatic hormone, insulin, and other factors, the liver regulates the amount of glucose in the circulating blood. When the level of glucose in the blood is high, as it is after digestion of a meal rich in carbohydrates, the liver synthesizes large amounts of glycogen. When the amount of blood glucose is low, as it would be following intensive exercise, glycogen is depolymerized to yield glucose. Thus the liver, under the influence of insulin and other hormones, keeps the level of glucose in the blood relatively constant. The digestion and further metabolism of carbohydrates can be summarized as follows:

$$\text{polysaccharides} \xrightarrow{\text{digestion}} \text{monosaccharides} \xrightarrow{\substack{\text{absorption} \\ \text{through} \\ \text{intestinal} \\ \text{wall}}} \text{monosaccharides} \longrightarrow CO_2 + H_2O$$
$$\downarrow$$
$$\text{glycogen}$$

The concentration of glucose in the blood is extremely important for proper functioning of the human body. The normal fasting level of glucose is between 70 and 90 mg/100 ml. (Some methods of analysis which measure other compounds in addition to glucose give slightly higher values.) A concentration of glucose below the normal range is called *hypoglycemia*, and one higher than the normal level is called *hyperglycemia*. (See Fig. 38-2.)

Extreme hypoglycemia produces a series of reactions known as shock. The symptoms may include the trembling of muscle, a feeling of weakness, and a whitening of the skin. Serious hypoglycemia can cause unconsciousness and lowered blood pressure, and may result in death. Loss of consciousness is probably due to the lack of glucose in the brain which depends on this sugar for most of its energy. Extreme hypoglycemia is usually due

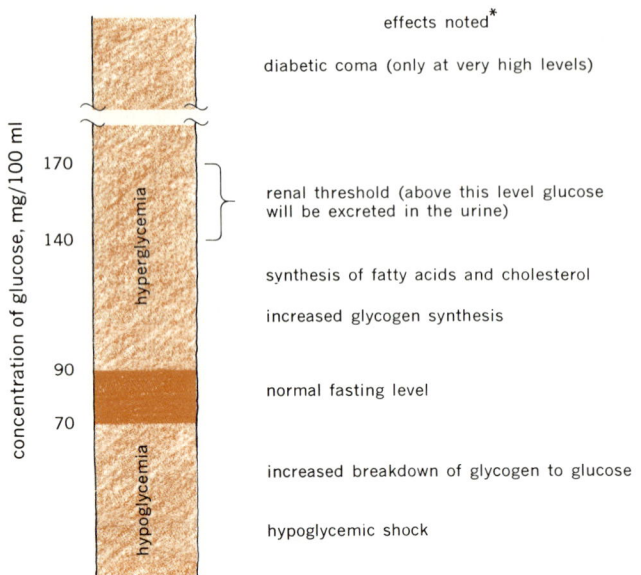

Fig. 38–2 Effects noted at various levels of blood glucose.

to the presence of excessive amounts of insulin. Some individuals who suffer from hyperinsulinism (overproduction of insulin) are subject to hypoglycemia if they eat a meal rich in carbohydrates. The most common cause of severe hypoglycemia is probably the injection of excessive amounts of insulin by diabetics; most diabetics carry candy or other glucose-rich substances with them to eat if they feel the onset of hypoglycemia. It is interesting to note that insulin shock has been used in treatment of mental diseases.

Research on hyperglycemia has shown that at levels higher than normal, the conversion of glucose to glycogen is stimulated. At these higher levels, fatty acids and cholesterol are synthesized from glucose. At some level, between 140 and 170 mg/100 ml for most people, glucose is excreted in the urine. The concentration of glucose necessary for the appearance of glucosuria (glucose in the urine) is called the *renal threshold*. Actually, glucose is always filtered into the urine in the kidneys but is reabsorbed before the urine leaves the kidney. Thus glucosuria represents not so much an abnormal filtering out of glucose but rather an inability of the kidney tubule cells to reabsorb the glucose. Extremely high levels of blood glucose are observed in cases of diabetic coma; factors other than hyperglycemia are probably responsible for the coma, however.

A variety of hormones controls the level of blood glucose. Insulin (from the pancreas) lowers blood glucose and enhances the conversion of glucose to glycogen. Adrenalin (epinephrine), one or more pituitary hormones, and a second pancreatic hormone—glucagon—all tend to raise the concentration of glucose in the blood. All these factors, acting together, keep the blood glucose level within certain limits beyond which the animal could not survive.

38-4 GLYCOGEN SYNTHESIS AND BREAKDOWN

The major intermediates in the conversion of glucose to glycogen are glucose 6-phosphate and glucose 1-phosphate. The same two intermediates are involved in the conversion of glycogen to glucose, but there are at least two instances where the conversions take different pathways. We can think of the situation as being similar to going between city A and city D with cities B and C being intermediate points, where there are oneway roads for part, but not all, of the way.

> The phosphate required to convert glucose to glucose 6-phosphate, as well as the energy required for the reaction, is provided by ATP. The reaction may be catalyzed by an enzyme known as hexokinase. Glucose 6-phosphate is then converted to glucose 1-phosphate. In glycogen synthesis, glucose 1-phosphate next reacts with uridine triphosphate (UTP) to give the compound UDP-glucose. Then UDP-glucose condenses to give glycogen, and the UDP is set free. The synthesis of glycogen requires a small amount of preformed glycogen to act as a pattern or primer, and also requires another enzyme or set of enzymes to make the 1-6 branches which are important in the glycogen molecule.
>
> When glycogen is converted to glucose 1-phosphate in the process of glycogenolysis, the enzyme catalyzing this reaction is different from the glycogen synthetase which catalyzed its formation. The enzyme catalyzing the degradation is a phosphorylase, i.e., an enzyme that uses phosphoric acid to cleave a larger molecule. Although this enzyme can catalyze the synthesis of glycogen, equilibrium lies in the direction of the formation of glucose 1-phosphate. This glucose phosphate is then isomerized to glucose 6-phosphate which can then enter a variety of metabolic sequences.
>
> The transformation of glucose 6-phosphate to glucose involves an enzyme called a phosphatase. The reaction gives glucose and phosphoric acid. It does not result in the synthesis of ATP from ADP and phosphate, so the reaction is not the reverse of that catalyzed by hexokinase. (The symbol ⓟ is used to indicate a phosphate group.) These reactions may be summarized as follows:

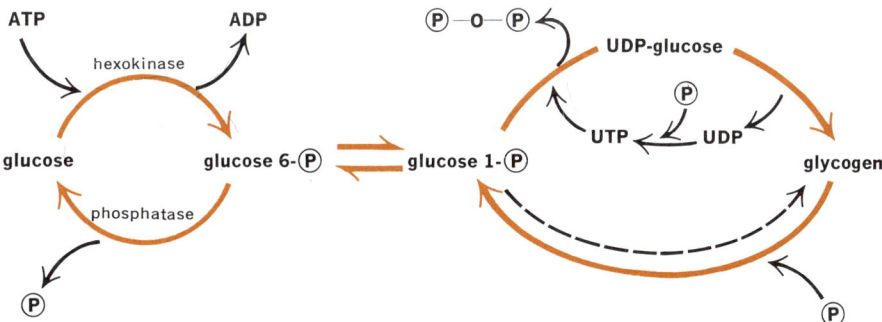

> The effects of deficiencies of the enzymes that catalyze these reactions are discussed in Section 38–11.

The irreversibility of parts of the scheme for the reversible conversion of glucose to glycogen is an illustration of the generalization that synthetic and degradative pathways may use the same general intermediates but do not follow the same paths. A further reference to this is made in Section 40–5.

38–5 DEGRADATION OF GLUCOSE

The conversion of glucose to carbon dioxide and water, with the release of much of the energy contained in the glucose molecule, is essentially an oxidation. The amount of free energy released in the complete oxidation of one mole of glucose is about 690 kcal.

glucose + oxygen → carbon dioxide + water + 690 kcal

This amount of energy is much greater than the amount that can be stored effectively in a molecule of ATP. From the discussion of the previous chapter, we would expect that the degradation and oxidation of glucose proceeds in a series of steps, some of which are capable of yielding energy that can be stored in ATP. This is true, and we know a great number of intermediates involved in the process. For convenience, we generally think of one series of reactions as leading from glucose to either pyruvic* or lactic acid. We call this series of reactions **glycolysis**. The second series of reactions involves the conversion of pyruvic acid or lactic acid to acetic acid, and finally to carbon dioxide and water. This series is referred to as **aerobic metabolism** and involves the **citric acid cycle** or the *Krebs cycle*.

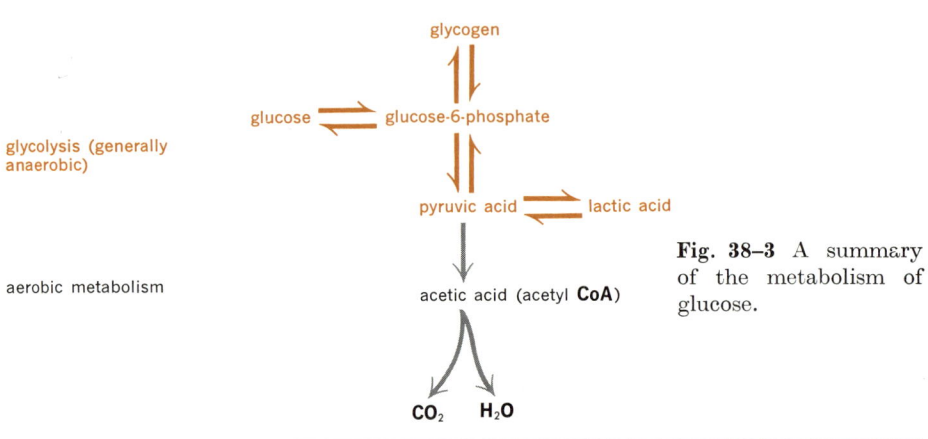

Fig. 38–3 A summary of the metabolism of glucose.

The term *aerobic* means in the presence of air. Actually, it is the oxygen of the air that is required for aerobic reactions. Anaerobic reactions do not require oxygen. In some cases, oxygen inhibits anaerobic processes, while in others the presence of oxygen has no effect on the reactions. The degradation of glucose is summarized in Fig. 38–3.

---

* Lactic, pyruvic, and acetic acids are weak acids and ionize only slightly; however, at the pH of the body, these substances are present as lactate, pyruvate, and acetate ions. Acetate is present in most biological systems as acetyl coenzyme A. Throughout the biochemistry section of this book, the term acid should be taken to mean an equilibrium mixture of an acid and its anion. Often the anion is present to a greater extent than the undissociated acid.

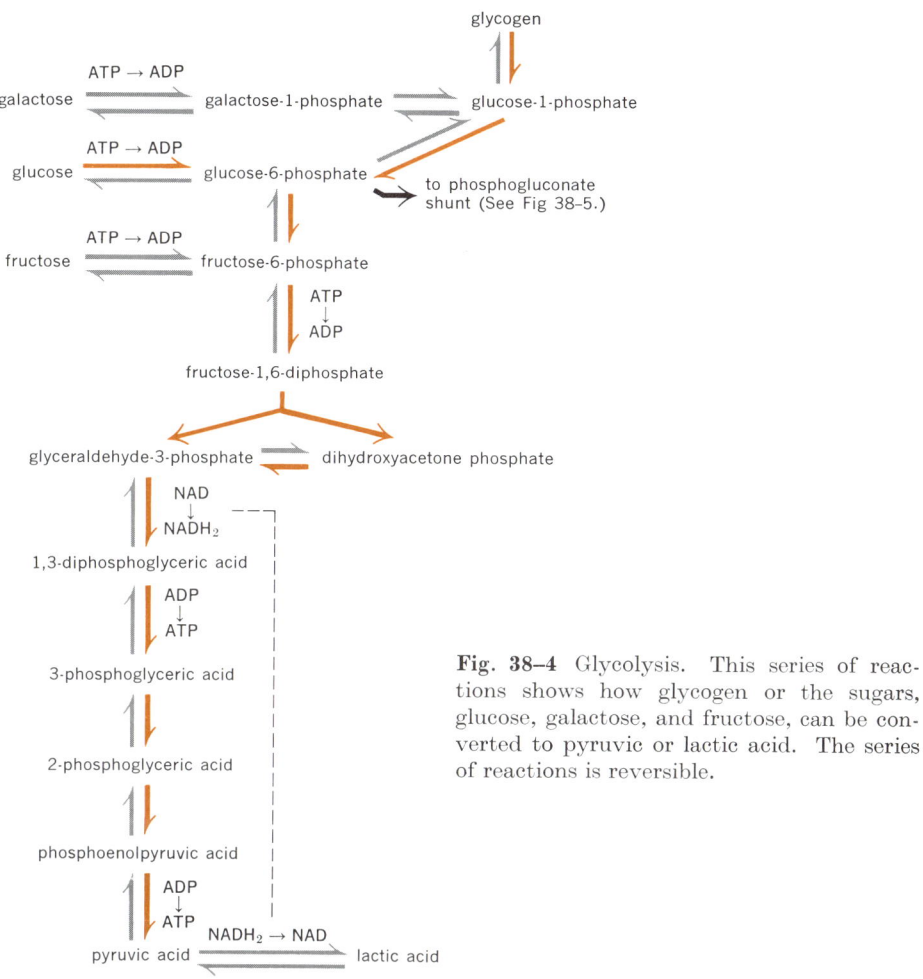

Fig. 38-4 Glycolysis. This series of reactions shows how glycogen or the sugars, glucose, galactose, and fructose, can be converted to pyruvic or lactic acid. The series of reactions is reversible.

38-6 GLYCOLYSIS

Since the prefix *glyco-* is used to designate carbohydrates generally and not only glycogen, the term glycolysis means the lysis or degradation of carbohydrates. The reactions of glycolysis can proceed without oxygen to yield lactic acid. They may also proceed in the presence of oxygen, and in this case the end product is pyruvic acid.

glucose
glycogen } **glucose 6-phosphate** — with oxygen → **pyruvic acid**
— without oxygen → **lactic acid**

Glycolysis results in the splitting of a 6-carbon compound into two 3-carbon compounds, and the synthesis of ATP. The reactions of glycolysis are summarized in Fig. 38-4 and discussed in detail in the following paragraphs.

Metabolism of Carbohydrates

We may consider glycolysis as starting with either glucose or glycogen. The first step in either case involves addition of a phosphate group. Ordinary phosphate is sufficient for phosphorylation of glycogen, but ATP is required to phosphorylate glucose. Glucose 6-phosphate is formed in one step from glucose, and after two steps when glycogen is the starting material.

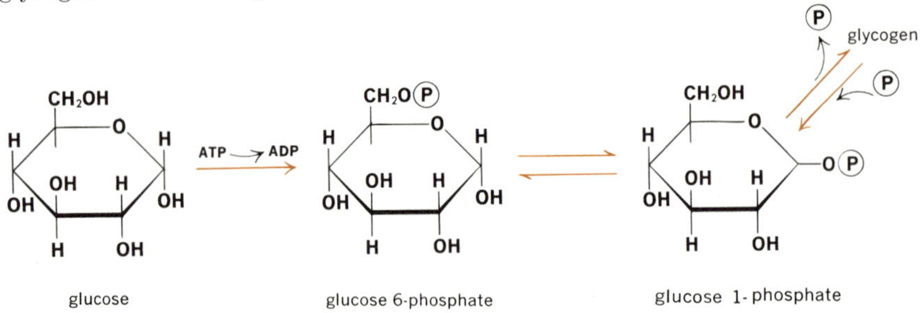

glucose →(ATP → ADP) glucose 6-phosphate ⇌ glucose 1-phosphate ⇌ glycogen

By an isomerization, glucose 6-phosphate is converted to fructose 6-phosphate: This reaction is best shown using straight-chain or Fisher formulas.

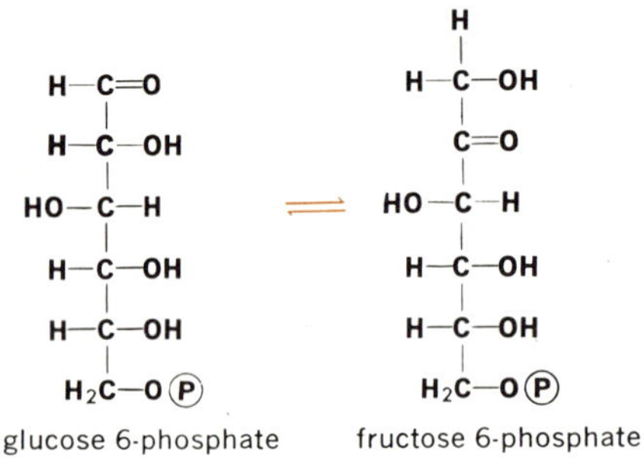

glucose 6-phosphate ⇌ fructose 6-phosphate

At this point another molecule of ATP is required to transform fructose 6-phosphate to fructose 1,6-diphosphate. This diphosphofructose is then split to give two 3-carbon compounds, glyceraldehyde 3-phosphate, and dihydroxyacetone phosphate.

fructose 1,6-diphosphate → dihydroxyacetone phosphate + glyceraldehyde 3-phosphate

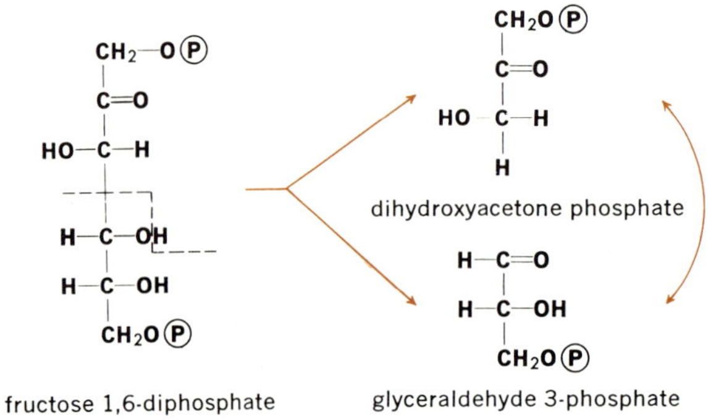

Dihydroxyacetone phosphate represents a dead end in this metabolic pathway; only the fact that the enzyme triosephosphate isomerase can catalyze the conversion of dihydroxyacetone phosphate to glyceraldehyde 3-phosphate prevents the loss of half the energy of the glucose molecule. We can say that the net effect of the splitting of one molecule of fructose diphosphate is the formation of two molecules of glyceraldehyde 3-phosphate. So far the glycolytic scheme has required an energy input in the form of two molecules of ATP, but has not yet released any of the energy of glucose.

The next step in glycolysis, the conversion of glyceraldehyde 3-phosphate to 1,3-diphosphoglyceric acid, is a complex reaction. It involves the addition of a phosphate and the oxidation of the molecule from an aldehyde to an acid. This oxidation results in the formation of reduced NAD ($NADH_2$).

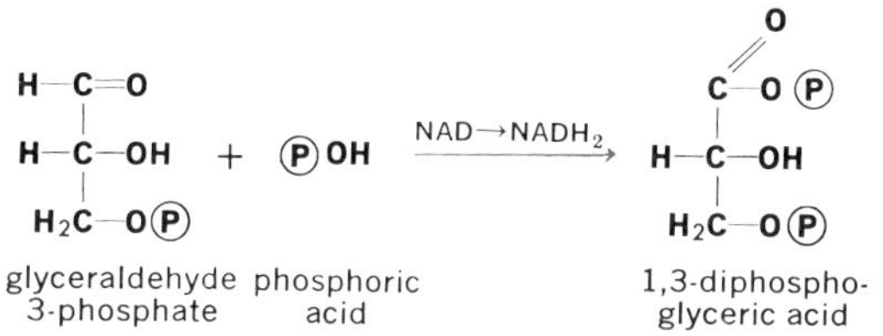

$$\begin{array}{c} H-C=O \\ | \\ H-C-OH \\ | \\ H_2C-O\,\textcircled{P} \end{array} + \textcircled{P}OH \xrightarrow{NAD \to NADH_2} \begin{array}{c} O \\ \parallel \\ C-O\,\textcircled{P} \\ | \\ H-C-OH \\ | \\ H_2C-O\,\textcircled{P} \end{array}$$

glyceraldehyde 3-phosphate   phosphoric acid   1,3-diphosphoglyceric acid

We will discuss the fate of this reduced NAD later, since we wish to follow the general course of glycolysis which involves the further reaction of 1,3-diphospho-glyceric acid.

One bond between the phosphoric acid and glyceric acid in 1,3-diphosphoglyceric acid is an anhydride linkage, and as such its rupture provides sufficient energy to synthesize a molecule of ATP from ADP and phosphate. The other product of the hydrolysis of 1,3-diphosphoglyceric acid is 3-phosphoglyceric acid.

$$\begin{array}{c} O \\ \parallel \\ C-O\,\textcircled{P} \\ | \\ H-C-OH \\ | \\ H_2C-O\,\textcircled{P} \end{array} + ADP \to \begin{array}{c} O \\ \parallel \\ C-OH \\ | \\ H-C-OH \\ | \\ H_2C-O\,\textcircled{P} \end{array} + ATP$$

1,3-diphosphoglyceric acid   3-phosphoglyceric acid

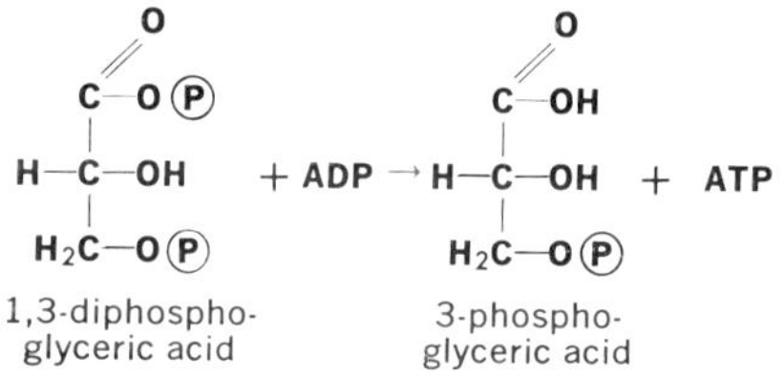

The compound 3-phosphoglyceric acid is converted to its isomer, 2-phosphoglyceric acid. Removal of a molecule of water from the latter compound gives the enol form of phosphopyruvic acid.

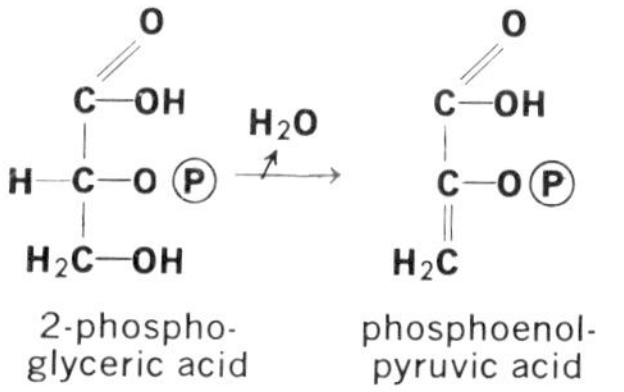

$$\begin{array}{c} O \\ \parallel \\ C-OH \\ | \\ H-C-O\,\textcircled{P} \\ | \\ H_2C-OH \end{array} \xrightarrow{H_2O} \begin{array}{c} O \\ \parallel \\ C-OH \\ | \\ C-O\,\textcircled{P} \\ \parallel \\ H_2C \end{array}$$

2-phosphoglyceric acid   phosphoenolpyruvic acid

Hydrolysis of phosphoenolpyruvic acid again gives both the phosphate and sufficient energy to result in the synthesis of ATP from ADP.

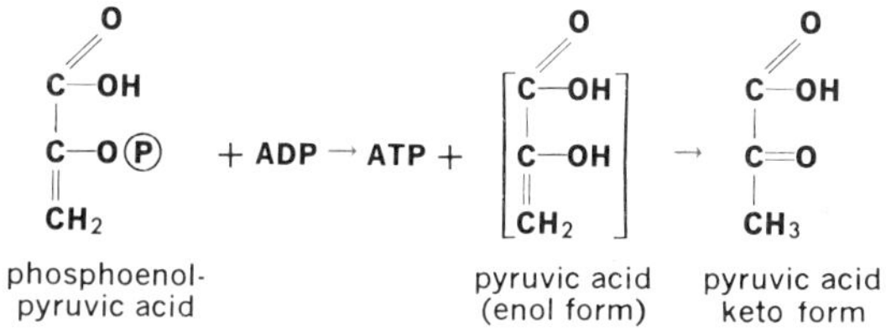

$$\begin{array}{c} O \\ \parallel \\ C-OH \\ | \\ C-O\,\textcircled{P} \\ \parallel \\ CH_2 \end{array} + ADP \to ATP + \left[\begin{array}{c} O \\ \parallel \\ C-OH \\ | \\ C-OH \\ \parallel \\ CH_2 \end{array}\right] \to \begin{array}{c} O \\ \parallel \\ C-OH \\ | \\ C=O \\ | \\ CH_3 \end{array}$$

phosphoenolpyruvic acid   pyruvic acid (enol form)   pyruvic acid keto form

With the removal of the phosphate we are left with the nonphosphorylated compound pyruvic acid. It is of interest to note that phosphorylated carbohydrates are found almost always totally within cells, and the nonphosphorylated compounds, glucose, pyruvic acid, and lactic acid are found both in cells and in the blood stream.

The splitting of glucose to two 3-carbon compounds results in the formation of reduced NAD. Pyruvic acid is the end product of glycolysis if oxygen is present and the reduced NAD (formed in the conversion of glyceraldehyde 3-phosphate to 1,3,-diphosphoglyceric acid) is oxidized through an electron-transport chain to give water. When glycolysis occurs in the absence of oxygen, it is necessary that some other agent be found for oxidizing the reduced NAD. The oxidized form of NAD is an essential catalyst; without it, glycolysis would stop with the accumulation of glyceraldehyde 3-phosphate. The compound which oxidizes NAD in anaerobic glycolysis is pyruvic acid itself. In this process, the pyruvic acid is reduced to give lactic acid. Thus we say that lactic acid is the normal end product of anaerobic glycolysis.

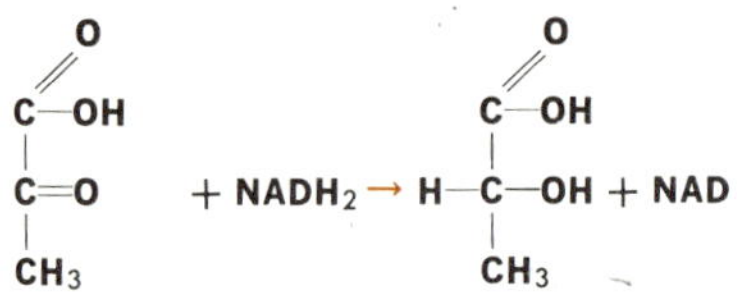

$$\begin{array}{c}O\\\parallel\\C-OH\\|\\C=O\\|\\CH_3\end{array} + NADH_2 \rightarrow \begin{array}{c}O\\\parallel\\C-OH\\|\\H-C-OH\\|\\CH_3\end{array} + NAD$$

pyruvic acid — lactic acid

Monosaccharides other than glucose are also metabolized in glycolysis. Fructose makes up half of the sucrose molecule, and is a product of digestion of any sucrose-containing foods. Fructose is phosphorylated to give fructose 6-phosphate, which is a normal glycolytic intermediate. Lactose, commonly known as milk sugar, produces one molecule of glucose and one of galactose when hydrolyzed in the digestive process. Full utilization of the energy of lactose requires that galactose be metabolized.

The route by which galactose enters the glycolytic pathway involves phosphorylation to give galactose 1-phosphate. This compound is then converted to its isomer glucose 1-phosphate by a change in configuration of the alcoholic OH group on carbon number 4.

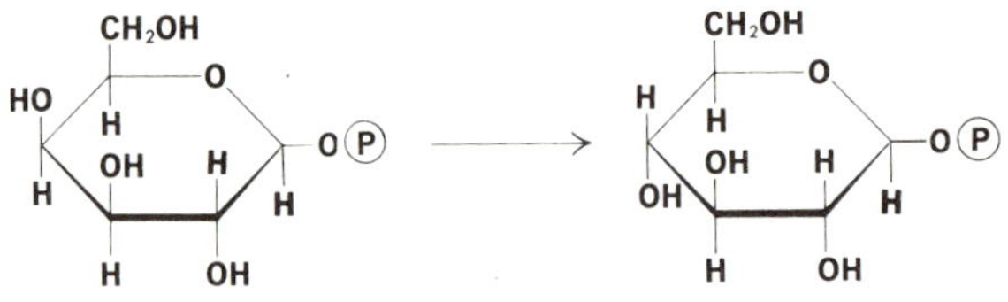

galactose 1-phosphate → glucose 1-phosphate

The glucose 1-phosphate so formed cannot be distinguished from that formed from glycogen or glucose, and it is metabolized via the glycolytic reactions.

The glycolysis of one molecule of glucose requires the investment of two molecules of ATP, and results in the synthesis of four molecules of ATP. (There are two reactions which yield ATP, but one molecule of glucose gives two molecules of each of the reactants, and hence two molecules of ATP at each of the steps.) Thus the net energy gain in anaerobic glycolysis is the formation of two molecules of ATP per molecule of glucose. The $NADH_2$ which is generated is oxidized to NAD by the conversion of pyruvate to lactate.

Aerobic glycolysis gives more ATP synthesis since the oxidation of each $NADH_2$, with the formation of H_2O, give an average of three ATP's (Section 37–3); there are two reduced NAD molecules per molecule of glucose, so we get six ATP's from this source. When we add these to the two that are obtained whether or not oxygen is present, the net energy trapping in aerobic glycolysis is eight molecules of ATP per molecule of glucose.

The free energy resulting from complete oxidation of some glycolytically significant compounds is

glucose, 690 kcal/mole   pyruvic acid, 275 kcal/mole   lactic acid, 315 kcal/mole.

From these data we see that there is more potential energy in lactic acid than in pyruvic acid. Since two moles of either pyruvic or lactic acid are formed from one mole of glucose, the energy differences are

690 − 2(275) or 140 kcal/mole for the conversion of glucose to pyruvic acid, and
690 − 2(315) or 60 kcal/mole for the conversion of glucose to lactic acid.

From this calculation we see that 140/690 or about 20% of the energy of glucose is released when glycolysis is aerobic and the end products are pyruvic acid and water. Only 60/690 or about 9% of the energy is released when glycolysis is anaerobic and lactic acid is formed as the only end product.

We can calculate the efficiency with which the released energy is trapped in ATP as follows. The formation of pyruvic acid and water results in the conversion of 8 molecules of ADP to ATP. If we multiply 8 by 12 kcal, which is the energy stored when a mole of ADP is converted to one of ATP, we get 96 kcal of energy stored. Since 140 kcal of energy is released and 96 are trapped, the efficiency is about 96/140 or 70% for aerobic glycolysis. Anaerobic glycolysis results in the synthesis of only 2 ATP's per molecule of glucose, so we have a storage of only 24 kcals of energy. The amount of the energy of glucose released is 60 kcal, so the efficiency of trapping of the released energy in anaerobic glycolysis is only 24/60 or 40%. Thus we can see that not only does anaerobic glycolysis release less of the energy of glucose than does aerobic glycolysis, but that it also traps the released energy much less efficiently.

Glycolysis results in the conversion of the energy of glucose to ATP. Only a small part of the energy of glucose is released; most of it is still in the lactic acid or pyruvic acid which is left at the end of the process. However, this amount of energy is apparently sufficient to support life and growth of anaerobic bacteria. Some of these bacteria are characterized as lactobacilli because of the lactic acid they produce. Human muscles, particularly those working with inadequate amounts of oxygen, can use anaerobic glycolysis as a source of energy and they, too, produce lactic acid. Lactic acid in the muscles of animals may be the cause of cramps. A more beneficial scheme would result in the release of a greater percentage of the energy of the glucose and would end with products such as carbon dioxide and water that were nontoxic and easy to eliminate. Such a system exists and will be discussed, after a few digressions, in the section on aerobic metabolism.

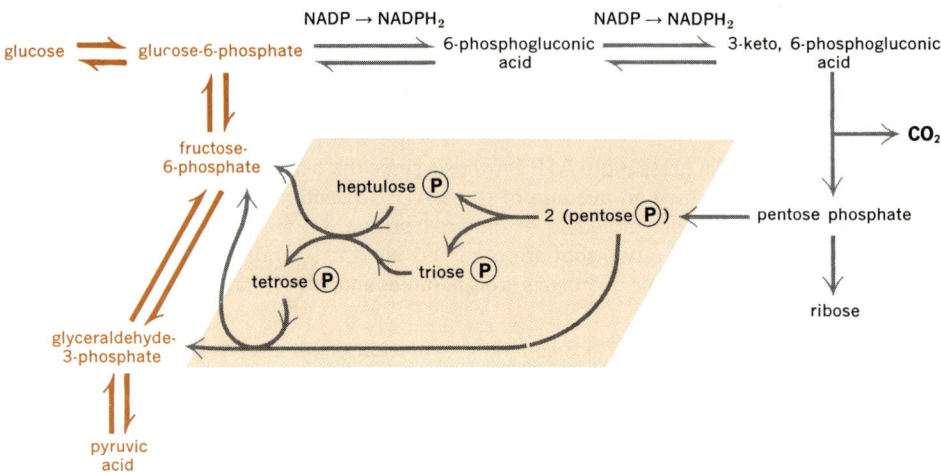

Fig. 38-5 The phosphogluconate shunt. This series of reactions is an alternative to those of glycolysis. The shaded area represents various possible condensations that may be used to convert pentose phosphate molecules to glyceraldehyde phosphate or fructose 6-phosphate. Glycolytic reactions are also shown for reference.

38-7 THE PHOSPHOGLUCONIC ACID SHUNT

Not all glucose 6-phosphate is metabolized by being converted to fructose 6-phosphate. An alternative pathway involves the oxidation of glucose 6-phosphate to 6-phosphogluconic acid. The series of reactions resulting from this conversion is called the *phosphogluconate shunt*. The term shunt indicates a bypass, and in some ways this pathway is a bypass to glycolysis. The phosphogluconate pathway is shown in Fig. 38-5.

The pathway provides oxidations in the first two reactions and thus provides energy. The energy here is transferred to nicotinamide adenine dinucleotide phosphate (NADP), a compound much like NAD but containing an additional phosphate group. Reduced NADP is an important energy source for some biosynthetic reactions, particularly for lipid synthesis. Thus it is not surprising that enzymes for the phosphogluconate pathway are found in tissues that synthesize lipids. In addition to producing $NADPH_2$, this pathway results in the synthesis of pentoses, and is probably the major source of the ribose and deoxyribose needed for nucleic acid synthesis.

Many tissues, particularly the liver, have enzymes for both glycolysis and the phosphogluconate series of reactions. These tissues can metabolize glucose by either pathway, and the pathway chosen probably depends on conditions such as the availability of oxygen and the level of $NADPH_2$ or of pentoses in the cells or in the circulating blood that moves through the tissues. Muscle and brain tissue apparently use the glycolytic pathway exclusively for metabolism of carbohydrates, since they do not contain the enzymes necessary for the phosphogluconate pathway. Some bacteria use reactions similar to this pathway. It is interesting to note that the photosynthesis of carbohydrates involves many of the intermediate compounds found in the phosphogluconate reaction sequence. It also involves NADP.

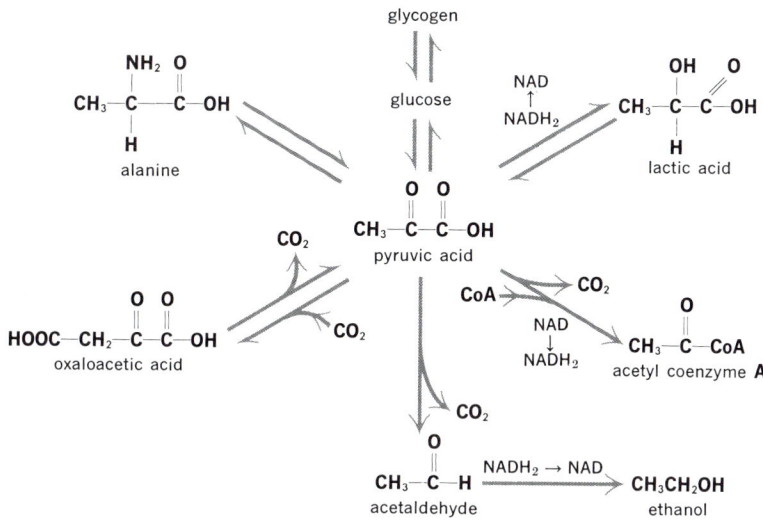

Fig. 38-6 Metabolism of pyruvic acid. Pyruvic acid represents a major branching point in metabolism. Some of the important products derived directly from pyruvic acid are shown.

38-8 METABOLISM OF PYRUVIC ACID

Pyruvic acid may undergo any of several alternate reactions, and thus represents a major branching point in metabolism. Some of these reactions lead to other series of reactions, and some represent metabolic dead ends. Figure 38-6 summarizes these reactions.

Addition of an amino group transforms pyruvic acid to the amino acid alanine. The reverse reaction also occurs and represents one method for conversion of amino acids into carbohydrates. Addition of carbon dioxide, perhaps not to pyruvic acid itself but to phosphoenolpyruvic acid, gives the dicarboxylic acid oxaloacetic acid whose significance will be shown later. The reversible conversion of pyruvic acid to lactic acid has been discussed previously. This conversion represents a metabolic dead end, and lactic acid, if it is to be further metabolized, must first be changed back to pyruvic acid. In normal muscle tissue when there is an adequate supply of oxygen, it is doubtful that any lactic acid is formed. Pyruvic acid may, by a reversal of glycolysis, be converted to glucose or glycogen.

Pyruvic acid may lose a molecule of carbon dioxide in either of two ways. These two different methods lead to quite different pathways and represent a major difference in metabolism in different kinds of organisms. In the simpler process, carbon dioxide is lost and acetaldehyde is formed:

$$CH_3-\underset{\substack{\| \\ O}}{C}-COOH \rightarrow CH_3-\underset{\substack{\| \\ O}}{C}-H + CO_2$$

pyruvic acid → acetaldehyde

The acetaldehyde may then be reduced to yield ethanol. This pathway is found primarily in yeasts, and is the basis of alcoholic fermentation. In this process yeasts metabolize glucose through the anaerobic glycolytic pathway with the formation of pyruvic acid. They do not, however, oxidize the reduced NAD by transforming pyruvic acid to lactic acid. Instead, they use the reduced NAD to convert acetaldehyde to ethanol:

$$CH_3-\underset{}{\overset{O}{C}}-H + NADH_2 \rightarrow CH_3-\underset{H}{\overset{H\quad OH}{C}} + NAD$$

acetaldehyde   ethanol

Thus the alcoholic fermentation of carbohydrates yields as end products carbon dioxide and ethanol. The series of reactions may be summarized as

$$C_6H_{12}O_6 \rightarrow 2C_2H_5OH + 2CO_2$$

The process does not involve oxygen, and the amount of energy trapped by the yeasts is small (two ATP molecules per molecule of glucose), but is sufficient to support their growth and reproduction. Eventually the ethanol formed by the yeasts results in a medium that will not support further growth, and they are thus inhibited by one of the products of their metabolism.

> Alcoholic fermentations have been known and widely used since prehistoric times. The widespread availability of both carbohydrates and yeasts has led to a variety of alcoholic beverages. Starches of wheat and barley yield beer, sugar of grapes gives wine, fermented honey gives mead—the list of conversions is almost endless. When alcohol is removed from the fermentation mixture by distillation, we get vodka, saki, brandy, whiskey, and other beverages. While all these beverages contain large amounts of alcohol, the differences between them are due to compounds other than alcohol which either distill over with the alcohol or are added later.
>
> The process of fermentation has been explained in many ways. The Greeks described it as the god Dionysius acting upon grape juice; Pasteur realized that the fermenting agents were yeasts; we now describe fermentation as being the enzymatically catalyzed conversion of carbohydrates to ethanol and carbon dioxide. The process of alcohol production has always fascinated man. However, he has undoubtedly been more concerned with the effects achieved by consumption of the products than he has with understanding the process of the formation of alcohol.
>
> An interesting abnormality, recently reported in the press, involved a Japanese man who was repeatedly accused of drunkenness even though he maintained that he had not consumed any alcoholic beverages. He was found to have, as the result of a previous injury, an extra pouch in his stomach. This pouch, which was protected from the effects of the acid gastric juice, was harboring a colony of yeasts which were efficiently producing ethanol. He then, quite unconsciously, absorbed the alcohol with the usual results. It is not likely, however, that similar pouches are present in the stomachs of others who protest in the face of overwhelming evidence to the contrary that they have not consumed alcohol.

The other way pyruvate may lose carbon dioxide is by a complex reaction. In this reaction or series of reactions, carbon dioxide is formed. The resulting acetaldehyde-like intermediate is indirectly oxidized by an enzyme using NAD as a hydrogen acceptor, and acetic acid in the form of acetyl coenzyme A (acetyl CoA) is formed. This series of re-

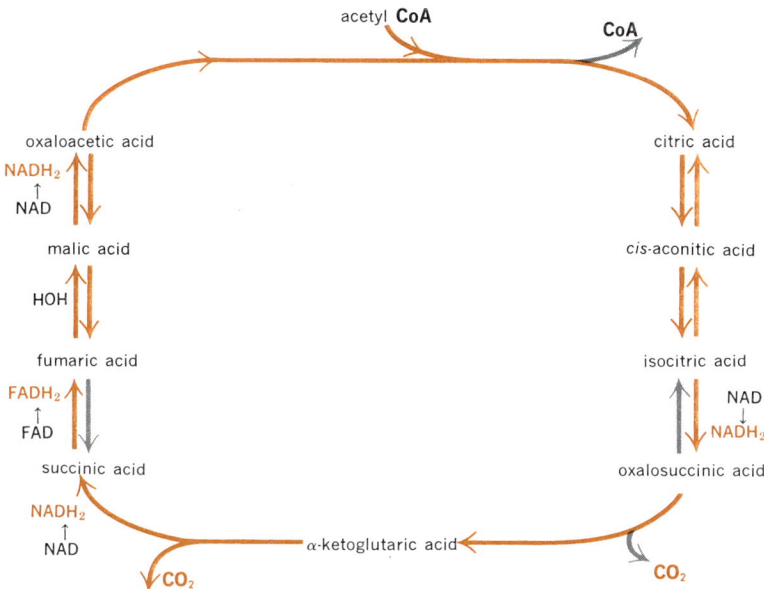

Fig. 38-7 The citric acid cycle. This series of reactions, which is also known as the Krebs cycle, converts an acetate group to carbon dioxide and water. The water is formed indirectly by the oxidation of $NADH_2$ and $FADH_2$ through the electron transport system. The isomers of citric acid and oxalosuccinic acid contain 6 carbon atoms each, α-ketoglutaric has 5, and succinic, fumaric, malic, and oxaloacetic contain 4 carbon atoms each.

---

actions is not reversible: in other words, acetyl CoA is not carboxylated to yield pyruvic acid. There are several possible fates for acetyl CoA; the most common one is the series of reactions known as the citric acid cycle.

38-9 AEROBIC METABOLISM—THE CITRIC ACID CYCLE

The series of aerobic reactions in animals is variously called the *Krebs cycle*, *tricarboxylic acid cycle*, and the *citric acid cycle*. We will use the last name since it seems to be the most descriptive. This series of reactions is not restricted to the metabolism of carbohydrates; it is also important in the metabolism of both lipids and proteins. Any substance that can be converted to acetyl CoA may enter the cycle. In this cyclic series of reactions a 2-carbon unit, the acetyl group, is completely converted to carbon dioxide and water. The cycle requires oxygen to oxidize (through the electron transport-system) the $NADH_2$, $NADPH_2$, and $FADH_2$ which accepted hydrogen atoms (electrons) from the intermediates in the cycle. Although there are minor variations, the cycle is used by almost all plants and animals that have been studied. Anaerobic organisms, of course, do not use it.

The citric acid cycle is summarized in Fig. 38-7, and discussed in detail in the following paragraphs.

In the first reaction of the cycle an acetyl CoA unit is condensed with an oxaloacetate molecule to give citric acid.

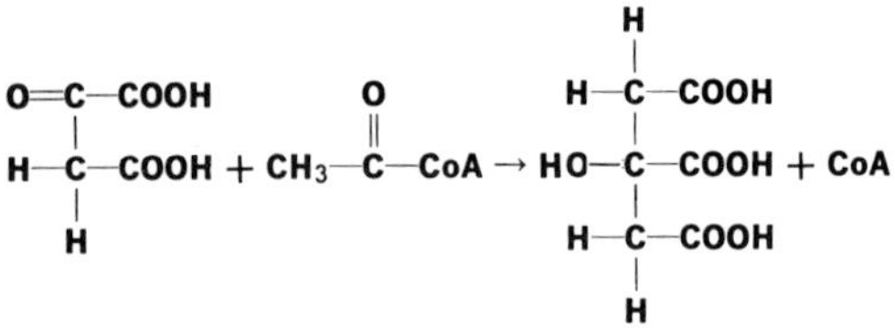

$$\underset{\substack{\text{oxaloacetic} \\ \text{acid}}}{\begin{array}{c} \text{O}=\text{C}-\text{COOH} \\ | \\ \text{H}-\text{C}-\text{COOH} \\ | \\ \text{H} \end{array}} + \underset{\text{acetyl CoA}}{\text{CH}_3-\overset{\text{O}}{\underset{\|}{\text{C}}}-\text{CoA}} \rightarrow \underset{\substack{\text{citric acid}}}{\begin{array}{c} \text{H} \\ | \\ \text{H}-\text{C}-\text{COOH} \\ | \\ \text{HO}-\text{C}-\text{COOH} \\ | \\ \text{H}-\text{C}-\text{COOH} \\ | \\ \text{H} \end{array}} + \text{CoA}$$

Actually, the citric acid formed is in equilibrium with *cis*-aconitic acid and isocitric acid, but the equilibrium favors the formation of citric acid.

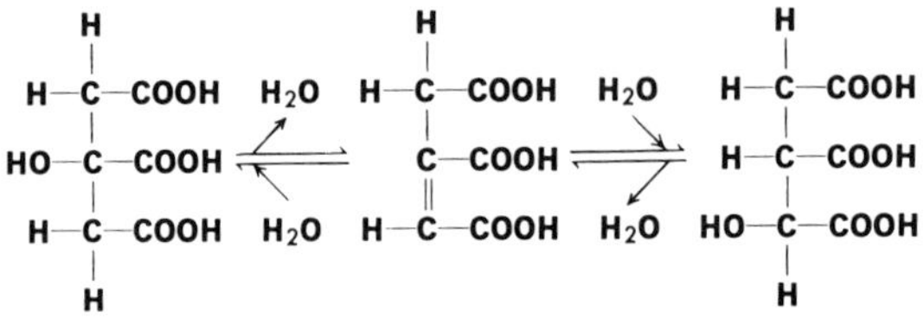

$$\underset{\substack{\text{citric} \\ \text{acid}}}{\begin{array}{c} \text{H} \\ | \\ \text{H}-\text{C}-\text{COOH} \\ | \\ \text{HO}-\text{C}-\text{COOH} \\ | \\ \text{H}-\text{C}-\text{COOH} \\ | \\ \text{H} \end{array}} \underset{\text{H}_2\text{O}}{\overset{\text{H}_2\text{O}}{\rightleftarrows}} \underset{\substack{\textit{cis}\text{-aconitic} \\ \text{acid}}}{\begin{array}{c} \text{H} \\ | \\ \text{H}-\text{C}-\text{COOH} \\ | \\ \text{C}-\text{COOH} \\ \| \\ \text{H}-\text{C}-\text{COOH} \end{array}} \underset{\text{H}_2\text{O}}{\overset{\text{H}_2\text{O}}{\rightleftarrows}} \underset{\substack{\text{isocitric} \\ \text{acid}}}{\begin{array}{c} \text{H} \\ | \\ \text{H}-\text{C}-\text{COOH} \\ | \\ \text{H}-\text{C}-\text{COOH} \\ | \\ \text{HO}-\text{C}-\text{COOH} \\ | \\ \text{H} \end{array}}$$

Because of this reversibility citric acid would seem to be a poor substance for the initiation of a series of reactions. However, the reaction by which it is formed from oxaloacetic acid and acetyl CoA is virtually irreversible, and there are no other significant biological reactions of citric acid. The dehydrogenation of isocitric acid to give oxalosuccinic acid causes an imbalance in the equilibrium. As would be expected from the law of mass action, the net effect is the series of reactions leading from citric acid to oxalosuccinic acid.

In the formation of oxalosuccinic acid, two hydrogen atoms are removed from isocitric acid to give reduced NAD (or NADP in some organisms) and oxalosuccinic acid:

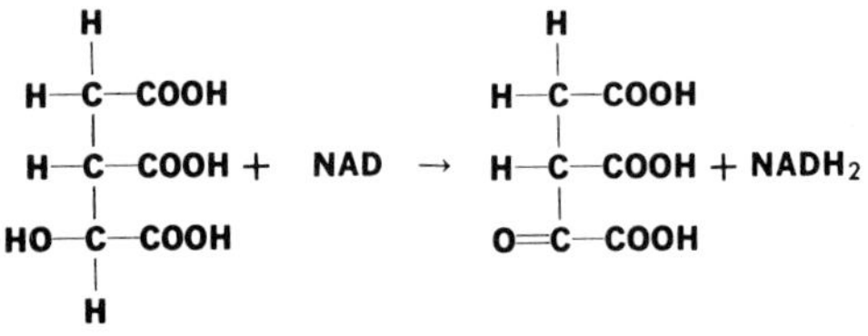

$$\underset{\substack{\text{isocitric} \\ \text{acid}}}{\begin{array}{c} \text{H} \\ | \\ \text{H}-\text{C}-\text{COOH} \\ | \\ \text{H}-\text{C}-\text{COOH} \\ | \\ \text{HO}-\text{C}-\text{COOH} \\ | \\ \text{H} \end{array}} + \text{NAD} \rightarrow \underset{\substack{\text{oxalosuccinic} \\ \text{acid}}}{\begin{array}{c} \text{H} \\ | \\ \text{H}-\text{C}-\text{COOH} \\ | \\ \text{H}-\text{C}-\text{COOH} \\ | \\ \text{O}=\text{C}-\text{COOH} \end{array}} + \text{NADH}_2$$

In the next reaction of the citric acid cycle, this compound loses carbon dioxide to give α-ketoglutaric acid:

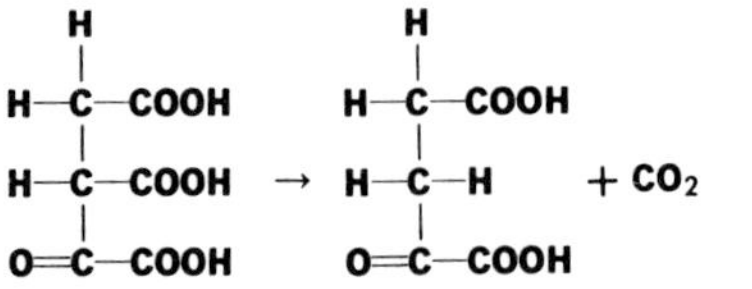

$$\underset{\substack{\text{oxalosuccinic} \\ \text{acid}}}{\begin{array}{c} \text{H} \\ | \\ \text{H}-\text{C}-\text{COOH} \\ | \\ \text{H}-\text{C}-\text{COOH} \\ | \\ \text{O}=\text{C}-\text{COOH} \end{array}} \rightarrow \underset{\substack{\alpha\text{-ketoglutaric} \\ \text{acid}}}{\begin{array}{c} \text{H} \\ | \\ \text{H}-\text{C}-\text{COOH} \\ | \\ \text{H}-\text{C}-\text{H} \\ | \\ \text{O}=\text{C}-\text{COOH} \end{array}} + \text{CO}_2$$

The following reaction, the conversion of α-ketoglutaric acid to succinic acid is complex, and includes both an oxidation and a decarboxylation. It is similar to the conversion of pyruvic acid to acetyl CoA, in that it requires both NAD and coenzyme A. This reaction, or perhaps to be completely accurate we should call it a series of reactions, is almost completely irreversible.

```
    H                          H
    |                          |
H — C — COOH              H — C — COOH
    |                          |
H — C — H     + NAD  →    H — C — H       + CO₂ + NADH₂
    |                          |
O = C — COOH              O = C — OH

α-ketoglutaric              succinic
    acid                      acid
```

Succinic acid is dehydrogenated under the influence of an FAD-containing enzyme to give fumaric acid. Fumaric acid is then hydrated to give malic acid which is dehydrogenated, in a reaction which requires NAD, to yield oxaloacetic acid.

```
  COOH                                           COOH              COOH
   |                                              |                 |
  CH₂                       H — C — COOH         H — C — OH         C = O
   |         FAD → FADH₂         ||                |     NAD → NADH₂ |
  CH₂       ─────────────→  HOOC — C — H    ──→   CH₂   ─────────→  CH₂
   |                                              |                 |
  COOH                          HOH              COOH              COOH

succinic                      fumaric            malic            oxaloacetic
 acid                           acid              acid               acid
```

The oxaloacetic acid unites with an acetate unit to give citric acid, and the cycle begins again. Of course, with a cyclic series of reactions there is really no beginning and no end, but the general availability of acetyl CoA and the significance of having sufficient oxaloacetate to react with it makes this an especially important reaction in the cycle, and a good point to start and stop any discussion of the cycle.

In the citric acid cycle a molecule of acetyl CoA, or more precisely of a 2-carbon unit is converted to two molecules of CO_2 and four of H_2O. The energy contained in the acetyl CoA is converted to ATP via the intermediates NAD (or NADP), FAD, and the cytochrome system.

One turn of the citric acid cycle results in the formation of three molecules of $NADH_2$ and one of $FADH_2$. This results in the synthesis of eleven molecules of ATP from ADP and phosphate (three from each $NADH_2$ and two from the $FADH_2$). In addition, the energy released in the conversion of α-ketoglutaric acid to succinic acid through the involvement of coenzyme A, is sufficient to synthesize another molecule of ATP. The total number of molecules of ATP synthesized per acetate unit is twelve.

We can summarize the amount of ATP synthesized in the complete oxidation of glucose to carbon dioxide and water as follows:

aerobic glycolysis	8 ATP
conversion of 2 moles of pyruvic acid to 2 moles of acetyl CoA	6 ATP
oxidation of 2 moles of acetyl CoA	24 ATP
total	38 ATP

Using 12 kcal as the energy stored in an ATP molecule, we have 12 × 38 or 456 kcal of energy stored in ATP. Since the total free energy change in the conversion of glucose to carbon dioxide and water is 690 kcal (Section 38–5), the series of reactions results in trapping 456/690 or 66% of the energy of glucose.

These calculations are based on an energy storage of 12 kcal per mole of ATP. It may be that the real value of this energy is only 8 kcal per mole. If such is the case, the efficiency of storage is only 44%, rather than 66%.

38–10 SUMMARY OF METABOLISM OF CARBOHYDRATES

Carbohydrates are usually ingested in the form of starch and cellulose. Small amounts of sucrose, glucose, and fructose are also present in many foods. Lactose (milk sugar) is an important component in the diet of babies, children, and many adults. The process of digestion degrades all digestible carbohydrates to monosaccharides, and the indigestible carbohydrates such as cellulose are excreted unchanged. The products of digestion, the monosaccharides, are absorbed into the bloodstream and carried to various organs where they are utilized. Utilization may involve anabolic reactions leading to glycogen in animals and starch or cellulose in plants. For the purpose of simplification, catabolism of carbohydrates is usually divided into several series of reactions: glycolysis, the phosphogluconate pathway, and the citric acid cycle. The latter serves for the metabolism of proteins and lipids as well as for carbohydrates.

Other series of reactions are known to involve metabolites of the carbohydrates, but the three metabolic schemes listed seem to be the most important ones. It is possible to list more intermediate compounds in each of the series of reactions given in the glycolytic, phosphogluconate, and citric acid pathways. The degree of detail given in Figs. 38–4, 38–5, and 38–7, however, is sufficient and perhaps extravagant for a basic understanding of the process.

Although remarkable similarity is observed in the reactions in organisms ranging from the yeasts and bacteria to man, there are some differences. Many microorganisms use anaerobic glycolysis and release only small amounts of the potential energy of the carbohydrates they ingest. Since they are essentially nonmotile, perhaps they need less energy. It is possible that formation of excess ATP might even be deleterious for them. The growth of these microorganisms is, in fact, inhibited by the products of their own metabolism—ethanol and lactic acid. Other microorganisms, the aerobes, not only get more energy from each molecule of glucose, but also produce nontoxic carbon dioxide and water as waste products.

Multicellular organisms, particularly those which move from place to place, often at rapid rates, use aerobic pathways. Certain of their tissues, such as the muscle tissues which in order to insure survival must often work under conditions of low oxygen concentration, retain the glycolytic system. This system uses NAD to allow oxidation reactions to proceed even in the absence of oxygen. When this happens, pyruvic acid is converted to lactic acid, and we say that the animal has incurred an "oxygen debt." The "debt" is repaid by rapid breathing, even after violent exercise has stopped. The rapid breathing results in (1) increased oxygenation of the blood flowing to the tissues, (2) the conversion of lactic acid to pyruvic acid and NAD to $NADH_2$, and (3) the oxidation of $NADH_2$ through the electron transport system.

38-11 ABNORMALITIES OF CARBOHYDRATE METABOLISM

Several diseases are caused by abnormalities of the carbohydrate metabolism. We will discuss briefly those that are best understood—*diabetes mellitus*, galactosemia, glycogen-storage disease, and glucose 6-phosphate dehydrogenase abnormalities. Adults generally have almost none of the enzyme lactase, which is required to digest lactose. Apparently they lose this enzyme with maturity, because almost all children have it. For further discussion, see the article by Kretchmer listed in the Suggested Reading.

Diabetes Mellitus

The word *diabetes* means excretion of large amounts of urine; *mellitus* means sweet. Therefore, the term *diabetes mellitus* signifies that a sweet urine—actually a urine containing glucose—is excreted in large amounts. Other forms of diabetes, such as *diabetes insipidus* are known, and it is not totally wrong to describe the increased excretion of urine following ingestion of large amounts of beer as a temporary diabetes.

The condition we know as *diabetes mellitus* was probably known to the ancient Egyptians. The Greeks scientist Celsus, who lived from 30 B.C. to 50 A.D., described a disease producing "polyuria without pain but with emaciation and danger," and a Chinese physician, about 200 A.D., described the condition as a "disease of thirst." About 1500, Paracelsus evaporated the urine of a diabetic and recovered "salt," which must have been a mixture of urea, salts, and sugar. Even in primitive societies it was known that ants were attracted to the urine of diabetics because of its sweetness.

From the dawn of civilization until the 1920's *diabetes mellitus* was fatal. No effective treatment was known. In severe cases, the patient could expect to live less than a year

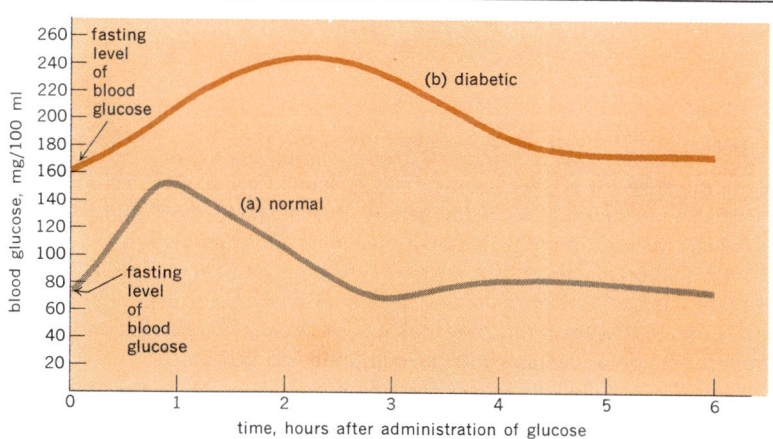

Fig. 38–8 Glucose tolerance curves. A glucose tolerance curve shows the amount of blood glucose at various intervals after an individual has consumed a large amount of glucose. Curve (a) is a normal curve. It shows a rise followed by a decrease and sometimes reaches a level slightly below the normal fasting level before returning to normal. Curve (b) represents results obtained when the test is made on a diabetic person. His fasting level is higher than normal and administration of glucose increases the blood glucose level to values that usually exceed the renal threshold. The typical "diabetic" curve falls off only slowly, and abnormally high values are found several hours after administration of the glucose.

after discovery of his condition. The outlook then for diabetics is analogous to the prognosis today for people who develop leukemia.

Diabetes mellitus is characterized by the presence of glucose and three other abnormal substances—acetone, acetoacetic acid, and β-hydroxybutyric acid in the urine. The glucose found in the urine is a reflection of the high level of glucose in the blood stream. The high level of blood glucose (hyperglucosemia) is due to the fact that the normal metabolism is severely impaired in the diabetic. He apparently cannot convert glucose either to glycogen or to carbon dioxide and water at anything like the normal rate. Since glucose cannot be utilized, it accumulates following digestion of a carbohydrate-containing meal, and is only slowly excreted. The variation in normal metabolism of glucose can be shown by a glucose-tolerance test. The person undergoing the test fasts for at least 12 hours, and then drinks a solution containing a large amount of glucose. Blood samples are withdrawn after specific time intervals and analyzed for glucose. While both normal and diabetic persons will show a rise in blood glucose under the conditions of the test, the concentration of glucose in the blood will be higher and the rate of decrease will be slower in the diabetic than in the normal person. Typical glucose tolerance curves are shown in Fig. 38–8.

The other unusual urinary components of diabetics, acetoacetic acid, β-hydroxybutyric acid, and acetone, are believed to accumulate because the diabetic is unable to oxidize acetyl CoA.

$$CH_3\overset{O}{\overset{\|}{C}}-CoA \rightarrow CH_3-\overset{O}{\overset{\|}{C}}-CH_2-COOH \rightarrow CH_3-\overset{\overset{OH}{|}}{\underset{H}{C}}-CH_2COOH$$

acetyl CoA   acetoacetic acid β-hydroxybutyric acid

$$CH_3-\overset{O}{\overset{\|}{C}}-CH_3 \quad + \quad CO_2$$

acetone

Thus we say that, in the absence of glucose metabolism, acetyl CoA tends to accumulate. It has been postulated that the real deficiency here is in oxaloacetic acid, and that this deficiency is caused by the lack of pyruvic acid which can lead to the formation of oxaloacetic acid by a carboxylation reaction. This possibility has not yet been definitely shown to be true, and perhaps it is best merely to say that three abnormal metabolites, all probably derived from acetyl CoA are present in the urine of diabetics. Acetone is also excreted with the air expired from the lungs of persons with severe, untreated diabetes.

In 1921, the Canadian scientists Frederick Banting and Charles Best (who was only a medical student at the time), proved that a substance that alleviated the symptoms of *diabetes mellitus* could be extracted from the pancreas. This substance, now known to be a hormone, is called *insulin*. Normally it is produced by special cells in the pancreas and released into the bloodstream and from the bloodstream absorbed by many tissues. Injections of insulin lead to the disappearance of all symptoms of *diabetes mellitus*.

How does insulin work? We really don't know. It has the net effect of promoting the metabolism of glucose. The best theory is that insulin promotes the absorption of glucose into cells. Until glucose is inside a cell, it cannot be metabolized. Absorption of glucose into a cell may involve its reaction with ATP to yield glucose 6-phosphate, but it is not certain that this is the reaction influenced by insulin.

Insulin is now extracted from pancreases of beef, swine, and sheep, and made available in a highly purified form. It must be injected into the diabetic, since ingestion as a pill would result in the digestion of the polypeptide, insulin, and reduce it to its component amino acids. Insulin is a treatment, not a cure, for diabetes. A cure would have to increase the ability of the pancreas to manufacture and release insulin.

Mild cases of diabetes, those that develop in later life, can be treated with various chemical substances which can be administered orally. These drugs do not contain insulin, and are effective only in mild cases—those in which the patient is able to produce some but an insufficient amount of insulin. These drugs, whose trade names are Orinase and Diabinese, apparently function by stimulating the secretion of the insulin that the diabetic person can still make. However, the fact that these drugs can produce bad side effects has led many physicians to stop using them for the treatment of diabetes.

$$CH_3-\bigcirc-SO_2-\underset{H}{N}-\underset{O}{\overset{\|}{C}}-NH-(CH_2)_3-CH_3$$

tolbutamide (Orinase)

$$Cl-\bigcirc-SO_2-\underset{H}{N}-\underset{O}{\overset{\|}{C}}-\underset{H}{N}-(CH_2)_2-CH_3$$

chlorpropamide (Diabinese)

Diabetes mellitus can be characterized as a disorder of carbohydrate metabolism. The exact nature of the defect is not known. The treatment of diabetes did not come from an understanding of the chemistry of the disease, but as the result of careful observation and a hunch on the part of Dr. Banting. In other disorders, treatment has been suggested only after the biochemical nature of the disease was understood.

Galactosemia

Galactosemia is a disease of infants. It is characterized by a high level of galactose in the blood. Galactose is also found in the urine. This condition becomes evident soon after the birth of affected children. They vomit when milk is fed to them, fail to gain weight, their liver enlarges, and in severe cases, they may die. Untreated survivors of this condition tend to be dwarfed, mentally retarded, and may have cataracts in their eyes. Fortunately, this disease is relatively rare and can be treated effectively.

When it was realized that the disease was caused by the inability to metabolize the galactose which was derived from the lactose present in milk, the treatment was obvious—restriction of the intake of the milk and milk products which contain lactose. If this treatment is begun during the first few weeks of life, all symptoms disappear and there are no long-term effects.

It is now known that galactosemia is caused by a hereditary deficiency of the enzyme responsible for the conversion of galactose 1-phosphate to glucose 1-phosphate. As the children grow older they usually develop an alternate pathway for metabolizing galactose, and thus the need to restrict milk is not permanent.

With the understanding of the cause and treatment of this condition, the problem becomes one of detection. Unfortunately, the early symptoms of galactosemia are similar

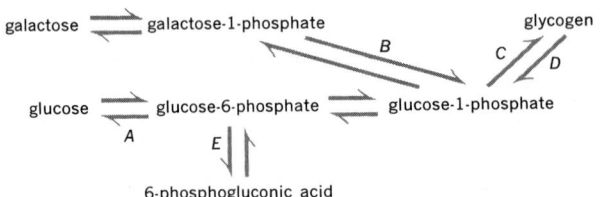

Fig. 38-9 Site of enzymatic deficiencies in disease of carbohydrate metabolism: *A*, site of defect in von Gierke's disease; *B*, site of defect in galactosemia; *C*, site of defect in diffuse glycogenolysis (Type III); *D*, site of defect in glycogen storage disease of liver and muscle (Type IV); *E*, site of defect in drug-induced hemolytic anemia and favism.

to those of many other diseases of infants. Careful observation and examination of the urine and blood for the presence of galactose can lead to prompt diagnosis and treatment. Ignorance is the major reason for the prevalence of this disorder.

Glycogen Storage Disease

A much more serious situation is involved in glycogen storage disease. Actually, several forms of this disorder are known, each probably involving the failure of a specific reaction in the reversible conversion of glucose to glycogen. The disease is characterized by an accumulation of glycogen in the liver, heart, or skeletal muscle. These conditions are hereditary and rare, but are usually fatal, with death often the result of decreased resistance to infections.

One form of glycogen-storage abnormality, known as Von Gierke's disease, involves the absence of the enzyme necessary to convert glucose 6-phosphate to glucose. No treatment has been found to be really effective. Other forms of these diseases involve the enzyme responsible for forming branches or for debranching the glycogen molecule—that is, a lack of the enzyme involved in the formation or disruption of 1-6 bonds between glucose units. A summary of the blocks involved in various types of glycogen storage disease is given in Fig. 38-9.

Disorders Involving Glucose 6-phosphate Dehydrogenase

The administration of certain drugs, namely phenacetin and some of the sulfonamides, causes the rupture of blood cells (hemolysis) and a subsequent anemia in some persons. In severe cases, half of the hemoglobin may be destroyed in this way. Although hemoglobin is normally synthesized by these persons, and blood transfusions may be given to persons with this defect, this disorder is still a serious one. A related condition is caused when sensitive persons eat fava beans. The defect in these disorders involves the enzyme glycose 6-phosphate dehydrogenase, the enzyme involved in the entrance of glucose 6-phosphate into the phosphogluconate shunt. It is quite likely that persons having these disorders are somewhat more resistant to malaria than are normal persons, so the condition is not without its advantages. Since this trait is widely found in Africans, it has apparently helped them to survive in a malarial area and has consequently been transmitted to their offspring. We say such traits have "positive survival value."

The hereditary diseases discussed above seem to be rare. There is some tendency for *diabetes mellitus* to be inherited, although the mode of inheritance is not completely understood. It is highly likely that other hereditary diseases involving irregularities in carbohydrate metabolism will be found. In some cases, such as in galactosemia, understanding the disease has led to an obvious and effective treatment. In others, as in the glycogen storage diseases, no effective treatment has yet been found. The study of defects in metabolism or "metabolic" diseases has only begun, and represents a real challenge for the medical biochemist. This type of disease is discussed further in Chapter 42, "The Chemistry of Heredity."

IMPORTANT TERMS

Adrenalin
Aerobic
Amylase
Anaerobic
Citric acid cycle (Krebs cycle)
Diabetes mellitus
Disaccharidase
Galactosemia
Glucagon
Glucosuria
Glycolysis
Hyperglycemia
Hypoglycemia
Insulin
Phosphogluconate shunt
Phosphorylase
Phosphorylation

WORK EXERCISES—Part A

1. How does the metabolic oxidation of glucose differ from the reaction when a small amount of sugar is burned in a flame? In what ways are the reactions similar?
2. Which glycolytic intermediates have no phosphorus in their structure?
3. Describe how the level of glucose in the blood is kept within a narrow range.
4. What is the source of blood glucose in an animal that has eaten no carbohydrates for a long period of time?
5. Summarize the digestion of a molecule of a highly branched polysaccharide.
6. Does glucosuria always mean that a person has *diabetes mellitus?*
7. Compare and contrast the three major metabolic pathways by which a molecule of glucose can be metabolized—glycolysis, phosphogluconate shunt, and the citric acid cycle.

In solving problems 8, 9, and 10, assume that the free energy "stored" when a mole of ADP is converted to one of ATP is 12 kcal.

8. The complete oxidation of acetic acid to carbon dioxide and water using a bomb calorimeter yields approximately 200 kcal of free energy per mole of acetic acid. Compare this energy with the amount of energy stored as ATP when acetic acid is oxidized via the citric acid cycle.
9. Fluoroacetic acid occurs in the leaves of poisonous plant (*Dichapetalum cymosum*). The toxicity of this substance is due to the fact that it combines with oxaloacetic acid to form fluorocitric acid which acts as an inhibitor of the dehydration of citric acid in the citric acid cycle. Assuming that the further reactions of citric acid are completely blocked, but that no other reactions are blocked, how many ATP molecules would you estimate could be produced per molecule of glucose fed to an animal poisoned with fluoroacetic acid? How does this amount of ATP compare to that normally obtained in an unpoisoned animal?
10. Malonic acid is poisonous. Apparently the toxicity is due to the fact that it inhibits the dehydrogenation of succinic acid in the citric acid cycle. Assuming (for ease of

calculation) that the free energy of the conversion of glucose to CO_2 and H_2O is 700 kcal/mole, estimate what percentage of the free energy of the glucose would normally be trapped as ATP in an animal poisoned with malonate. How does this figure compare with the amount of energy produced by glucose in an unpoisoned animal?

The Use of Isotopes in Biochemistry

The reactions by which the carbohydrates are metabolized have been studied by using isotopes. In these studies, a labeled compound (one containing the isotope) is administered to an animal either by injection or by feeding, and later some part of the animal or his waste products is analyzed for the isotope. Individual enzymatic reactions may be studied by adding isotopic substrates to purified enzyme systems.

The radioactive isotope carbon-14 (^{14}C) is probably the most widely used, but the non-radioactive isotopes deuterium (^2H), nitrogen-15, and oxygen-18 are often used instead. The radioactive isotopes tritium (^3H), phosphorus-32, and iodine-131 can also be used.

Some isotopic compounds are uniformly labeled (for example, all the carbon atoms may contain ^{14}C) while in other compounds only one carbon atom is isotopic. Using an asterisk to indicate labeled carbon atoms (which should not be confused with the use of an asterisk for indicating asymmetry in optically active compounds), we can differentiate between uniformly and specifically labeled pyruvic acid as follows:

$$H_3C*-C*-C*OOH \qquad H_3C-C-C*OOH$$
$$\overset{O}{\underset{\|}{}} \qquad \overset{O}{\underset{\|}{}}$$

uniformly labeled pyruvic acid   carboxyl-labeled pyruvic acid

In performing and interpreting experiments involving isotopes, we make the following assumptions:

1) The labeled compound is metabolized in the same manner as a nonlabeled compound.
2) The label stays with the atom which was originally labeled, and this atom maintains its position in a complex molecule. Care must be taken when labeling hydrogen atoms since the fact that the hydrogen of acids dissociates from its original molecule makes interpretation of results difficult.
3) When symmetrical compounds are formed, the results are the same as if the isotopic label were distributed between similar atoms. Thus if succinic acid is labeled in only one of the noncarboxylic carbons ($HOOC-CH_2-C*H_2-COOH$), it behaves as if the label were distributed over the two similar carbons ($HOOC-C*H_2-C*H_2-COOH$). Metabolism of this succinic acid to give the nonsymmetrical malic acid (in the citric acid cycle) would produce both

$$HOOC-C*H_2-CH-COOH \quad \text{and} \quad HOOC-CH_2-C*H-COOH$$
$$\qquad\qquad\qquad | \qquad\qquad\qquad\qquad\qquad\qquad\qquad | $$
$$\qquad\qquad\qquad OH \qquad\qquad\qquad\qquad\qquad\qquad\qquad OH$$

and analysis of such a mixture would give results indicating that the labeling was spread over the noncarboxyl carbons:

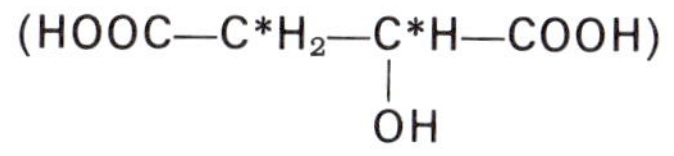

(HOOC—C*H₂—C*H—COOH)
           |
           OH

The carboxyl carbons, however, would not be labeled. There is one important exception to this rule about labeling as it applies to symmetrical compounds. Although citric acid is symmetrical, research indicates that the enzyme that catalyzes its metabolism in the citric acid cycle treats the citric acids as if it were nonsymmetrical, and

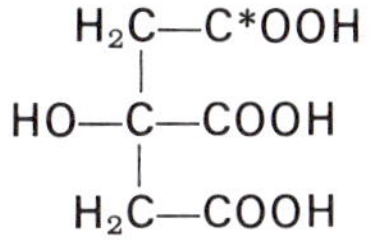

H₂C—C*OOH
|
HO—C—COOH
|
H₂C—COOH

gives α-keto glutaric acid labeled only in one carboxyl group,

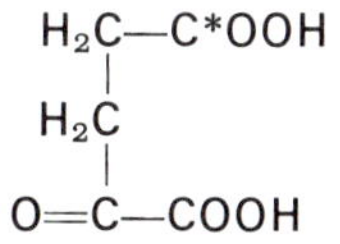

H₂C—C*OOH
|
H₂C
|
O=C—COOH

WORK EXERCISES—Part B

On the basis of the foregoing discussion, do the following work exercises:

11. List five radioactive compounds that might be isolated from a rat or his excretion products following the injection of pyruvic acid with the carboxyl carbon labeled with carbon-14. Locate the position of the radioactive carbon in each.

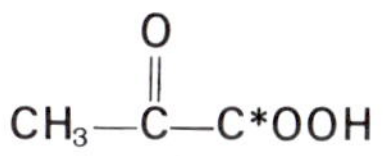

         O
         ‖
    CH₃—C—C*OOH

12. What compounds which are normal metabolites of pyruvic acid would you expect to have no radioactivity when carboxyl-labeled pyruvic acid was administered to rats?

13. If carboxyl-labeled pyruvic acid had been added to a culture of yeast, which products of yeast fermentation would be radioactive and which products would not contain carbon-14?

14. One of the major reactions of pyruvate is its conversion to acetyl CoA. This reaction could be studied by the following experiments:
    The enzymes and cofactors known to be necessary for the conversion of pyruvic acid to acetyl coenzyme A are used in both experiments. In Experiment 1, the addition of pyruvate labeled in the methyl or carbonyl carbon gave acetyl CoA containing radioactivity. In Experiment 2 the addition of acetate labeled in both carbon atoms did not result in the formation of labeled pyruvic acid.
    What can you conclude about the reversibility of this reaction?

15. Glucose was synthesized to contain carbon-14 in the fourth carbon and later given to a white rat. The following compounds were later isolated from the animal: glucose,

fructose-1,6-diphosphate, pyruvic acid, acetyl CoA, α-ketoglutaric acid. Indicate the position of the carbon-14 (if present) in each of these compounds.

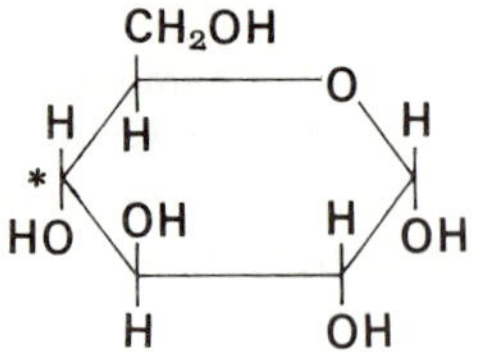

SUGGESTED READING

Kretchmer, N., "Lactose and Lactase," *Scientific American*, Oct. 1972. Whereas children have the enzyme lactase, which is necessary to digest milk sugar, most adults don't. The significance of this finding and its racial and genetic aspects are discussed in this interesting article.

Levine, R., and M. S. Goldstein, "The Action of Insulin," *Scientific American*, May 1958. This article discusses glucose metabolism and the function of insulin.

Wolf, G., *Isotopes in Biology*, Academic Press, New York, 1964. This paperback book gives a wealth of information on the use of isotopes for the study of biochemical processes.

METABOLISM OF PROTEINS 39

39-1 INTRODUCTION

The digestion of proteins is necessary in order to convert them into amino acids which can be absorbed across the intestinal wall. Undigested proteins are potent antigens and bring about the production of antibodies (that are themselves proteins) which produce allergic reactions. These body defenses against foreign proteins means that absorption of proteins without digestion would produce severe allergic shock reactions following each meal; however, it is possible that some proteins are absorbed through the intestine without bad effects during the first few days of a baby's life. In this way the mother's milk probably provides the child with preformed antibodies which will serve until he can manufacture his own. The permeability of the intestine is rapidly lost within a few days after birth.

Following digestion and absorption, the amino acid of proteins may either be catabolized to yield energy or built into one of the thousands of proteins necessary for the life and well-being of the organism. The process of protein synthesis is extremely complex. Each protein contains at least a hundred amino acids in a highly specific sequence. The regular misplacement of just one amino acid can lead to serious consequences. At the same time that the exact structure is being formed, energy must be provided for this anabolic process.

This chapter discusses first the digestion of proteins and then the reactions of the amino acids that result either in their degradation to yield energy or in their synthesis into larger molecules, especially the proteins which play a vital role in the life of any organism.

39-2 PROTEIN-DIGESTING ENZYMES

The digestion of proteins to yield amino acids is catalyzed by a variety of enzymes all of which may be called **peptidases,** since the bond hydrolyzed is the peptide linkage between the amino acids. The word **proteolytic** is also used to describe these enzymes. The reaction may be written as follows:

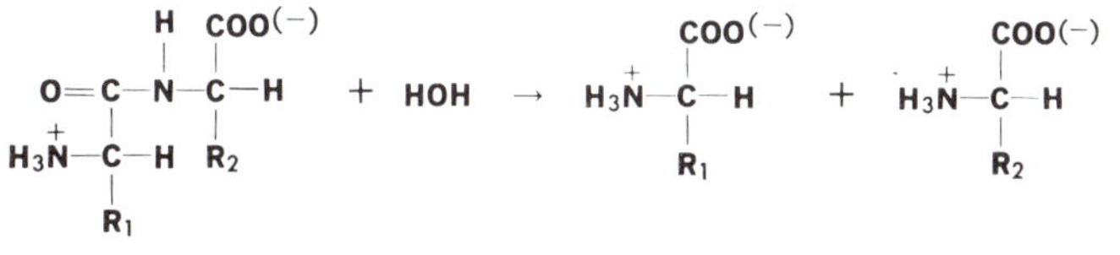

$$\text{a dipeptide} + \text{water} \rightarrow \text{amino acid} + \text{amino acid}$$

There are many varieties of peptidases. **Exopeptidases** act on peptide bonds nearest to the end of a long protein chain. They may be either **aminopeptidases** or **carboxypeptidases** depending upon whether or not they exert their action nearest the end of the protein (or polypeptide) chain having the free amino or carboxyl group. **Endopeptidases** catalyze the hydrolysis of peptide bonds that are not near the end of a chain. **Dipeptidases** act only upon dipeptides. Many of the peptidases are specific, for example, endopeptidases might catalyze hydrolyses only at places where a tyrosine residue was present in the polypeptide chain; others seem to work for only a limited number of similar amino acids.

All the proteolytic enzymes that have been carefully studied to date are secreted in inactive forms. Pepsin, for instance, is produced as an inactive protein **pre-pepsin** (also called pepsinogen*). Its conversion to pepsin involves the hydrolysis of some of its own peptide bonds. This hydrolysis is catalyzed by hydrogen ions or by pepsin itself. Two questions are immediately suggested: "How does hydrolysis of peptide bonds convert an inactive protein into an active enzyme?" and "Why are these enzymes secreted in an inactive form?" "How" questions are much easier to answer than "Why" questions, but answers have been proposed for both. Conversion of inactive pre-pepsin to pepsin probably involves the removal of an inhibitor and perhaps a "masking" group. The "masking" group is a part of the pre-pepsin molecule that covers or masks the active site of the enzyme much as a sheath masks a knife—although in this case the masking group is much smaller than the active agent. Pre-pepsin has a molecular weight of 42,600 and pepsin of 34,500, indicating that the inhibitor and other peptides removed have a weight of approximately 8000.

The answer to the question concerning the reason for production of inactive precursors (sometimes called **zymogens,** or enzyme generators) is suggested by the realization that the cells that manufacture these enzymes are themselves proteins. It would be inefficient or even disastrous if these proteolytic enzymes destroyed the very tissues that made them. The basic question "Why doesn't the stomach or intestine digest itself?" has plagued scientists for some time. Apparently the answer is that most of the stomach and intestinal wall is protected by a lining of polysaccharide-protein complex (a mucopolysaccharide) called mucus which is relatively indigestible but which does, however, slough off continuously—much as do the outer layers of our skin. The second part of the answer is that the really unprotected part of the stomach, the glands that manufacture pepsin, are protected because they make an inactive form of the enzyme. The same explanation can be used to describe the manufacture of pre-trypsin and pre-chymotrypsin by the pancreas.

39–3 PROTEIN DIGESTION

Protein digestion does not start until the food reaches the stomach. Here the combined action of HCl and the endopeptidase, **pepsin,** initiates protein digestion. Complete action by pepsin results in the hydrolysis of about 10% of the bonds of typical proteins and leaves particles having molecular weights of between 600 and 3000. Such a mixture is also produced *in vitro* (using pepsin, of course) and sold as *peptone* for use in bacterial growth media.

---

* The name pepsinogen is most commonly used. The International Union of Biochemistry recommends pre-pepsin. They make similar recommendations for the precursors of trypsin and chymotrypsin.

In the small intestine, the partially digested proteins are exposed to the action of the endopeptidases **trypsin** and **chymotrypsin,** and carboxypeptidases from the pancreas. Aminopeptidases and dipeptidases are probably secreted from the intestinal wall. Trypsin and chymotrypsin are secreted as pre-trypsin and pre-chymotrypsin, and it is probable that the exopeptidases are also secreted as inactive zymogens. As the result of the action of the various peptidases, proteins are converted into amino acids and these are absorbed through the intestinal wall into the blood stream.

Some proteolytic enzymes are extracted from their natural sources and used for various purposes. Papain, which comes from the papaya fruit, is used in meat tenderizers where its action partially digests the meat. Other proteolytic enzymes have been used to aid in freeing the lens of the eye prior to its removal in surgery for cataract and to facilitate regrowth of nerve cells that have been cut.

39–4 METABOLISM OF AMINO ACIDS

The digestion and the eventual fate of amino acids absorbed into the blood stream may be summarized as follows:

proteins → **amino acids**
1. synthesis into protein,
2. removal of amino group to give a keto acid, which may:
   a) be converted to glucose or glycogen,
   b) be converted to CO_2 and water,
3. participation in a pathway peculiar to a given amino acid

Protein Synthesis

The synthesis of protein is an extremely rapid process when one considers the complexity of the protein being synthesized. Only minutes after the injection of radioactive amino acids into an animal, radioactive protein can be found in that animal, indicating that the synthesis occurs rapidly. We also know from other studies with isotopes that there is a continual degradation and resynthesis of all the proteins in any organism. The white rat replaces half of its proteins in 17 days. The human takes 80 days to do the same. These figures are somewhat misleading, however, since not all proteins are degraded and resynthesized (the term protein turnover is used to describe the process) at the same rate. Half of the proteins in the blood serum are turned over in 10 days, liver proteins require only 20 to 25 days, but the turnover in the protein of bone (collagen) is slow.

The problem of exactly how an organism synthesizes protein has intrigued scientists for many years. How does a cell put together just the right amino acid, in the right sequence to give a protein that may serve as an enzyme, an antibody against bacteria, or perhaps a hormone? Since the number of possible arrangements of the 20 amino acids in even a small protein is greater than the number of atoms in the known universe, protein synthesis cannot be just a random synthesis, but must be under strict control from some kind of "director."

The realization that the kinds of proteins an organism synthesizes depends upon its heredity led to speculation that DNA was probably involved in some way and might serve as the "director" or source of information. This assumption we now know to be true. In fact, both DNA and RNA are involved in the synthesis of proteins. Three kinds of RNA each having different physical characteristics, are involved in protein synthesis. These three forms are called (1) **soluble (transfer) RNA,** (2) **ribosomal RNA,** and (3) **messenger (template) RNA.** Each plays a specific role in protein synthesis.

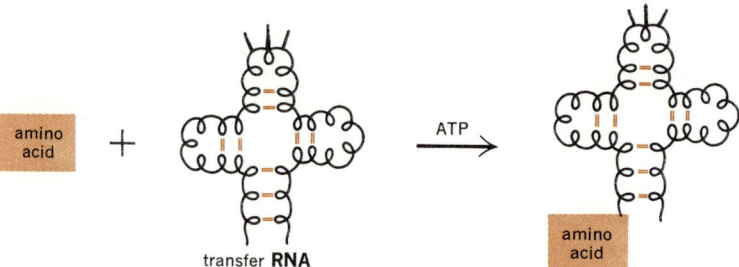

Fig. 39-1 The activation and binding of an amino acid to transfer RNA. In the reaction shown above, a single enzyme is responsible for both activating an amino acid and binding it to the specific transfer RNA that will carry it to the ribosome. The structures given here are only rough diagrammatic approximations of the molecules they represent.

The general scheme for protein synthesis has been worked out only during the past ten to fifteen years. There are still unanswered questions about this highly detailed process and research is continuing at a voluminous rate. In the discussion that follows, we have attempted to give a broad description of the process. It is beyond the scope of this book to discuss all the details of, and exceptions to, the general process.

The first step in the synthesis of proteins involves a reaction of a given amino acid with ATP. This reaction produces a compound having a structure that can be summarized as adenine-ribose-phosphate-amino acid. The amino acid is said to be "activated" in this process. (See Fig. 39-1.) This activated amino acid is then transferred to a specific kind of soluble RNA. This kind of RNA is known as transfer RNA. It is now certain that there are several kinds of transfer RNA for each amino acid but that they all have certain characteristics in common. See Fig. 39-2 for the structure of a molecule of transfer RNA that binds to the amino acid alanine.

The transfer-RNA-amino-acid complex then travels to the ribosome (a small organelle within the cell) where the amino acid is added to other amino acids to form polypeptide chains and eventually proteins. The transfer RNA then leaves the ribosome and picks up another molecule of its particular kind of amino acid. The order in which the amino acids which are brought to the ribosome by transfer RNA are linked together in a peptide chain is specified by messenger RNA.

Messenger RNA, as well as other known kinds of RNA, is believed to be synthesized in the nucleus of the cell under the influence of DNA. Apparently DNA not only duplicates itself (as described in Chapter 36) but also serves as a pattern for the synthesis of a specific kind of messenger RNA. In DNA duplication, a guanine in one strand of DNA serves as a guide for adding a cytosine to a forming chain of DNA, and adenine is the guide for thymine. Since RNA contains uracil instead of thymine, an adenine in a DNA molecule directs the placement of a uracil (actually a uridylic acid molecule) in a chain of messenger RNA. (See Fig. 39-3.) Thus, we can summarize the synthesis of RNA from DNA as follows:

DNA chain: —G—C—A—G—G—C—T—A—·····

RNA chain: —C—G—U—C—C—G—A—U—·····

Messenger RNA then travels from the nucleus to the ribosome, carrying in its structure the message describing, in a type of code, a particular protein molecule. Messenger RNA and

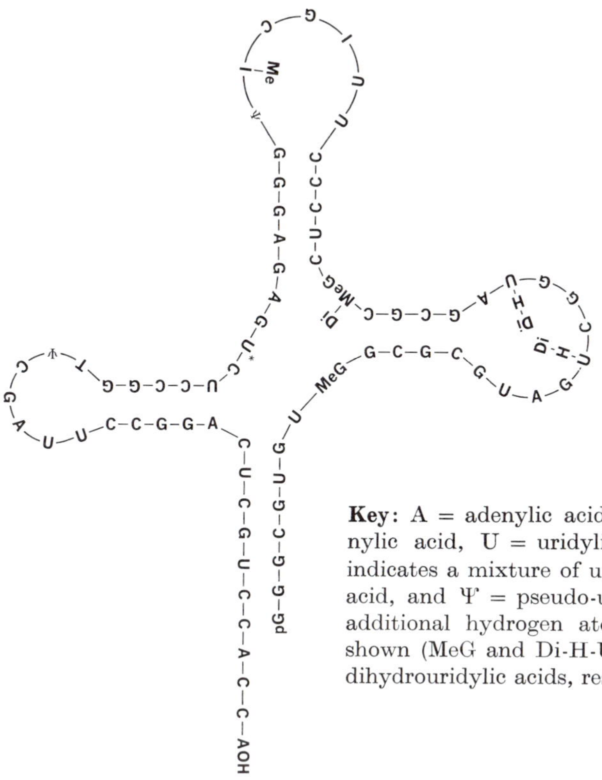

Fig. 39-2 Model of the soluble RNA that transfers alanine. Regions of hydrogen bonding between strands are indicated by the closeness of the strands. Note that in most RNA there are bases other than the four found in the greatest quantity. The triplet I-G-C or C-G-G may be the part of the molecule that matches with messenger RNA. [Adapted from R. W. Holley et al., "Structure of a Ribonucleic Acid," *Science* **147**, 19 Mar. 1965, pp. 1462–1465. Copyright 1965 by the American Association for the Advancement of Science.]

Key: A = adenylic acid, C = cytidylic acid, G = guanylic acid, U = uridylic acid, I = inosinic acid, U* indicates a mixture of uridylic acid and dihydrouridylic acid, and Ψ = pseudo-uridylic acid. The presence of additional hydrogen atoms or methyl groups is also shown (MeG and Di-H-U indicate methyl guanylic and dihydrouridylic acids, respectively).

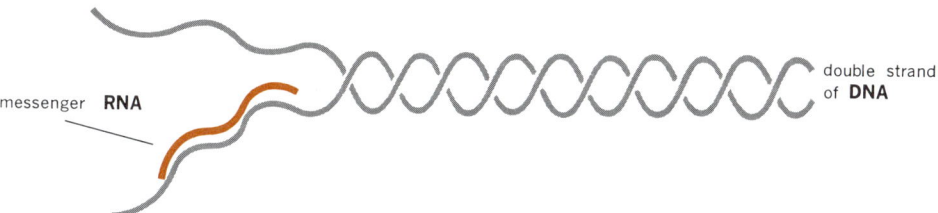

Fig. 39-3 Synthesis of messenger RNA. It is believed that DNA strands in the nucleus of the cell unwind sufficiently for a given strand to serve as a pattern for the synthesis of messenger RNA. The messenger RNA then carries the instructions contained in the DNA to the ribosomes.

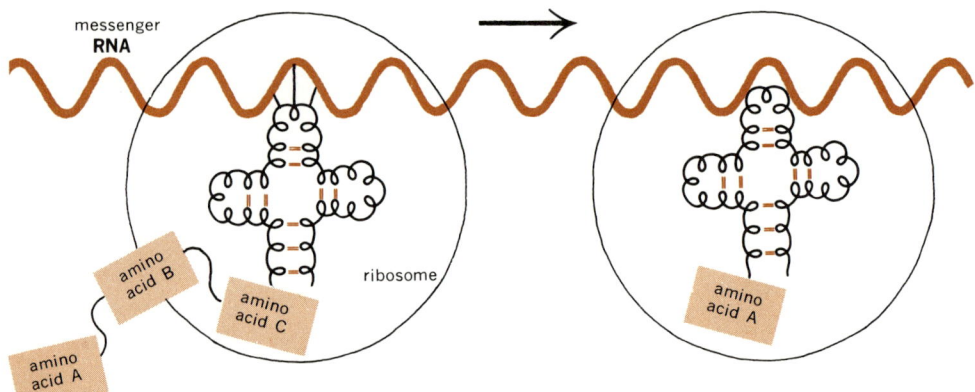

Fig. 39-4 Synthesis of protein at the ribosome. The ribosom apparently has two major active areas. In one, messenger RNA is held so that transfe. RNA having the correct sequence of nucleotide bases can match it. When the matching takes place, the activated amino acid is positioned over the area where peptide bonds are generated between amino acids to give the protein chain.

---

the transfer-RNA-amino-acid combination meet at the ribosome (which itself is made of RNA and protein). In some way messenger RNA specifies the kind of amino acid to be placed at each part of a forming protein chain. Presumably it does so by matching some specific part of the transfer RNA. In the illustration shown below only a small area of the soluble RNA exactly matches the messenger RNA.

```
                        ribosome
messenger RNA:  —C—G—U—( C—C—G )—A—U—·····

transfer RNA:   —A—A—U—( G—G—C )—A—G—·····
```

From experimental evidence, we are quite certain that the code requires three nucleotides (bases) to specify an amino acid. We call each of these three bases sequences in messenger RNA a codon. In the previous illustration, the codon —C—C—G— is being matched with the complementary anticodon —C—C—G— located on the transfer RNA which is carrying the amino acid. Since we believe that the codon C—C—G is specific for the amino acid proline (see Table 39-1), a molecule of this amino acid is added to the growing protein chain whenever its codon is specified in messenger RNA. Figure 39-4 shows this process diagrammatically.

The establishment of the list of codons that specify each amino acid is one of the triumphs of modern biochemistry. It came about when experiments were carried out in which synthetic RNA molecules were added to a mixture of ribosomes, transfer RNA, amino acids, and other factors. It was found, for instance, that when polyuridylic acid [represented as —U—(U—U)$_x$—U—] was added to the mixture, only polyphenylalanine was formed. It was surprising to find that there is also another codon for phenylalanine, U—U—C, but this repetition proved to be the general rule. That is, there is more than one codon for most of the amino acids, but each known codon is specific; it governs the addition of only one kind of amino acid molecule to the growing peptide chain. Other codons probably serve to terminate the growing protein molecule.

TABLE 39-1 The genetic code

This table gives the trinucleotides (codons) that are believed to contain the information for incorporating a given amino acid into a polypeptide chain. Although the code is degenerate in that more than one trinucleotide codes for a given amino acid, there is a general similarity in two of the bases in the codons for an amino acid. The codons UAA, UAG, and UGA serve the purpose of stopping peptide synthesis, and thus are called *terminator* codons. The data in the table are from the reports of a number of investigators.

Amino Acid	Codons	Amino Acid	Codons
alanine	GCA, GCC, GCG, GCU	leucine	CUA, CUC, CUG, CUU, UUA, UUG
arginine	CGA, CGC, CGG, CGU, AGA, AGG	lysine	AAA, AAG
asparagine	AAC, AAU	methionine	AUG
aspartic acid	GAC, GAU	phenylalanine	UUC, UUU
cysteine	UGC, UGU	proline	CCA, CCC, CCG, CCU
glutamic acid	GAA, GAG	serine	UCA, UCC, UCG, UCU, AGC, AGU
glutamine	CAA, CAG		
glycine	GGA, GGC, GGG, GGU	threonine	ACA, ACC, ACG, ACU
histidine	CAC, CAU	tryptophan	UGG
isoleucine	AUC, AUU	tyrosine	UAC, UAU
		valine	GUA, GUC, GUG, GUU

Evidence indicates that messenger RNA from one kind of organism can be used with ribosomes and transfer RNA amino acids from another, and that the kind of protein synthesized is characteristic of the messenger RNA. That is, the process of protein synthesis is universal. This may indicate the process by which nucleic acids from viruses can "take over" the synthetic systems of the cell. It would also explain why viruses require whole cells—not just an assortment of nutrients if they are to grow and reproduce.

Messenger RNA molecules are depolymerized after being used only once or, at most, a few times. The messenger RNA of higher plants and animals seems to be more stable than that of bacteria. While this destruction of messenger RNA seems wasteful, it provides a sensitive means of controlling the kind of protein synthesized. Bacteria usually need to respond to more environmental changes than do the cells of higher plants or animals. Bacteria that stop making a protein they no longer need are more efficient and tend to be successful in their never-ending struggle to survive and reproduce.

Amino Acid Degradation

The basic biochemical reactions of amino acids were described in Chapter 34 as peptide synthesis, transamination, deamination, and decarboxylation. We have seen that the process of peptide bond formation with the synthesis of proteins is a significant one for all organisms. The extent to which the other reactions play a role in any given organism is an important aspect of the specific biochemistry of that animal.

To determine the extent to which transamination actually occurs in an animal, we can feed him any one amino acid that has been labeled with the nonradioactive isotope nitrogen-15 in the amino group. Later when we examine the various tissues of the animal we find that nitrogen-15 is present in the amino acids, both free and in proteins, with the exception of lysine. If we make our examination only a short time after feeding the animal

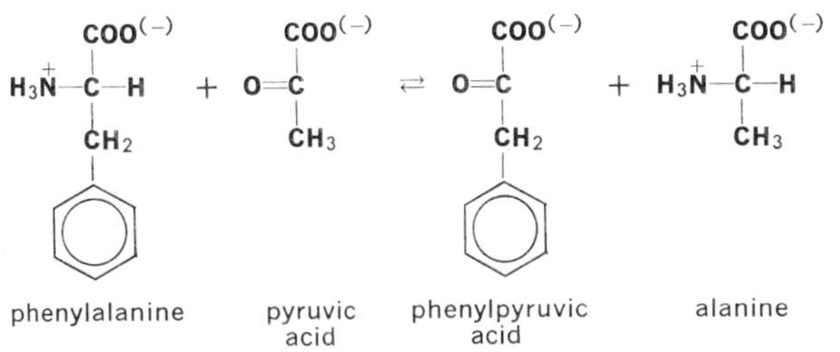

$$\underset{\text{phenylalanine}}{\overset{\text{COO}^{(-)}}{\underset{\underset{\bigcirc}{\overset{|}{\text{CH}_2}}}{\overset{|}{\underset{|}{\overset{+}{\text{H}_3\text{N}}-\text{C}-\text{H}}}}}} + \underset{\substack{\text{pyruvic}\\\text{acid}}}{\overset{\text{COO}^{(-)}}{\underset{\text{CH}_3}{\overset{|}{\underset{|}{\text{O}=\text{C}}}}}} \rightleftarrows \underset{\substack{\text{phenylpyruvic}\\\text{acid}}}{\overset{\text{COO}^{(-)}}{\underset{\underset{\bigcirc}{\overset{|}{\text{CH}_2}}}{\overset{|}{\underset{|}{\text{O}=\text{C}}}}}} + \underset{\text{alanine}}{\overset{\text{COO}^{(-)}}{\underset{\text{CH}_3}{\overset{|}{\underset{|}{\overset{+}{\text{H}_3\text{N}}-\text{C}-\text{H}}}}}}$$

Fig. 39-5 A transamination reaction in which the amino group of the amino acid phenylalanine is transferred to pyruvic acid to yield phenylpyruvic acid and alanine. The reverse of this reaction would result in the synthesis of phenylalanine if sufficient phenylpyruvic acid were present from another metabolic pathway.

---

the isotopic amino acid, we find that the greatest amount of ^{15}N is still in the amino acid administered, with lesser amounts in glutamic acid and aspartic acids, and still lesser amounts in the other amino acids, except lysine, of course, which contains no ^{15}N. These findings lead to the conclusion that the keto acids corresponding to the amino acids, glutamic acid and aspartic acid, are the most important "amino acceptors" in transamination reactions. It is not surprising to find that these keto acids, α-ketoglutaric and oxaloacetic acids, are found as reactants in the citric acid cycle.

Another approach to finding the extent of transamination is made by feeding an animal the keto analogs of amino acids. We know that a supply of certain amino acids is essential for a given animal, a white rat, for example, because the animal cannot synthesize these amino acids at a sufficient rate to allow him to grow normally. When we feed the keto acids corresponding to the essential amino acids to this animal, we find that all the keto analogs can be substituted for the essential amino acids with the exception of those of lysine (which the previously described experiments had shown not to be transaminated) and threonine. Since our other experiments indicated that threonine could be transaminated, we must conclude that the failure of the keto analog to substitute for dietary threonine may be due to the fact that it is metabolized so quickly that it cannot be transaminated. A typical transamination reaction is shown in Fig. 39-5.

These experiments show one of the important uses which an animal makes of transamination reactions. If the basic carbon skeleton is present in the body from some other source, the amino acid can be synthesized. Thus a requirement of an animal for an essential amino acid is really only a requirement for a certain type of carbon chain with a keto acid on it. A list of essential amino acids is given in Chapter 41, and it is interesting to see that they represent complex or unusual compounds whose carbon chains or benzene rings are not present in metabolites of the carbohydrates or lipids.

Not only are transamination reactions necessary for conversion of keto acids to amino acids, but the enzymes that catalyze these processes have been used to aid in diagnosing various diseases. After a heart attack there is a sharp rise in the amount of the **transaminase** that catalyzes the reaction between glutamic acid and oxaloacetic acid present in the blood. A determination of the amount of this enzyme in blood serum is one of the best methods of diagnosing such a heart attack. Other transaminases are used in diagnosing hepatitis.

The conversion of an amino acid to a keto acid may also occur by a process of **deamination.** Many organisms contain the keto analogs of the amino acids; however, there is little evidence that oxidative deamination occurs to any extent in the bodies of humans and other higher animals.

With certain important exceptions, the amines that would be produced by the **decarboxylation** of amino acids are not found in the human; consequently, we conclude that decarboxylation of amino acids is not an important process in humans. Bacteria in the intestine are responsible for some decarboxylations, and some of the odor of normal feces is due to these vile-smelling amines.

The production of keto acids, primarily by transamination, gives molecules that can then enter either the glycolytic or citric acid pathways. The conversion is direct for alanine, glutamic acid, and aspartic acid, yielding pyruvic acid, α-ketoglutaric acid, and oxaloacetic acid, respectively. Additional reactions are required to convert other amino acids into compounds of one of the major metabolic schemes, but these reactions are known for all the amino acids. Thus we can say that the carbon chains of amino acids can be converted via the reverse of glycolysis to glucose or glycogen, or through the citric acid cycle to carbon dioxide and water, with an attendant release of energy.

39-5 EXCRETION OF NITROGENOUS WASTE PRODUCTS

The conversion of amino acids to keto acids leaves an amino group. Although the amino group may be passed from compound to compound, the net utilization of amino acids for energy or glucose and glycogen synthesis means that the amino group must be eliminated. The metabolism of this group is extremely important since very low levels of ammonia in the blood are extremely toxic to all animals. A chronic alcoholic often dies from ammonia toxicity; his liver, which is the source of the enzymes that catalyze conversion of the amino group to the less toxic urea, has been so damaged by cirrhosis that he is unable to get rid of the poisonous ammonia. Other animals, such as fish, can apparently tolerate slightly higher levels of ammonia in their blood, but they too are poisoned by relatively low levels of this substance. One of the specific biochemical differences in different species involves the manner in which they excrete the nitrogen from proteins.

There are two major sources of nitrogenous waste products—amino acid metabolism and the metabolism of the purine and pyrimidine bases which are either consumed in the diet as nucleic acids or synthesized in the body. In the human, approximately 95% of the nitrogenous excretion products come from amino acid metabolism. Nitrogenous waste products are excreted primarily via the kidneys in man and other mammals. Not all animals do so; for instance, fish use their gills for excretion of ammonia. Most animals excrete more than one nitrogen-containing product, but there is often a predominance of one substance.

Amino Acid Excretion

Some simple animals excrete amino acids. This loss is essentially wasteful, since there is still a great deal of energy in amino acids. Humans excrete minor amounts of some amino acids in the urine, and some persons having hereditary abnormalities excrete relatively large amounts of specific amino acids. **Cystinuria** is the condition in which cystine is excreted in relatively large amounts in the urine. Since cystine is quite insoluble, it may form "stones" in the bladder or kidneys of persons who are cystinurics. In fact, the first chemical characterization of cystine was made from a bladder stone of a person who probably was a cystinuric individual.

Ammonia Excretion

Levels of only 5 mg of ammonia per 100 ml of blood are toxic to humans, and the normal concentration of this metabolite is only 1 to 3 micrograms per 100 ml of human blood. Animals that live in water, from simple one-celled organisms to the fish, excrete free ammonia. Even fish, however, do not have concentrations of ammonia greater than 0.1 mg (100 micrograms) per 100 ml of blood. How do these animals excrete ammonia when there is so little of it in the blood? Apparently the ammonia is usually combined with glutamic acid to give glutamine. Glutamine is then carried to a membrane next to the surrounding water (the gills for fish and certain other animals) where the ammonia is released.

Humans excrete ammonium ions (NH_4^+) in the urine, and presumably the ammonia is formed from glutamine in the kidneys. The amount of ammonium ions excreted is not very large, and there is no detectable excretion of ammonia by normal healthy humans. The reaction can be described as follows:

$$\text{glutamine} + \text{water} \rightleftarrows \text{glutamic acid} + \text{ammonia}$$

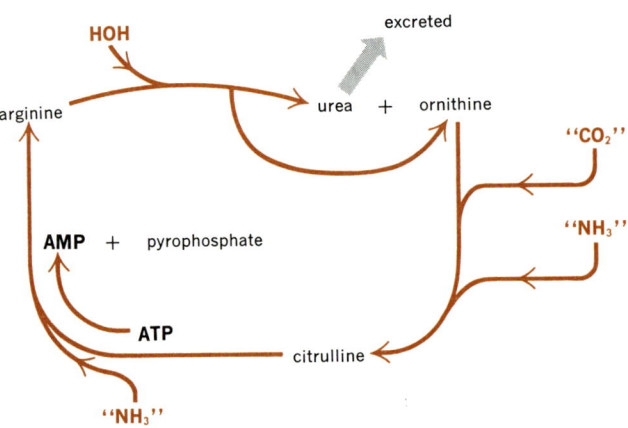

Fig. 39-6 The urea cycle. By the series of reactions shown above, urea is produced by the hydrolysis of the guanido group of arginine. Arginine is then regenerated in a series of reactions so that the cycle can continue.

Urea Excretion

Many animals, including man, excrete urea as the major nitrogen-containing metabolite derived from amino acid metabolism. Urea is much less toxic than ammonia. Levels between 18 and 38 mg/100 ml of blood are considered normal in a human. Sharks have as much as 2% (2 g/100 ml of blood) of urea in their blood. In fact, shark hearts will not beat if urea is not present. The fact that freshwater sharks have much less urea in their blood (0.6% instead of 2%), leads us to the conclusion that one function of the urea in sea sharks is that of balancing the osmotic pressure of the salty sea water.

Energy is required to synthesize urea. Apparently the production of the relatively non-toxic urea is sufficiently advantageous to compensate for the extra energy required for the synthesis. The series of reactions by which urea is synthesized in mammals is called the urea cycle. This series of reactions is summarized in Fig. 39–6 and discussed in detail in the following section.

In this series of reactions, arginine is hydrolyzed to give urea and a molecule of the amino acid ornithine.

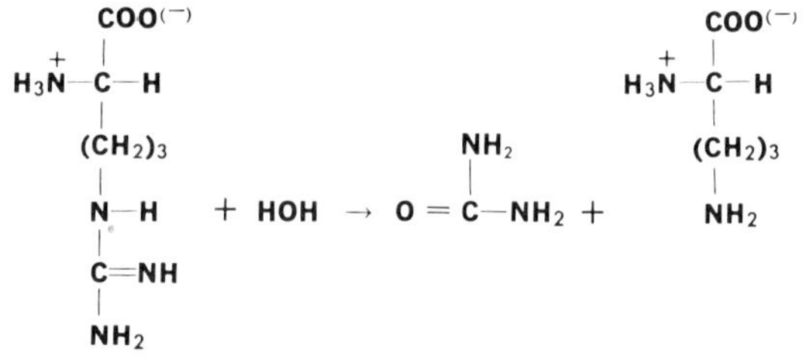

$$\text{arginine} + \text{water} \rightarrow \text{urea} + \text{ornithine}$$

Ornithine then combines with one unit of CO_2 and one of NH_3. Actually, neither CO_2 nor ammonia is present as such, but they are added in the form of carbamyl phosphate. Carbamyl phosphate is synthesized in the liver of mammals by the following reaction:

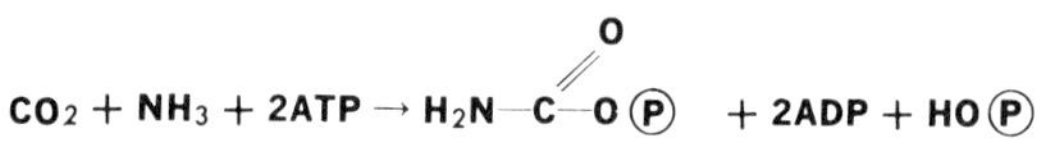

$$CO_2 + NH_3 + 2ATP \rightarrow H_2N-\overset{O}{\underset{\|}{C}}-O\,\text{\textcircled{P}} + 2ADP + HO\,\text{\textcircled{P}}$$

The addition of carbamyl phosphate to ornithine gives citrulline.

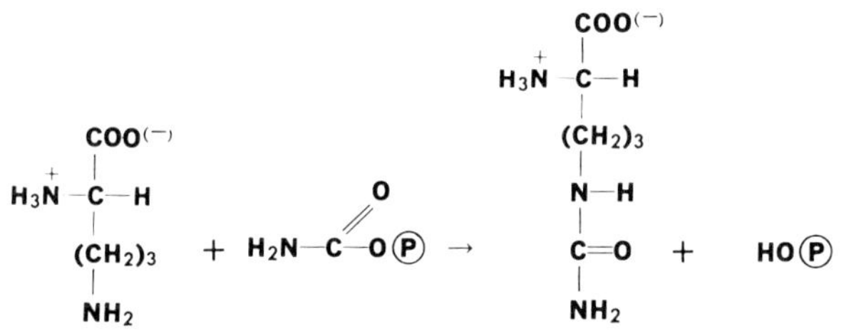

ornithine + carbamyl phosphate → citrulline + phosphoric acid

Citrulline acquires an amino group from aspartic acid. The final product of the reaction is arginine and fumaric acid.

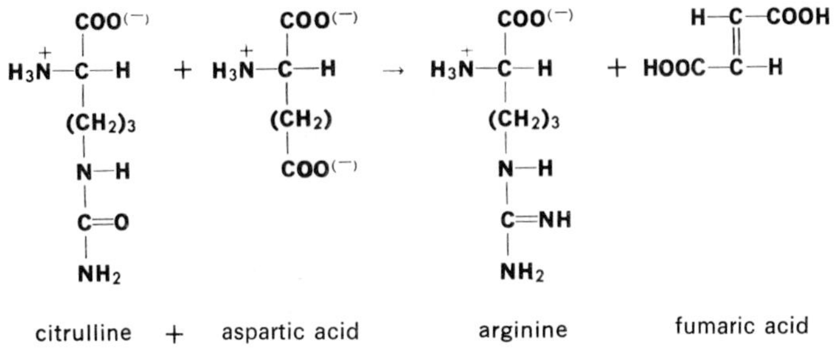

$$\begin{array}{c} COO^{(-)} \\ | \\ H_3\overset{+}{N}-C-H \\ | \\ (CH_2)_3 \\ | \\ N-H \\ | \\ C=O \\ | \\ NH_2 \end{array} + \begin{array}{c} COO^{(-)} \\ | \\ H_3\overset{+}{N}-C-H \\ | \\ (CH_2) \\ | \\ COO^{(-)} \end{array} \rightarrow \begin{array}{c} COO^{(-)} \\ | \\ H_3\overset{+}{N}-C-H \\ | \\ (CH_2)_3 \\ | \\ N-H \\ | \\ C=NH \\ | \\ NH_2 \end{array} + \begin{array}{c} H-C-COOH \\ \| \\ HOOC-C-H \end{array}$$

citrulline + aspartic acid → arginine + fumaric acid

The arginine is then hydrolyzed to yield another molecule of urea and one of ornithine, and the cycle repeats itself. The fumaric acid can be oxidized via the citric acid cycle.

In this series of reactions we can see that ATP is used to provide the energy for the synthesis of urea. In mammals, this series of reactions takes place only in the liver. Extensive liver damage leads to the accumulation of ammonia, and can lead to death.

Uric Acid Excretion

In animals that excrete urea, the urea is dissolved in the urine. It can be said that the excretion of urea actually requires water. Animals which must conserve water (at least during parts of their life cycle), such as the reptiles, birds, and many insects, excrete uric acid. This substance is excreted in a solid form mixed with, but not dissolved in, small amounts of water. The synthesis of uric acid is a complex, energy-requiring process and we must assume, as we did in the case of urea synthesis, that the advantages of conservation of water are sufficient to justify the excretion of a relatively energy-rich compound which the animal must synthesize.

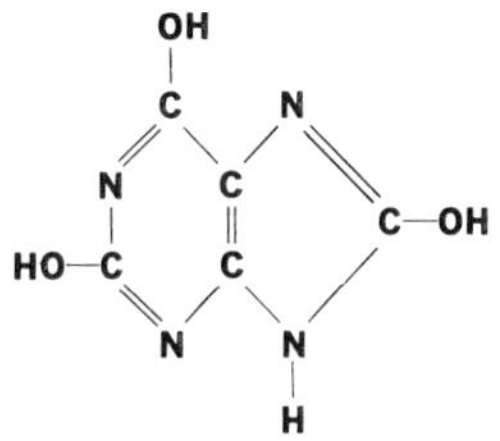

uric acid

It is difficult to determine what is cause and what is effect, but it is true that many reptiles are found in deserts where conservation of water is of primary importance to survival. The condition which is shared by birds, reptiles, and some insects is that the embryo develops within an egg which is enclosed in a shell that is impervious to water. Thus all the water the developing embryo will be able to use is within the egg at the time it is deposited by the female. The sequence of events observed in the development of the chicken embryo is interesting. During the first four days of incubation of the egg, ammonia is produced. From the fifth to the fourteenth day, urea is produced instead of ammonia, and indeed some of the previously produced ammonia is converted to urea. From the fourteenth

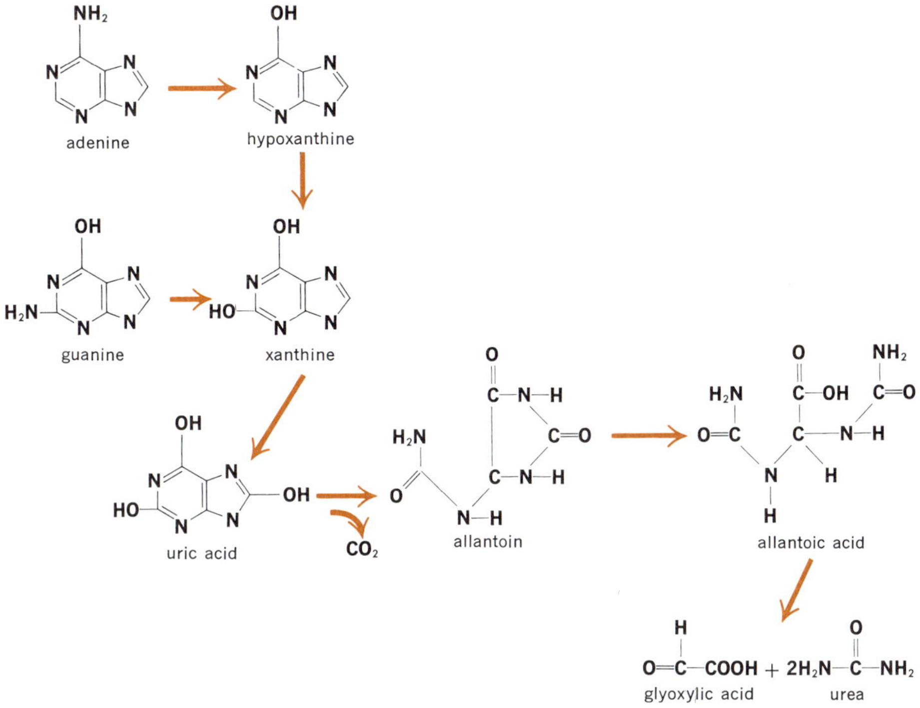

Fig. 39-7 Metabolism of purine bases. In this series of reactions adenine and guanine are oxidized to uric acid. Man and other primates excrete uric acid, since they lack the enzyme necessary for the formation of allantoin. Excretion products of other animals indicate that they carry the series of reactions further.

day until hatching and throughout the rest of its life, uric acid is synthesized by the chicken. Animals that develop from eggs laid in water generally excrete ammonia—as do the adults of these species. However, when amphibians such as frogs begin to grow legs, their metabolism switches, and they begin to synthesize urea—apparently in anticipation of their coming conversion to land-dwelling animals. Philosophically, it is probably preferable to interpret the change in excretion products as being the factor that makes life on dry land possible, rather than to make the teleological interpretation that a change in excretion anticipates a change in environment.

Humans also excrete uric acid; however, it is not synthesized as a method of excreting ammonia but represents the end product of the metabolism of the purine bases adenine and guanine. With minor exceptions, only humans and other primates (apes) excrete uric acid as the normal end product of purine metabolism. Most mammals excrete allantoin, and simpler animals excrete alantoic acid or urea as the end products of purine metabolism (see Fig. 39-7 for this series of reactions). In this series of reactions, animals generally regarded as simpler than humans are actually more complex in that they have more enzymes and can tap more of the energy of the purines. The "loss" of these enzymes, especially of uricase, can be blamed for some of the ills that humans suffer. Some types of painful arthritis and gout result from deposits of uric acid in the joints.

The "loss" of enzymes by higher species leads us to wonder about the advantages of losing the enzyme uricase and the attendant accumulation of uric acid in the blood and other parts of the body. Some scientists have speculated that perhaps man's superior intelligence is due to the high level of uric acid in his blood. While this is a tempting hypothesis, the fact that men normally have a slightly higher level of uric acid in their blood than women (4.5 mg/100 ml in men as against 3.5 mg/100 ml in women) has led to the rejection of this hypothesis by approximately half of the population. The fact that some fowls (chickens and turkeys) have an even slightly higher level of uric acid in their blood (5 mg/100 ml), and that insects have as much as 20 mg/100 ml again argues against the hypothesis!

Miscellaneous Nitrogenous Excretion Products

Ammonia, urea, and uric acid are major nitrogen-containing excretion products of amino acid metabolism. Allantoin and allantoic acid are also excreted by many animals but these come primarily from purine metabolism. Trimethylamine oxide is excreted by some marine animals, but the source of this compound has not been definitely established. Swine and spiders both excrete guanine, the swine because they have no guanase. Presumably the same is true of spiders whose metabolism has not yet been extensively investigated. In spite of the fact that he has the enzyme uricase, the Dalmatian dog excretes some uric acid. Apparently this results from a slightly different kidney function and not from an intent to imitate human metabolism!

39-6 SPECIFIC METABOLIC PATHWAYS FOR AMINO ACIDS

In addition to the general reactions of amino acids which have been previously discussed, each amino acid has a series of specific reactions in which it participates. In some cases, similar amino acids share at least parts of a specific metabolic pathway. As a result of these special pathways, the amino acid either is converted into a compound that can enter the glycolytic pathway or citric acid cycle, or it is converted into a special metabolite. There are many of these special metabolites. Some of the more important ones are summarized in Table 39-2. The specific pathways of phenylalanine and tyrosine are described in Chapter 42 in connection with a discussion of abnormalities in these pathways. A complete discussion of these special metabolic pathways is beyond the scope of this book, a general biochemistry textbook (see the list at the end of Chapter 32) should be consulted by those wishing further details.

39-7 DISORDERS OF PROTEIN METABOLISM

Several instances are known in which a lack of the correct enzyme has led to an abnormal accumulation of metabolites. From our knowledge of protein synthesis, we believe that failures in synthesis of proteins, especially where the condition is hereditary, are due to defects in nucleic acids. It is probably incorrect to describe these enzyme deficiencies as disorders of protein metabolism, even though they represent abnormalities in synthetic pathways. Numerous other conditions are known in which lack of enzymes leads to abnormal metabolism of amino acids. The most striking example involves the metabolism of phenylalanine and tyrosine. These defects are discussed in Section 42-4.

TABLE 39-2 Important products of the metabolism of specific amino acids

Amino Acid	Product	Importance of Product
alanine	pyruvate	glycolytic intermediate
aspartic acid	oxaloacetic acid	important intermediate of citric acid cycle
arginine	ornithine, citrulline	intermediates in urea synthesis
arginine	urea	important nitrogen-containing excretion product
glutamic acid	α-ketoglutaric acid	important intermediate in citric acid cycle
methionine	S-adenosyl methionine	important source of methyl group in biosynthetic reactions
cysteine	taurine	combines with cholic acid to give bile salts
histidine	histamine	important in allergic reactions
tryptophane	nicotinic acid	vitamin
phenylalanine, tyrosine	thyroxine, adrenalin	important hormones
tyrosine	melanin	pigment of skin and hair

39-8 SUMMARY

Proteins are digested under the influence of a variety of enzymes to yield amino acids. These amino acids can be used to synthesize new protein or other simpler molecules such as the hormones thyroxine and adrenalin. The process of protein synthesis is not only complex but amazing, because it causes complex molecules to have a precisely specific sequence of amino acids. Directions for the synthesis are found in DNA, but various types of RNA are required to carry out the process. Synthetic reactions are remarkably similar whether the organism is as simple as the bacterium *E. Coli* or as complex as man. The energy of these anabolic reactions is provided by ATP. Amino acids not used for synthetic purposes can be degraded to yield energy. When this happens, the carbon, hydrogen, and oxygen are converted to carbon dioxide and water by reactions of the citric acid cycle and other common metabolic sequences. The nitrogen may be eliminated in a variety of compounds; different species use different reaction sequences to convert the amino acid nitrogen into a waste product, depending on the habitat and state of evolution of the animal. Thus throughout the animal kingdom, the metabolism of amino acids follows some pathways that are similar and others that are distinctive.

IMPORTANT TERMS

Aminopeptidase
Carboxypeptidase
Chymotrypsin
Cystinuria
Dipeptidase
Endopeptidase
Essential amino acid
Exopeptidase
Messenger (template) RNA
Pepsin
Peptidase
Pre-pepsin (pepsinogen)
Proteolytic
Ribosomal RNA
Soluble (transfer) RNA
Transaminase
Trypsin
Zymogen

WORK EXERCISES

1. The metabolism of the amino acid histidine can be summarized as follows:

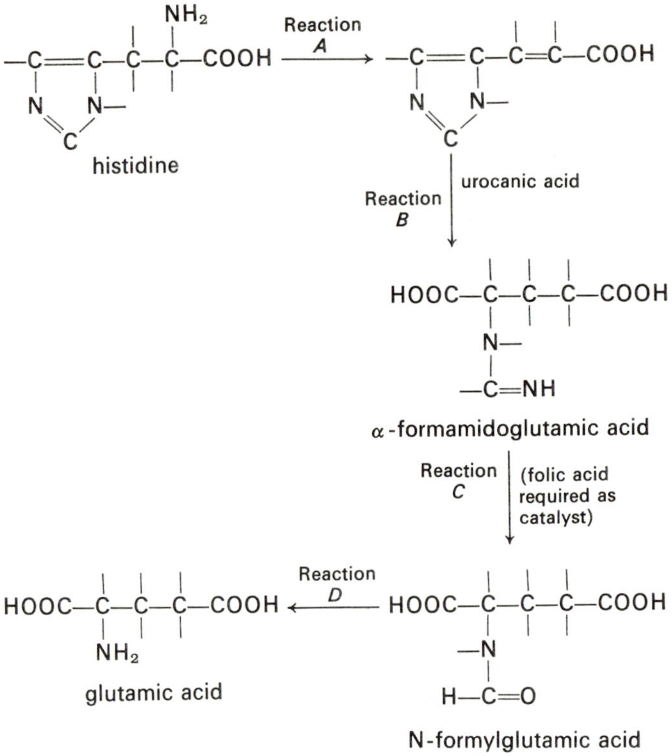

$$\text{histidine} \xrightarrow{\text{Reaction A}} \text{urocanic acid} \xrightarrow{\text{Reaction B}} \alpha\text{-formamidoglutamic acid} \xrightarrow[\text{(folic acid required as catalyst)}]{\text{Reaction C}} \text{N-formylglutamic acid} \xrightarrow{\text{Reaction D}} \text{glutamic acid}$$

Reaction C has been shown to require the vitamin folic acid as a coenzyme. Answer the following questions with respect to the metabolism of histidine:

   a) If histidine containing ^{14}C in the carboxyl carbon is given to an animal, what compound in the citric acid cycle would you expect to be labeled first? (Give the structural formula and show the position of the label.)
   b) What clinical test might be developed to indicate whether an individual had a deficiency of folic acid?

2. Give the experimental evidence that transamination occurs in an animal's body.

3. Write the reaction by which each of the following could be formed from an amino acid:
   a) glyoxylic acid (O=CH—COOH)   b) alpha ketoglutarate
   c) pyruvic acid   d) ethanolamine
   e) mercaptoethanolamine (H_2N—CH_2—CH_2SH)
   f) histamine   g) oxaloacetic acid

4. List the kinds of RNA involved in protein synthesis, and indicate the role of each in this process.

5. Citrulline containing ^{15}N in the position indicated below was given to a dog. What two compounds (other than citrulline) in the animal should contain the greatest amount of ^{15}N in a few minutes after administration of the labeled compound?

$$H_2N^*-\underset{\underset{O}{\|}}{C}-\underset{\underset{H}{|}}{N}-CH_2-CH_2-CH_2-\underset{\underset{NH_2}{|}}{CH}-COOH$$

6. The peptide alanylserylmethionylglutamyltyrosine was incubated with several different enzymes. The products of each specific reaction are given below. List the type or classification of the enzyme that catalyzed each reaction.

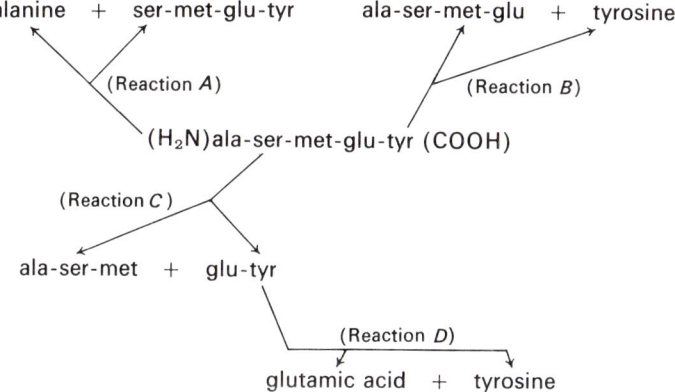

SUGGESTED READING

Asimov, I., *The Genetic Code*, a Signet Science Library Book (P2250), New American Library, New York, 1963.

Neurath, H., "Protein Digesting Enzymes," *Scientific American*, Dec. 1964 (Offprint #198). The article discusses the structure of chymotrypsin and other proteolytic enzymes, and relates the structure to their function as enzymes.

The following six articles from the *Scientific American* are excellent references on the subject of protein synthesis. The illustrations make a complex process quite understandable.

Clark, B. F. C., and K. A. Marcker, "How Proteins Start," *Scientific American*, Jan. 1968.

Crick, F. H. C., "The Genetic Code," *Scientific American*, Oct. 1962 (Offprint #123).

Crick, F. H. C., "The Genetic Code III," *Scientific American*, Oct. 1966.

Hurwitz, J., and J. J. Furth, "Messenger RNA," *Scientific American*, Feb. 1962 (Offprint #119).

Nirenberg M. "The Genetic Code II," *Scientific American*, Mar. 1963 (Offprint #153).

Nomura, M., "Ribosomes," *Scientific American*, Oct. 1969 (Offprint #1157). A well-illustrated discussion of these organelles and their role in protein synthesis.

METABOLISM OF LIPIDS   40

40–1 INTRODUCTION

The lipids are found in all parts of the human body and are especially important in the brain. Although the human body can synthesize most lipids if it is given sufficient raw material—principally carbohydrates—adequate diets should contain some lipids. The triglycerides represent the most important dietary lipids. Lipids from plants tend to contain more unsaturated fatty acids, and while they contain some sterols, contain no cholesterol as do animal products. The biosynthesis and many of the functions of the fatty acids, the triglycerides, and cholesterol are reasonably well understood; however, little is known about the way in which many lipids of the brain are synthesized, and almost nothing is known of the chemical role they play in the functioning of the brain and of the nervous system.

40–2 DIGESTION OF LIPIDS

The enzymes required for the digestion of lipids are given the general name of **lipases**. There is no digestion of lipids in either the mouth or stomach although lipase is found in the stomach. This enzyme is active only at pH values near 7, and thus it is ineffective in the stomach where the pH is normally between 1.5 and 2.5. The principal digestion of lipids occurs in the small intestine, where the combination of bile and the pancreatic and intestinal lipases catalyzes the hydrolysis of the ester bonds of triglycerides. See structure on page 621. A major problem faced in the digestion of triglycerides is that of attachment of the water-soluble lipases to water-insoluble lipid globules. The process is more efficient when the lipids are broken into small globules. This breakdown increases their surface area and consequently the area available for interaction with lipases. Consequently, lipids in an emulsified form (broken into small globules) are much more readily digested. The **bile** does not contain enzymes, but the bile salts are effective emulsifying agents. Thus, while bile is not absolutely necessary for digestion of lipids, it helps greatly in the process.

$$\underset{\substack{\text{palmitooleostearin}\\\text{(a triglyceride)}}}{\begin{array}{c}H_2C-O-\overset{O}{\overset{\|}{C}}-C_{15}H_{31}\\|\\C_{17}H_{33}-\overset{O}{\overset{\|}{C}}-O-C-H\\|\\H_2C-O-\overset{O}{\overset{\|}{C}}-C_{17}H_{35}\end{array}} + \underset{\text{water}}{HOH} \longleftrightarrow \begin{array}{c}\underset{\text{palmitic acid}}{C_{15}H_{31}COOH}\\\underset{\text{oleic acid}}{C_{17}H_{33}COOH}\\\underset{\text{stearic acid}}{C_{17}H_{35}COOH}\end{array} + \underset{\text{glycerol}}{C_3H_5(OH)_3}$$

There is some difference of opinion among biochemists as to whether triglycerides must be digested into glycerol and free fatty acids in order to be absorbed across the intestinal wall. If they are completely hydrolyzed, they are certainly resynthesized quite rapidly, since there are far more triglycerides than free fatty acids in the blood and in the lymph following digestion of a fatty meal. Lipids that are liquid tend to be more easily absorbed than are solid lipids.

Lecithinases (enzymes that catalyze the hydrolysis of the phospholipid lecithins) are present in the small intestine. Cholesterol esterases are also present. These latter enzymes catalyze the hydrolysis of cholesterol from the fatty acids with which it is often combined in natural products.

40–3 ABSORPTION OF LIPIDS

Glycerol, fatty acids, possibly the mono- or di-esters of glycerol and fatty acids, and cholesterol are absorbed across the intestinal wall into the blood and lymph. The lymph is a clear fluid much like blood, but without red corpuscles. It has a circulatory system similar to but separate from that of the blood (see Fig. 40–1). Following the digestion of a fatty meal, the lymph becomes cloudy from the presence of tiny globules of lipids, mostly in the form of triglycerides. Since the lymph flows into the blood, principally at the thoracic duct, the blood also becomes somewhat cloudy following the digestion of a fatty meal (this change can be observed only with serum or plasma since the red blood cells tend to obscure the effect in whole blood). Thus the net effect of absorption of lipids into the lymph is about the same as if the lipids were absorbed into the blood stream. Cholesterol is the only common sterol that is readily absorbed. Small amounts of the sterol-like compound, vitamin D, are also absorbed. Many of the substances classified as lipids are neither digested nor absorbed. Since these make up only a small part of the diet of animals, who consume principally the triglycerides, their indigestibility poses no real problems.

40–4 FURTHER METABOLISM OF LIPIDS

The major products of lipid digestion are the fatty acids and glycerol. These recombine immediately after absorption to form triglycerides. They circulate as small globules of triglycerides until they reach an organ which can rehydrolyze them and either convert the fatty acids and glycerol to a triglyceride characteristic of the species and of the specific organ doing the synthesis, or else the glycerol and fatty acids are oxidized to carbon dioxide and water with the attendant release of energy.

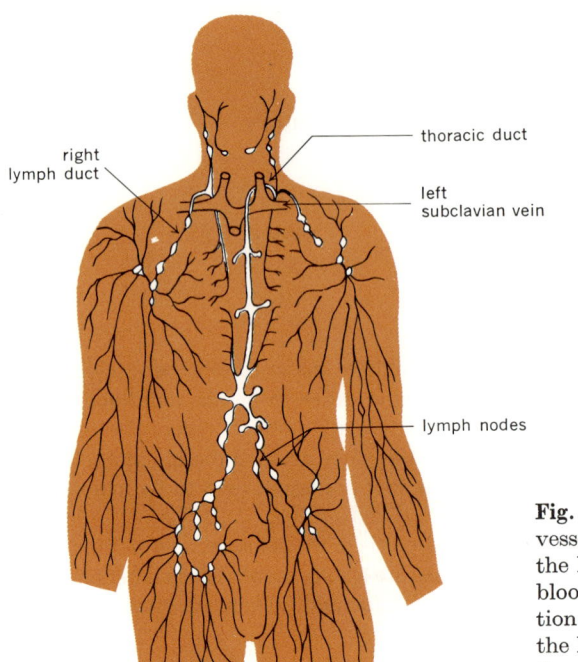

Fig. 40-1 The lymphatic system. The vessels which help in the circulation of the lymph and its connection with the blood system are shown here. In addition to its function in lipid absorption, the lymph returns valuable proteins to the circulatory system and is important in the response of the body to infections.

We can summarize the metabolism of triglycerides as follows:

$$\text{triglycerides} \downarrow$$
$$CO_2 + H_2O \leftarrow \text{fatty acids} \atop \text{(glycolysis)} \quad + \quad \text{glycerol} \Bigg\} \xrightarrow{\text{resynthesis}} \text{triglycerides}$$

Glycerol is metabolized by addition of phosphate from ATP and oxidation to yield glyceraldehyde-3-phosphate which is an intermediate in the glycolytic series of reactions (see Fig. 38-4).

Oxidation of Fatty Acids

The problem faced in the degradation of fatty acids is essentially that of breaking a carbon-to-carbon bond. In this case it is accomplished by dehydrogenation followed by a hydration, another dehydrogenation, and finally the rupture of the bond adjacent to the keto group formed in the process described above. Alternate carbon bonds are broken, giving two-carbon units. Coenzyme A (Fig. 36-14) is essential to the process. A schematic summary of the process is given in Fig. 40-2 and discussed in detail in the following paragraphs.

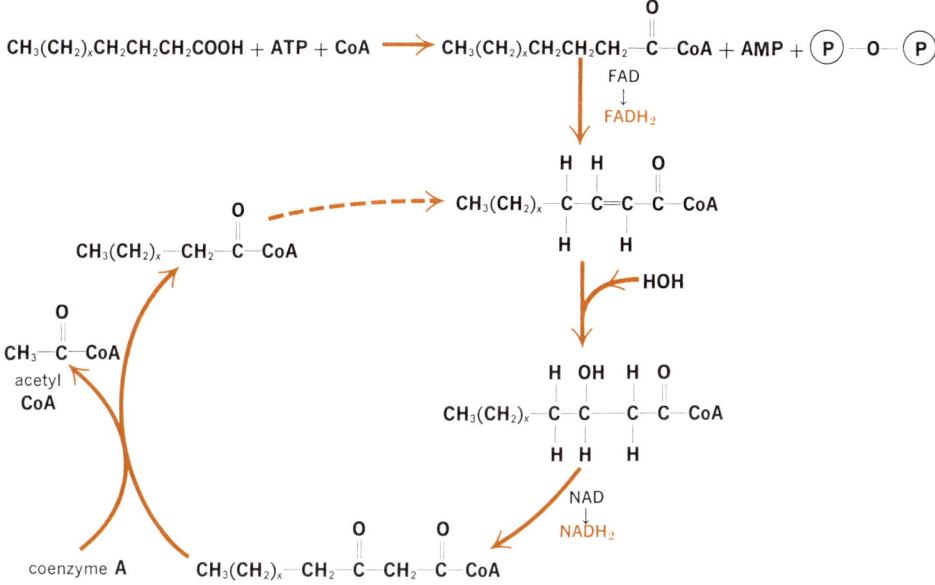

Fig. 40–2 The oxidation of fatty acids. The series of reactions given above shows how a two-carbon unit may be removed from a long-chain fatty acid. It should be noted that this is not truly a cyclic series of reactions since each reactant contains two fewer carbon atoms than it did in its previous trip through the series of reactions.

In the series of reactions shown in Fig. 40–2, we can see that the product of the oxidation of fatty acids is acetyl CoA. This product is then further metabolized to carbon dioxide and water via the citric acid cycle. The removal of an acetyl CoA unit leaves a fatty acid which is shorter by two carbon atoms. This shortened fatty acid goes through the cycle again and again until it is all converted to acetyl CoA. Since the starting material differs each time, the series of reactions is sometimes called a spiral—the **fatty acid spiral**. Each time a 2-carbon unit of the fatty acid is converted to a unit of acetyl CoA, 1 FAD is reduced to $FADH_2$ and 1 NAD is reduced to $NADH_2$. Thus a total of 5 ATP molecules can be synthesized each time a 2-carbon unit is released. The oxidation of 1 molecule of palmitic acid ($C_{15}H_{31}COOH$) requires that it go through the spiral 7 times to yield 8 molecules of acetyl CoA. Each turn of the spiral yields 5 ATP, so we get 5×7 or 35 ATP's. Remembering that oxidation of acetyl CoA via the citric acid cycle gives 12 ATP molecules per acetate unit, we get 12×8 or 96 ATP's. Of course, to be precise, we should subtract 1 ATP since it is required for the attachment of the original fatty acid to coenzyme A. (The attachment of the CoA to the acetate unit that is released at the end of the spiral does not require the energy of ATP.) Thus we get:

$5 \times 7 =$ 35
$12 \times 8 =$ 96
           ———
            131
            −1
           ———
130 ATP's/mole of palmitic acid

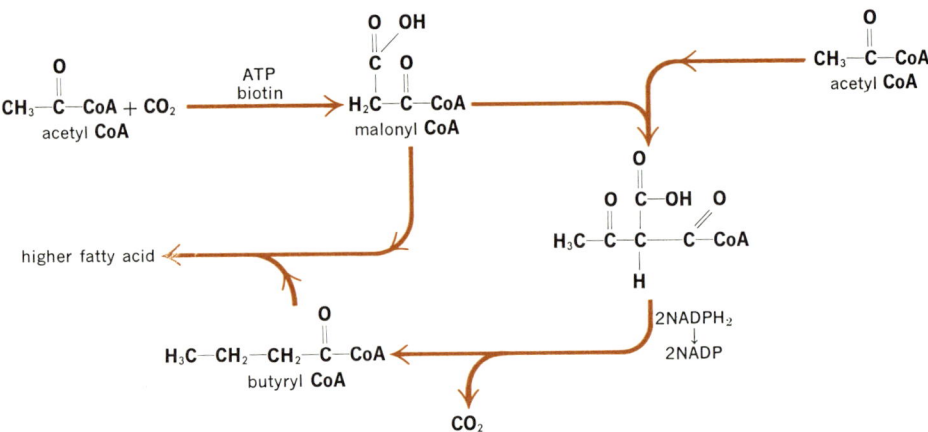

Fig. 40-3 The synthesis of fatty acids. The series of reactions represented above shows how a fatty acid may have a two-carbon unit added to it. By successive reactions, long-chain fatty acids can be synthesized.

The free energy available in the complete oxidation of 1 mole of palmitic acid as determined *in vitro* is about 2400 kcal. If the energy stored when 1 ATP is synthesized from ADP and phosphate is 12 kcal, we get 130×12 or 1560 kcal of energy stored, and the process is 1560/2400 or 65% efficient.

Since almost all the naturally occurring fatty acids have an even number of carbon atoms, the problem of metabolism of a 3-carbon fatty acid is not too significant. There are reaction sequences known, however, in which the 3-carbon acid adds a carbon dioxide to become succinic acid which is then oxidized via the citric acid cycle.

40-5 SYNTHESIS OF FATTY ACIDS

Most plants and animals can synthesize fatty acids. In some systems, synthesis seems to be primarily a reversal of the catabolic process; however, it now seems that this process is used primarily to add short units to fatty acid chains that are already formed. The hydrogen atoms (electrons) and the energy for fatty acid synthesis is provided by reduced NADP, in contrast to the degradation where FAD and NAD serve as hydrogen acceptors. It is interesting that most tissues in the human that synthesize fatty acids contain the enzymes for the phosphogluconate shunt (see Fig. 38-5) which produces reduced NADP.

Although acetyl CoA is used as the starting material, the process of synthesis is not just the addition of these 2-carbon units to one another. A carbon-dioxide molecule is added to acetyl CoA to form malonyl CoA, but the CO_2 unit is subsequently removed, and all the carbon atoms of the synthetic fatty acid are derived from acetate, none from carbon dioxide. The process is summarized in Fig. 40-3.

The minor differences in synthesis and degradation are apparently significant. In connection with these processes, Dr. David Green has stated, "Evidence is multiplying that life processes almost never follow the same pathway in building up and degrading a complex molecule. Readily reversible processes are rarely suitable for synthesis. The trick is to pick reactions that go almost exclusively in one direction."*

Some organisms, especially plants, synthesize fatty acids having double bonds. Humans are able to saturate or unsaturate one carbon-to-carbon bond in the C_{18} fatty acid molecule. Thus they can convert oleic acid to stearic acid or vice versa. Since humans require fatty acids containing more than one double bond, however, these fatty acids must be consumed in the diet, and we characterize them as essential fatty acids.

An animal or plant tends to synthesize triglycerides whose melting point is near the skin temperature of the animal or environmental temperature of the plant. Flax plants will grow at various temperatures, but the linseed oil from plants grown in colder temperatures is more highly unsaturated, that is, it will have a lower melting (freezing) point. Since we know that liquid lipids are more easily metabolized by animals it is not surprising that there is a need for fat deposits that are solid enough to be structurally sound but sufficiently soft to be readily metabolized. It is illogical to believe that a seal living in Arctic waters would deposit a triglyceride like beef—one that would be quite solid at the temperatures in which the seal lives.

We often divide the lipid stores of an animal into two categories, working or **tissue lipids** and storage or **depot lipids.** In any animal, tissue lipids are always present in about the same amount and are believed to be essential components of cells. Many of these tissue lipids are phospholipids. Since in man much of the lipid of the brain and nervous system is phospholipid, presumably his brain contains working and not just storage lipid. The amount of depot or storage lipid is variable and depends upon the nutritional state of the animal. The amount of depot lipid is high during periods of adequate or excessive food intake, and low following starvation.

There is a constant turnover of both kinds of lipids in the body, i.e., they are constantly being degraded and resynthesized. The metabolic half-life of the fatty acids in rat liver (mostly tissue lipids) is about 2 days, that of the brain and of the depot fats is about 10 days. Thus the depot lipids, while less active metabolically, are still being constantly degraded and resynthesized. In an experiment in which mice were maintained on a restricted diet so that they lost weight, they were given small amounts of isotopically labelled linseed oil. Their fats were soon found to contain the isotope. In fact, in 4 days, these animals made approximately 120 mg of depot fat and 130 mg of tissue lipid. Since they were losing weight, they were apparently catabolizing fats even faster than they were synthesizing them, but this did not mean that all synthesis stopped. This is only one example of the facts that lead us to believe that all components of the bodies of all organisms are in a state that has been described as **dynamic equilibrium.** Only when isotopes became available could scientists show that, although a person is maintaining his weight, the molecules in his body are constantly changing.

As we have said previously, an animal tends to accumulate lipids, particularly depot lipids that have a melting point near that of his skin temperature. Exceptions to this generalization are known. In an experiment designed to study the effect of diet on depot fat, dogs whose depot fat normally melted at 20°C, were fed either mutton tallow (highly saturated, high melting point) or linseed oil (highly unsaturated, low melting point). The melting point of the body fat of the dogs fed mutton tallow was found to increase to 40°C and that of those fed linseed oil was reduced to 0°C. Where large amounts of a particular type of lipid are ingested and when almost none of other types are available, an animal

---

* D. E. Green, *Scientific American*, **202**, No. 2, February 1960, p. 51.

is not always able to convert the fatty acids just as it would on a varied diet. Since plants synthesize their own fatty acids (starting originally with CO_2 and H_2O), they tend to produce more consistent types of fats or oils, with the exception noted previously that the degree of unsaturation may depend upon the climate in which the plant is grown.

Farmers who fatten hogs sometimes receive a lower price when they sell the animals if the fat is too soft—a condition called "soft pork." The condition is remedied if the diet is changed to include less unsaturated fat and more corn. Although the corn oil is relatively unsaturated, the starch in the corn is converted to fatty acids that are saturated, and consequently "solid" pork is produced, and a higher market price can be obtained.

40-6 THE SYNTHESIS OF CHOLESTEROL

Cholesterol is synthesized by the human and by most other animals. Although plants make other sterols, they do not synthesize cholesterol. The synthesis of the steroid nucleus starts with acetyl coenzyme A and involves nearly 40 distinct steps. Cholesterol and similar compounds are the precursors of the steroid hormones which are of great importance to the body.

40-7 DISORDERS OF LIPID METABOLISM

Cardiovascular disease is the leading cause of death in the United States. Actually, several conditions are included under this broad term, which really means diseases of the heart and blood vessels. The conditions we commonly call heart attacks, strokes, heart failure, and hypertension (high blood pressure) are all forms of cardiovascular disease. The underlying cause of cases of cardiovascular disease is **atherosclerosis**—a condition in which fatty deposits are formed in the inner lining of blood vessels. These deposits interfere with normal blood flow and may result either in depriving some areas of an adequate blood supply, in the formation of a blood clot, or in the rupture of blood vessels. The deposits formed in the blood vessels are composed largely of lipids in which cholesterol is found in the greatest quantity. There is no generally accepted theory as to the reason for formation of these fatty deposits in the blood vessels; however, it has been established that persons who have high levels of cholesterol in their blood are more likely to have heart attacks. This does not necessarily mean that the high level of cholesterol causes the attacks. Since the body receives some of its cholesterol from dietary sources and makes the rest, the extent to which cholesterol in the diet should be restricted is also a subject on which opinion is divided. One drug which successfully inhibited cholesterol synthesis in the body has been withdrawn from the market because it also produced undesirable side effects.

Research has shown that diets high in saturated fats tend to increase levels of cholesterol and that diets high in unsaturated fats tend to decrease blood cholesterol. Scientists do not agree on the reasons for this difference. It is also known that persons who are overweight are more likely to suffer from cardiovascular diseases. The explanation here probably depends on at least two factors—excess weight puts an extra strain on the heart and overweight persons probably have high blood lipid levels, particularly high cholesterol levels. The problems involved in understanding the cause and prevention of cardiovascular disease are much too complex to be adequately discussed in this text. They are problems about which the experts still disagree.

We might summarize what we do know about lipid metabolism by saying that anything that can be metabolized to yield acetyl CoA can be synthesized to either cholesterol or fatty acids. Therefore, excessive intake of either carbohydrates or lipids is not beneficial to health. While it has not definitely been established that unsaturated lipids, such as those found in various vegetable oils, are valuable in preventing cardiovascular disease, some of the normal dietary lipids should probably be in this form. There is insufficient evidence at this time to justify a person in excluding from his diet the animal fats, which include butter and eggs, as well as the fat found mixed in all meats. Indeed, it is difficult to design an adequate diet without the inclusion of milk and eggs. Probably the best policy a person can follow is to decrease the total number of calories eaten and to decrease the percentage of calories that are in the form of lipids, particularly the saturated lipids derived from animal fats. Americans derive from 40 to 45% of their calories from fats, whereas the Japanese, for example, derive only 10% of their calories from fats (and 78% from starches). When groups of men between the ages of 45 and 64 were compared, research showed that the mortality rate from cardiovascular disease for Japanese men is only 25% of that for American men. When Japanese move to the United States, however, and change their dietary habits to those of the Americans, their rate for heart attacks is similar to that for native Americans. We cannot, therefore, disregard the importance of the amount and type of lipid in the diet even though such factors as personality, sex, stress, and heredity are involved in cardiovascular disease. Much more research is necessary before we can give definite answers to problems associated with cardiovascular disease, and in particular, determine the part played by lipids.

40-8 SUMMARY OF LIPID METABOLISM

In spite of the evidence pointing to lipids as a possible cause of cardiovascular disease, lipids play many vital roles in the normal functioning of the body. One gram of lipid gives an average of 9 Cal (kcal) of energy while 1 g of carbohydrate or protein provides only 4 kcal/gram. Thus persons who work hard—who have a large demand for energy—must eat reasonable amounts of lipids. Lipids are stored in a water-free state, so a maximum amount of energy is stored with a minimum increase in body weight. The advantage of economic storage of energy was probably of great importance to our primitive ancestors whose survival depended upon both having energy stores and being mobile, expecially if they had to run from sabre-toothed tigers. It is interesting to note that the plants, which are essentially not mobile, accumulate carbohydrates in their tubers (potatoes) or seeds (wheat) instead of accumulating lipids. The plant lipids, which are mostly oils, however, are located in the most mobile part of the plant—the seeds.

Fats are stored by animals both subcutaneously and around delicate organs. Subcutaneous fat helps insulate the body from extremes of temperature and cushions it from the blows of external objects. Fatty deposits around the kidneys and other vital organs probably help protect them from physical shock.

In the instances given above, it is tempting to assume that the human body "knows" where to store fat and that it is better to store fat than protein or carbohydrate. It is certainly much more scientific to say that those animals who stored fats had a better chance for survival than did similar animals who accumulated carbohydrates or proteins—or maybe even some other kind of compound. We could conclude that since those who survived were, of course, the ones who reproduced, the most advantageous traits have persisted.

628 Metabolism of Lipids

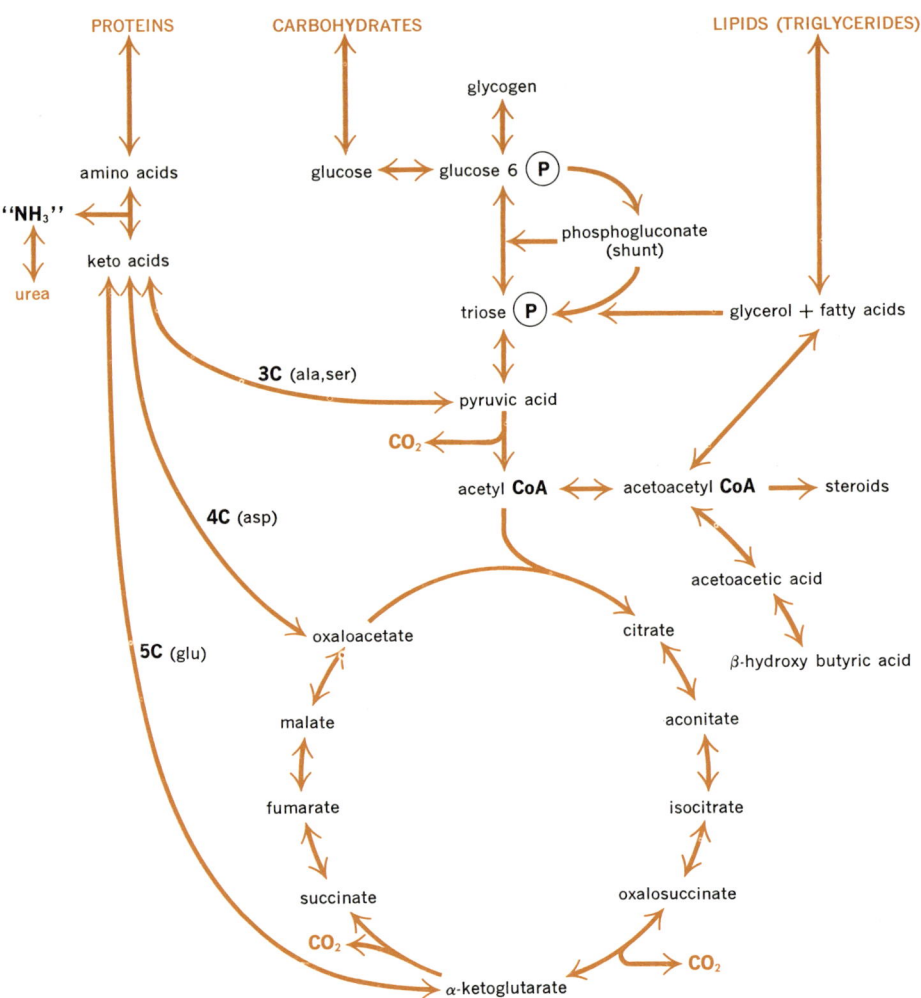

Fig. 40–4 Summary of intermediary metabolism. The reactions shown above represent the major metabolism of the basic biochemical compounds, the proteins, the carbohydrates, and the triglycerides.

Whether the prehistoric criteria for survival are still most important for our way of life may be open to question.

There is, however, a lot of nonsense in the currently preferred image of the American, especially the American female. The standard of leanness and absence of depot fat that is set by many fashion models probably represents a condition only slightly less sound from a physiological point of view than the picture of the overweight housewife. The mental and physical harm done both by the dread of being fat and by the attempts to diet is tragic. Reasonable amounts of depot fat are not so bad as we might be led to believe. A sensible approach to the problem of body weight requires a knowledge of nutrition—a topic which is discussed in the following chapter.

40-9 INTERRELATIONSHIP OF THE METABOLISM OF CARBOHYDRATES, PROTEINS, AND LIPIDS

Although proteins, carbohydrates, and lipids have distinctive metabolic pathways, there are several compounds that are common to two or more pathways. These represent important interconversion points. The citric acid cycle is used for the final oxidation of all three major types of compound, and both carbohydrates and lipids enter the cycle via acetyl CoA. A summary of intermediary metabolism is given in Fig. 40-4. Although the figure presents an abbreviated summary, it represents the major known pathways and shows the interrelationship of the metabolism of carbohydrates, proteins, and lipids.

From Fig. 40-4 we can see that carbohydrates can be converted to fatty acids and the lipids via acetyl CoA. The reverse is not true because the reaction of pyruvic acid to acetyl CoA is not reversible. The conversion of some amino acids to intermediates in the citric acid cycle and to pyruvic acid is also shown. Other amino acids can be converted to acetyl CoA or other intermediates by a relatively complex series of reactions. Many substances have been classified as being ketogenic or glucogenic. The ketogenic compounds will produce ketosis (excretion of acetoacetate, β-hydroxybutyrate, and acetone) when given to a diabetic animal. This reaction indicates that ketogenic substances are metabolized to yield acetate or acetoacetate. Glucogenic (antiketogenic) substances are those that are metabolized to yield a citric acid cycle intermediate, pyruvic acid, or some other glycolytic intermediate. A list of some common ketogenic and glucogenic substances is given below:

Ketogenic Substances	Glucogenic Substances
the amino acids leucine, isoleucine, phenylalanine, and tyrosine	the amino acids glycine, alanine, serine, cystine, methionine, aspartic acid, glutamic acid, proline, arginine, and histidine
all fatty acids	all carbohydrates
	glycerol

IMPORTANT TERMS

Atherosclerosis
Bile
Cardiovascular disease
Depot lipids

Fatty acid spiral
Lecithinase
Lipase
Tissue lipids

WORK EXERCISES

1. Summarize the digestion of a molecule of a triglyceride, of a lecithin.
2. List the series of reactions that is used to break a carbon bond in the metabolism of fatty acids.
3. What is the metabolic fate of the glycerol from the digestion of a triglyceride?
4. Why is the system for metabolism of fatty acids called a spiral rather than a cycle?
5. A table of food composition gives the following caloric values for 100 g of the edible portion of some sea foods: clams, 81; crabs, 104; oysters, 84; shrimp, 127; scallops, 78; salmon, 223; herring, 191. (All values are for the raw food; there is no significant

difference in water content of these foods.) On the basis of the material in this chapter, can you give a logical explanation and interpretation of the difference in caloric value of these foods?

6. Calculate the number of ATP's formed from ADP and phosphate in the complete oxidation of one mole of capric acid ($C_9H_{19}COOH$). If the theoretical free energy change in the complete oxidation of this compound *in vitro* is 1500 kcal/mole, and the change in free energy for the synthesis of ATP from ADP and phosphate is 12 kcal/mole, what is the percentage efficiency of this biological oxidation?

7. Alanine containing ^{14}C was given to a diabetic dog. Later β-hydroxybutyric acid containing ^{14}C was found in the urine of this animal. List the probable intermediates in the conversion of alanine to β-hydroxybutyric acid.

8. On the basis of the metabolic scheme for the amino acid histidine given in Problem 6 at the end of Chapter 39, explain why histidine is glucogenic.

9. It is known that a diet virtually free of lipids but high in carbohydrates can result in a person's accumulating large deposits of depot fat. Trace (in a general way) the pathway by which this conversion of carbohydrates to lipids can occur.

10. What major metabolic reaction of carbohydrates is irreversible? What is the significance of this fact?

11. Consult a general biochemistry text for the reactions involved in the synthesis of cholesterol by animals.

SUGGESTED READING

Dawkins, M. J. R., and D. Hull, "The Production of Heat by Fat," *Scientific American*, Aug. 1965 (Offprint #1018). This article discusses the significance of "brown" fat deposits in heat production.

Green, D. E., "The Metabolism of Fats," *Scientific American*, Jan. 1954 (Offprint #16).

Green, D. E., "The Synthesis of Fats," *Scientific American*, Feb. 1960 (Offprint #67). This article and the one above, covering the degradation and synthesis of fatty acids, are excellent.

Sprain, D. M., "Atherosclerosis," *Scientific American*, Aug. 1966. A recent discussion of this leading cause of death, and the relationship of diet to the disease.

NUTRITION 41

41-1 INTRODUCTION

There are two reasons why a plant or an animal must receive nourishment. Each organism must have a supply of energy for the many processes which must occur if life is to continue. In green plants most of the energy requirements are provided by the plant's ability to capture and use the radiation from the sun. Other plants, the fungi, not only cannot use the sun's radiation, but are killed or have their growth slowed by sunlight. This is especially true of bacteria which are one type of fungi. Although humans like to acquire a suntan, they do not utilize the sun's radiation as a source of energy; in fact, the tanning reaction probably serves to protect humans from the bad effects of overexposure to the sun. The energy an animal uses to maintain basic life processes and to perform work comes from the many different kinds of food it eats.

The second reason a living organism must receive nutrition is to obtain necessary chemical substances which the organism itself cannot synthesize. For most plants, the nutritional requirements are merely a source of carbon dioxide, some nitrogen other than the elemental nitrogen found in the air, and a variety of mineral elements—the most important of which are phosphorus, potassium, and sulfur. Animals require a more highly reduced form of carbon—the carbohydrates and lipids. Usually nitrogen must be provided in the form of amino acids or proteins. Mineral elements are necessary for animals just as they are for plants. Animals also have a special requirement for small amounts of chemical substances called vitamins. Green plants do not require preformed vitamins because they can synthesize these substances; many bacteria, however, have definite requirements for vitamins. This chapter will discuss the nutrition of animals, principally of man, in terms of both energy requirements and essential nutrients. Although neither plants nor animals can exist for any length of time without water, it is not usually listed as a nutrient.

41-2 ENERGY REQUIREMENTS

The energy requirements of an animal are ordinarily expressed in terms of **Calories.**

The Calorie we use in nutritional discussions is actually the kilocalorie—the amount of energy required to raise the temperature of one kilogram of water one degree centigrade.

The total number of Calories a person requires depends a great deal on his activity. Men doing heavy labor expend more energy and consequently require more energy from their food than do desk workers. Even a person who spends the day in a hospital bed is doing some work—his heart is beating, there are peristaltic movements of the digestive tract, and nerves are sending messages. Since the amount of work depends upon the amount of muscles, the size of the heart, and other factors, caloric requirements also depend upon body weight. One reasonable estimate of the energy expended just to keep a person alive is that it requires 1 Cal/kg of body weight per hour. Thus a person weighing 50 kg (about 110 lb) requires 50×24 or 1200 Cal/day as a minimum. For this person doing average work the energy requirements are probably from 2000 to 2400 Cal/day. A large man doing heavy work may require up to 5000 or 6000 Cal/day. These figures are only a general indication of caloric requirements. The efficiency of digestion and the basal metabolic rate of a person have a great deal to do with the number of calories required.

We customarily say that carbohydrates and proteins provide 4 Cal/g and lipids (the triglycerides) give 9 Cal/g, although for specific foods the number of calories differs slightly from the averages given. These figures are for the digestible members of the groups listed. For example, although cellulose contains energy, cellulose is not digested by humans and therefore provides no calories for them. The opposite is true for ruminants, who can utilize cellulose (see Section 38–2). Actually, proteins and amino acids give slightly more energy than 4 Cal/g when they are oxidized *in vitro*. However, their energy is not completely used, since urea is excreted rather than oxides of nitrogen. As a result, the average energy actually available from the metabolism of proteins is reduced to 4 Cal/g. Those who proclaim that calories do not count are talking about the excess calories of a poorly balanced diet. This topic will be discussed further in Section 41–9.

Most persons are aware that they feel somewhat warmer after they have eaten a meal. Those who are aware of different kinds of foods may have noticed that the increase in heat production is greater following a meal rich in protein-containing foods. This increased heat production which follows eating is called the **specific dynamic action** of foods. It represents the conversion of some of the caloric content of food into heat. The amount of this conversion is about 4% for lipids, 6% for carbohydrates, and an amazing 30% for proteins.

This conversion of the stored energy of foods into heat over and above that required for the maintenance of body temperature represents a "waste" of energy. This loss is probably due to the failure of the body to trap the food energy which is released in the digestion and further metabolism of the food. This untrapped energy is then dissipated as excess heat. In practical terms, then, 30% of the caloric value of protein foods is dissipated as heat, and while the calories from proteins do count, the amount must be discounted by about 30% in calculating the total energy provided by this class of foods.

This so-called energy waste is a bonanza for "calorie watchers." By increasing the protein portion of the diet and reducing the fat and carbohydrate portion they can lose weight on a caloric intake that might otherwise cause an increase in weight.

41–3 THE ESSENTIAL NUTRIENTS

An essential food is defined as a *chemical substance which cannot be synthesized by a given animal at a rate equal to his need*. Strictly speaking, carbon dioxide is an essential nutrient for a plant. When we use the term, **essential nutrient,** especially with respect to man, we mean that the omission of this one nutrient would cause deficiency symptoms even though the individual had a varied diet containing sufficient energy sources. The deficiency symptoms might be quite indefinite, perhaps shown only in a failure to grow normally, or they might be specific as they are in the diseases known as rickets and pellagra.

41-4 CARBOHYDRATES IN NUTRITION

None of the carbohydrates are essential, in the terminology of the nutritionist. However, for the majority of the people on earth more than half of the diet consists of carbohydrates. Rice, wheat and wheat flour, bread, potatoes, macaroni, and similar products are over 75% carbohydrate. In the more prosperous countries, proteins and lipids usually account for a large portion of the total food intake, but in poorer countries only small quantities of protein are available to most people. In these countries the land is used to produce cereal grains. Because of the expense involved and the continuous demand for grain, it is not fed to cattle or swine which would convert the calories contained in the carbohydrate grains to protein. Animals are poor calorie converters—the food they supply as meat or milk represents only 10 to 25% of the calories they consume.

41-5 PROTEINS IN NUTRITION

The proteins are significant in nutrition because of the amino acids they contain. Ten of the common amino acids are not synthesized by the human at a rate sufficient to maintain normal growth. These essential amino acids are given in Table 41-1. It will be seen that these amino acids are the ones having unusual carbon chains.

TABLE 41-1 Essential amino acids (for the human and the white rat)

arginine	leucine	phenylalanine
histidine	lysine	threonine
isoleucine	methionine	tryptophan
		valine

Most proteins derived from animals—such as beef and pork products, milk, and eggs—are good sources of the essential amino acids. Although some plants contain appreciable amounts of protein, many plant proteins lack one or more of the essential amino acids. Soybean protein contains only small quantities of methionine, and corn protein is deficient in both lysine and tryptophan. Proteins lacking one or more essential amino acids are considered poor quality proteins. For those whose diet consists primarily of plant products, and this means for much of the world's population, a variety of different plant proteins is necessary if they are to receive a balanced diet. Adequate nutrition for vegetarians is easier to achieve if eggs or milk products are eaten. Fowl and eggs are good protein sources, and in countries located near the sea fish is available as a protein source.

A good general rule is that a person should eat one gram of good quality protein each day for each kilogram of body weight. Since most protein foods are about 70% water, this means that a 70-kg (154-lb) person should eat a minimum of approximately 230 g (about half a pound) of high-protein foods each day. This rule-of-thumb assumes that the food is 30% protein, which is true only of high protein foods like meat, eggs, and cheese. If the protein is provided by a plant that contains only 5% protein, the intake must be increased, of course.

41-6 LIPIDS IN NUTRITION

Lipids are good sources of food energy. In most instances an animal can make all the lipids it requires, provided it is given a diet containing a sufficient amount of carbohydrates.

White rats, however, require small amounts of the highly unsaturated fatty acids, linoleic, linolenic, and arachidonic acids. The formulas for these essential fatty acids are given in Table 41–2.

TABLE 41-2 The essential fatty acids

Name	Formula	Location of Double Bonds
Linoleic acid (octadecadienoic acid)	$C_{17}H_{31}COOH$	9–10, 12–13
Linolenic acid (octadecatrienoic acid)	$C_{17}H_{29}COOH$	9–10, 12–13, 15–16
Arachidonic acid (eicosatetraenoic acid)	$C_{19}H_{31}COOH$	5–6, 8–9, 11–12, 14–15

No definite symptoms of deficiency due to a lack of these fatty acids are known in the human. It is logical to assume, however, that they are necessary nutrients since the nutrition of white rats and of humans is remarkably similar. There is also considerable evidence that lipids containing unsaturated fatty acids are beneficial in preventing heart attacks (see Section 40–7). It is wise, therefore, for people to include in their daily diet unhydrogenated vegetable oils which contain these highly unsaturated fatty acids.

41-7 MINERAL ELEMENTS IN NUTRITION

There is no simple, unequivocal listing of the mineral elements required for adequate human nutrition. In the first place, there is some question as to just which elements are mineral elements. The generally accepted definition is that **mineral elements** *are those elements not normally found in carbohydrates, proteins or lipids*. This conception eliminates the elements carbon, hydrogen, oxygen, nitrogen, sulfur, and phosphorus since these are commonly found in the three basic nutrients. Enough phosphorus for the human body's requirements cannot always be supplied from the three major sources of nutrients. Some inorganic phosphate must also be included in the diet. For this reason some authorities regard phosphorus as a mineral element. An adult requires about one gram of phosphorus per day either from inorganic phosphates or other compounds.

With the elimination of carbon, hydrogen, oxygen, nitrogen, and sulfur from the list of mineral elements, the major question remaining is "Which of the other chemical elements are essential for normal health and growth?" The question is not easily answered. More than 50 elements are found in the human body. Some are present in large quantities, and a deficiency produces specific deficiency symptoms. A lack of iron, for example, produces anemia. Other elements such as aluminum and silicon are probably present primarily because they are so abundant in the soil of the earth that it would be extremely difficult not to eat them along with normal food. It is, in fact, virtually impossible to place a human on a completely aluminum- or silicon-free diet. Consequently, we have no experimental evidence as to whether or not these elements are essential. A summary of the mineral elements which are known to affect humans is given in Table 41–3.

We can list several general functions of the mineral elements. Many are found in the structure of parts of the body, as are calcium and phosphates in bone. Others are probably

TABLE 41-3 The mineral elements in nutrition

elements needed daily in large quantities:	sodium, calcium, magnesium, potassium (phosphorus)
elements needed in smaller quantities:	iron, copper, chlorine, iodine, cobalt, manganese, zinc, molybdenum
mineral elements whose need is questionable:	bromine, fluorine, selenium, tin, silicon, arsenic, boron, vanadium
mineral elements which are toxic in relatively small quantities:	tin, arsenic, barium, beryllium, bismuth, cadmium, lead, mercury, selenium, silver, tellurium, thorium

less obviously structural; iron and iodine are parts of the molecules of hemoglobin and thyroxine, respectively. The mineral elements are also important in maintaining the correct osmotic pressure of body fluids. The maintenance of a definite osmotic pressure is necessary to prevent dehydration or overhydration of body cells and to keep proteins in their active forms. Nerves and muscle cells also require certain of the mineral elements for normal function. In cases in which either of two or perhaps three elements has an effect, it is logical to assume that the effect is due to osmotic regulation. A requirement for a specific mineral element indicates a specific effect.

One of the most important functions of mineral elements is the part they play in enzymatic catalysis. When a mineral element (probably in an ionic form) is required for the maximum function of an enzyme, we call the mineral element an "activator." The biochemist H. R. Mahler has said, "There probably does not exist a single enzyme-catalyzed reaction in which either substrate, product, enzyme, or some combination within this triad is not influenced in a very direct and highly specific manner by the precise nature of the inorganic ions which surround and 'modify' it."*

41-8 IMPORTANCE OF SPECIFIC MINERAL ELEMENTS

The following paragraphs discuss briefly the specific functions of some of the more important mineral elements.

Calcium

Calcium is found in the form of calcium phosphate in bones and teeth. It is also necessary for blood clotting. The removal of calcium ions by addition of citrate or other chelating agents keeps blood from clotting. Calcium activates some lipases and ATP-ase. Most persons require about 1 g of calcium per day. Milk is one of the best sources of calcium. For pregnant women the calcium requirement is appreciably higher than normal because of the amount of calcium being deposited in the bones of the growing child before his birth. Pregnant women who have insufficient calcium in their diets often deplete the calcium in their own bones and teeth, with resulting tooth decay and fragility of bones. Milk production requires large amounts of calcium in the diet. The old saying that a mother loses a tooth for every child she bears may have been true at one time, but there

---

* H. R. Mahler, *Mineral Metabolism*, Academic Press, New York.

is no need for such damage with our modern knowledge of nutrition and the availability of milk and calcium supplements.

Magnesium

Magnesium is present in small quantities in bones, but is found primarily in body fluids. It activates many enzymes and is found as part of the chlorophyll molecule. Low levels of magnesium in the diet (a condition which is hard to produce experimentally since magnesium is present in almost all natural water) can cause hyperirritability. Mice suffering from magnesium deficiency have running fits when given stimuli as simple as being exposed to a blast of air. High levels of magnesium in the blood produce anesthesia; however, many anesthetics are much more effective. Humans need about 250–500 mg of magnesium per day.

Sodium and Potassium

Sodium ions account for 93% of the cations of the blood. Sodium ions are found in all body fluids, but most cells have little sodium inside the cell wall. Instead, there is a relatively high concentration of potassium ions within cells. While in a resting state, nerve cells contain potassium ions. When a nerve impulse passes, these potassium ions flow out into the surrounding medium and sodium ions flow into the nerve cell. In order for another impulse to be conducted by that nerve cell, the resting conditions must be re-established, sodium must be pumped out, and potassium ions must enter. The whole process happens in milliseconds.

Potassium is present in relatively large quantities in plants, but sodium is not. Normally, humans consume from 2 to 4 g of potassium per day (in an ionic form, of course). Sodium ions are consumed primarily in the form of NaCl. While consumption varies widely, 5 g/day of sodium ions is a good average value. Herbivores (animals that eat only plants) often crave sodium chloride and will travel long distances to find a source of salt. Anthropologists have pointed to the fact that ancient civilizations developed in areas near either the sea or dried up seas and lakes where good supplies of salts, particularly NaCl, were available. When the human body contains abnormally large amounts of sodium ions it tends to retain water, and we see the condition known as edema. Persons who develop edema—often as a result of certain heart conditions—are restricted from consuming NaCl, and are given medicines to aid them in excreting sodium ions.

Chlorine

Chloride is the principal anion of the body. No specific function is known for chlorine, and we assume that it is merely the most readily available and convenient anion to balance the cations Na^+ and K^+ in body fluids.

Iron

Iron is part of the important body substances, hemoglobin, myoglobin, and the cytochromes. Although a human contains from 3 to 5 g of iron, he usually needs only small amounts, perhaps 10 mg/day, in his diet. The fact that females lose some blood at each menstrual cycle means that they require a slightly higher intake of iron than do males. The body re-uses iron, however, and although the iron-containing hemoglobin is turned over relatively rapidly in the body, iron is not excreted as are other parts of the hemoglobin molecule. It is retained and re-used in the synthesis of a new molecule of hemoglobin.

Nevertheless, iron deficiencies are known and are one of the more common causes of anemia.

Manganese

Manganese is known to activate many enzymes, among them arginase and many phosphatases. A deficiency has been shown to cause sterility in male rats.

Copper

Copper serves as an activator for many enzymes, particularly oxidative enzymes. Copper which is part of the pigment of the respiratory system of crabs, shrimp, and some other animals serves much the same function that iron serves in the hemoglobin of humans. This pigment found in those creatures is blue and has the name *hemocyanin*.

Cobalt

Cobalt is part of the structure of vitamin B_{12}; in fact, the name of the vitamin, cobalamin, reflects this fact. Deficiencies of cobalt in the food of ruminants leads to deficiencies of cobalamin and to anemia. Humans cannot synthesize cobalamin but must ingest the vitamin which already contains the cobalt.

Fluorine

It is known that large amounts of fluoride ions are poisonous. It is also true that small amounts may be incorporated into teeth, and that these small amounts make teeth less subject to decay. The addition of fluoride ions to water supplies has led to a great deal of controversy.

Iodine

Iodine is part of the thyroid hormones thyroxine and triiodothyronine. Lack of iodine in the diet is one of the causes of an enlargement of the thyroid gland which is called goiter. The use of iodized salt, which contains some sodium iodide in addition to sodium chloride, eliminates goiters caused by iodine deficiency.

Other Mineral Elements

Molybdenum and zinc serve as enzyme activators. Zinc is also found as part of the structure of the hormone insulin. Bromide has a sedative effect when present in relatively high concentrations in the blood. An appreciable amount of aluminum in the diet may lead to poor bone structure because aluminum prevents the absorption of the phosphates normally present in the diet and necessary for bone formation. This condition apparently comes about because aluminum phosphate is quite insoluble. Recent studies on animals maintained in germ-free conditions indicate that they require fluorine, silicon, tin, and vanadium for growth. Boron and zinc are required for normal plant growth although high levels, especially of zinc, may be toxic to plants.

41–9 VITAMINS

At the beginning of the twentieth century it was generally believed that an animal given a diet containing sufficient calories, adequate amounts of carbohydrates, proteins, lipids, and mineral elements would be well nourished. Additional research, however, with dietary components that were purified further indicated that something more than these nutrients was needed. Animals given highly purified diets soon got sick and, if not treated with less highly purified foods, they died. Gradually nutritionists realized that the animals on these purified but inadequate diets had symptoms similar to diseases seen in humans. The addition of certain substances to the purified diets cured the "deficiency" diseases in experimental animals, and it was not long until similar additions were made to diets of humans.

Placing experimental animals on given diets whose composition was well established has developed into an important experimental procedure for establishing exact nutritional requirements. The first new food substance to be identified was an amine—an important or vital amine. From these two words the name vitamine, which we now spell *vitamin*, was coined.

We now know that these vital food substances are not all amines. The current definition of a **vitamin** is that it is *an organic compound occurring in natural foods, either as such or as a utilizable precursor, which is required for normal growth, maintenance, or reproduction.* Vitamins are essentially catalysts and are not sources of body energy. The vitamins were originally named for letters of the alphabet. When a vitamin which was thought to be only one substance was found to be two or more different substances, both letters and numbers were used. Now that chemical structures are known for the vitamins, it is better to use descriptive names, although the old letters and numbers are still widely used. Another classification of vitamins is based on their solubility. Vitamins A, D, E, and K are insoluble in water and are often called the fat-soluble vitamins. It is better to call them lipid-soluble vitamins. Substances classified as one of the vitamins B or vitamin C (ascorbic acid) are water-soluble. Although these two classifications (lipid- and water-soluble) were developed merely for convenience in classifying vitamins, there is an underlying significance in the classification. The lipid-soluble vitamins are stored in the fats of the body; the water-soluble vitamins are not stored to any extent. Thus, the water-soluble vitamins must be ingested daily, while stores of vitamins A, D, E, and K may be sufficient to keep deficiencies from developing even after long periods on vitamin-deficient diets. Now that vitamins are readily available in supplements as well as in foods, physicians are beginning to see cases of **hypervitaminoses**—an excessive amount of a vitamin. The only vitamins for which true hypervitaminoses are known are the lipid-soluble vitamins—the ones the body can accumulate, sometimes to its own detriment.

A vitamin must always be defined with respect to a given type of animal. Ascorbic acid (vitamin C) is a necessary substance for all animals; however, most animals synthesize this compound. Man, other primates, and the guinea pig, however, cannot synthesize ascorbic acid. For these animals ascorbic acid is a vitamin—for others it is not.

Any real understanding of the vitamins must involve some idea of their structure and their general functions.

Vitamin A

The formula for this vitamin is given below:

vitamin A (retinol)

It can be seen that vitamin A (chemically known as retinol) is a hydrocarbon with one —OH group; the compound is a lipid and would be classified as lipid-soluble. Vitamin A is similar to β-carotene (see Chapter 35), and can be derived from β-carotene in the diet. The vitamin itself is yellow and its precursor, β-carotene, is orange, so that small amounts of it give a yellow color. Thus many foods that are good sources of the vitamin are yellow or orange (carrots, squash). Many green foods also contain vitamin A, but its color is masked by the more brilliant chlorophyll. The vitamin can be destroyed by heating and oxidation.

Vitamin A is necessary for growth, maintenance of epithelial tissues, and proper vision. Lack of the vitamin causes night blindness, poor dark adaptation, and a disease of the eyes called xerophthalmia. Although vitamin A undoubtedly plays many roles in the animal body, the one best understood is its function in vision. The visual cycle can be summarized as follows:

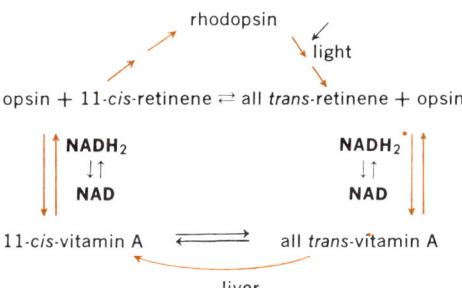

This cycle involves rhodopsin (visual purple) which is a combination of a protein called opsin and retinene (the aldehyde corresponding to vitamin A). Since there are five double bonds in the side chain of vitamin A, each one could exist in either the *cis* or *trans* configuration. Vitamin A is usually found in the all *trans* form; however, in rhodopsin there is the *cis* configuration between carbon atoms 11 and 12. This 11-*cis* configuration is necessary for the binding of retinene to the protein opsin. When a quantum of light hits a molecule of rhodopsin it converts the 11-*cis* form of the vitamin to the all-*trans* form. This change of configuration apparently changes the shape of the retinene so much that it cannot stay bound to opsin. The other reactions of the series serve to regenerate the 11-*cis*-retinene so that it can recondense with opsin and again react with light.

Rhodopsin is not the only visual pigment. It is most important in vision in dim light; lack of rhodopsin due to a dietary deficiency of vitamin A (or carotene) causes a condition known as "night blindness." Eating liver, which is a good source of the vitamin, and carrots, which contain β-carotene, prevents or cures night blindness.

In the United States today there may be more excesses of vitamin A than deficiencies. The observation that Eskimos generally do not eat polar bear's liver correlates with the fact that it contains extremely high levels of Vitamin A and when eaten produces symptoms of hypervitaminosis A. These cases of hypervitaminoses produce skin irritations and a lack of appetite. In many areas of the world, however, vitamin A deficiencies create serious problems, particularly since they cause the disease xerophthalmia. It was recently estimated that, in India, at least one million persons are blind because of vitamin A deficiencies. Yet the materials to prevent these deficiencies would cost only a few pennies a year per person.

The B Vitamins

At one time it was believed that there was a substance called vitamin B. However, further study indicated that this substance was really a mixture of more than one chemical compound. Further fractionation proved that there were many compounds, all water-soluble, most containing at least one nitrogen in a ring structure, and having somewhat similar functions. These vitamins were given names such as B_1, B_2, and B_3. Some of those originally given a numerical classification proved either to be one of the other vitamins or else compounds that were not vitamins; only those called B_1, B_2, B_6, and B_{12} have retained the letter-number designation. These designations are gradually being replaced with names which reflect the chemical structures. Other substances now classified as B vitamins are known primarily by their chemical names. For purposes of discussion we will group these various substances under the general heading of B vitamins.

All the B vitamins are necessary components of all living cells. They usually function as coenzymes, usually with a nucleotide, or at least a phosphate, attached. Since many organisms can synthesize these chemical compounds, they are not truly vitamins for these organisms. In many animals, bacteria in the gastrointestinal tract synthesize appreciable amounts of the B vitamins, and the host organism may use the products of intestinal synthesis. In fact, some vitamin deficiencies in some animals cannot be produced without destroying their gastrointestinal bacteria.

All B vitamins are involved in promotion of the appetite, growth, and general well-being of the animal which requires them. Foods that are good sources of one B vitamin as a rule are rich in the others also. The classical deficiency diseases which have been attributed to a deficiency of a specific B vitamin were probably multiple deficiency diseases. Actually, we know much more about the specific deficiency diseases in the white rat and other experimental animals than we do in humans, since our knowledge of human deficiencies has come primarily from reports of doctors who were much more interested in curing the disease than prolonging the condition in order to study it. We may feel some displeasure about the lack of precise knowledge concerning human nutrition, but when we realize that definitive knowledge could come only from experimentation on humans, we are willing to accept the limitations.

Thiamine (Vitamin B_1)

The disease beriberi has been characterized as a thiamine-deficiency disease. Beriberi was prevalent in the Orient where a large part of the diet is composed of polished rice. The simple way to cure this disease is to restore to the diet the materials from rice hulls which are removed in polishing the rice. While the classical human beriberi may have been a multiple deficiency disease, many of the symptoms are similar to those shown in rats on diets deficient only in vitamin B_1.

thiamine chloride

Thiamine deficiency causes abnormalities in the nervous system. A neuritis of the nerves in muscles leads to muscular weakness. Anxiety and mental confusion are also seen in humans. In some cases the heart does not function normally.

Thiamine functions as a coenzyme in decarboxylation reactions. A biochemical deficiency symptom is the finding of high levels of pyruvic and lactic acids in the blood, apparently due to the failure of a thiamine-deficient animal to decarboxylate pyruvic acid to acetyl CoA. We do not yet know how the biochemical abnormality—lack of decarboxylation—leads to the physiological symptoms.

Thiamine was the "vital amine" which resulted in the name vitamin. It is one of the most important vitamins; the complete absence of this substance from the diet will produce death in experimental animals in a week or two. A human requires about 1 mg of thiamine per day. Yeast, pork, fresh fruits, vegetables, and whole grains are good sources of thiamine. Thiamine is now added to white bread and polished rice. The vitamin is hydrolyzed to an inactive form when heated in neutral or alkaline solutions, however. Consequently, prolonged cooking of foods containing thiamine, particularly if sodium bicarbonate has been added, as it often is to retain the fresh green color of vegetables, destroys the vitamin.

Some interesting experiments he did with white rats led Dr. R. J. Williams to reach some conclusions regarding alcoholism and thiamine. When white rats were given a choice of either water or a dilute solution of ethanol, each rat established an individual drinking pattern. Some preferred water, others drank both water and alcohol. When the diet of the rats was altered so that the thiamine was decreased, all the rats started drinking large amounts of alcohol. This kind of alcoholism could be cured by adding thiamine to the deficient diets. It is known that human alcoholics often suffer from thiamine deficiencies. Apparently the intake of alcohol diminishes the desire for food so that these people develop a slight thiamine deficiency which leads to further alcoholism, more serious vitamin deficiency, and so on. The administration of thiamine is effective in treating some cases of alcoholism; however, on the basis of current evidence we cannot say that adequate nutrition is a cure for all cases of alcoholism, since many psychological factors are also involved. Relatively large doses of thiamine have been touted as a preventive or cure for a "hangover." This prescription is probably as effective as most cures, other than abstinence, that have been proposed. To our knowledge this area has not been subjected to rigorous scientific experimentation.

Riboflavin (Vitamin B_2)

The deficiency of riboflavin can be said to produce generally poor health. White rats given diets deficient in vitamin B_2 fail to grow normally, have a poor coat of hair, a tendency toward skin irritations, and can be described as "looking sick." Human deficiencies of riboflavin that are not complicated by other deficiencies are not well characterized. The vitamin in the form of FAD (flavine adenine dinucleotide) and FMN (flavine mononucleotide) serves as an important coenzyme (see Fig. 37–8 for the formula of FAD).

Humans require about 1.5 mg/day of riboflavin. The vitamin is found especially in yeast, liver, and wheat germ. Milk and eggs, although containing smaller amounts, probably represent the major sources of the vitamin in the average American diet. Green leafy vegetables are also good sources of riboflavin.

Nicotinamide (Niacin)

This vitamin is related to nicotinic acid (which is only remotely related to nicotine, see Section 28–6A and B). Nicotinic acid can be converted to the vitamin by the human.

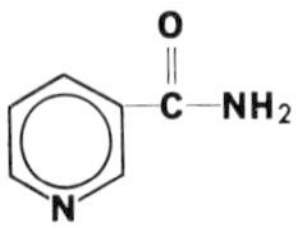

nicotinamide

The classical deficiency disease of this vitamin is pellagra. Until 40 or 50 years ago many of the residents of the southeastern United States and of Spain suffered from this condition. It was characterized by a dermatitis, mental symptoms, and diarrhea. The addition of nicotinamide to the diets of persons suffering from pellagra cures the disease; so does the consumption of a diet rich in good quality protein. This originally led to confusion as to the real cause and cure of pellagra. When it was found that the amino acid tryptophan could be converted to nicotinamide by a series of reactions, the link between the two cures for pellagra was found. Those areas in which pellagra was widespread were areas in which corn was a major part of the diet. In fact, the diet of a typical pellagrin consisted principally of the fatty part of pork or "side meat", cornmeal mush, and molasses. Thus, the caloric needs were satisfied, but no good source of either nicotinamide or tryptophan was ingested. The protein present in corn is notably lacking in the amino acid tryptophan. The story of the discovery of the cause and the treatment of pellagra by Dr. Goldberger of the United States Public Health Service is a striking example of nutritional detective work. One dramatic account of this story is in the book *Hunger Fighters* by Paul DeKruif. (See the "Suggested Reading" at the end of the chapter.)

Nicotinamide is found in the coenzymes NAD and NADP. (For formulas see Fig. 36–13.) These coenzymes function in a great number of oxidation reactions and provide an important link in the conversion of the energy of foods to the energy stored in ATP.

The requirement for this vitamin depends upon the intake of tryptophan in the diet; a human requires probably 10 to 16 mg/day. The vitamin is found in a variety of plant and animal foods. Meat products, especially liver, are good sources of nicotinamide.

Pyridoxal (Vitamin B_6)

Pyridoxal is the functional form of the substance also known as vitamin B_6. The closely related compounds pyridoxine and pyridoxamine can also serve as sources of the vitamin.

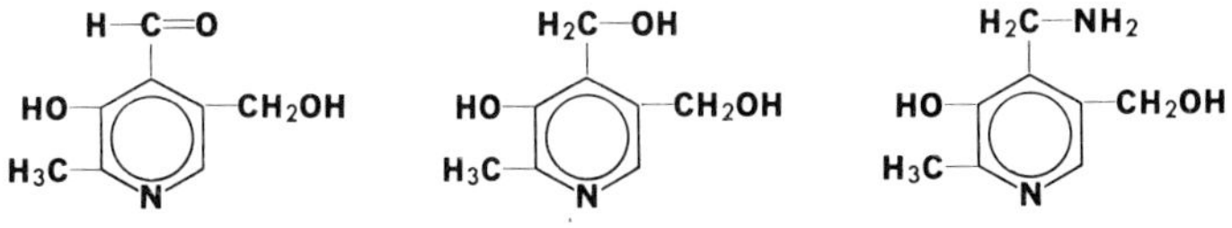

pyridoxal   pyridoxine   pyridoxamine

No human diseases due to a deficiency of this vitamin are known, although convulsions have been reported in infants who had been fed a formula deficient in sources of pyridoxal. Rats on a diet deficient in vitamin B_6 develop a dermatitis and fail to grow normally. The fact that the amounts of urinary excretion products of pyridoxal are greater than the amount known to be ingested has led to the conclusion that pyridoxal is synthesized in the human body. Whether this synthesis is due to bacteria in the intestinal tract or whether some part of the body synthesizes the compound is not known. If the latter is found to be true, pyridoxal may no longer be considered a vitamin. Whole grains, fresh meats, vegetables, and milk are good sources of pyridoxal.

Pyridoxal phosphate is the coenzyme for many of the reactions of amino acids. Transaminations are among the most important reactions catalyzed by pyridoxal-containing enzymes. Decarboxylation and racemization (conversion from the D- or L-form to a DL mixture) of amino acids are also catalyzed by B_6-containing enzymes.

Pantothenic Acid

Pantothenic acid is part of the structure of coenzyme A. (See Fig. 36–14 for the formula.) Human deficiency diseases are not known. In rats deficiencies produce a decreased rate of growth, impaired reproduction, and graying of hair. The latter is not observed, of course, unless the rat used is a dark-haired one instead of the usual albino white rat used in nutritional research. No relationship has been found between pantothenic acid, and graying of human hair. As coenzyme A, the vitamin functions in a great number of reactions, especially those involving acetic or other fatty acids. Because of its widespread occurrence in foods, pantothenic acid deficiencies are rare. Liver and yeast are rich sources of the vitamin.

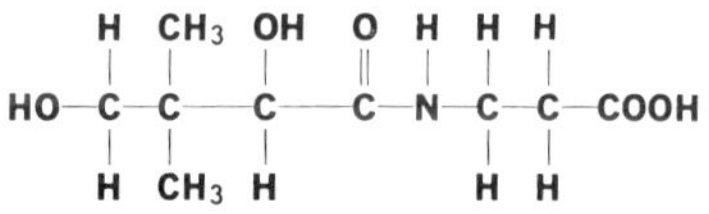

```
    H  CH₃ OH   O  H H H
    |   |   |   ||  |  |  |
HO—C—C———C———C—N—C—C—COOH     pantothenic acid
    |   |   |       |  |
    H  CH₃  H       H  H
```

Biotin

Biotin is widely distributed in foods. In most animals the excretion of metabolites of biotin exceeds the intake, and we assume that intestinal synthesis provides sufficient amounts of this compound.

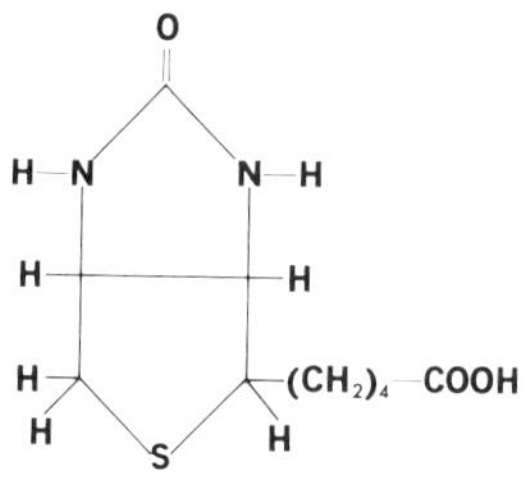

```
              O
              ||
         /    \
     H—N      N—H
     H—|      |—H
     H—|      |—(CH₂)₄—COOH    biotin
        \    /
      H  S  H
```

Biotin serves as a coenzyme for reactions in which carbon dioxide is added to a substrate. The synthesis of fatty acids is one of the more important of such reactions. The conversion of pyruvate to oxaloacetate requires biotin, as does the incorporation of CO_2 into urea via the urea cycle.

Folic Acid

Deficiencies of folic acid produce various kinds of anemias. Such deficiencies in the human are usually due to a malfunction of the digestive tract rather than to an actual dietary insufficiency. The vitamin is widespread in nature, and most diets contain appreciable quantities of it. It is also synthesized by bacteria in the intestine.

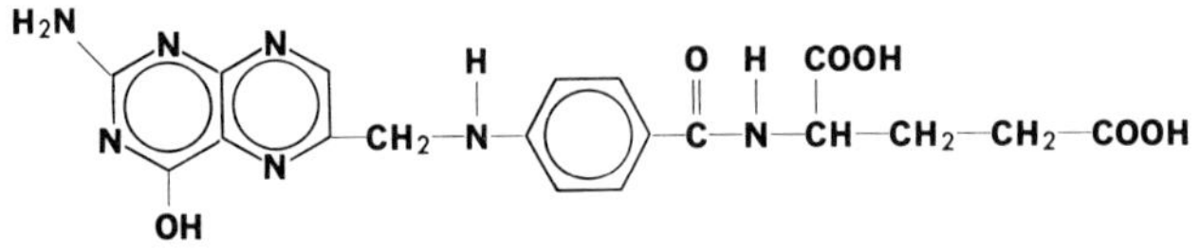

[Structure of folic acid shown] folic acid

Folic acid functions in the metabolism of one-carbon compounds, particularly formyl and hydroxymethyl groups. It is certainly a necessary component of the body, although the daily requirements are not known.

Cobalamin (Vitamin B_{12})

This vitamin was one of the last ones to be studied and characterized. Deficiency of vitamin B_{12} causes the condition known as pernicious anemia. As with folic acid, most deficiencies are due to failure to absorb the vitamin which is present in food. The vitamin is widespread in animal products, but is not found in most green plants. Apparently the human body can store some amounts of this vitamin and since the daily requirement is small, perhaps only 1 μg (microgram)/day, B_{12} is not especially significant in normal nutrition. The formula for B_{12} is given in Fig. 41–1.

Ascorbic Acid (Vitamin C)

Scurvy has been known for hundreds of years. It was especially seen among sailors. Scurvy was characterized by sore gums and loss of teeth, swollen and tender joints, small hemorrhages located just below the skin, and anemia. The British learned that lemons or limes which could be kept on long voyages would not only cure but prevent this condition. From this discovery came the practice of carrying limes and the name "Limey" to describe a British sailor. Later developments led to the identification of ascorbic acid as vitamin C.

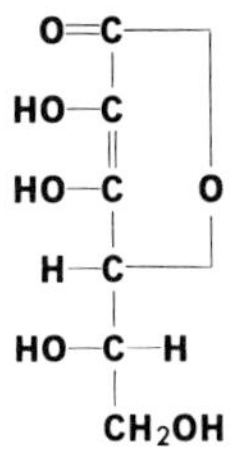

[Structure of ascorbic acid shown] ascorbic acid

Most animals can synthesize ascorbic acid from glucose, but man, the apes, and the guinea pig cannot. The human requirement has not been definitely established, but is apparently in the range between 30–50 mg/day. The vitamin is present not only in citrus fruits, but in most fresh fruits and even in potatoes. Ascorbic acid is relatively easily oxidized, however, and consequently its effectiveness as a vitamin is lost when foods containing it are cooked.

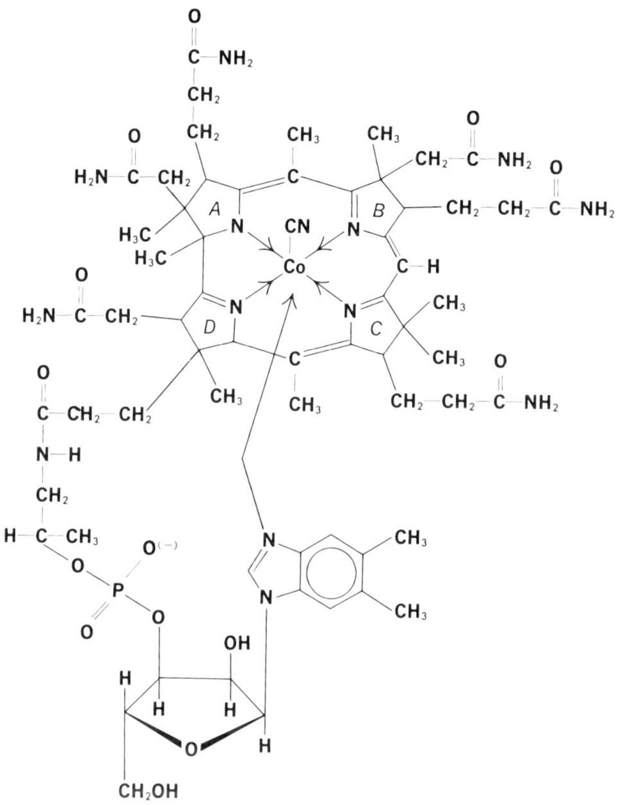

Fig. 41-1 Structure of vitamin B$_{12}$ (cobalamin). Note the cobalt atom in the center of the pyrrolidine rings. These rings are in a plane with the cobalt, with cyanide group, and the dimethylbenzimidazole groups above and below the ring. Note also that rings *A* and *D* do not have a carbon bridge between them as was found in hemoglobin and chlorophyll.

The exact function of ascorbic acid in human metabolism is not known. It presumably functions in oxidation-reduction reactions, and many such reactions have been shown to be aided by ascorbic acid. For most of these systems other easily oxidizable and reducible substances also function. No reaction is known for which ascorbic acid is a specific coenzyme. Despite much publicity, there is little scientifically valid evidence that vitamin C cures or prevents colds Such evidence may be found, but large doses of the vitamin seem not to be warranted at present. On the other hand, vitamin C is water soluble and is not stored; so overdoses are not so bad as those of lipid-soluble vitamins.

Vitamin D

Several substances have the function of Vitamin D. They are steroids differing only in the side chain (R group) at position 17. Children who have inadequate amounts of the vitamin in their diet develop rickets. The disease is now rapidly disappearing, but cases are prevalent in the poorer countries of the world. Rickets is characterized by the failure of the bones to harden normally. Children with rickets usually have bowed legs and are stunted in growth. The vitamin promotes the absorption of calcium and phosphorus through the intestinal wall and functions in depositing these two elements in the protein matrix of bone.

Vitamin D is sometimes called the sunshine vitamin because the ultraviolet radiation of the sun can convert an inactive precursor in the skin (7-dehydrocholesterol) to vitamin D. (See Section 35–6.)

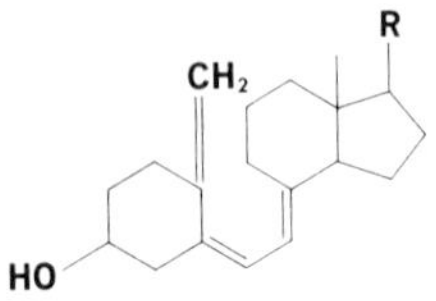

[structure of vitamin D]

Children need vitamin D during the time they are growing and, to a lesser extent, mothers require the vitamin during pregnancy and lactation. There is no demonstrable vitamin D requirement for adult males. Vitamin D is readily available, and since it is lipid-soluble and can be stored, cases of hypervitaminoses have been reported. In these cases bones are actually demineralized, and the calcium and phosphate deposited in many soft tissues. Deposits may also form in the kidney with serious consequences. Deaths have been reported due to kidney failure brought on by excessive intake of vitamin D. Considerations such as these make one consider whether the addition of vitamin D to milk, which has proved to be an excellent preventive for rickets, may not also have unfavorable consequences.

Vitamin E

The substance α-tocopherol and related compounds are necessary for normal reproduction in the rat.

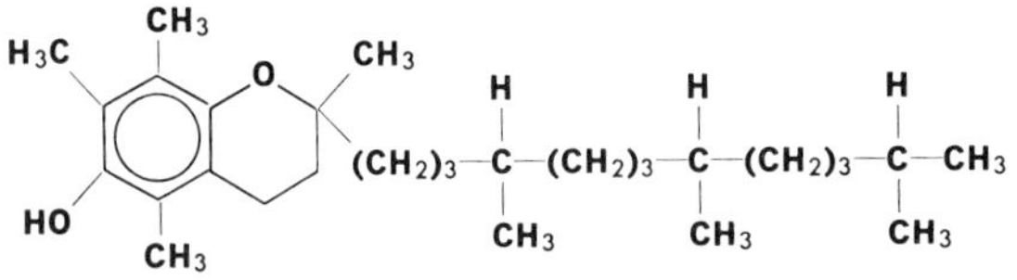

[structure of α-tocopherol]

Also, the absence of these compounds, which are collectively known as vitamin E, causes a serious muscular dystrophy in rabbits. Although there are other indications that the vitamin is necessary for several species of animals, there is no convincing evidence that vitamin E is essential for humans. It is ineffective in treating human muscular dystrophy and sterility. On the other hand, since other species do require the vitamin, it is possible that there is sufficient dietary intake and storage of vitamin E in humans so that symptoms of deficiency do not appear. The best natural sources of vitamin E are the oils derived from plants; wheat germ oil and cottonseed oil are good sources. The lipids of the leaves of green plants also contain vitamin E.

Excessive claims are often made for the effectiveness of vitamin E in helping certain heart conditions and in improving muscle action. Some athletic coaches advise the use of plant oils that are rich in vitamin E. To the best of our knowledge, at the present time, such uses are unwarranted. The ever-present danger of hypervitaminoses of the lipid-soluble vitamins should keep one from indiscriminate use of this substance.

Vitamin K

Vitamin K is necessary for proper clotting of blood; deficiencies are characterized by hemorrhages. Since the vitamin is widely found in foods and is probably synthesized by intestinal bacteria, deficiencies are usually due to poor absorption.

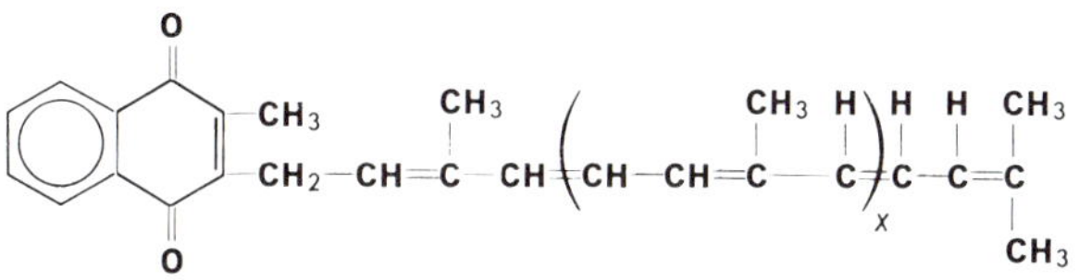

There are several forms of vitamin K, each differing in the length of the side chain. The widespread occurrence of the vitamin and its production by intestinal synthesis mean that normal persons need not be concerned with the amount of this nutrient in their diets.

In cases where clotting of the blood is damaging to an individual, extensive use is now made of substances which oppose the action of the vitamin, reducing the tendency of the person's blood to clot. These substances are structurally similar to vitamin K, but sufficiently different so that they interfere with the action of the vitamin. Dicoumarol was originally isolated from spoiled clover hay which caused the death of cattle due to internal hemorrhages. Now both dicoumarol and a synthetic analog, warfarin, are given to persons who have had heart attacks caused by blood clots. Warfarin is also used now as a rat poison. It causes death by producing internal hemorrhages. Continued consumption of the poison is required, however, before it is lethal.

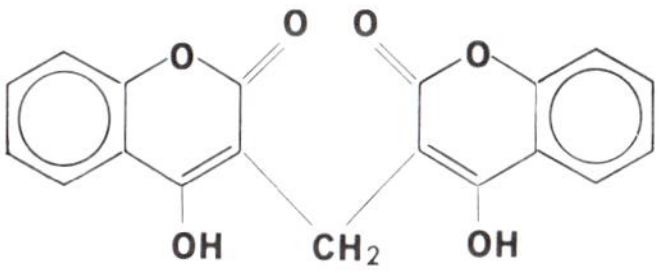

41–10 ON BEING WELL FED

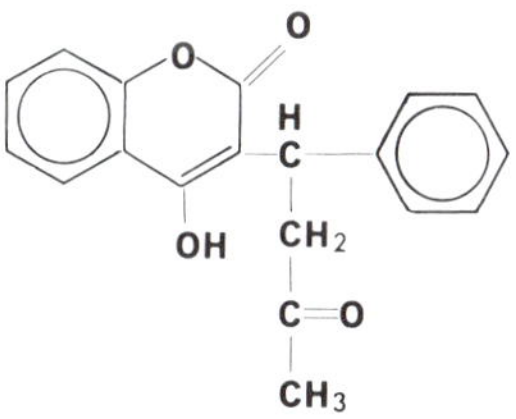

An adequate diet should contain foods that provide sufficient energy. Either carbohydrates, proteins, or lipids can provide this energy. Evidence from studies of persons having cardiovascular disease indicates that in the normal American diet the percentage of calories derived from lipids is probably too high. Many nutritionists recommend that lipids should provide no more than 15% of the calories a person consumes.

When the consumption of energy-producing foods exceeds the energy requirements of an individual, his body usually converts the energy of the food to energy stored in his triglycerides, and he gets fat. The simplest way to lose weight or to keep from gaining weight is to decrease the number of calories consumed.

A good diet, however, must also include all essential nutrients. It should contain the essential amino acids which are derived from proteins, the essential fatty acids, the mineral elements, and the vitamins. To meet these requirements, a varied diet should include high quality protein, vegetables, and fruits, some of which should be uncooked, and some plant oils. Vitamin pills are not necessary for normal persons who are able to have a varied diet as outlined above. Large amounts of the water-soluble vitamins are probably not harmful, but serious consideration should be given before including large amounts of the lipid-soluble vitamins, particularly in the diet of adults. Eggs, milk and other dairy products are excellent foods. Both eggs and milk are "specifically designed" for the nutrition of a developing individual—the designer, of course, being the process of evolution which assures the non-survival of the inadequately nourished.

Many Americans are overweight, and many others are obsessed with the idea that they must lose weight. They keep hoping that there is some easy way to eat all they want and still be thin. Well-meaning but ignorant persons and charlatans are only too ready to take advantage of this situation. Many good types of diet are known, but others, even some that are widely publicized, are so grossly unbalanced that continued adherence to them would have serious consequences. For persons who wish to lose weight the following suggestions may be helpful.

1) Remember that calories do count, and don't eat high calorie foods. Many vegetables contain large amounts of water and cellulose neither of which add any calories to your diet. The green leafy vegetables, in addition to being low in calories, have high mineral and vitamin content.
2) If you need to decrease food consumption drastically, it may be advisable to include vitamin and mineral supplements.
3) The proteins in your diet should be maintained at a high level. You can decrease the amounts of lipids and carbohydrates in your diet—within the limits of discretion, of course.
4) For serious dieters it is advisable to get a chart listing not only the caloric, but also the protein, mineral, and vitamin content of foods.

This chapter serves only as a general introduction to the science of nutrition. There are exceptions to most of the generalizations presented. The presentation is intended to give our readers an understanding of normal nutrition, not to provide sufficient information for proposing drastic diets. Any prolonged dietary restrictions should be undertaken only under the supervision of a competent physician.

41–11 WORLDWIDE NUTRITIONAL PROBLEMS

It is often said that half the world's people go to bed hungry every night. Although the truth of such a claim is hard to establish, it is a fact that far too many people suffer from malnutrition, either from lack of calories or from lack of a specific essential nutrient. This problem exists because fewer people are dying from infectious diseases and more of them are living to help aggravate the rapid rate of population growth in many countries.

Lack of sufficient protein containing the essential amino acids is probably second only to general lack of food as a worldwide problem. Although plant proteins are poor sources of several of the essential amino acids, many people cannot afford to feed plant materials to animals in order to secure the higher quality, hence more desirable, animal protein. Since animals convert only a relatively small percentage of the protein in their food into body protein, it is highly questionable as to whether we should continue the wasteful practice of feeding high protein foods to cattle. The situation is different when cattle or sheep eat grass or other food unavailable for direct human use.

There are ways of making a diet of plant protein more nutritious. One is by carefully mixing plant proteins so that the deficiences of one food are met by the amino acids supplied by another. Instructions for doing this are given in the book *Diet For a Small Planet* by Lappé which is listed at the end of the chapter. Another procedure for increasing the quality of protein food is by breeding new kinds of plants that have either higher protein content or a higher content of an important amino acid. The article by Harpstead in the list of suggested reading tells of the development of a high lysine corn. This and similar breeding experiments combined with better methods of production are generally known as the "Green Revolution." It shows promise of meeting some of the world's food problems.

We will probably never solve our worldwide nutrition problems until we find ways of dealing with population growth and with the increased consumption of food by the developed countries. It is tragic to contemplate that dog and cat food, so much seen in every supermarket, contains fish protein that children in many parts of the world badly need. Disputes over territorial fishing rights show how countries value the fish caught off their shores.

While science can help solve some of the worldwide food problems, social problems are involved also. Only when awareness and concern are coupled with scientific and social knowledge will we have a chance of solving the crucial problems of the world.

IMPORTANT TERMS

Basal metabolic rate   Essential nutrient   Nutrient   Specific dynamic action
Calorie   Hypervitaminosis   Nutrition   Vitamin
Deficiency disease   Mineral element

WORK EXERCISES

1. Why is tryptophan an essential amino acid?
2. Why must vegetarians eat a variety of types of plant food?
3. Calculate your basic daily requirements for calories and protein.
4. Keep a record of all foods eaten for a day. Using a dietary chart, compute your intake of carbohydrates, proteins, lipids, and total calories.
5. Why should you include nonhydrogenated vegetable oils in your diet?
6. What are the basic functions of mineral elements in the human body?
7. What are the biological functions of each of the following mineral elements?
   a) iron   b) chlorine   c) calcium   d) cobalt   e) iodine
8. What vitamin deficiency is the primary cause of each of the following diseases?
   a) beriberi   b) rickets   c) scurvy   d) pellagra

9. What vitamin is found in each of the following coenzymes?
    a) NAD   b) FAD   c) coenzyme A
10. Examine a carton of milk to find the amount of vitamin D that has been added. Check in a grocery or dairy store to see whether adding vitamin D is widespread.
11. Examine the label on a loaf of bread to determine the amount of vitamins added to the bread. Most breakfast cereals also have a listing of their vitamin content. How do they compare with bread as a source of vitamins?
12. Which is the "sunshine" vitamin? Why is it so named?
13. How do the requirements of certain animals for vitamin C correlate with the generalization (Chapter 39) that animals higher in the evolutionary scale often lack enzymes that simpler animals have?
14. List the vitamins which contain the following elements: N, S, Co.

SUGGESTED READING

Champagnat, A., "Protein from Petroleum," *Scientific American*, Oct. 1965. The author discusses the possibility of using microorganisms that grow on hydrocarbons to synthesize edible proteins.

DeKruif, P., *Hunger Fighters*, Harcourt, New York, 1928. Also available as a paperback book.

Dowling, J. E., "Night Blindness," *Scientific American*, Oct. 1966. The role of vitamin A in night blindness is discussed.

Frieden, E., "The Biochemistry of Copper," *Scientific American*, May 1968. The role of this metallic element in the life processes of plants and animals, particularly its role in the function of proteins, is discussed in this article.

Frieden, E., "The Chemical Elements of Life," *Scientific American*, July 1972. An excellent discussion of the functions of mineral elements in life processes, particularly interesting is its report on the elements fluorine, silicon, tin, and vanadium, which have recently been shown to be essential for normal growth of some animals.

Harpstead, D. D., "High-Lysine Corn," *Scientific American*, Aug. 1971 (Offprint #1229). This article tells of the breeding of strains of corn that contain this essential amino acid, so important in world nutrition.

Hubbard, Ruth, and A. Kropf, "Molecular Isomers in Vision," *Scientific American*, June 1967. A well-illustrated discussion of the role of vitamin A in vision.

Lappé, F. M., *Diet for a Small Planet*, Friends of the Earth/Ballantine, New York, 1971. This book discusses world protein problems and gives specific instructions and recipes for making plant protein diets more nourishing.

Loomis, W. F., "Rickets," *Scientific American*, Dec. 1970 (Offprint #1207). The author discusses this disease in terms of lack of sunlight rather than as a vitamin D deficiency.

Meyer, L. H., *Food Chemistry*, Reinhold, New York, 1960. A chemically-oriented book about the foods we eat.

Pirie, N. W., "Orthodox and Unorthodox Methods of Meeting World Food Needs," *Scientific American*, Feb. 1967. A discussion of new sources of foods for our growing population—some of the proposals are exotic and exciting.

Young, V. R., and N. S. Scrimshaw, "The Physiology of Starvation," *Scientific American*, Oct. 1971 (Offprint #1232). The authors discuss what happens biochemically when people are starved. The social implications are also discussed.

THE CHEMISTRY OF HEREDITY

42-1 INTRODUCTION

Since the times of the ancient Greeks, and probably long before, people have been aware of the fact that a young animal resembles its parents. Characteristics regarded as hereditary have included eye color, stature, and even a fiery temper. Occasionally heredity seemed to fail when, for example, normal parents produced an albino offspring. As with many biological theories, those concerning heredity have become more and more detailed in order to explain observations. Underlying the details, however, we can now see a few simple generalizations which help us to explain heredity and to predict what will happen in many cases.

When it was realized that all living things were produced by parents similar to them, and that lice, frogs, and mice were not generated spontaneously from decaying matter, mud, or old rags, theories of heredity became important. Part of the way in which characteristics are transmitted was deduced by the Austrian monk, Gregor Mendel, whose studies of pea plants were the basis for adding to the theory of heredity the concepts of specific unit characteristics which do not blend. Figure 42-1 illustrates one of Mendel's experiments involving tall and dwarf peas. The development of Mendel's theories led to the belief that there were at least a pair of factors called **genes** for each hereditary characteristic. Some characteristics, such as skin color, are now believed to be controlled by more than one pair of genes. For each hereditary characteristic or for each pair of genes, both dominant and recessive varieties are known.

With sweet peas, a plant produced by crossing one having red flowers with one having white flowers produces only red flowers, not pink. Since the new plant presumably has genes for both red flowers and white flowers, we conclude that the gene for red flowers is dominant. If we cross the red-flowering pea plants from mixed parents with each other, their offspring produces red or white flowers, but no pink ones. In the illustration of the tall and dwarf peas (Fig. 42-1) we see that plants were either tall or dwarf, and that tallness is dominant. An individual having two different genes for a single trait is called a *hybrid* or a **heterozygote.** A pure-bred individual, one having a pair of identical genes, is homozygous—or is a **homozygote**—for that trait.

Our present theory about inheritance can be summarized as follows. There is a pair of genes for each hereditary characteristic in most cells of each individual. These genes may be identical, in which case the individual is homozygous for that characteristic, or they

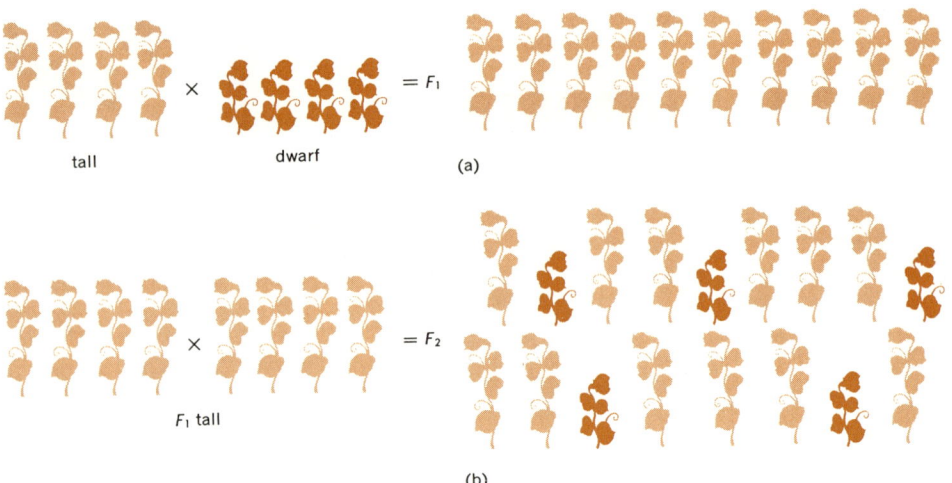

Fig. 42-1 Mendel's experiments with tallness in pea plants. Diagram (a) represents Mendel's first cross of tall and dwarf peas. All the offspring (F_1 generation) were tall. Diagram (b) shows the results of matings between the members of the F_1 generation. When Mendel performed the experiment he found 787 tall plants and 277 dwarf plants. [Adapted from Baker and Allen, *The Study of Biology*, Addison-Wesley, Reading, Mass., 1967.]

may be different, in which case the individual is heterozygous for that characteristic. When an individual forms reproductive cells, whether in the parts of a flower or in the testes or ovary of a mammal, only one member of the pair of genes goes into an individual reproductive cell. (The general name given this reproductive cell is *gamete*.) A homozygote produces only one kind of gametes; a heterozygote, two. When the gametes from a male (sperm cells) meet with and fertilize the gametes (eggs or ova) from a female, various combinations are possible. If the number of offspring of the mating of a single male and a female is large, the proportions predicted by laws of probability are found. In cases where the number of offspring is small, the lack of an adequate sample may result in some deviation from predictions. The situation is analogous to tossing a coin only a few times and having the percentage of "heads" not equal to that of "tails."

An analysis of the hereditary factors involved in the tall and dwarf pea plants of Mendel is given in Fig. 42-2. (We use T to represent the dominant characteristic—tallness—and t to represent the dwarf characteristic.) The male (♂) produces gametes containing either T or t, and so does the female (♀). Chances are 1 in 4 that the offspring will receive TT or tt, and 2 in 4 that the new individual will be heterozygous (Tt). If T is completely dominant, 3 of 4 individuals (those having TT and Tt) will show the dominant characteristic, and only 1 in 4 will show the recessive character. If B represents brown hair color in rats, and b represents an albino condition, the mating of two heterozygotes, both of whom had brown hair, would be expected to produce 25% albino offspring and 75% normal-appearing brown-haired offspring. However, 50% of the total offspring (the heterozygotes) of such matings would be normal-appearing carriers of the albino trait.

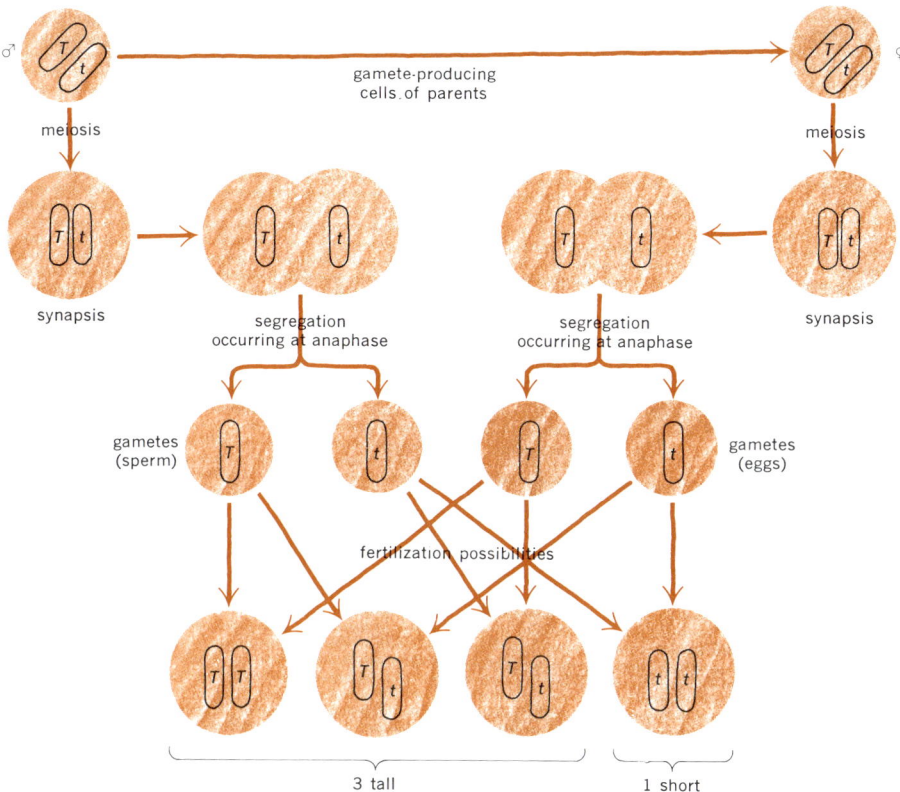

Fig. 42–2 A diagram showing the mechanism of inheritance of tallness in pea plants. T indicates the gene for tallness, and t the gene for the dwarf condition. [Reprinted from Baker and Allen, *The Study of Biology*, Addison-Wesley, Reading, Mass., 1967.]

The theory and examples given above are derived from observation and statistical interpretation. No special instruments or chemical techniques are required for these observations and their interpretation. Other observations about heredity involve the use of the microscope, biochemical information, and specialized techniques.

42–2 GENETICS FROM A CELLULAR LEVEL OF OBSERVATION

Information about the cellular mechanism of heredity is based on microscopic observation of the behavior of individual cells. During part of their life cycle, cells are seen to contain long threadlike bodies called chromosomes in their nuclei (see Fig. 42–3). These chromosomes are normally present in pairs. During normal cell division, each member of the pair of chromosomes splits to form two chromosomes, one of which goes to each daughter cell. In this form of cell division, which is called **mitosis** and represents an asexual form of reproduction of cells, the parent and offspring have exactly the same kind of chromo-

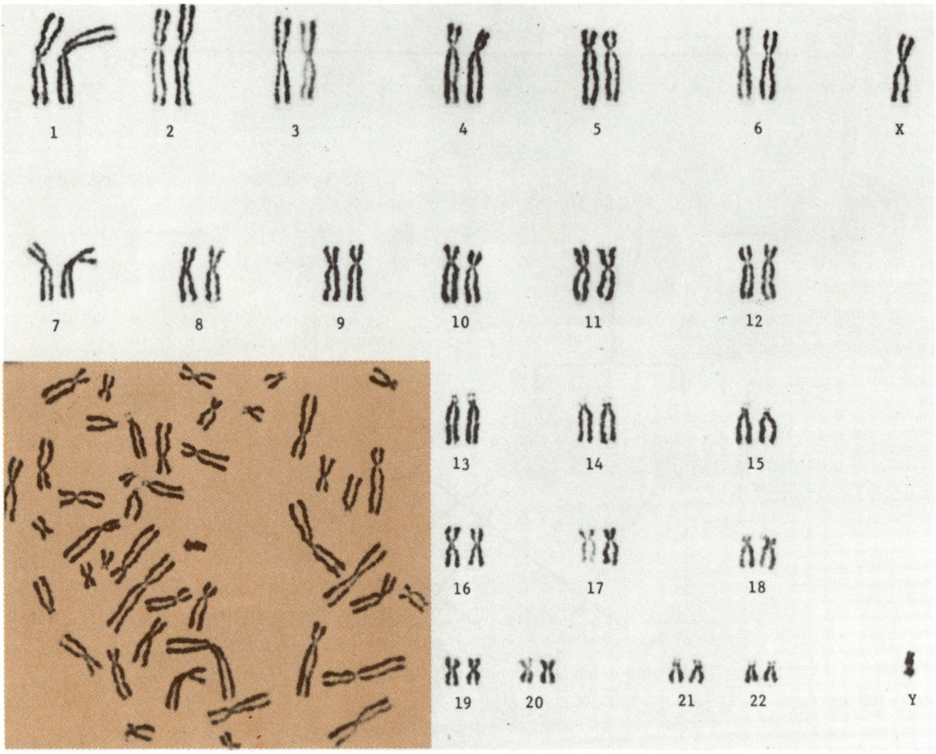

Fig. 42-3 Human chromosomes. When the contents of a human cell are treated in certain procedures and mounted on a microscope slide, we can take a picture such as that shown in the lower left segment shown above. If the individual chromosomes are cut out and arranged in the special form shown in the photograph, we have what is known as a karyotype. The karyotype given here is that of a normal human male. It contains one X and one Y chromosome. Females have two X chromosomes and no Y chromosome in each cell. The cells of some people contain more or less than 46 (23 pairs) of chromosomes. In most cases, these abnormalities in chromosome number have serious consequences. Persons with Downs syndrome, who are commonly called Mongoloid idiots, have three chromosomes 21. [Photograph courtesy of James German, M.D.]

somes. Since the chromosomes contain the hereditary material—the units a geneticist would call genes—each offspring has exactly the same genes as its parents.

In the process of sexual reproduction, **gametes** (sperm and ova) are formed which contain half the number of chromosomes contained in the normal body cells of the parents. This is accomplished by means of **meiosis** (Fig. 42-4), a type of cell division in which only one member of the pair of chromosomes goes to each developing gamete. Thus, on a cellular level, the observations correlate nicely with the theory derived from observation of individual organisms. The genetic information is contained in the chromosomes. The production of a new individual involves first the separation and then the recombination of chromosomes. We postulate that each chromosome contains the units we call genes. While we cannot see the genes, we picture them as being specific areas of a chromosome.

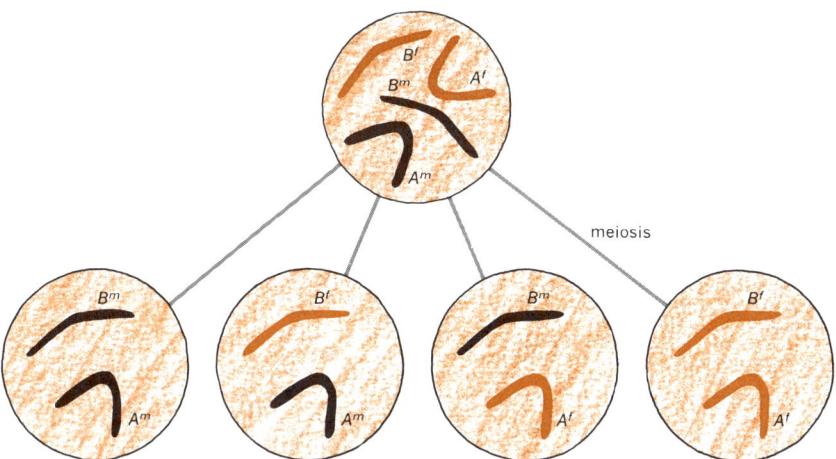

Fig. 42-4 Meiosis. The diagram above represents the random assortment of chromosomes from the mother (A^m and B^m) and the father (A^f and B^f) during meiosis. Note that each of the cells formed contains only half as many chromosomes as the parent cell. [Adapted from John W. Kimball, *Biology*, Addison-Wesley, Reading, Mass., 1965.]

42-3 GENETICS FROM A BIOCHEMICAL VIEWPOINT

The **chromosomes** seen in the microscopic examination of cells are found, upon chemical analysis, to be composed of deoxyribonucleic acid (DNA) and protein. While either the protein or DNA could be the chemical carrier of hereditary information, most evidence supports the fact that DNA is the hereditary substance. Some of the evidence for this conclusion is given below.

1) While the amounts of all other constituents of a cell vary widely, the amount of DNA per cell is generally the same in all cells of all animals in any given species.

2) The amount of DNA per cell in different species is different.

3) When DNA from certain bacteria having one characteristic is added to similar organisms lacking the characteristic, the characteristic is acquired by the deficient organism. For example, resistance to certain antibiotics can be transferred from a culture of bacteria that *is* resistant to another culture of the same organism that is *not* resistant merely by adding DNA of the resistant culture to the growth medium of the nonresistant bacteria. This type of transfer of DNA from one organism to another appears to occur only in microorganisms.

4) Viruses, which can be said to alter the heredity of the cells they attack, are known to inject only DNA (in some cases RNA), not protein, into the cell they infect.

When the structure of DNA was shown to be a double stranded helix which had the capability of replicating itelf (see Chapter 36), a mechanism for the duplication of the chromosomes was suggested. Whether each individual strand represents different information or whether both of them carry the same information, one in reverse order to the other, has not been established.

DNA is believed to be the basic hereditary substance—the molecules that are transferred from parent to offspring. It is found primarily in the nucleus of the cell. However, the major synthetic work of the cell is done outside the nucleus, in the cytoplasm. Thus the transmission of the information contained in DNA requires a messenger. This function is fulfilled by the kind of RNA called messenger RNA (see Chapter 39). DNA controls the synthesis of messenger RNA to transfer the information to the cytoplasm of the cell concerning types of protein to be synthesized. At the ribosomes where protein synthesis occurs, the message of DNA is actually put into action. The theory assumes that all hereditary characteristics are somehow dependent on the synthesis of a specific kind of protein. Thus DNA has the specifications for synthesis of the proteins of muscle, hemoglobin, the protein hormones, and most significantly, the enzymes. Since biochemical reactions do not proceed at any appreciable rate without enzymatic catalysis, the kind of enzymes present in a cell probably controls the general course of metabolism within that cell. For example, hereditary differences between species with respect to the kind of waste products made from amino acids (which was described in Chapter 39) are dependent not on the presence of the amino acids but on the presence of specific enzymes.

Since the kind and amount of DNA in any cell of an organism is identical with that of every other cell in the organism, we are faced with a major problem in explaining differences in cell function. DNA having the information for making the enzymes necessary for urea synthesis is in every cell of the human body, yet only liver cells actually carry out the synthesis of urea. If we compare the total information in each cell to a large book of instructions, the question becomes, "How does each cell know just which page of instructions to follow?" Answers to this question are now being proposed, and they involve a complex group of molecules called repressors and de-repressors. (See the article by Changeux in the list of references at the end of this chapter for a discussion of this theory.) Current research supports this theory.

We can summarize the biochemical or molecular interpretation of heredity as follows: Males produce sperm cells which consist largely of a head which contains DNA surrounded by protein and a long tail-like structure called a flagellum. Egg cells contain lipids and proteins, in addition to DNA (see Fig. 42–5). When a sperm cell fertilizes an egg cell, a chromosome from the sperm cell pairs up with a similar chromosome from the egg cell. In order for these chromosomes to form matched pairs, the male and female must have the same kind of chromosomes—that is, be of the same species.

At the moment of fertilization (conception), the zygote, or fertilized egg cell, has all the hereditary information it will ever receive. As the zygote divides to form 2, 4, 8, 16 cells, and finally the millions or trillions of cells found in the new individual, some cells begin to assume different functions in a process appropriately known as differentiation. Differentiation is believed to be the result of directions from repressors and de-repressors which control exactly which molecules will be synthesized of the many thousand possible kinds for which instructions are available in the DNA. As the organism grows and matures, it will produce sperm or egg cells containing only the kinds of DNA (genes) that it received from its parents. Occasionally some agent, perhaps some form of radiation, changes the DNA in a gamete, and this gamete produces a mutant organism having different hereditary characteristics. Such mutations are rare. The probability of mutations is expressed as mutations per gene per generation. In humans this probability is between 1 in 100,000 and 1 in 1,000,000 with different genes having different rates of mutation.

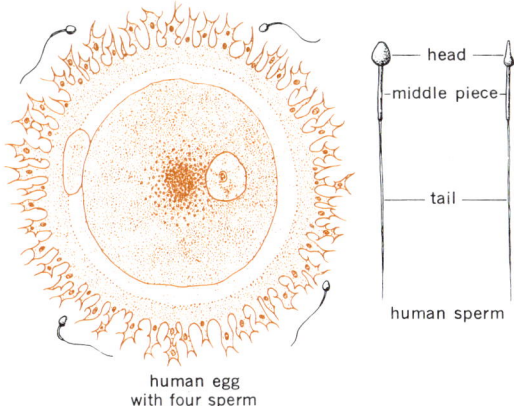

Fig. 42–5 A human egg with four sperm cells. This drawing shows the relative size of human sperm and egg cells. In the process of fertilization one sperm cell penetrates the egg. [From Baker and Allen, *The Study of Biology*, Addison-Wesley, Reading, Mass., 1967.]

Thus the long-range preservation of a species is assured by the transfer of potentially "immortal" DNA from parent to offspring. Sufficient variety is introduced by formation of hybrid individuals and by mutations. Most mutants are probably less able to survive than are the normal parents, but on extremely rare occasions a mutant is able to do things its parents could not do and in the struggle for survival, the mutant wins and becomes the first of a new kind of individual. A change in environmental conditions may help a mutant which is in competition with unmutated individuals.

The development of our understanding of the process of heredity is a striking example of the way in which science functions. Earlier civilizations could talk about heredity only in terms of individuals, and could offer no real explanation of the processes involved. This general observation was followed by refinement to the cellular and finally the biochemical level of observation and explanation. Our present way of looking at and thinking about heredity has had two major consequences. First, we have, of course, a better understanding of just how the process works. While new knowledge is one of the motives for scientific research, the second major consequence of our progress is that we can predict, and in certain cases control, some of the effects of heredity. Control always brings problems, but the benefits are so great that we must find solutions to these problems. Some examples of biochemical understanding and control of hereditary abnormalities are given in the remainder of this chapter.

42–4 INBORN ERRORS OF METABOLISM

In 1906, the English physician Garrod wrote a book describing several abnormal patterns of excretion. He recognized that these conditions were inherited, and called them inborn errors of metabolism. We now know more than a hundred such conditions. The following

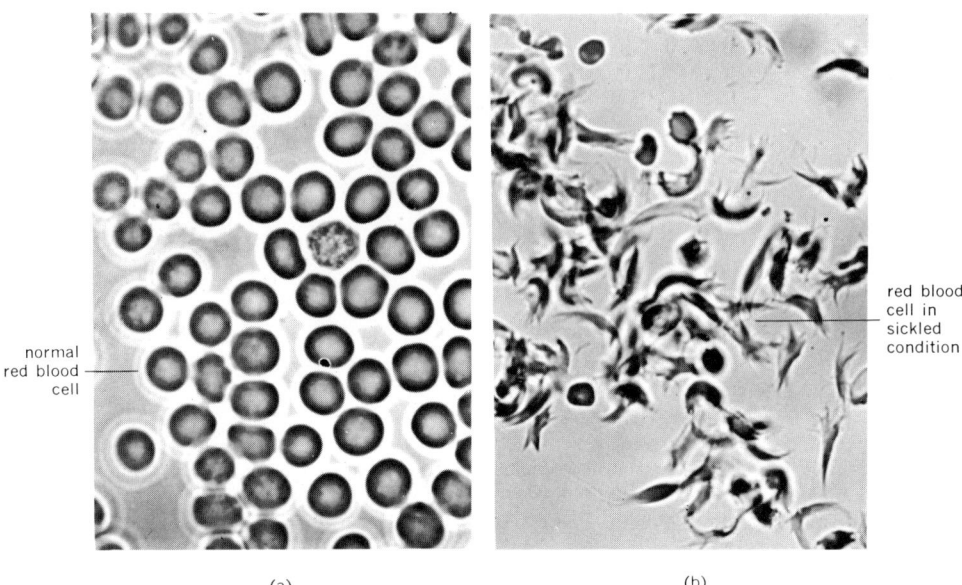

Fig. 42–6 Normal red blood cells and sickle cells: (a) normal human red blood cells; (b) a similar preparation that was made of the red blood cells of a patient with sickle cell anemia. The sickle-like form of the blood cells in (b) gives the disease its name. [From Baker and Allen, *The Study of Biology*, Addison-Wesley, Reading, Mass., 1967. Photograph courtesy of Dr. A. C. Allison.]

sections discuss abnormalities in protein structure and enzyme formation that are known to be hereditary.

Recently techniques have been developed to test whether a developing human embryo has certain types of inborn errors. This technique, which is discussed in the article by Friedmann listed at the end of the chapter, can lead to a decision for an abortion. Such a decision poses serious moral questions. Other research involves the possibility of "genetic engineering" by attempts to replace inadequate DNA with a type that is better by the device of "infecting" an individual with viruses carrying the DNA for the better genes. Other types of genetic engineering are also being proposed. There are tremendous technical difficulties involved, but the processes are probably not impossible.

Sickle Cell Anemia

More biochemical information is available about sickle cell anemia than about any other inherited condition. The correlations of chemistry, physiology, and genetics are amazing. In fact, studies of sickle cell anemia were responsible for some of our current theories about the chemical nature of heredity.

If we take blood samples from persons with this form of anemia, we find that the red blood cells, when placed under conditions where the oxygen concentration is low, form peculiar sickle shapes (see Fig. 42–6). We find that some of the red blood cells of certain other apparently healthy persons will also sickle when similarly treated. If we study the hemoglobin from these persons we find that they have two different kinds. When the mixture is analyzed by electrophoresis at pH 6.9, normal hemoglobin migrates in one direction and the abnormal hemoglobin—called sickle cell hemoglobin—migrates in the other direction. Persons with sickle cell anemia, a serious condition, have only sickle cell hemoglobin (abbreviated Hgb S), and those who have two kinds of hemoglobin have both Hgb S and normal hemoglobin (abbreviated Hgb A). These latter persons are characterized as having the sickle cell trait.

The tendency to form sickle cell hemoglobin is hereditary. Those having sickle cell anemia are homozygous, and those with the sickle cell trait are heterozygous. Thus we have a hereditary condition that is directly related to the kind of protein that an individual makes.

Studies of Hgb A and Hgb S show that they have different solubilities in weak salt solutions (similar to the salt solution of the blood). When carrying oxygen, both forms of hemoglobin are about equally soluble. However, when the oxygen is removed (a process that occurs as blood cells pass from the lungs to other parts of the body) Hgb S is only 1/50th as soluble as Hgb A. Apparently this decreased solubility is the cause of the formation of the "sickles." Hgb S actually crystallizes out of solution, and the crystals then either stretch or break the membrane of the red blood cells. Such sickled cells cannot perform the function of normal red blood cells, and the person with sickle cell anemia must either manufacture red blood cells at a faster rate or suffer from the effects of a deficiency in red blood cells. The sickled cells also tend to block the small blood vessels, the capillaries, and this failure of blood circulation is the apparent cause of the pain these people suffer. Thus sickle cell anemia is a disease caused by a difference in solubility of an abnormal protein. This disease is one where the physical or chemical cause is known.

The next questions are: "Why is Hgb S less soluble than Hgb A, and why do the two forms migrate differently in an electrical field?" The answer to both questions is the same, and is found in the chemical structure of the protein hemoglobin. This protein contains two kinds of subunits, with two of each kind combined to give the hemoglobin molecule (Fig. 34–6). In one of the chains there is a glutamic acid residue in Hgb A and a valine in Hgb S. Glutamic acid has the side chain $-CH_2-CH_2-COO^-$ and valine has $-CH-(CH_3)_2$. This difference might seem to be minor, since both have three carbons in the side chain. The effect, however, is due to the charge on the carboxyl group of glutamic acid which makes that group much more soluble in water than the hydrocarbon-like valine. The fact that one has a negative charge and the other no charge also explains the difference in electrophoretic mobility. Thus the difference on only one amino acid in a chain of about 150 amino acids can result in a life of weakness, pain, intermittent illness, and the constant threat of death from those infections that a normal person handles easily.

Why do some people make Hgb S and others only Hgb A? Since this condition is a hereditary trait, we assume that one person produces DNA that has the pattern for making messenger RNA which makes normal hemoglobin (Hgb A), and that the other has DNA that leads to the production of Hgb S. How different are the "codes" for glutamic acid and valine? From Table 39–1 we find that the codes for glutamic acid are GAA and GAG, and two of the possible codes for valine are GUA and GUG. The difference between pairs is only in one nucleotide. It is logical to assume that a single mutation may be responsible for the change. Why has this mutation persisted? The answer is also fascinating.

Since persons with sickle cell anemia often do not live to an age where they can have children, why hasn't the condition died out? Why hasn't the evolutionary principle of survival of the fittest eliminated persons with the gene for sickle cell anemia? The answer is related to the disease malaria. While persons with sickle cell anemia (homozygotes) die young and are eliminated, those with the sickle cell trait (heterozygotes with both Hgb S and Hgb A) have an increased resistance to malaria. The parasite that causes malaria lives in red blood cells, and apparently it can't find as good a place to live and reproduce if its host's red blood cells contain half Hgb S. The gene for Hgb S is found almost totally among blacks. Approximately 8% of American blacks have some Hgb S. The occurrence of this abnormal hemoglobin is as high as 45% in some African tribes—which may lead us to ask whether we can really call this an "abnormal" hemoglobin.

Thus the story of sickle cell anemia and the sickle cell trait relates physics and chemistry (the solubility and electrophoretic behavior of Hgb S) to physiology (anemia and decreased supply of oxygen to the tissues) to genetics where it shows how recessive traits may be present but not obvious. This line of reasoning supports the theory that genes influence protein synthesis. The sickle cell anemia story further shows how a normally deleterious condition can persist in a given environment, and we have a picture of natural selection (evolution) in action. The difference in incidence of the sickle cell trait in Africa and America indicates that, in the absence of malaria, the trait is less common.

In no other instance are all aspects of a condition so well known and the interrelations so well established. Unfortunately, in the case of sickle cell anemia, understanding does not lead to curing the condition. Some medications seem to increase the survival rate of persons with sickle cell anemia, but in the long run, the only hope seems to lie in preventing the birth of children doomed to lives of suffering and to early death from sickle cell anemia. Control will require widespread examinations of the hemoglobin of prospective parents followed by genetic counselling of individuals who have this trait, so that they will not want to bring afflicted children into the world. One of the aims of science in general, and biochemistry in particular, is to reach a similar understanding of all diseases.

Hereditary Abnormalities in Phenylalanine Metabolism

The metabolism of the amino acid phenylalanine is summarized in Fig. 42–7. This summary indicates that phenylalanine from proteins in the diet may be metabolized in a variety of ways. It may be resynthesized into protein; it may be transformed into the thyroid hormone thyroxine, into the adrenal hormone epinephrine (adrenalin), or into the basic pigment of human hair and skin, melanin; it may also be oxidized completely into carbon dioxide and water with the release of an appreciable amount of energy. Tyrosine, which can be derived from phenylalanine, follows the same general pathways except that it cannot be converted to phenylalanine. In terms of nutrition, phenylalanine is an essential amino acid; tyrosine is not.

The understanding of the metabolism of phenylalanine assumed a new importance when it was observed that a certain type of mental defectives—particularly idiot children —excreted an unusual substance in their urine. This compound was identified as phenylpyruvic acid. It was found that about 1% of the feeble-minded persons in institutions excreted this abnormal metabolite. The questions immediately suggested are: "Does this excretion have anything to do with the feeblemindedness?" and "If so, can an understanding of the condition lead to a cure?" Further observation and study indicated that the persons who excreted phenylpyruvic acid (they are called phenylketonurics) often came from marriages of cousins. This type of observation is a strong indication that the disease may be associated with the inheritance of recessive genes.

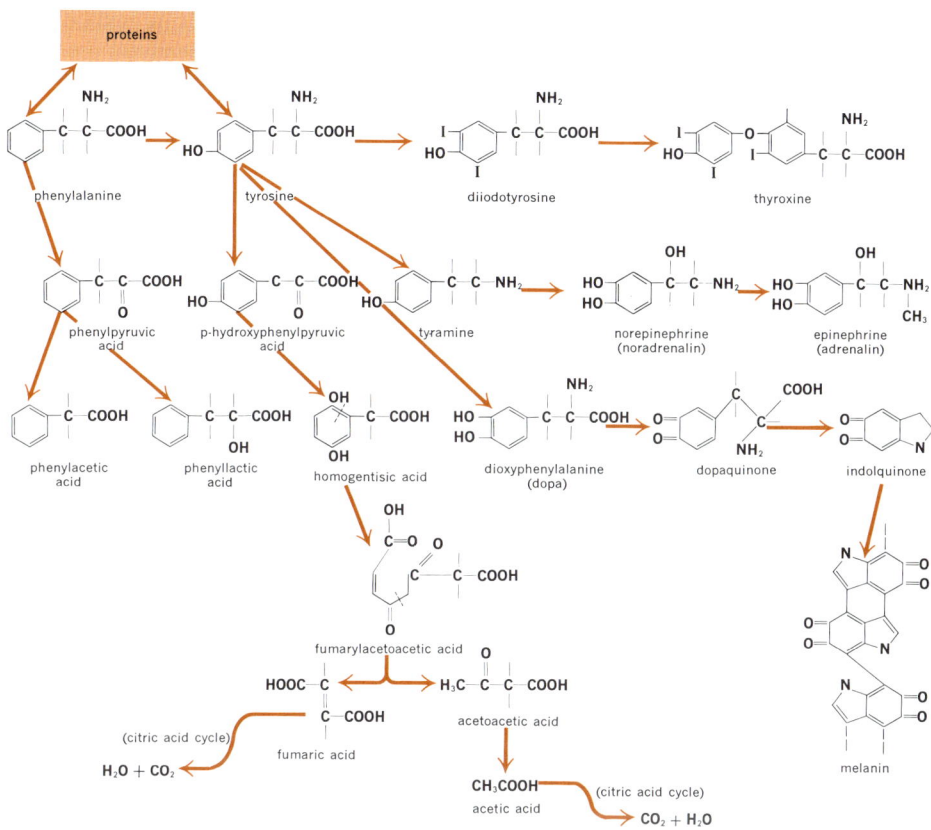

Fig. 42-7 The metabolism of phenylalanine. The series of reactions shown above represents some of the possible products and intermediates that can be formed from the amino acid phenylalanine. The consequences of failure to produce some of these reactions are discussed in the accompanying text.

We now know that the condition is inherited and that the reason phenylketonurics excrete this abnormal compound is that they cannot convert phenylalanine to tyrosine. When this reaction is blocked, the phenylalanine which is not used for protein synthesis has only one pathway left—the conversion to phenylpyruvic, phenyllactic, and phenylacetic acids. Why is the conversion of phenylalanine to tyrosine blocked? It is blocked because the phenylketonuric does not make the necessary enzyme for conversion. In genetic terms, the DNA of the phenylketonuric lacks the information required for synthesis of phenylalanine oxidase. Whether he makes a similar protein that is just different enough to lack catalytic activity is not known, although our theory suggests that he does.

Since most parents of phenylketonurics have essentially normal intelligence and do not excrete phenylpyruvic acid, they must be heterozygotes having one gene (or type of DNA) that has the information for synthesis of phenylalanine oxidase and one that doesn't. The next question that presents itself is: "Can we find these hybrid carriers of the defect?" Previously, geneticists believed that the only way a recessive characteristic could be detected in a heterozygote was by letting the individual produce offspring and examining them for the trait. This is a tragic procedure when human lives are involved. Reasoning that perhaps the heterozygous carriers of the recessive gene for phenylketonuria (abbreviated as PKU) might have less than normal amounts of phenylalanine oxidase, research scientists tested parents of phenylketonuric children. While the test has not been entirely successful, it does show that these heterozygous carriers excrete larger amounts of phenylpyruvic acid when given a large dose of phenylalanine than do normal persons. Thus the understanding of the metabolism of phenylalanine and the biochemistry of a hereditary defect has led to the development of a test for a hidden recessive characteristic. Results of such tests can be used for counselling individuals related to phenylketonurics as to the possibility of their having children with this disorder. If both a husband and wife are given such tests, the probability of having children with this genetic defect can be predicted. Procedures of this sort offer great promise in improving the genetic qualities of humans. Control would also pose a great threat in the hands of unscientific or inhumane persons.

The prediction of the possibility of having phenylketonuric children may be an interesting development but what about the baby born with the disease? Is this infant doomed to develop no further than the level of an idiot or imbecile? Fortunately the understanding of the biochemistry of the defect has led to an effective treatment—not to prevent the defect, but to help children who have the disease.

One treatment that might be proposed would be to give the phenylketonuric child some of the enzyme he lacks. Since enzymes are digested into their component amino acids in the gastrointestinal tract, administration would presumably have to be made by injection. This technique works for the administration of insulin to diabetics, but would pose two more problems in the case of phenylketonuria. The first would be that of getting the large enzyme molecule from the blood into the cells where it must work. The second problem is even more formidable—the human body generally responds to the injection of a foreign protein by forming antibodies to the protein. These antibodies neutralize any effect of the protein.

Fortunately, an alternate procedure can be used, and, in fact, is used successfully. This treatment consists of restricting the amount of phenylalanine in the diet of the phenylketonuric. It has been found that dietary restriction of this sort allows the infant with PKU to develop normally and that after the age of five or six, perhaps even earlier, the diet may be discontinued without bad effects. Thus the condition, while it can never be cured, can be treated.

The problem of phenylketonuria now becomes one of discovering the babies who are potential phenylketonurics. Most states now require that the urine, or better yet, the blood, of all newborn babies be tested for the presence of phenylpyruvic acid or phenylalanine. The blood tests require only a small sample of blood and are probably best, since the concentration of phenylalanine in the blood increases some time before urinary excretion of phenylpyruvic acid is readily detectable. Thus treatment can be initiated at the earliest possible moment. There is now no reason for children with this defect to grow up facing life as mental defectives.

The question, "Why does an excess of phenylalanine or phenylpyruvic acid in the blood cause mental defects?" has not yet been definitely answered. It is probable that either phenylpyruvic acid or phenylalanine interferes with the synthesis of serotonin—a compound involved in brain function.

Phenylketonuria is not the only hereditary defect related to the metabolism of phenylalanine. Albino individuals lack the ability to synthesize melanin. Apparently their systems lack the information for making one of the enzymes for the conversion of tyrosine to melanin. Since this condition represents the lack of a metabolic product, melanin, and not an excess of a metabolite (as in PKU), there is no known treatment for albinism. If tests for detection of normal carriers of the recessive gene are developed, control of the incidence, if not treatment, might become possible. The consequences of albinism, while serious, are not nearly so tragic as those of PKU.

Lack of an enzyme responsible for oxidizing homogentisic acid leads to the excretion of this product in the urine. This condition is hereditary and presents another inborn error in the metabolism of phenylalanine and tyrosine. The condition is usually not serious, and no treatment is given. Another abnormality, the excretion of p-hydroxyphenylpyruvic acid, is traced to a blockage in the oxidation of tyrosine. The incidence of this condition apparently is rare, and it has few, if any, bad consequences.

The metabolism of phenylalanine was chosen to illustrate inborn errors of metabolism. Although errors are known in the metabolism of other substances, more have been found in the metabolism of phenylalanine than in that of other substances. This fact may mean that there really are more such errors in phenylalanine metabolism, but it probably means only that the abnormal metabolic products of phenylalanine tend to be obvious because of their distinctive odors or color.

42-5 THE ONE-GENE-ONE-ENZYME THEORY

The principles of hereditary control of enzyme formation which are so well illustrated by the metabolism of phenylalanine and its abnormalities were first formulated on the basis of work with the red bread mold, *Neurospora*. From their studies with this organism, Doctors Beadle and Tatum formulated the *one-gene–one-enzyme theory* which should perhaps now be characterized as the *one-gene–one protein theory*. A good summary of this theory, which has led to a much better understanding of heredity, was given by Dr. E. L. Tatum in his Nobel Prize Lecture of 1958. He stated that:

1) All biochemical processes in all organisms are under genic control.
2) These overall biochemical processes are resolvable into a series of individual, stepwise reactions.
3) Each single reaction is controlled in a primary fashion by a single gene, or, in other terms, in every case a 1:1 correspondence of gene and biochemical reaction exists, such that:
4) Mutation of a single gene results only in an alteration in the ability of the cell to carry out a single primary chemical reaction. . . . As has been repeatedly stated, the underlying hypothesis, which in a number of cases has been supported by direct experimental evidence, is that each gene controls the production, function, and specificity of a particular enzyme.*

The understanding of any chemical process leads to the question of control. If we understand the mechanism of heredity, can we then control it? The answer to this question is that we cannot now control the basic biochemical processes involved in

---

* E. L. Tatum, "A Case History in Biological Research," *Science* **129**:1711–1715 (26 June 1959), copyright by the American Association for the Advancement of Science.

heredity. On the basis of biochemical understanding and perhaps some chemical analyses, we can only counsel persons about the possibilities of producing offspring of a definite type. Recent research by Kornberg and his associates (Section 36–4) may be the first of many steps that may lead to some kind of chemical control of heredity.

The other problem in genetic control is a social and moral one. Presuming that we learn how to control the biochemistry of heredity, should we? If controls are to be instituted, who will initiate them? These problems are formidable, and are beyond the scope of this book. They cannot and must not be discussed without a basic understanding of the biochemical, as well as cellular and organismic basis of heredity.

IMPORTANT TERMS

Chromosome	Heterozygote	Mitosis
Gamete	Homozygote	Mutation
Gene	Inborn error of metabolism	Ribosome
Genetic code	Meiosis	

WORK EXERCISES

1. Explain what is meant by the one-gene–one-protein theory. What kind of evidence supports this theory?
2. List all examples of inborn errors of metabolism that have been discussed in this text. List the specific protein missing or altered in each case.
3. What might be the consequences of an inborn error of metabolism that prevented the synthesis of thyroxine from tyrosine?
4. Differences in chemical structure in several kinds of hemoglobin are listed in Table 42–1. Check the genetic code as given in Table 39–1 to determine the differences in the code for the normal and abnormal amino acids in these structures.

TABLE 42–1 A summary of amino acid differences in some hemoglobin variants

Type of Hemoglobin	Position Number in B-Chain					
	6	7	26	63	67	121
normal	Glu	Glu	Glu	His	Val	Glu
Hgb S (sickle cell)	Val					
Hgb C	Lys					
Hgb G San José		Gly				
Hgb E			Lys			
Hgb M Saskatoon				Tyr		
Hgb M Milwaukee					Glu	
Hgb D Punjab						Glu·NH$_2$
Hgb Zurich				Arg		
Hgb O Arabia						Lys

SUGGESTED READING

Beadle, G. W., "The Genes of Men and Molds," *Scientific American*, Sept. 1948 (Offprint #1). Dr. Beadle discusses some of the experiments that led to the one-gene–one-enzyme theory and the Nobel Prize.

Beermann, W., and U. Clever, "Chromosome Puffs," *Scientific American*, Apr. 1964 (Offprint #180). This article tells the fascinating story of the relationship of puffs in chromosomes and of new protein synthesis.

Cairns, J., "The Bacterial Chromosome," *Scientific American*, Jan. 1966. A discussion of the shape and methods of replication of DNA in some bacteria.

Changeux, J.-P., "The Control of Biochemical Reactions," *Scientific American*, Apr. 1965 (Offprint #1008). The control of enzyme activity and production by feedback is discussed.

Dobzhansky, T., *Heredity and the Nature of Man*, Signet Science Library Book (P2837), New American Library, New York, 1966. This paperback book and the article below give some of the views of this important geneticist and humanitarian.

Dobzhansky, T., "The Genetic Basis of Evolution," *Scientific American*, Jan. 1954 (Offprint #6).

Friedmann, T., "Prenatal Diagnosis of Genetic Disease," *Scientific American*, Nov. 1971 (Offprint #1234). A discussion of techniques used to determine, even before birth, whether a child will have a genetic defect. The social implications are discussed.

Hanawalt, P. C., and R. H. Hayes, "The Repair of DNA," *Scientific American*, Feb. 1967. A description of the ways bacteria can repair damage to their DNA.

Horowitz, N. H., "The Gene," *Scientific American*, Oct. 1956 (Offprint #17). A discussion of this basic unit of heredity that no one has ever seen—yet.

Hotchkiss, R. D., and Esther Weiss, "Transformed Bacteria," *Scientific American*, Nov. 1956 (Offprint #18). A report on transfers of DNA from one bacterium to another.

Ingram, V. M., "How Do Genes Act?" *Scientific American*, Jan. 1958 (Offprint #104). A discussion of the genetics of sickle-cell anemia.

Jacob, F., and E. L. Wollman, "Viruses and Genes," *Scientific American*, June 1961 (Offprint #89). This article tells about the biochemistry of viral infection of cells. Significant both for the study of heredity and of infection.

Kellenberger, E., "The Genetic Control of the Shape of a Virus," *Scientific American*, Dec. 1966. The article discusses the relationship of heredity to shape of virus particles.

McKusick, V. A., "The Royal Hemophilia," *Scientific American*, Aug. 1965. This article traces an inherited trait back to Queen Victoria of England.

McKusick, V. A., "The Mapping of Human Chromosomes," *Scientific American*, April 1971 (Offprint #1220). The article is complex but understandably written by an expert in human heredity.

Muller, H. J., "Radiation and Human Mutation," *Scientific American*, Nov. 1955 (Offprint #29). This and the following article by Puck treat radiation effects from the organismic and cellular levels, respectively.

Puck, T. T., "Radiation and the Human Cell," *Scientific American*, Apr. 1960 (Offprint #71).

Zuckerkandle, E., "The Evolution of Hemoglobin," *Scientific American*, May 1965 (Offprint #1012). The author uses the structures of hemoglobin in various species to trace evolution.

BIOCHEMISTRY OF PLANTS   43

43–1 INTRODUCTION

Most people think of plants as being distinctly different from animals; the differences are obvious, of course, if we compare a horse with a tree. But those who are familiar with the sea anemone, which is an animal that looks like a flower, or with microscopic plants and animals realize that the differences may not be so distinct. Some one-celled organisms such as the euglena, which moves with the aid of a whiplike flagellum but contains chlorophyll, have characteristics of both plants and animals. Assignment of these organisms to either classification is sometimes arbitrary.

Biochemically, plants and animals are similar. Both contain the same saccharide units, amino acids, and lipid constituents. These basic units are sometimes assembled differently in plants than they are in animals, however. The starch of plants is somewhat different from glycogen of animals, and the beta-linked cellulose of plants is significantly different from both starch and glycogen even though all are made only of glucose units. Plant lipids tend to be more unsaturated than animal lipids because they contain more of the highly unsaturated fatty acids.

The metabolism of plants is also highly similar to that of animals. The critic acid cycle is common to both plants and animals as are many other reactions. In general, the green plants have greater synthetic capability than do the colorless plants such as bacteria, yeasts, and other fungi. In those cases which have been carefully investigated, as with the mold *Neurospora*, the fungi seem to have a greater ability to synthesize compounds than do animals. There are, however, some bacteria whose dependence upon preformed nutrients is even greater than that of animals.

A major difference between a self-sufficient green plant and a parasitic fungus is in the presence of **chlorophyll** (Fig. 43–1). This green pigment is sometimes obscured by other pigments, but its presence in a plant indicates that the plant can synthesize its own components if it has access to the energy of the sun and elements or simple compounds for building blocks. This synthesis under the influence of light—photosynthesis—is the distinguishing characteristic of plant life, and probably represents the most important series of reactions that occur on the surface of the earth. The general equation for photosynthesis may be written as:

$$6CO_2 + 6H_2O \xrightarrow{\text{light}} C_6H_{12}O_6 + 6O_2$$

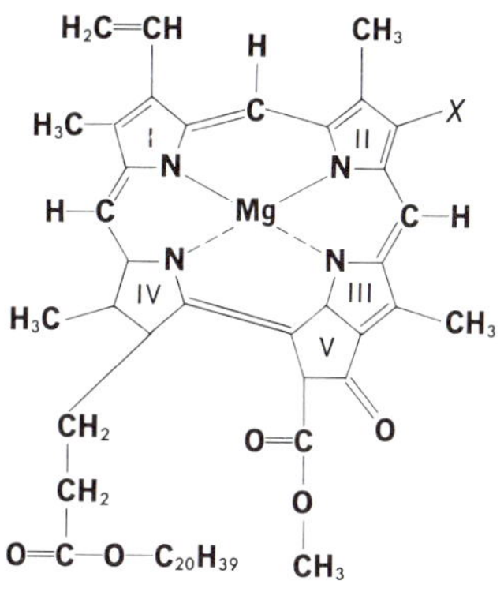

Fig. 43-1 The chlorophyll molecule. In addition to the four pyrrolidine rings, there is a fifth ring in chlorophyll. At position X, chlorophyll a has an ethyl group, and chlorophyll b has a CHO group. The chlorophyll molecule may be compared with similar structures in cytochrome c (Fig. 37-9) and vitamin B_{12} (Fig. 41-1).

43-2 PHOTOSYNTHESIS

Green plants, when given a source of light, CO_2, H_2O, and the necessary catalysts, can produce carbohydrates. If a source of nitrogen is added, amino acids are made. The addition of sulfate and phosphate for the sulfur-containing amino acids and the many phosphorus-containing building blocks and catalysts makes possible the synthesis of almost all major components of the plant. Mineral elements such as iron, potassium, magnesium, and many others are also required by plants either as minor structural constituents or as catalysts.

The details of the processes by which plants synthesize these many compounds, particularly of photosynthesis, were not understood until about twenty-five years ago. Only when the radioactive isotope carbon-14 became available could scientists investigate the pathway by which carbon dioxide becomes incorporated in carbohydrates. Before research into the way in which a photon of light is trapped and used, techniques had to be developed for studying extremely fast reactions, since these processes occur in milliseconds or microseconds.

For simplicity, the complex process of photosynthesis is divided into those processes that require light—called the **light reaction**—and those that can proceed in the absence of light—the **dark reaction**. Both reactions are really series of reactions, and while they are separated for purposes of discussion we must always remember that they occur at the same time and use some of the same intermediate compounds, whenever a living green plant is exposed to sunlight.

The Light Reaction

We can think of photosynthesis as starting when a quantum of light is absorbed by chlorophyll in a green plant. The energy absorbed by chlorophyll causes one of the chlorophyll electrons to become excited; it contains more energy than other electrons. This high-energy electron is then transferred to other molecules much as in the process of electron transport (Section 37-5). The electron loses some of its energy at each step. This energy is stored in the compounds ATP and reduced NADP ($NADPH_2$), which are used in the dark reaction. Some of the energy absorbed by a plant is also used to break water into oxygen, electrons and protons. This **photolysis** of water is the source of the molecular oxygen which is one of the products of photosynthesis. Photolysis provides the electrons to replace the excited electrons lost by chlorophyll and the protons that are bound to $NADPH_2$, which is the energy source in the fixation of CO_2 by the green plant.

$$2H_2O \begin{cases} \rightarrow O_2 \; [\overset{..}{O}::\overset{..}{O}] \\ \rightarrow 4H^+ \\ \rightarrow 4 \text{ electrons } [::] \end{cases}$$

$$2[H:\overset{..}{O}:H]$$

The processes in which radiation is absorbed occur very rapidly—probably in 10^{-15} to 10^{-9} sec. We do not have good methods for studying such reactions, and consequently we know very little about the first stages of photosynthesis. Nor do we know much about the mechanics of the photolysis of water. Biochemists and biophysicists are currently doing research on the mechanism by which the energy of the chlorophyll is trapped by compounds similar to NAD, FAD, and the cytochromes of the electron transport system.

The Dark Reaction

Those processes of photosynthesis which can occur in the dark, provided sources of energy are available, are reasonably well understood. They involve the conversion of CO_2 to carbohydrates and other compounds. They are catalyzed by enzymes, and although they occur rapidly, methods have been developed to study this conversion of gaseous CO_2 to nongaseous compounds. The process is given the general name of **fixation**.

Many compounds are present in a green leaf or in a single-celled green alga. If we want to study the process by which CO_2 is converted into these compounds we must look for the compound which first reacts with the CO_2. To do so we add radioactive CO_2—carbon dioxide containing carbon-14, symbolized $^{14}CO_2$—to a container of thousands of algae, separate the various compounds found in the algae, and then determine which ones contain carbon-14. Since such experiments always give many compounds containing carbon-14, the reactions of photosynthesis must be stopped a short time after the addition of the radioactive carbon dioxide. This can be done by dropping the culture of algae into boiling methanol.

Using this technique, research chemists have found that only 90 seconds after addition of the $^{14}CO_2$, 90% of the CO_2 had been fixed and at least 14 radioactive compounds were present. When the time interval was cut to 5 seconds, only 5 compounds were found to be radioactive—malic acid, aspartic acid, pyruvic acid, 3-phosphoglyceric acid, and 2-phosphoglyceric acid; 65% of the radioactivity was in the latter two compounds. These experiments indicated that there was a 2-carbon compound in the green plant that was reacting with CO_2 to give either the 2- or 3-phosphoglyceric acid. The search for this 2-carbon compound, however, proved fruitless. Instead, a 5-carbon compound, ribulose diphosphate, was found which first combined with CO_2 and then immediately split to give two molecules of a phosphorylated glyceric acid. The problem of working out the details of the dark reaction of photosynthesis then became one of showing how both this 5-carbon compound, ribulose diphosphate, and the normal

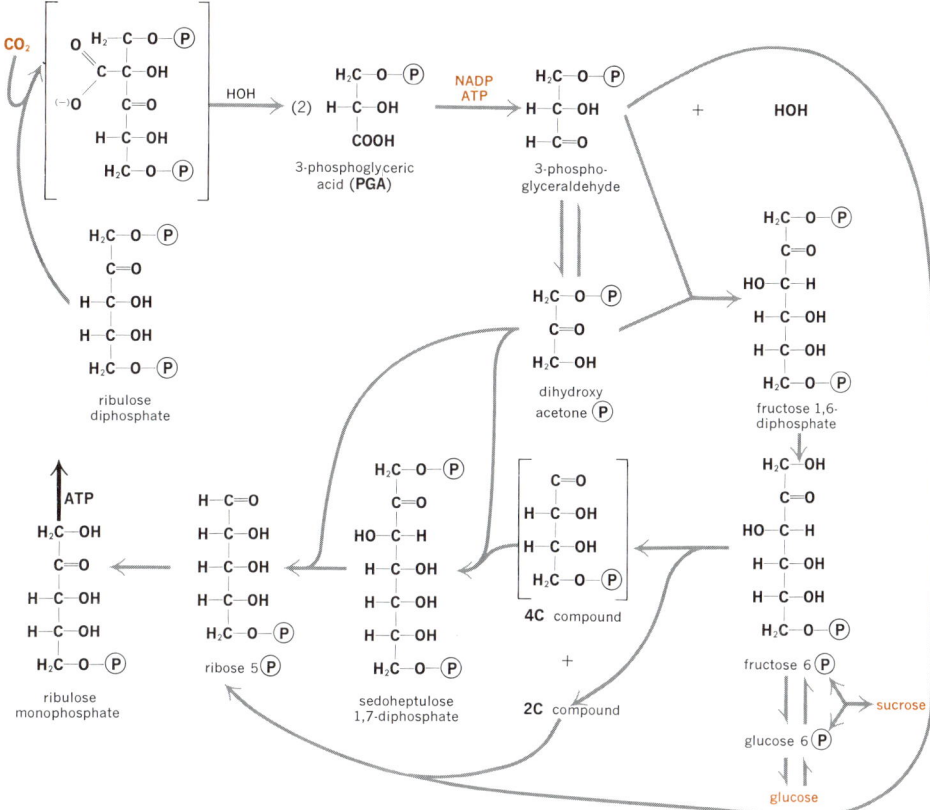

Fig. 43-2 The dark reaction of photosynthesis. This series of reactions which was proposed by Calvin and Bassham shows how carbon dioxide can combine with ribulose diphosphate to yield the sugars glucose or sucrose or to regenerate ribulose diphosphate for further reactions.

---

product of photosynthesis, the sugar glucose, could be made at the same time. The scheme of reactions that was developed by Dr. Melvin Calvin and his coworkers on the basis of their experiments summarized above is illustrated in Fig. 43–2. This series of reactions shows how glucose, fructose, or sucrose can be generated while ribulose diphosphate is still available for continuation of the process. There is no simple conversion of one molecule of one substance to one molecule of another. The scheme may be summarized as follows:

$6CO_2 + 18ATP + 12NADPH_2 \rightarrow C_6H_{12}O_6 + 6H_2O + 18ADP + 12NADP + 18\text{\textcircled{P}}-OH$

All substances other than glucose, hydrogen atoms, and CO_2 can be regarded as catalysts in this highly complex series of reactions.

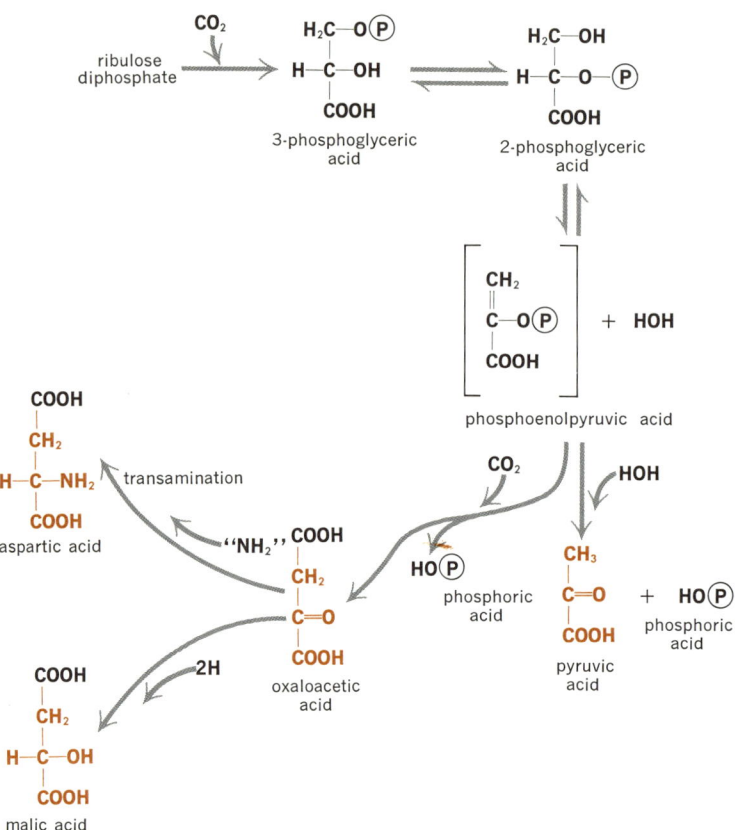

Fig. 43-3 Synthesis of pyruvic acid, oxaloacetic acid, malic acid, and aspartic acid by photosynthesis. In addition to the reactions shown in Fig. 43-2, the reactions given above show how other known products of photosynthesis may be synthesized from phosphoglyceric acid.

A scheme for the synthesis of malic acid, aspartic acid, and pyruvic acid, which were also found after short exposure of algae to the CO_2, is shown in Fig. 43-3. The conversion of pyruvic acid to alanine and serine explains the formation of these amino acids, which some investigators have found to be early products of photosynthesis. Since the reactions of pyruvic acid leading to acetyl CoA are well known, the basic starting materials for the synthesis of fatty acids are also indicated. The relation of photosynthesis to glycolysis and the citric acid cycle is shown in Fig. 43-4.

Variations of Photosynthesis

The photosynthesis described above is essentially what happens in all green plants—from the unicellular algae to spinach to the giant redwood trees. Other kinds of photosynthesis exist, particularly among the bacteria. Some bacteria use H_2S instead of H_2O in photosynthesis, and instead of releasing O_2 to the air, these bacteria deposit sulfur. Other

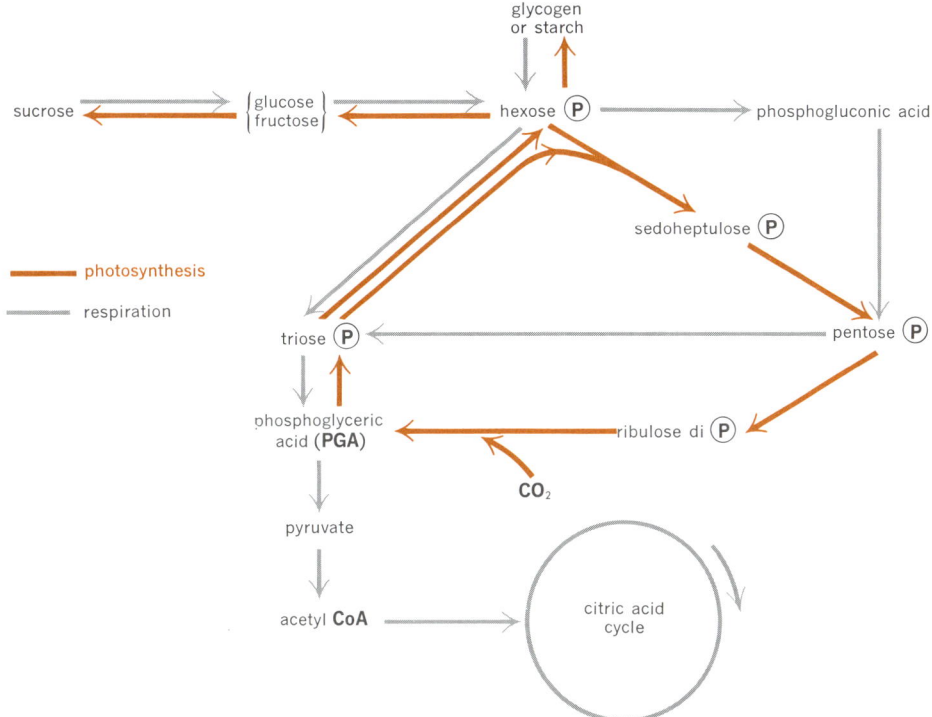

Fig. 43-4 The relationship of photosynthesis to respiration (glycolysis). This diagram shows some of the compounds important both in oxidation of carbohydrates and in photosynthesis. Arrows indicating photosynthetic reactions are shown in color and those of respiration are in black.

---

bacteria use H_2 in place of H_2O or H_2S and release neither oxygen nor sulfur. These and other variations in the basic process of photosynthesis indicate that perhaps the best definition is one given by Dr. Martin Kamen: "Photosynthesis is a series of processes in which electromagnetic energy is converted to chemical free energy which can be used for biosynthesis."*

43-3 NITROGEN FIXATION IN PLANTS

Although about 80% of the atmosphere is nitrogen, many plants suffer from nitrogen deficiency. Plants cannot use free nitrogen, but must have a combined or "fixed" form. One of the essential components of fertilizers is nitrogen in the form of liquid ammonia, nitrates, urea, or other compounds. Many nitrogen-containing compounds are found in manure.

Long before they understood why, farmers knew that some plants such as peas, beans, and clover seemed to enrich the soil, particularly if these plants were not harvested but were plowed under. These plants—the legumes—were found to have nodules on their

---

* Martin D. Kamen, *Primary Processes in Photosynthesis*, Academic Press, New York, 1963.

roots, and the nodules were found to contain bacteria. Such bacteria, we know now, can fix nitrogen—that is, convert nitrogen from its unuseable gaseous form to a nutritious compound such as a nitrate or nitrite. Thus the green plant not only fixes carbon dioxide from the air, but by the help of other plants, the bacteria, can also fix nitrogen from the air.

43–4 OTHER PLANT PRODUCTS

In addition to their basic function of fixing carbon dioxide and nitrogen, plants make a variety of compounds that are not synthesized by animals. Plants, but not animals, can synthesize compounds containing benzene rings. As a consequence, any benzene derivative that is required by an animal must be obtained from some plant source. Particularly important in this respect are the amino acids phenylalanine and tryptophan, which humans get either from plant products or from the meats of animals which have eaten the plants.

Many plants have widespread medicinal use. Especially important in this respect are the alkaloids, which were discussed in Section 28–8 B. The antibiotics (Section 44–5) are also important products of certain kinds of plants, particularly the molds.

43–5 SUMMARY

From the preceding discussion we can see that green plants serve as the manufacturers of all the naturally occurring organic compounds on the earth. Animals can take the basic units from plants and reassemble them into different compounds, but animals cannot exist without green plants. Animals, in turn, supply large amounts of carbon dioxide for photosynthesis. An earth with only green plants would probably soon have an excess of oxygen and various plant components and a deficiency of CO_2 and H_2O.

$$CO_2 + H_2O \underset{\text{animals}}{\overset{\text{plants}}{\rightleftarrows}} \text{carbohydrates} + O_2$$

It has been speculated that an earth devoid of animal and fungal life would have oceans full of a thick soup of dead algae. A deficiency of CO_2 in the air would probably result in less trapping of the energy of the sun (for discussion of this, see Section 45–2), and all life on the planet might cease. Thus a balance between the photosynthesis of green plants and oxidative processes of animals is required to maintain the equilibrium we need.

It has been proposed that animal life rose from plant life—or perhaps from a simpler plant-animal form. Animals became the mobile, energy-dissipating kind of life while plants remained, or perhaps became, the stable, energy-trapping synthesizers. In acquiring their motility the animals gave up some of their manufacturing ability. It is perhaps more logical to assume that those animals who became the most mobile, those who lost some storage and manufacturing ability were best able to survive in a struggle for existence. It is probable that because these mobile animals were able to find more food and could catch and eat their slower-moving competitors, they survived and reproduced. With this sort of differentiation have come the varied forms of life, both plant and animal, we see in the world around us—each different from, but also each highly dependent upon, other plants and animals. Although perhaps we should not extend analogies from the biological world to the social, differences and mutual dependence are certainly important in both.

IMPORTANT TERMS

Chlorophyll Fixation Photolysis
Dark reaction Light reaction Photosynthesis

WORK EXERCISES

1. What metal is found in chlorophyll?
2. Supposing that some disaster destroyed all animal life on the earth but that the green plants survived and were able to reproduce. Speculate on conditions on the earth 100 years after the disappearance of animal life.
3. What metabolic reactions have been discussed in this book, other than those of photosynthesis, in which carbon dioxide is added to a molecule?
4. If you added carbon dioxide containing carbon-14 to a green plant, what compound would you expect to be the first to contain radioactivity? In which carbon atom would you expect to find most of the radioactivity? If sucrose were isolated from this plant, which carbon atoms would you expect to be labeled?

SUGGESTED READING

Fogg, C. G., *The Growth of Plants*, Pelican Books, A-560, Penguin Books, New York, 1963. A paperback book on the general subject of plants.

Goodenough, U. W., and R. P. Levine, "The Genetic Activity of Mitochondria and Chloroplasts," *Scientific American*, Nov. 1970 (Offprint #1203). The authors suggest that mitochondria may have descended from simpler unicellular organisms.

Hendricks, S. B., "How Light Interacts with Living Matter," *Scientific American*, Sept. 1968. The author discusses how light acts in photoperiodism, vision, and photosynthesis. The discussion of the last topic is especially good.

Levine, R. P., "The Mechanism of Photosynthesis," *Scientific American*, Dec. 1969 (Offprint #1163). A good, up-to-date discussion of photosynthesis, especially of the "light reaction."

Van Overbeek, J., "The Control of Plant Growth," *Scientific American*, July 1968 (Offprint #1111). The effect of plant hormones in increasing and decreasing growth is discussed in this well-illustrated article.

The following three articles represent excellent reports by persons involved in unravelling the various aspects of photosynthesis. The article by Bassham is an excellent summary.

Arnon, D. I., "The Role of Light in Photosynthesis," *Scientific American*, Nov. 1960 (Offprint #75).

Bassham, J. A., "The Path of Carbon in Photosynthesis," *Scientific American*, June 1962 (Offprint #122).

Rabinowitch, E. I., and Govindjee, "The Role of Chlorophyll in Photosynthesis," *Scientific American*, July 1965 (Offprint #1016).

THE BIOCHEMISTRY OF DISEASE AND THERAPY 44

44-1 INTRODUCTION

Attempting to discuss, in a single chapter, the biochemistry of disease and therapy (the treatment of disease) is a major task. Whole books are devoted to the subject, and in many cases they discuss only a single disease. Nevertheless, in this chapter we will present some generalizations about diseases and their relationship to biochemistry. We believe that the information is essentially correct, that it will give the reader a better understanding of some medical practices, and will illustrate some practical applications of biochemical theories.

For the purposes of this discussion we will define **disease** as any departure from health or, in other words, any failure of the body to function normally. From the viewpoint of the biochemist the functioning of the body depends upon chemical reactions. A breakdown in these reactions leads to abnormal functioning which we can call disease. To simplify our discussion, we have divided the diseases into several categories. These are arbitrary categories, of course, under which we attempt to describe processes which we do not completely understand. Some diseases may belong in two or more categories and others do not really belong in any of those listed.

The major division is between (1) infectious diseases (those caused by bacteria, viruses, and other organisms) and (2) noninfectious diseases. Noninfectious diseases are further classified as (a) nutritional diseases, (b) mental diseases, (c) diseases of the endocrine glands, and (d) the metabolic or degenerative diseases. Cancer and arthritis, will be discussed under the last classification. Cardiovascular disease, which also belongs with the degenerative diseases, was discussed in Chapter 40. Those conditions or diseases known as inborn errors of metabolism were discussed in Chapter 42.

Forms of therapy have two major aims, to relieve the symptoms and to cure the disease. As will be evident from the following discussion, there is too little of the latter and far too much of the former. Our knowledge of the causes and cures of diseases is still so inadequate that often all we can do is to make the patient as comfortable as possible. In many cases this simple treatment is all that is needed because the natural defenses of the body can take care of the diseased condition. In others, treatment of symptoms only prolongs the time a patient must wait before his death from an incurable disease.

44-2 THE INFECTIOUS DISEASES

"At some early stage of evolution, possibly before plants and animals had become clearly differentiated from each other, one or more species must have made the discovery that a very convenient source of food consists of the tissues of other organisms. In such tissues may be found materials, already preformed, suitable as energy sources and for building one's own tissues . . . Among some of the species which had discovered the importance and desirability of eating other organisms as a source of food, the expedient of parasitism developed. After all, it is not economical to destroy the source of your food material completely, although some micro-organisms, perversely enough, do this to their hosts. Instead, it is easier to live in or on the host, making use of the food supply which he provides directly or indirectly. This is parasitism of which infection is an example." *

Man has a clear idea of what the parasites are: the bacteria, viruses, and simple animals such as lice. From the viewpoint of many plants, humans and indeed all animals are potential parasites. The animals that have survived have developed various forms of defense against their enemies. In addition to weapons and fortifications against large foes, humans have developed several defenses against the microscopic agents of infection.

44-3 DEFENSES AGAINST INFECTION

The human body has several natural defenses against infection. The tough, relatively impermeable skin is a major defense. For those invaders that do manage to get around or through this barrier, the **phagocytes,** those cells (such as some kinds of white blood cells) that engulf or "eat" invaders are potent antagonists. Biochemists are most interested in two other natural defensive agents—the antibodies and some of the hormones. **Antibodies** are produced by the body in response to the invasion of various types of foreign molecules. The human can make antibodies against materials as varied as egg white and viruses. The process of antibody formation is not completely understood, but apparently each human has a recognition system that can tell when a substance in his body is foreign. If the foreign particle is of sufficient size and proper character, the body manufactures antibodies that are apparently shaped in such a way as to combine with the antigen (invading foreign particle) and negate its effects. The antibodies are primarily protein, although many also contain carbohydrates in their structure. Some of the most important antibodies are present in the blood serum and can be separated in a group called the gamma globulins. Vaccination is effective because it stimulates the production of antibodies. It is assumed that the pooled gamma globulin from many individuals contains antibodies to many specific diseases, and thus it is often given to help the body fight specific diseases.

Some of the steroid hormones produced in the cortex of the adrenal gland also help the body fight infection. These substances enable the body to "wall off" the foreign invaders and thus keep them from spreading throughout the body and doing damage. A large percentage of the American population has had a mild infection of the tubercule bacillus, has developed antibodies to it (as shown by skin tests), and may also have "walled off" groups of these bacteria in small nodules in the lungs. Some adrenal cortical hormones help the body produce these inflammatory walls. The adrenal cortex also

---

* B. S. Walker, W. C. Boyd, and I. Asimov, *Biochemistry and Human Metabolism,* 3rd ed., Williams and Wilkins, Baltimore, 1957.

produces hormones which decrease the reaction of the body to foreign substances by suppressing antibody formation. These hormones, the best known of which is cortisone, are effective in diminishing inflammation in many parts of the body. They are widely used in the treatment of arthritis and other diseases. In the early days of cortisone therapy, some persons who received this medication developed tuberculosis, apparently because the tuberculosis bacteria that had been "contained" were again set free. Physicians prescribing cortisone or other similar substances now take measures to prevent such consequences.

The transplanting of organs such as the heart and kidney from one human to another are made possible by the use of substances that repress the responses of the body to foreign tissue. Many persons given such transplants died later from pneumonia or other diseases. Their deaths were due, at least in part, to the fact that the drugs given to keep the body from rejecting the transplanted organ also keep the body from rejecting or fighting the bacteria that cause pneumonia and other diseases. In the healthy individual there is a delicate balance of the pro-inflammatory and anti-inflammatory substances.

44-4 CHEMOTHERAPEUTIC AGENTS

Against some diseases the natural body defenses can be aided by chemical substances either synthesized in the laboratory or extracted from other organisms. These substances are classified as synthetic **chemotherapeutic agents** and **antibiotics,** respectively. Although antibiotics can now be synthesized in the laboratory, they were originally isolated from microorganisms, particularly the molds. Even in the laboratory, synthesis of antibiotics often relies on cultures of molds to perform certain steps in the synthesis. In most cases, laboratory synthesis is not economical; it is less expensive to grow the molds and let them do the entire job of producing the antibiotic.

A perfect chemotherapeutic agent would be totally effective against a kind of bacteria or other infectious agent and would have no bad effects on the host. The action has been characterized as being like a magic bullet that hits and affects only infectious agents. Since the biochemistry of all organisms from the bacteria to man is essentially similar, what is poisonous to one organism is usually poisonous to the other. However, sufficient exceptions to this general rule exist to make chemotherapeutic agents useful.

One of the first synthetic therapeutic agents was the compound called Salvarsan, or Dr. Erlich's compound 606, which was effective against syphilis. (The story of Dr. Erlich is dramaticaly told in the book *Microbe Hunters* by Paul de Kruif.) Probably the most important synthetic chemotherapeutic agents are sulfanilamide and its derivatives, known as the sulfonamides, or popularly as the sulfa drugs. The effectiveness of these drugs led to the proposal of a basic theory of chemotherapeutics. It was postulated that sulfanilamide was effective against bacteria because it was an antimetabolite to *para*-aminobenzoic acid (PABA). An **antimetabolite** is defined as a compound that is sufficiently similar in chemical structure to a normal metabolite so that it can replace it in an organism. However, the antimetabolite does not function as does the metabolite, a circumstance which often leads either to failure in growth or reproduction and thus to the eventual death of the organism. Bacteria use PABA in the synthesis of folic acid. It was argued that because of similarity in structure between PABA and sulfanilamide, organisms tend to incorporate sulfanilamide and compounds related to it into a molecule that is similar to folic acid but which cannot perform the functions of folic acid.

sulfanilamide **PABA** **folic acid**

This deficiency would lead to decreased ability to grow and reproduce. The test of a true antimetabolite is that its effects must be reversed by the addition of large amounts of the normal metabolite. Thus bacteria that are destroyed by sulfanilamide or another sulfonamide (actually perhaps just kept from reproducing, with the natural body defenses doing the destroying) should be able to live and reproduce in the presence of the sulfonamide when sufficient folic acid is supplied to them. While many sulfonamide-sensitive bacteria can meet this test, others cannot. It now seems most likely that the sulfa drugs are not true antimetabolites but that they do inhibit a specific step in folic acid synthesis. However, the antimetabolite theory has led to the design of many chemotherapeutic agents specifically designed to be antimetabolites for a chemical substance. The compound aminopterin, which is successful, although only for a limited time, in the treatment of leukemia, is another folic acid antimetabolite.

Why do sulfa drugs inhibit the growth of bacteria but have little effect on humans? The answer may be twofold. First, humans require folic acid as such, and do not synthesize it—it is a vitamin. Thus interruption of its synthesis or synthesis of a sulfonamide-containing folic acid are not serious matters. Secondly, humans can survive even if deprived of all folic acid for a period of a day or two, but this period of time represents several lifetimes (times between divisions) to bacteria. Whenever the bacteria are prevented from reproducing at a normal rate, the natural defenses of the human body can usually effectively combat the invaders.

44-5 ANTIBIOTICS

The first antibiotic to be widely used was penicillin. Its discovery came about as the result of an accidental contamination of a bacterial plate by a particular kind of blue-green mold. When Dr. Alexander Fleming observed that the colony of mold was surrounded by a zone in which no bacteria (a staphylococcus culture) were growing, although the rest of the plate was covered with bacteria, he realized that the mold was secreting something which was inhibiting the growth of bacteria in the surrounding medium. To overcome the difficulties in extracting, testing, purifying, and producing in quantity the bacterial inhibiting substance required a great deal of research. It was many years before the substance which we call penicillin became available for general use.

Many other workers, notably Dr. Selman Waksman, have found hundreds of antibiotics. The general procedure is to make culture plates of the organism that we wish to control and expose them to extracts from other organisms—particularly the molds. An extensive search has been conducted for rare kinds of molds and yeasts that might produce antibiotics. Although hundreds of antibiotics are known, only a limited number are useful in treating diseases. Most of the antibiotics are toxic not only to the infectious agent but to the humans or other animals being treated. Penicillin has the advantage of being both effective against certain infections and the least toxic of the common anti-

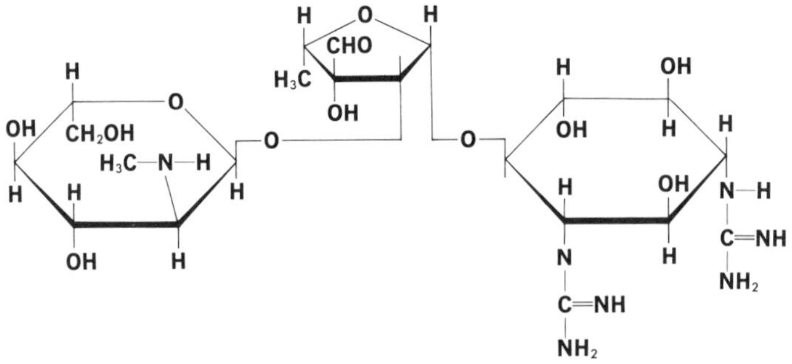

streptomycin

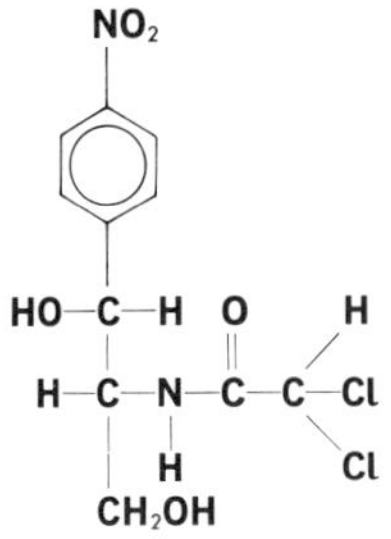

chloramphenicol

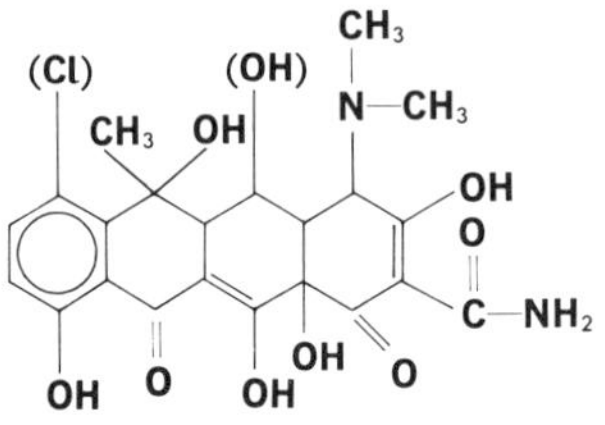

tetracycline

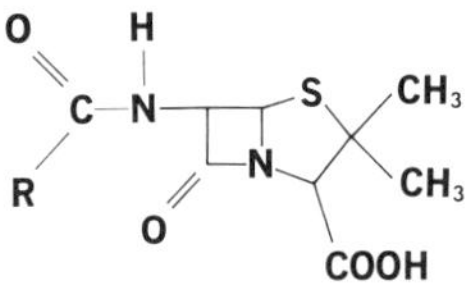

penicillin

Fig. 44–1 The structure of some common antibiotics. The form of penicillin known as penicillin G has a benzyl group at R. Other forms have the *p*-hydroxybenzyl, pentenyl, *n*-amyl, or *n*-heptyl R groups. The tetracycline molecule is known as Achromycin. Oxytetracycline which has an OH as indicated on ring number 3 is known as Terramycin, and chlorotetracycline, which has a chlorine atom on the first ring, is known as Aureomycin.

---

biotics; some individuals however, have acquired a sensitivity to penicillin so that administration causes serious consequences. Apparently what has happened is that these penicillin-sensitive persons have produced antibodies to penicillin itself, since it, too, is a foreign molecule. In cases of this kind the natural defenses of the body are acting in a way which is bad for its general health. This reaction of the body against foreign substances is also a serious problem in the surgical substitution or transplanting to replace various parts of the human body.

The formulas for some of the more important antibiotics are given in Fig. 44–1.

Because of the wide use of vaccination and the development of effective chemotherapeutic agents, and perhaps also because of the generally good state of nutrition of most persons in the United States, deaths caused by infectious diseases have greatly decreased in the past few decades. This is not true of all countries in the world, and infectious diseases still take a large toll, especially of children. Part of the solution to the problems of infectious diseases lies in education and part in the provision of vaccines, chemotherapeutic agents, and nutritious foods. The role of adequate sanitation in the prevention and spread of infectious diseases is also important.

44-6 THE NONINFECTIOUS DISEASES

Diseases in this classification, as far as we know, do not result from the presence of an infectious agent. It is possible, however, that some of them are caused by agents which have not yet been associated with the disease. This is especially true of certain kinds of cancer which may be caused by viruses.

Nutritional Diseases

Malnutrition is a minor problem in the United States today; however, among the poor it still prevails. But there are definite evidences that some diseases are aggravated, if not caused, by overeating or "excessive" nutrition which some would classify as another form of malnutrition. It is estimated that 15% of the American population, some 30,000,000 persons, are overweight and that at least 5,000,000 of these are pathologically overweight. Quite the opposite is true of many other countries of the world. In Africa, and in other under-developed countries throughout the world, kwashiorkor, a disease caused by inadequate protein nutrition, is prevalent. Pellagra, beriberi, and other deficiency diseases are still present in large areas of the world.

Many nutritionists believe that the laziness, stupidity, and other characteristics which are often attributed to the natives of poor countries are really due to prolonged nutritional deficiencies. Controlled research on the effects of starvation on human volunteers supports this conclusion. Recent studies with experimental animals and of malnourished children indicate that malnutrition of infants produces mental defects that can never be overcome. Offspring of malnourished mothers show similar irreversible mental deficiencies.

Individuals vary in their utilization of food and in their needs for specific nutrients. While not much is known about this subject, we do know that there is wide variability in all measurable human characteristics, including the requirements for vitamins. With respect to nutritional requirements, we are too often prone to remember the generalization and forget the individual. Screening of people on a massive scale and a great deal of research will be required to determine the individual variations. However, this lack of refinement should not keep us from acting upon the best available generalizations as to normal requirements. Malnutrition in the poorer countries of the world can be combated by education in nutrition and by increasing food supplies. The latter problem is becoming more and more serious as a result of the rapid increase in population. This increase is due in a large part to the control of infectious disease and to better nutrition. Thus the solution to problems of disease and malnutrition may lead to even more complex problems. It may be said that in treating diseases man has interfered with the balance of nature, and this is, of course, true. The alternative to this interference is continued infant mortality, starvation, and disease. Another alternative, which many persons regard as another interference with nature, is birth control. The problems raised by the population explosion are not insurmountable. Some countries have reversed the increase. Oral contraceptives show great promise for those who can accept use of "the pill" (see Chapter 35-6).

Mental Diseases

Various forms of mental illness may affect as many as ten percent of the population of the United States at some time during their lives. Some fifty percent of all the patients in United States hospitals are victims of mental diseases. These estimates lead many persons to say that mental diseases are the primary health problem in the United States at the present time. It is true, of course, that the high percentage of mental patients in hospitals is due to the fact that hospital stay is much longer for these patients than for the patients in general hospitals. Other authorities consider the greatest problem is in the control of cardiovascular disease or cancer, which rank highest as causes of death.

One difficulty in discussing mental diseases is the lack of a good definition of the diseases and objective reports of the effects of drugs used in their treatment. Mental illness is usually categorized as a psychosis or neurosis, and the victims of such conditions are known as psychotics or neurotics. The psychotic is generally thought of as a person who differs seriously from normal persons in his idea of reality and contact with his environment. He may be in a stupor, he may be wildly irrational, he may see and hear things no one else does, or he may become uncontrollably violent. A neurotic is described as a person whose mental illness does not involve loss of contact with reality. He may be overwhelmed by persistent fears, haunted by anxiety, subject to unexplainable aches and pains, and perhaps unable to sleep. Definite borderlines between psychotics, neurotics, and normal persons are not easily drawn. It has been said that the mentally ill are just like normal persons except that some tendencies are exaggerated.

For many years biochemists have looked for an abnormal metabolite in the blood or urine of psychotics. This search has led to a number of compounds that can produce some of the symptoms of psychoses when given to experimental animals or humans. However, there are no clear indications that these substances are present in the blood or urine of psychotics in sufficient quantities to account for their illness. The search for abnormal metabolites is continuing and may yet be successful. There is evidence that some forms of schizophrenia are hereditary, and this lends support to the search for an inherited biochemical defect.

While the search for metabolic abnormalities in the mentally ill has been largely unsuccessful, therapy in treating mental illness has made major advances in the past twenty years. The use of tranquilizing drugs such as reserpine and chlorpromazine has transformed mental hospitals from madhouses or snake pits with straitjackets and padded cells to quieter, calmer places. These potent tranquilizers alleviate the symptoms, but do not cure mental diseases. They make it possible for a psychiatrist to establish contact with a psychotic patient and thus help him treat the disease by psychotherapy. These drugs have also made possible the release of many patients from mental hospitals.

Reserpine can be isolated from an extract of the roots of the plant *rauwolfia serpentina*. This plant has been used by medicine men of India for centuries for the treatment of a variety of conditions ranging from snakebite to madness. Purified extracts of *rauwolfia* were first used in the United States in the treatment of high blood pressure. The tranquilizing effects were noted, and the most active of the many substances (more than 25 are known) in *rauwolfia* extract was found to be reserpine (see Fig. 44–2). Chlorpromazine was synthesized to be used as an antihistamine, but its usefulness as a potent tranquilizer was soon observed. Both reserpine and chlorpromazine were first generally obtainable in 1954. Since then many chemical compounds closely related to chlorpromazine have been synthesized and are used as tranquilizers (see Fig. 44–2).

The milder tranquilizers such as meprobamate, which is known by its trade names Miltown and Equanil, are used by many people who would be described only as neurotic or perhaps just "overanxious normals." Meprobamate is both a muscle relaxant and a tranquilizer. Although it has been widely used, reports of bad side effects have decreased use of this drug. Other similar mild tranquilizers are available.

Many other chemical substances have an effect on the mind—or to use more modern

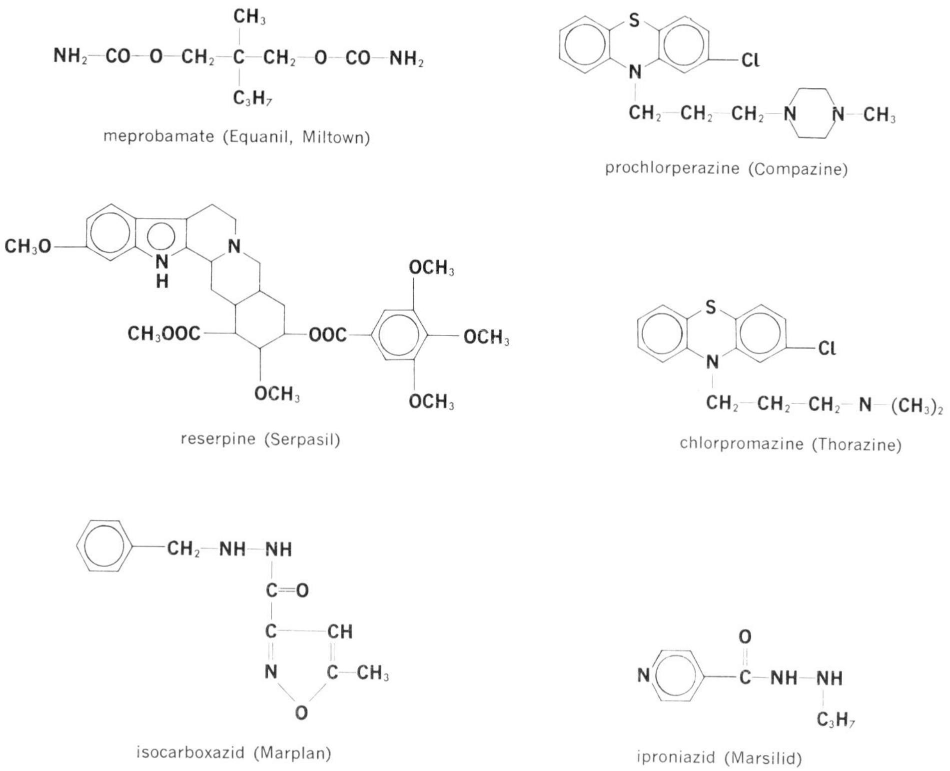

Fig. 44-2 Drugs that influence mental states. Chlorpromazine, prochlorperazine, reserpine, and meprobamate are generally described as tranquilizers, and iproniazid and isocarboxazid are known as psychic energizers.

---

terms, alter mental states. These include antidepressants or "psychic energizers" such as iproniazid and isocarboxazid (Fig. 44-2). These substances counteract depression—a sense of despair, fear, or guilt. These drugs relieve depression without making a patient jittery or irritable. Iproniazid was originaly used in the treatment of tuberculosis. When it was observed that patients taking the drug became elated and sometimes so happy they began dancing in the hospital rooms, the mood-elevating aspects were obvious. Unfortunately, many bad side effects of iproniazid became evident, and it was withdrawn from general use as an antidepressant (and more effective drugs were found for treating tuberculosis). Since iproniazid is a derivative of hydrazine (H_2NNH_2), chemical synthesis of similar compounds has produced a variety of effective and less toxic antidepressants. These drugs inhibit the enzyme monoamine oxidase (MAO) that is important in the oxidation of various amines (norepinephrine, for example, or serotonin). These amines are important in the functioning of the brain and the nervous system. Apparently this inhibition of MAO is responsible for the antidepressant properties of the hydrazine derivatives.

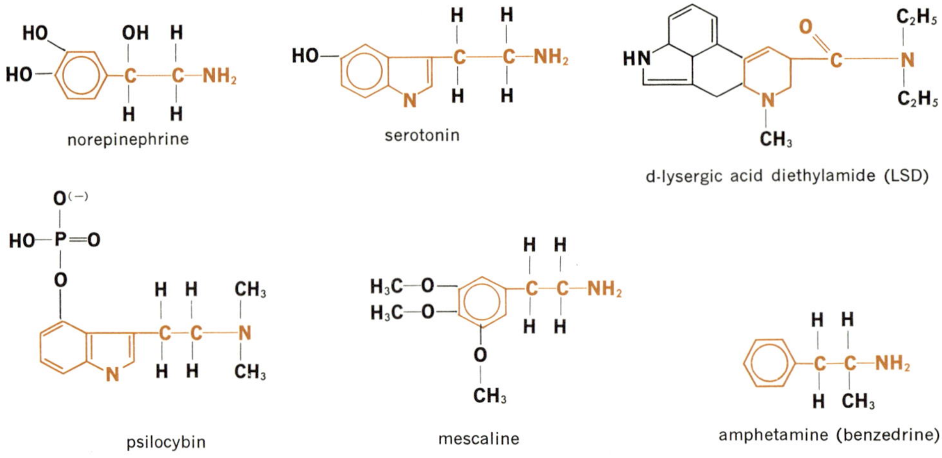

Fig. 44-3 Drugs that affect the central nervous system. Norepinephrine and serotonin are involved in transmission of nerve impulses. LSD, psilocybin, and mescaline produce altered mental states and may be called psychedelic or hallucinogenic substances. Amphetamine is a stimulant. Structural elements that are similar are shown in color.

The amphetamines (Benzedrine and Dexedrine) are also used as antidepressants. They have the general disadvantage of producing a nervous irritability and depressing the appetite. The latter effect, however, makes them useful for weight reduction. Since these are potent drugs, they are legally available only on a doctor's prescription. Although these antidepressants are used by college students who want to stay up all night to finish a term paper or study for an examination, there is little evidence that such use is warranted. They can keep a student awake and may elevate his spirits, but they probably do not elevate his grades.

Another class of mind-influencing substances which are variously described as hallucinogenic, psychotogenic, or psychedelic drugs are the object of much interest today. The first of these substances to be chemically characterized, lysergic acid diethylamide (LSD) (see Fig. 44-3), has been the object of much experimentation, both scientific and unscientific. The drug is relatively nontoxic in the dosages used, but it can have profound effects on mental states and on the personality. Psilocybin, mescaline, and tetrahydrocannabinol (from marijuana) are reported to have effects similar, but not identical, to those of LSD. The effects of these drugs are largely not describable in objective, scientific terms but are reported in subjective, personal language. As a result, scientific evaluation of these substances is difficult. It is to be hoped that further, well-controlled experimentation with these potent substances will lead to better evidence concerning both their usefulness and their potential danger. Further discussions of these drugs are listed in the suggested reading at the end of this chapter.

An examination of the formulas of the compounds given in Fig. 44-3 shows that many of the substances affecting the mind are similar in structure to norepinephrine and serotonin. These latter substances, as we have already noted, are important in the functioning of the brain and nervous system. It is also known that reserpine and chlorpromazine influence the levels of serotonin in the brain. The best-established fact on the biochemical effect of any of these drugs is that the hydrazine derivatives inhibit the enzyme monoamine oxidase which is important in the metabolism of norepinephrine and serotonin. Thus the evidence is mounting that these two naturally occurring amines are very important in influencing mental states. LSD is known to block the metabolism of serotonin; however, other similar substances that also block serotonin metabolism do not have the mind-altering effects of LSD. These observations are intriguing and suggest possible hypotheses about the action of mind-altering substances. However, no consistent theory has yet been able to stand the challenge of explaining all the experimental data.

Diseases Caused by Malfunction of the Endocrine Glands

The endocrine glands are characterized by the fact that they secrete chemical substances called hormones directly into the blood stream. The hormones travel throughout the circulatory system and eventually affect many parts of the body. The endocrine glands are the pituitary (or hypophysis), located just below the brain; the thyroid, located in the neck; the parathyroids, located in the thyroid gland; the pancreas, located in the abdominal region; the adrenals, located on top of the kidneys; and either the ovaries, located abdominally in the female, or the testes located in the scrotal sac in the male. (See Fig. 44-4.) Each endocrine gland produces at least one hormone, and many of them produce several chemically distinct hormones. The deficiency or oversupply of any of the hormones has serious consequences for the individual. We can give examples of only a few of these conditions for the purposes of illustration.

Diabetes mellitus, which was discussed in Chapter 38, is caused by a deficiency of the pancreatic hormone, insulin. Treatment by insulin does not cure diabetes, but continued use of it allows the diabetic to lead an essentially normal life. Control of diabetes represents one of the most successful instances of correction of a hormone deficiency. Purified hormones from other animals or synthetic products have been used successfully to treat endocrine insufficiencies. The administration of thyroxine for hypothyroidism (underactivity of the thyroid gland), the adrenocorticotropic hormone (ACTH) to compensate for underactivity of the adrenal cortex, and either male or female hormones to persons who fail to mature normally are other examples of treatment for hormone deficiencies.

Treatment is more difficult in cases of overproduction of a hormone or hyperactivity of an endocrine gland. In many cases, surgery is used to remove the overactive gland. This may be followed by replacement of the hormones—at a normal level—from a synthetic or natural source. In other instances, specific substances are known which can decrease the secretion of a gland. For instance, propylthiouracil decreases the output of thyroxine in a hyperthyroid person. In other cases, the action of one hormone is opposed to that of another, and administration of the antagonist may be helpful. An example of such antagonism exists between the male and female hormones and in the inflammation-promoting and inflammation-suppressing hormones of the adrenal cortex.

Hormones are sometimes used to treat diseases not directly attributable to the endocrine glands. Male hormones are effective in treating some forms of cancer of the breast in the female, and cortisone is widely used to treat various diseases characterized by inflammatory reactions.

There are no real cures for diseases of the endocrine other than surgical removal of overactive glands and possibly the surgical transplantation of glandular tissue to com-

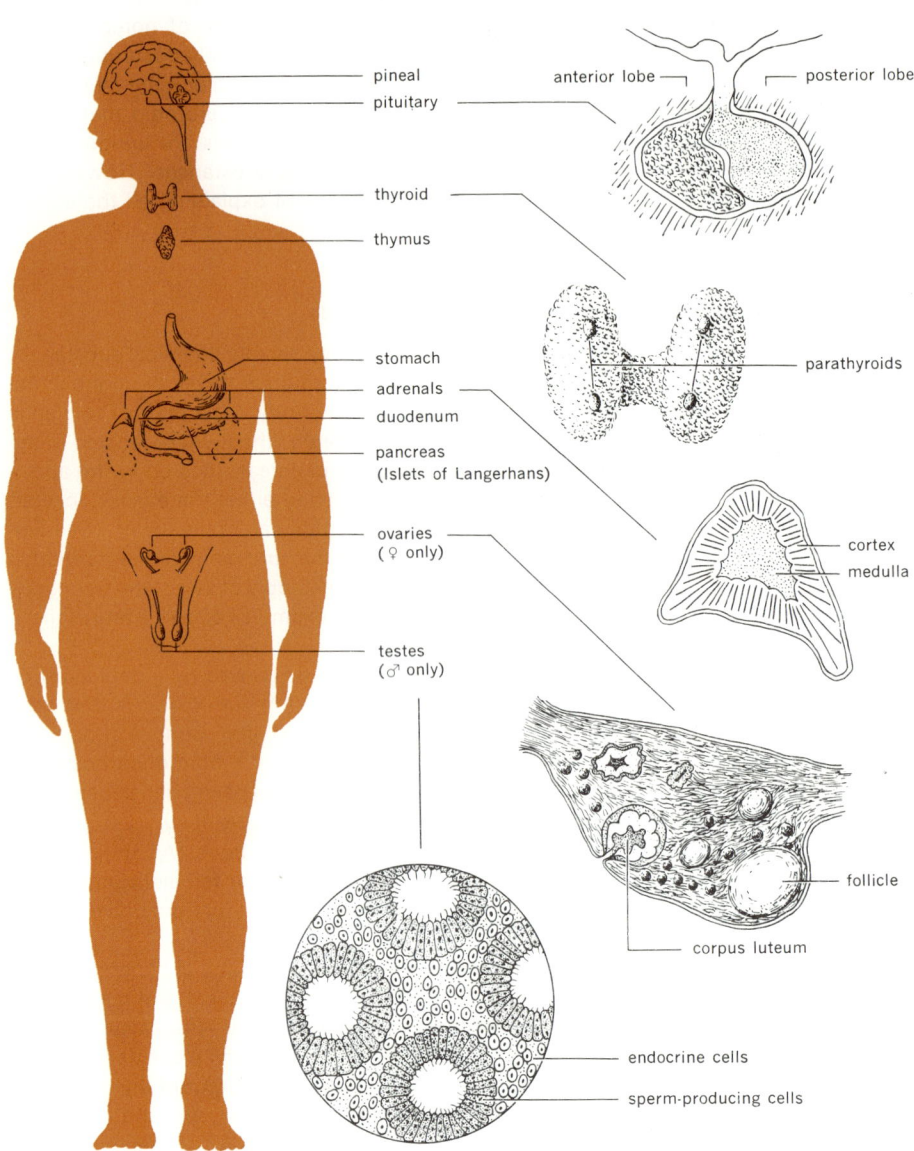

Fig. 44-4 The endocrine glands of the human. The location of the glands is shown at the left, and drawings of the detailed structure of the glands at the right. [Reprinted from John W. Kimball, *Biology*, Addison-Wesley, Reading, Mass., 1965.]

pensate for underactive glands. The purification and laboratory synthesis of hormones has, however, led to treatment of many endocrine disorders. Our knowledge of just how the hormones work is minimal. It is to be hoped that as we find out more about the reactions that hormones influence, other means of treatment and perhaps control of endocrine diseases will emerge. (See Section 37–8 for a discussion of hormones.)

Metabolic or Degenerative Diseases

The terminology used to describe these diseases is not totally satisfactory. From one aspect, all diseases are metabolic since they influence reactions in the human body. The term degenerative diseases is perhaps better, yet that is not totally accurate, either.

Cancer. The term cancer is commonly used to describe a variety of conditions all of which are characterized by a nonphysiological growth or multiplication of cells. This growth may remain restricted in area and have little tendency to recur after removal. In a case like this we describe the growth, or neoplasm, as benign. In common usage, benign neoplasms are not cancerous. When a tumor grows unrestrictedly and spreads through the blood or lymph to other parts of the body to form new growths, it is described as malignant and called a cancer. Medical men use more descriptive terms and speak of *carcinoma*, a malignant tumor of epithelial cells; *sarcoma*, a malignant tumor of the muscle or connective tissue; or *leukemia*, which is characterized by excessive numbers of leucocytes or white blood cells.

Biochemists and other medical researchers have attempted for years to find some abnormal metabolite in cancerous cells, but so far the search has been unsuccessful. Cancerous cells seem to be similar to others except that they have lost the ability to stop reproducing. Malignant cells tend to lose their differentiation or specialization and revert to a more primitive type of cell, and in some cases this reversion involves the use of anaerobic glycolysis even in the presence of oxygen. These cells apparently do not use the reactions of the citric acid cycle.

More is known about the causes of cancer than about the changes it brings about in the metabolism at a cellular level. Ingestion of certain organic chemicals such as dimethylbenzanthracene and methylcholanthrene,

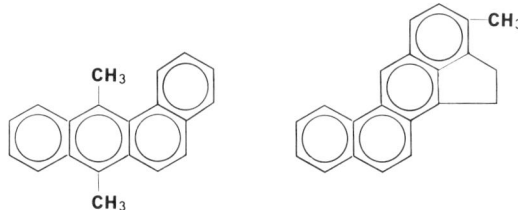

dimethylbenzanthracene   methylcholanthrene

can produce cancer in rats and other experimental animals. Closely related compounds found in cigarette smoke are also effective **carcinogens** (producers of cancer). Cancer can also be produced by radiation. Ultraviolet irradiation of the skin can lead to skin cancer. X-rays and gamma radiation from radioisotopes penetrate deeper and produce cancer in other organs. The radioactive substances strontium 90 and radium, which the body tends to treat like their sister element, calcium, are found in bones and are likely to produce cancer of the bone or bone marrow.

Evidence is accumulating that some forms of cancer are caused by viruses. In fact, some authorities believe that all cancer is caused by viruses. If this proves to be true, we will have to reclassify cancer as an infectious disease, although it will certainly be different from other such diseases.

One theory has recently been proposed that links all the bits of evidence concerning the causes of cancer. This theory must still be tested, but it presents an interesting way of regarding the cancerous condition. According to this theory, viruses may infect cells and remain in the cells in an essentially inactive form. When the proper stimulus comes along, either radiation or a chemical carcinogen, the virus becomes active and produces cancer. A similar theory has proposed that viruses are not necessary but that carcinogenic chemicals or irradiation affect the normal DNA in a cell in such a way that it produces malignancy. (See Section 36–7.)

There are various treatments for cancer. Probably the best is surgical removal at an early stage of development of the neoplasm. In some cases irradiation with X-rays destroys the cancerous cells along with normal cells. Thus X-rays can both cause cancer and be used to treat it. The probability that they will cure a known neoplasm is greater than the probability of their producing a new one, so X-ray therapy is useful. Significant progress is being reported toward finding antibodies to cancerous cells—presumably these antibodies would have to come from someone other than the victim of the cancer, since the fact that he has cancer indicates that he cannot make sufficient antibodies. The form of cancer therapy most closely related to biochemistry is the use of antimetabolites. Antimetabolites of many of the components of the nucleic acids have been effective in treating malignancies. Mercaptopurine and fluorouracil,

mercaptopurine   fluorouracil

have been used for this purpose. Treatment with antimetabolites often prolongs the life of the victim of cancer, but it does not offer a cure. Over a period of time the cancerous cells seem to become resistant to most antimetabolites. The antimetabolites eventually begin to affect the normal cells of the body to such an extent that treatment must be stopped.

It is to be hoped that further studies of the basic defect, causes, and treatment of cancer will lead to either a cure or an effective treatment of this dread disease.

Arthritis. Arthritis means, literally, inflammation of the joints. In any condition so generally described, there are probably many causes of the symptoms which characterize the condition. Two types of arthritis have been distinguished, gouty arthritis and rheumatoid arthritis. In the former, deposits, largely of uric acid and urates, form at various places in the body and are most noticeable at the joints. Deposits at places other than joints are classified as gout. This condition represents an abnormal deposition of a normal metabolite. The disease is apparently due to the synthesis of excessive amounts of uric acid in these individuals.

Rheumatoid arthritis is one of a group of diseases called the collagen diseases. In these there are inflammatory changes in connective tissue which can produce lesions in joints, skin, muscle, heart, and blood vessels. The cause of these inflammatory changes is not definitely known, but in some cases there are antibodies in the serum which react with other components of the host's own tissues. We can characterize this disease as one in which an individual is immune to himself, and say he has an auto-immune disease.

Both forms of arthritis represent abnormalities in essentially normal body constituents. Gout and gouty arthritis can be treated with colchicine and allopurinol, both of which apparently decrease the synthesis of uric acid. It is interesting that allopurinol, which is a purine antimetabolite, was first synthesized to be used as a chemotherapeutic agent against cancer. Both forms of arthritis, or perhaps it is best to say both diseases, can be treated with anti-inflammatory drugs such as cortisone.

Cardiovascular disease. This disease, which was discussed in Chapter 40, is also included with the degenerative diseases. It is, in fact, probably the most important of these diseases, at least with respect to the number of persons affected.

44–7 AGING

It is probably wrong to classify the process of aging as a disease, but it has much in common with metabolic or degenerative diseases. We are all aware of the general symptoms of aging; we know that the skin wrinkles, the hair turns gray and may fall out, sight is impaired, the bones become brittle, and the body no longer can stand so much stress as previously.

The biochemist is interested in three basic aspects of aging: What is the biochemical nature of the changes accompanying aging? What is the cause of these changes? What treatments or procedures may offset or delay the onset of these changes? At the present time there are only fragmentary answers to these questions, but research is proceeding, and we will undoubtedly have better answers in the future.

The search for the nature of the biochemical changes in aging has centered upon the connective tissue—the material between the cells that helps hold them together. The protein, collagen, is the most abundant component of connective tissue. Upon aging, the collagen becomes more rigid, less soluble in phosphate and citrate buffers, and assumes an almost crystalline form. This increased rigidity is probably due to increased cross linkage between collagen molecules. Other components of the connective tissue, elastin and various mucopolysaccharides (combinations of protein and polysaccharides) change in ways that may make nutrition and elimination of waste products of the cells, which are surrounded by the connective tissue, much more difficult. It is known that with aging many cells die and are replaced with connective tissue.

A brown lipoprotein pigment called lipofucsin accumulates in many cells with increasing age. Neither the source nor significance of this pigment is known. However, in persons who live to be 100 years old, six to ten percent of the volume of some heart cells is made up of this pigment.

While there is no lack of theories to explain why persons age, most of the theories lack any real verification. Since many of the cells of the body (nerve and muscle cells) never divide, they may just "wear out." We must ask, then, what wearing out means in biochemical terms. It has been suggested that the DNA is disrupted by chance irradiations so that it no longer has all the "information" a cell needs for survival. The amount of RNA found in the nucleus of cells decreases with age as does its rate of turnover. Perhaps toxic products produced in the cell disrupt some part of the cellular machinery. Other theories of aging point to the fact that many older persons do not eat well-balanced nutritious diets, and thus some of the effects of aging may be due to malnutrition.

Few valid treatments have been found to delay the onset of aging. In experiments with rats, a drastically reduced caloric intake with maintenance of the levels of vitamins and proteins increased the life span from a normal value of 1000 days to 1400 days. In severely starved animals the life span was doubled. In countries where starvation is common, the average life span is not increased, but these countries also have poorer nutrition and more infectious diseases, so perhaps the comparison is not valid. It is certainly true that obesity is not consistent with long life.

Various people have advocated the consumption of yoghurt or the implantation of monkey glands (especially the testes) as a means for preventing aging. Others who have been somewhat more flippant about the whole idea have suggested that the way to live to be old is to advertise for a pair of long-lived parents. It is known that heredity is important in longevity. Another facetious statement is the quip attributed to G. B. Shaw to the effect that the one comforting thought about the process of aging is its alternative!

In a more serious vein—research into the cause of aging and possible ways of slowing down the process should prove fruitful in the next several years. Dr. Hans Selye has said: "Among all my autopsies (and I have performed quite a few), I have never seen a man who died of old age. In fact, I do not think anyone has ever died of old age yet . . . To die of old age would mean that all the organs of the body would be worn out proportionately, merely by having been used too long. This is never the case. We invariably die because one vital part has worn out too early in proportion to the rest of the body."*

Some may question the desirability of prolonging life for persons who feel alienated from society even while young. The prospect of a prolonged existence filled with mental anguish or boredom is not inviting. It would seem, then, that various areas must be explored for ways of making life more meaningful as well as longer.

IMPORTANT TERMS

Antibiotic	Carcinogen	Infection
Antibody	Chemotherapeutic agent	Metabolic (degenerative) disease
Antigen	Disease	Phagocyte
Antimetabolite	Endocrine gland	Therapy

WORK EXERCISES

1. Distinguish between an antibiotic and an antibody. How are they different, and in what ways are they similar?

2. Distinguish between carcinoma and sarcoma, between a chemotherapeutic agent and an antibiotic, between a tranquilizer and an antidepressant, and between the effects of benzedrine and iproniazid.

3. The enzyme monoamine oxidase (MAO) probably catalyzes the metabolism of serotonin and norepinephrine. What would be the net chemical effect of inhibiting MAO?

4. Discuss the relationship of radiation to cancer.

---

* H. Selye, *The Stress of Life*, New York, McGraw-Hill, 1956.

SUGGESTED READING

Allison, A., "Lysosomes and Disease," *Scientific American*, Nov. 1967. The role of these subcellular organelles in disease is discussed in a strikingly illustrated article.

Barron, F., *et al.*, "The Hallucinogenic Drugs," *Scientific American*, Apr. 1964. This article gives structures and discusses effects of such alkaloids as LSD, psilocybin, and mescaline.

Best, J. B., "Protopsychology," *Scientific American*, Feb. 1963 (Offprint #149). This article discusses learning in planarian worms and the effects of diet—particularly a diet consisting of trained worms—on their learning. Some scientists who tried to repeat some of these experiments were unsuccessful. This casts doubt on the validity of some of the experiments reported in this article.

Braun, A. C., "The Reversal of Tumor Growth," *Scientific American*, Nov. 1965. A discussion of the types of cells that change from malignant to normal.

Collier, H. O. J., "Aspirin," *Scientific American*, Nov. 1963 (Offprint #169). A discussion of ways in which this long-used "miracle drug" works.

Comfort, A., *The Process of Ageing*, Signet Science Library, New American Library, New York, 1964. Dr. Comfort discusses many aspects of aging, in an engaging although essentially pessimistic book.

Cooper, L. Z., "German Measles," *Scientific American*, July 1966. This article tells the story of the development of a vaccine which may eventually control this disease.

DeKruif, P., *Microbe Hunters*, Harcourt, Brace and World, New York, 1932. A classic popularization of the lives of many great scientific microbe hunters. Also available in paperback.

Frei, E., III, and E. J. Freireich, "Leukemia," *Scientific American*, May 1964. A discussion of the disease and some treatments being used for it.

Gorini, L., "Antibiotics and the Genetic Code," *Scientific American*, Apr. 1966. A discussion of the effects of streptomycin and related drugs on protein synthesis.

Hammond, E. C., "The Effects of Smoking," *Scientific American*, July 1962 (Offprint #126). Present data is even more frightening, but this older report, along with its basic biological approach to the problem, is still good reading.

Hilleman, M. R., and A. A. Tytell, "The Induction of Interferon," *Scientific American*, July 1971 (Offprint #1226). Interferon is an important body defense against viral infections. This article describes ways to make cells produce more of this virus fighter.

Hirschorn, N., and W. B. Greenough, III, "Cholera," *Scientific American*, Aug. 1971. Although cholera can be treated by replacing body fluids, it continues to kill, particularly in under-developed countries. Treatment and possible antidotes are discussed in this article.

Langer, W. L., "The Black Death," *Scientific American*, Feb. 1964. This article discusses the plague that killed one-fourth of the people in Europe in the years 1348 to 1350.

Lasagna, L., *The Doctors' Dilemmas*, Collier Books (BS177v), New York, 1963. Dr. Lasagna discusses quacks, the drug industry, legal matters, and other controversial aspects of medicine.

Li, C. H., "The ACTH Molecule," *Scientific American*, July 1963 (Offprint #160). An excellent summary of the functions of the anterior pituitary gland, and a specific discussion of the polypeptide ACTH, its structure and function.

Linder, F. E., "The Health of the American People," *Scientific American*, June 1966. A discussion of techniques used and the results obtained from a survey of the health of many Americans.

Macalpine, I., and R. Hunter, "Porphyria and King George III," *Scientific American*, July 1969 (Offprint #1149). Did the English king who was at least partly responsible for the American Revolution suffer from the hereditary metabolic disease porphyria? The evidence is considered in this article, which shows a relationship between biochemistry and history.

Nossal, G. J. V., "How Cells Make Antibodies," *Scientific American*, Dec. 1964 (Offprint #199). A well-illustrated discussion of the involvement of genes in antibody synthesis.

Selye, H., *The Stress of Life*, McGraw-Hill, New York, 1956. This book gives Dr. Selye's fascinating story of the experiments that led to his theory of stress. He also discusses stress, aging, and philosophical implications of his theories.

Sharon, N., "The Bacterial Cell Wall," *Scientific American*, May 1969 (Offprint #1142). This article discusses both the chemical structure of bacterial cell walls and the action of the antibiotic penicillin in destroying bacteria.

Shock, N. W., "The Physiology of Aging," *Scientific American*, Jan. 1962. This article and the one below by Verzar treat aging from a physiological viewpoint and from the chemical-structural changes in the protein collagen which accompany aging.

Smith, I. M., "Death from Staphylococci," *Scientific American*, Feb. 1968. The changes in glucose, glycogen, and ATP in mice dying from staphylococcal infections led the author to speculate about the biochemical cause of death.

Speirs, R. S., "How Cells Attack Antigens," *Scientific American*, Feb. 1964 (Offprint #176). Discussion of the functions of specialized cells that protect the body from invasion by antigens.

Taussig, Helen B., "The Thalidomide Syndrome," *Scientific American*, Aug. 1962. A report of the effects on unborn children of the supposedly "safe" sedative.

Verzar, F., "The Aging of Collagen," *Scientific American*, Apr. 1963.

Zinsser, W., *Rats, Lice, and History*, Bantam Pathfinder paperback, New York, 1960. The subtitle of the book is "Being a study in biography, which, after twelve preliminary chapters indispensable for the preparation of the lay reader, deals with the life history of typhus fever." A fascinating book.

Up-to-date, nonprejudiced discussions of marihuana, LSD, and other similar drugs are hard to find. Those listed below are scholarly and authentic.

Dishotsky, N. I., et al., "LSD and Genetic Damage," *Science*, 30 April, 1971.

Grinspoon, L., "Marihuana," *Scientific American*, Dec. 1969 (Offprint #524).

Hollister, L. E., "Marihuana in Man: Three Years Later," *Science*, 2 April, 1971.

CHEMISTRY OF THE ENVIRONMENT 45

45-1 INTRODUCTION

Most of us have been fascinated by the voyages of the astronauts across vast reaches of inhospitable space. One of the more interesting aspects of these voyages is the fact that the astronauts are totally dependent on their spacecraft for all the resources that maintain life—and the spacecraft have only limited amounts of resources. There is a striking analogy between space vehicles and our earth. Every person on earth is totally dependent on our present supplies of oxygen, water, and other materials.

The earth gains and loses only insignificant amounts of matter each year, and our supplies of all materials (oxygen, water, metals) are only those we have "on board." Although we have a great number of atoms of oxygen and the other elements on our spaceship, they exist not only in a free state but also in various compounds that may or may not be useful to us.

To live, we must breathe free molecular oxygen in the air; carbon dioxide suffocates us and carbon monoxide poisons us. We are "using up" some elements, such as iron. That is, we are using them and then failing to return them to the environment in a form that we can use again. Our material resources are not inexhaustible. It is quite possible for us to "run out" of an element that our civilization depends on, such as copper. It is even possible to decrease an element such as free molecular oxygen to a level that would not support human life.

Although the specific number of atoms on the earth is fairly constant, we cannot make the same statement about the amount of energy. Every day we receive vast amounts of energy from our star, the sun. Since we know that the temperature of the earth is relatively stable, we can deduce that the earth loses energy at a rate that is approximately equal to that at which we receive energy from the sun. The relatively stable temperature is possible only because of a delicate balance between gain and loss of energy. We know that at times this balance has been upset, and there have been drastic changes in the climate of the earth. There is evidence that there have been five or six ice ages. Careful measurements show that the average surface temperature of the earth was increasing throughout the past century but that, since 1950, a cooling trend has started, and our temperature has decreased a few tenths of a degree celsius. Although the overall increase since 1850 seems slight, relatively small increases can trigger processes that have tremendous consequences. A warming of only a few degrees could melt most of the polar ice caps, raise the sea level, and flood many of the major cities of the world.

What is causing the current decrease in temperature? Is this the beginning of another ice age? How is the temperature decrease related to the large amounts of energy released when we burn fossil fuels—coals, natural gas, petroleum? What are the possibilities of oxidizing these fuels so rapidly that we will run out of free molecular oxygen and suffocate in an excess of carbon dioxide? These are important questions for everyone concerned with the continuance of life on earth. Their answers, although complex, are based on the chemical and physical principles discussed in this book. Some of these principles have already been presented. We will discuss both the global aspects of the problem and the less widespread but more obvious local and regional problems of resource exhaustion and pollution.

Spaceship earth does not have a mother planet to return to in order to replenish her supplies. If life as we know it is to survive, we must learn to use and re-use the materials we have. Many of the resources we depend on are currently being depleted, while others (plants, animals) are being threatened by a complex of chemical substances and processes created by modern society. These substances and processes could, in time, destroy our basic resources. They could even destroy man himself; already other species have been wiped out or are doomed to extinction in the near future.

We must be aware of the complex relationships of the various forms of matter and energy and the factors influencing conversion of one substance into another. These relationships, the factors that control them, and the general principles involved, are discussed in this chapter on the chemistry of the environment.

45-2 OUR GLOBAL ENVIRONMENT

The earth is surrounded and protected by the atmosphere. Its temperature and composition affect all of us, and its fluidity means that changes in any part of it can quickly be transferred to distant parts of the globe.

The composition of the atmosphere, with its accompanying effects on temperature and climate, is held in balance by an interesting series of interrelationships. We will explore some of them in this section.

It is important to remember that although temperatures and climate have varied widely in the past, these fluctuations have been due primarily to natural causes—not the works of man. There is little we can do about natural changes other than accept them. However, with the increase in the number of people on our planet and with growth of technological knowledge, man now has the power to initiate drastic changes. Although there are powerful natural forces serving to keep things in a state of dynamic equilibrium, many of them are delicately balanced, and only slight disruptions of their balance can lead to greatly amplified consequences.

Oxygen in the Atmosphere

It is difficult to explain exactly why we have the quantity of oxygen we find on our planet earth. We do not find this gas in the atmospheres of other planets, and apparently it was not present when the earth was formed. The oxygen we now find in the air came from various types of rocks—the silicates and perhaps carbonates, and from water. The photolysis of water, particularly in the upper atmosphere, to yield hydrogen and oxygen gases has been important and still continues. Some of the hydrogen formed in this process is lost into space, but most of the oxygen is retained in our atmosphere. The other major source of oxygen today is the process of photosynthesis, in which plants take in water and carbon dioxide and release free oxygen into the atmosphere. We can summarize the process by the following equation:

$$6CO_2 + 6H_2O \underset{\text{oxidation}}{\overset{\text{photosynthesis}}{\rightleftarrows}} C_6H_{12}O_6 + 6O_2$$
$$\text{(a simple sugar)}$$

The process is reversible and yields water and carbon dioxide when the products of photosynthesis are burned in an animal body or in a fire. In the distant past, huge amounts of the products of photosynthesis were converted to coal, petroleum, and natural gas. This process involved the removal of oxygen from the molecules found in the plants, leaving hydrocarbons in the case of petroleum and natural gas, and carbon in coal. Some of the oxygen in our atmosphere is there only because these fossil fuels were stored in a reduced (nonoxidized) state.

We are burning gasoline, fuel oil, natural gas, and coal at a rate infinitely more rapid than the rate at which they are being produced on the earth today. While this fact has serious consequences in terms of exhausting sources of stored energy, it could have even more serious consequences in terms of the amount of oxygen and carbon dioxide in our atmosphere. At the present time, the level of oxygen in our atmosphere seems to be staying constant at a 20.946% in spite of the amount of fossil fuels we are burning. This balance is due in part to the fact that green plants are producing large amounts of free oxygen. In fact, as predicted by the law of mass action, the presence of increased amounts of carbon dioxide increases the rate of photosynthesis by green plants. Although we are aware of trees and grasses as green plants, we are likely to forget, unless reminded, those one-celled green plants, the algae, which are found both in fresh water and in the oceans. These unicellular plants are responsible for about a third of the oxygen produced by green plants. Thus, we decrease our supplies of atmospheric oxygen not only when we cut down forests and convert farm or grass land to housing areas and asphalt parking lots, but also when we allow the pollution of streams, lakes, and the seas to inhibit the growth of algae. Although the evidence indicates that we have not *yet* exceeded the power of natural processes to regulate the level of oxygen, it is not a matter we can ignore.

Carbon Dioxide in the Atmosphere

In contrast to the constant level of oxygen, that of carbon dioxide in our atmosphere has been increasing at a slow but steady rate. It is estimated that the amount of carbon dioxide increased from 290 to about 330 parts per million (ppm) from 1900 to about 1950. As the result of definite measurements, we know that the level of carbon dioxide has been increasing about 0.2% or 0.7 ppm each year since 1958.

Some scientists are concerned about this increase, because it is occurring in spite of several processes that tend to keep the level constant. We have already mentioned the effects of the law of mass action on photosynthesis, which is one important control. Another homeostatic force is the solution of carbon dioxide in water. The oceans of the world are huge reservoirs of dissolved carbon dioxide, not because it is present in any great concentration but because the volume of the oceans is so great. The higher the concentration of carbon dioxide in the air, the more we would expect to be dissolved in the oceans. Another factor that influences the solubility of a gas in water is the temperature, and we would expect any increase in the average earth temperature to decrease the amount of carbon dioxide in the seas and a general cooling to add to the amount dissolved.

There is another relationship between the amount of carbon dioxide in the atmosphere and temperature. This gas absorbs more of the heat the earth would normally lose to space than either of the other major gases of the atmosphere, oxygen and nitrogen. Thus an increase in the concentration of carbon dioxide in the air would tend to increase the temperature of the earth. The higher temperature would cause the oceans to release more carbon dioxide to the atmosphere, and this would, in turn, tend to further increase the temperature, resulting in the release of more carbon dioxide. Thus we have a self-catalyzing process in which a small change could result in major consequences.

Since we know from careful measurements that the amount of carbon dioxide in the atmosphere is increasing, we would suppose that the temperature of the earth is increasing. However, careful measurements of temperature show the opposite and this decrease in temperature is happening even though we are releasing great amounts of heat by burning fossil fuels. Therefore we must look for a compensating factor to account for the decrease in temperature. It is possible that the amount of radiation being emitted by the sun is decreasing; however, it seems more likely that the amount of this radiation being absorbed by the earth's atmosphere is decreasing. This tendency may be due to the fact that we are constantly increasing the amount of particulate matter (ash, soot) that is being discharged into the air. This not only reflects the sun's energy back into space, but also forms nuclei around which water particles can condense and form clouds. Clouds also reflect the sun's radiation back into space. One has only to recall the difference in temperature on sunny and cloudy days to become aware of the tremendous effect of cloud cover on temperature. Although we know that the amount of particulate matter in our atmosphere is increasing, we have inadequate information about the amount of cloud cover on the whole globe on an annual basis. The increasing use of weather satellites should give us more data in the near future.

There is, under natural conditions, very little circulation or deposition of particles in our upper atmosphere. It is estimated that in the stratosphere (elevations above 11 km), particles may persist for as long as five years. Since supersonic planes are now being designed to fly at these elevations, many people are concerned about the effect of the water (ice) particles as well as other exhaust substances from these high flying aircraft. Our information is inadequate to accurately estimate the results of such particles, but one possibility is that they could form long-lasting clouds that would decrease still further the amount of solar radiation that reaches the earth.

From the foregoing discussion, you can see that the levels of just two of the gases of the atmosphere, oxygen and carbon dioxide, are the result of several related processes. These relationships are represented in Fig. 45–1. Although the amount of carbon in each of its various forms is important, the rate of conversion of one compound to another is even more important. We use the concept of **feedback** in describing the rate and control of these changes (see Fig. 45–2). We say that there is positive feedback when the production of one factor increases the production of itself or of another substance. Such a relationship is seen in the case of atmospheric carbon dioxide increasing the temperature of the earth's surface, which in turn increases the rate of release of carbon dioxide from sea water. An example of negative feedback is found in the fact that high levels of atmospheric oxygen depress the rate of photosynthesis. If we are to understand and perhaps control a complex series of physical and chemical changes, it is necessary to know not

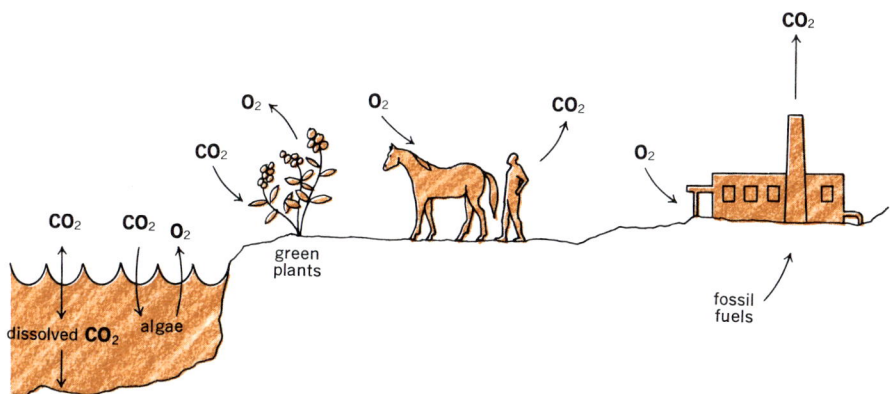

Fig. 45-1 The carbon and oxygen cycle. Many processes are active in maintaining the amounts of oxygen and carbon dioxide in our atmosphere. The most important ones are shown above. Photosynthesis in green plants on land and in the sea consumes carbon dioxide and produces free oxygen, while oxidation of plant material (wood, coal, etc.), whether in an animal's body or in a fire, consumes oxygen and releases carbon dioxide. On a global scale the amount of carbon dioxide dissolved in seawater is a significant factor; this process is highly sensitive to temperature.

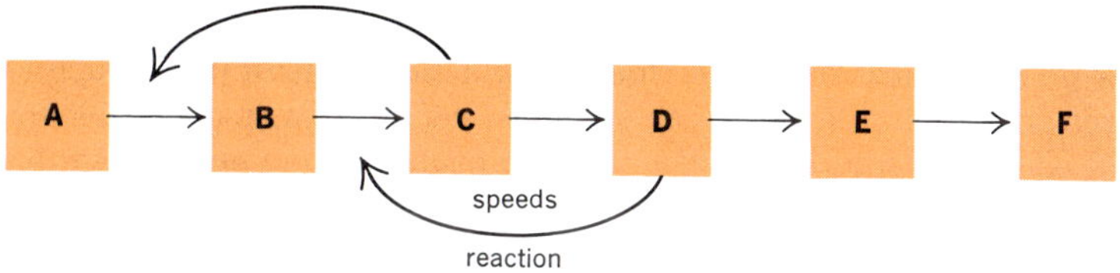

Fig. 45-2 Feedback. The basic process is the conversion of substance A to substance F. If one of the intermediates or products in the overall series of reactions inhibits a step in the process, we say there is negative feedback. If an intermediate or product helps the reaction proceed at a faster rate, we call this positive feedback, or amplification. In the illustration above, C exerts negative feedback and D positive feedback on the conversion of A to F.

---

only the products and reactants involved but also the rate of change and the effects of many factors on these rates of change.

45-3 EFFECTS OF MAN ON LOCAL ENVIRONMENTS

Although the net effect of man on his global environment seems to have been relatively small so far, we must realize that this fact is due to the extremely large size and volume of our spaceship earth. Natural processes are powerful, but man now has the potentiality to interfere with them, and minor changes may become major if positive feedback occurs.

The degree of alteration of our local environments is quite different, however. Over each of the major cities of our world there is a pall of dirty air. Sometimes the level of air pollution is so extreme that children are not allowed to go out to run and play because exercise would cause them to breathe more of the polluted air. Streams that once carried water so pure that it could be used for drinking purposes are now so contaminated that even swimming is forbidden. Like global problems, these can be explained in chemical and physical terms, and their solutions require scientific knowledge.

Three air pollution disasters have received worldwide attention. In 1952, a black fog that persisted for several days over London was responsible for at least 4000 deaths. In Donora, Pennsylvania, in 1948 and in the Meuse Valley of Belgium in 1930 scores of deaths and a great deal of intense discomfort were caused by persistent fogs that became loaded with pollutants from nearby factories. Many of the victims of these disasters were older people, and the cause of death was usually respiratory or heart failure.

Since these disasters produced immediate death and widespread discomfort, they created public concern. Probably much more dangerous are the effects of continuous breathing of polluted air over a period of months or years. These effects are difficult to attribute to pollution since people die from a variety of lung diseases and heart conditions even in the absence of air pollution. Only extended studies will show the extent to which various causes of death are influenced by long-term exposure to relatively low levels of pollutants. However, current evidence concerning the increase of the lung disease emphysema, as well as other lung disorders, is alarming. Further research will probably uncover more of the relationships between pollutants and health problems.

45-4 THE AIR WE BREATHE

More than 3000 foreign substances have been found in our air. Some of these, such as the odor of pine trees and pollen from a variety of plants, we regard as natural. Others, such as the oxides of nitrogen, carbon monoxide, and the ozone we find near the earth's surface, can be directly attributed to man and his works—especially the internal combustion engine in his automobiles. Many of these substances are present in only a few parts per million even in the air over our most polluted cities on a bad day. While these levels seem low, it is known that a concentration of only 6 ppm of ozone will kill any laboratory animal so far tested, and it does so within a period of four hours.

Ever since he first used fire, man has polluted the air. The concentration of people in cities and the introduction of factories and automobiles have greatly complicated the problem. Another complicating factor is the presence of **inversion layers** in our atmosphere. Normally, when we produce a pollutant, it rises in the air and is scattered by wind currents because warm air is lighter and rises through the denser cold air. If pollutants reach a region of **temperature inversion,** that is, a region where the air is warmer and less dense, they stop rising and concentrate below the inversion zone (see Fig. 45-3). Although there is little evidence that cities create these inversions, many cities are located in areas where they form easily. The tragic incidents in London, Donora, and the Meuse Valley were probably due to inversion layers which trapped gases being discharged into the atmosphere.

Most coal and petroleum contain sulfur. When burned, they produce a pungent gas, sulfur dioxide. Although the sulfur content of petroleum can be reduced by refining, this is not done with coal. In general, soft coal contains more sulfur than hard coal.

Sulfur dioxide is the oxide of a nonmetal, and when it is combined with water it forms sulfurous acid (H_2SO_3). If sulfur dioxide is oxidized by oxygen or other substances in air, it forms sulfur trioxide, which reacts with water to form sulfuric acid. Although the

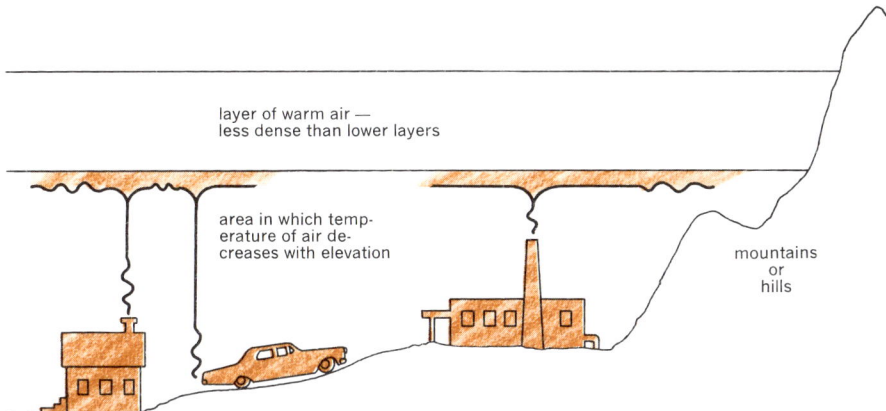

Fig. 45-3 Diagram of an inversion layer in the atmosphere. Normally, warm, polluted air rises and is scattered by the wind. When an inversion layer forms, polluted air can no longer rise and disperse. Inversion layers form only in the absence of wind and are broken up when the wind returns.

---

amount of this acid in the air is small, it is a strong acid capable of causing major lung damage and death, as well as the disappearance of nylon hose when ladies wear them on days when pollution is high. Sulfur dioxide was the major culprit in the killer fogs of London, Donora, and the Meuse Valley.

Another by-product from the burning of coal is the mixture of carbon particles we call soot. Lungs of people who have lived in London or other major cities where coal has been burned are usually a dirty gray, due to the continued inhalation of soot. Along with these dark particles are other coal derivatives, many of them colorless, but nonetheless deadly. One of the first correlations of occupations with cancer was observed in chimney sweeps. Something in the soot and other debris they continually encountered appeared to be causing cancer of the skin and lungs. We are now quite certain that many of the compounds that are found not only in coal smoke but also in the smoke of cigarettes and even on charcoal-broiled steaks can cause cancer. These substances, which are commonly known as tars, do not cause cancer immediately and not everyone exposed to them will die from cancer, but the cause-effect relationship is well established.

Although coal is the source of two major pollutants, it is not the major cause of the carbon monoxide, ozone, oxides of nitrogen, and hydrocarbons we find in the air above major cities today. Here the culprit is the internal combustion engine. This engine is a device to derive power from oxidation of the hydrocarbons in gasoline and similar fuels. While it does this efficiently enough to power automobiles and trucks, it also produces carbon monoxide, oxides of nitrogen, and unburned hydrocarbons which are usually eliminated in the exhaust. When these substances react, particularly in the sunlight, with the oxygen in the air and with each other, they produce the mixture we call *smog*. The name is really inappropriate, since it should indicate a mixture of smoke and fog, neither of which is present in cities located in relatively dry, sunny areas such as Los Angeles and Denver. The word *smaze* has been coined for a mixture of smoke and haze. Whatever we choose to call this polluted air, it represents a major cause of concern for everyone living in an urban area.

Carbon monoxide is produced whenever carbon-containing substances are burned in insufficient oxygen. Carbon monoxide has an affinity for hemoglobin that is more than 200 times that of oxygen, and molecules of hemoglobin which have reacted with carbon monoxide (carboxyhemoglobin) are unable to combine with oxygen. If a sufficient percentage of molecules of hemoglobin are so tied up, death occurs. We are all aware of the deaths of people who operate automobiles in closed garages. If the automobile operates outside, the same carbon monoxide is being produced, but it is being widely scattered until it reaches a nontoxic level. If thousands of cars are producing carbon monoxide in a city under an inversion layer, we have increased the size of the "garage" but must expect bad consequences. It is probable that many automobile accidents are partially due to drivers forced to breathe too much carbon monoxide. Continued breathing of air containing 0.1% or 1 part per thousand of carbon monoxide for 90 minutes may be fatal. (In discussing toxic effects of gases, we must specify not only a level of poison but also the period of exposure, since it is the total amount of poison absorbed that is significant).

Those who smoke cigarettes and particularly those who inhale are also exposed to relatively large concentrations of carbon monoxide and probably expose those in their immediate vicinity also. It has been shown that a person who smokes one pack of cigarettes per day and inhales deeply has about 6% of his hemoglobin tied up as carboxyhemoglobin. Experiments have shown that this level can affect eyesight and time discrimination. Other effects that have been connected with carbon monoxide toxicity are impairment of vision and time-interval discrimination, fatigue, headaches, irritability, drowsiness. If the levels of carboxyhemaglobin are high enough, coma and death result. Experiments with animals have shown definite heart damage due to relatively low levels of carbon monoxide.

The state of California lists 30 ppm of carbon monoxide for eight hours as an adverse level, and 120 ppm for one hour as a serious level of pollution. The latter level has been measured on the streets of Los Angeles. The present maximum allowable atmospheric concentration for occupational exposure to carbon monoxide is 50 ppm for eight hours. This limit was set at 100 ppm until 1964.

The human body gradually excretes carbon monoxide that has combined with hemoglobin and makes new hemoglobin to replace it, so that the effects of the poison can be reversed. However, the impaired function of the body operating under decreased supplies of oxygen may produce permanent damage, especially to the heart. As with many poisons, children are more susceptible to carbon monoxide than are adults.

Although carbon monoxide is oxidized to carbon dioxide in the atmosphere and certain soil organisms also remove carbon monoxide, these processes are affected by many variables. If the conversion of CO to CO_2 is decreased it is conceivable that we could begin to accumulate high levels of CO in our global air supply. This is another substance whose concentration should be monitored more carefully on both a local and global scale.

The two most abundant substances in air are oxygen and nitrogen. Under ordinary conditions these two gases do not react with each other. However, when heated under pressure in the presence of catalysts—conditions present in automobile engines—they react to form nitrogen oxide (NO). This gas, commonly called nitric oxide, is readily oxidized to yield nitrogen dioxide (nitrogen(IV) oxide). These and other possible oxides

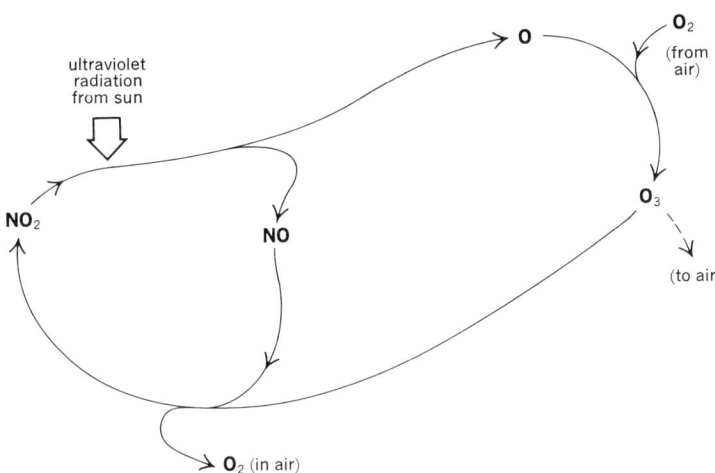

Fig. 45-4 Interrelationship of ultraviolet radiation, oxides of nitrogen, ozone, and oxygen. This diagram shows the effects of ultraviolet radiation on production of some of the components of photochemical smog. If some part of the cycle is interrupted, one of the substances (e.g., ozone) may accumulate or react with substances not included in this cycle.

of nitrogen make up another significant and deadly component of air pollution. It is nitrogen dioxide that is responsible for the brown color of much air pollution.

We can summarize the reactions of oxygen and nitrogen as shown below:

$$O_2 + N_2 \rightarrow 2NO$$
$$2NO + O_2 \rightarrow 2NO_2$$

People exposed to 500 ppm of NO_2 in the air die almost immediately. Lower concentrations produce lung damage in experimental animals; in fact, nitrogen dioxide has been characterized as the only gas that invariably produces the lung disease emphysema in rats. Scientists disagree about levels that can be tolerated by humans. The American Conference of Governmental and Industrial Hygienists gives 5 ppm as a tentative threshold (allowable) dose, whereas Russian scientists report that a level of 3 ppm has produced emphysema. Levels of nitrogen dioxide reach 0.9 ppm in Los Angeles on days of high pollution.

In addition to its primary effects, nitrogen dioxide serves as a facilitator in the production of ozone, as shown by the following reaction:

$$NO_2 + O_2 \xrightarrow{\text{ultraviolet radiation}} O_3 + NO$$

This reaction accounts for the formation of ozone and explains the fact that ozone levels are higher on sunny days than on cloudy ones. The reaction is reversible and is further complicated by the fact that the nitrogen oxide formed can also react with normal oxygen to give more nitrogen dioxide which, in turn, reacts with oxygen to produce more ozone (see Fig. 45-4).

Ozone is not totally bad for life on earth. In the upper atmosphere, from 25 to 50 kilometers above the earth's surface, the reaction of oxygen with ultraviolet radiation from the sun produces ozone. This process prevents the ultraviolet radiation from reaching the earth's surface, where it would certainly increase the amount of sunburn that beach lovers get and might even destroy many forms of life.

Ozone is extremely toxic, with relatively large amounts producing immediate death, whereas smaller amounts cause respiratory diseases. So long as ozone is 15 miles (25 km) above the earth, only those who fly aircraft at this height need be concerned, but at the surface of the earth it becomes of great importance to all of us. It has been shown that 6 ppm of ozone kills laboratory animals in a short time. The effects of ozone are increased if the person exposed is exercising or if the temperature is high, and ozone is especially damaging to children. Ozone in the atmosphere is generally not measured as such, but it makes up about 90% of the general category of "oxidants" which are now measured in many locations on a continuing basis. In Los Angeles, school children are not allowed to play outside when the level of oxidants goes above 0.35 ppm.

Parenthetically, we should say that we make many references to Los Angeles not because conditions are necessarily worse there than in other places, but because the people there have become sufficiently concerned to find out the facts and make protective regulations. At the present time half of the air pollution experts in the United States work in the Los Angeles area, and it has become the source of accurate measurements which we need to speak precisely about levels and effects. Other cities in the United States and the world may have pollution as bad as or worse than Los Angeles; we just do not have the figures to cite. Tokyo and Mexico City have especially bad air pollution problems. Other major industrial cities are saved only because they receive large amounts of rain, which is one of the best remedies for air pollution—probably at the expense of the streams that carry this polluted rain water.

One of the most obvious effects of air pollution is its effects on the eyes. People living in polluted areas are highly aware of eye irritation that results in pain and excessive watering. Although a variety of chemical compounds are probably responsible, peroxyacetylnitrate (PAN) is probably the main tear gas or lacrimator in smog.

$$CH_3-\overset{\overset{\displaystyle O}{\|}}{C}-O-O-NO_2$$

peroxyacetyl nitrate (PAN)

It is produced by the reaction of unburned hydrocarbons (particularly those with a double bond in them) with ozone and the nitrogen oxides. Although eye irritation is painful, it does not seem to produce long-term permanent damage. Other pollutants, some of which our senses do not detect directly, are much more damaging. PAN, however, can permanently damage plants.

In addition to the many gases found in the air surrounding our cities, there are many kinds of small particles. Included in the value reported as "particulates" are particles of ash, soot, rubber, asbestos, and hundreds of other materials. Asbestos, which probably comes primarily from brake linings of cars, produces lung cancer. It does not do so immediately but has a latency period of several years, as is the case with many cancer-causing substances. Rubber particles from automobile tires are also suspected of producing cancer. Although ash and soot are decreasing in many cities as less coal is burned, the amounts of other particulate matter seem to be increasing.

So far we have concentrated on effects of air pollution on humans; however, widespread plant damage is also occurring near many cities. Spinach, orchids, and violets can no longer be grown in the Los Angeles area. Forests on the nearby mountains are also being affected. This damage is probably due to ozone and some of the unburned hydrocarbons, especially ethylene (C_2H_4). This sensitivity of plants is in contrast to their ability to grow in and concentrate certain other poisons. Plants growing near highways have been shown to contain high levels of lead. Although the lead does not seem to affect the plants, it can poison the animal that eats them. Other aspects of lead poisoning will be discussed in Section 45–6.

45–5 WATER POLLUTION

For hundreds of years, the rivers of the world have served as sewers for whatever soluble or suspendable waste man has wanted to discard. Many of the substances we dump into our waterways can be used for food by plants and animals living in the water. These organisms also destroy or inactivate many poisons. However, we are now overloading most of our streams both with nutrients and poisons, and the animals and many of the plants are dying. Other plants, especially the algae, are growing better than usual because of the increased nutrients, particularly phosphates, nitrates, and organic matter from human wastes. This increase in growth is called **eutrophication**. It provides more food for fish and other animals and releases oxygen into the air, but it also has disadvantages. If algae accumulate, they settle to the bottom of lakes or slow-running streams, where they decay. This process uses oxygen which comes from the water. When water is depleted

Fig. 45–5 Fish killed as a result of eutrophication. This process occurs when large amounts of nutrients, added to bodies of water, stimulate the growth of algae. If the algae are not eaten by fish, but instead fall to the bottom of the body of water, their decay consumes large amounts of oxygen. When the water is deprived of oxygen, the fish die. [Photo courtesy of the Environmental Protection Agency.]

of oxygen, fish and other animals die. The death of fish both from poisons and from lack of oxygen tends to increase the amount of algae, and many lakes now resemble pea soup. (See Fig. 45-5.) Some parts of Lake Erie are so full of pollutants that few things except algae live there now.

In many streams a precarious balance exists between the destruction caused by poisonous wastes, often from factories, and that caused by substances classed as nutrients. A more detailed discussion of these substances will be found in later sections of this chapter.

In addition to these life and death matters, the clear, fresh-smelling water that is so important in our enjoyment of lakes and rivers is seldom seen. The same problems apply to the mixture of fresh and salt water in bays opening to the sea. In many places, the oceans themselves are becoming polluted. They represent a large reservoir for dispersal of pollutants, but they are not inexhaustible. Many of the topics discussed in the following sections might also be classified as water pollution, since the pollutants are carried by water.

One of the more recently recognized kinds of water pollution is simply heat from industrial processes. Many factories, especially power plants, use large quantities of water for cooling. When this water is returned to streams and rivers at temperatures higher than normal, changes occur. The kinds of plants and animals that live in warm waters are different from those in cooler water, and so we see a change in ecological balance when the water is warmed. Since warming decreases the solubility of oxygen in water, thermal pollution (or calefaction) diminishes the oxygen available for fish and other animals.

45-6 POISONOUS ELEMENTS AND COMPOUNDS IN THE ENVIRONMENT

The ions of approximately 25 elements are highly toxic to man and other animals. Among the most toxic are the metals lead, mercury, arsenic, beryllium, cadmium, chromium, and nickel. The nonmetal fluorine is apparently needed in small concentrations, but it becomes toxic at higher levels. Although these elements occur widely in nature, they are usually found in compounds that are insoluble and therefore not likely to poison humans. It is when we concentrate and solubilize these elements that problems may arise. Many cases of acute poisoning have been caused by ingestion of relatively small quantities, but even more insidious are deaths and disabilities due to long-term ingestion of very small doses. Since the animal body does not excrete these substances very well, levels tend to increase: we say that these are cumulative poisons. Many of the effects of these poisons are similar to those of other diseases and disorders, and it is difficult to be certain whether they are the actual causes of death. We know from studies with experimental animals that deaths do occur after chronic poisoning, but we must make a test of the content of these elements in blood, bone, or other tissue before we can determine the extent to which they are responsible for death or illness of humans.

Lead

Lead was one of the elements known to ancient peoples. The ease with which it was separated from ores, its low melting point, the ease with which it was shaped, and its resistance to corrosion made it useful. Although the Greeks knew about the toxicity of lead, apparently the Romans did not. They used it for water pipes and for lining wine-storing vessels (in fact, the word *plumbing* and the chemical symbol Pb are derived from the Latin word for lead, *plumbum*). Examination of bones from Roman tombs shows concentrations of lead sufficiently high that we can be certain they suffered from lead poisoning. Since lead poisoning is known to affect mental processes, many have speculated that the erratic behavior of some of the latter-day Romans may have been due to lead poisoning, and others even go so far as to suggest that part of the reason for the fall of the Roman Empire was widespread lead poisoning among the ruling classes. Poor people did not use the expensive lead containers and were not so likely to be poisoned.

We have used lead for a variety of purposes in our civilization. While we no longer use lead pipes to carry drinking water, this change is comparatively recent. The insecticide lead arsenate (which contains arsenic in addition to lead) was widely used, particularly on tobacco plants, until comparatively recently. Both of these toxic elements were absorbed by plants and probably still persist in soils. Until about 1940 lead was widely used as a pigment in paints. It is used for glazes on pottery, and as recently as 1970 there were many episodes of lead poisoning due to drinking weakly acidic liquids, such as apple juice and cola beverages, from glazed pottery. Perhaps the greatest source of lead in our environment today is the tetraethyl lead used as an anti-knock ingredient in high octane gasoline. Examination of ice and snow in the region of the North Pole shows a steady increase in lead content from the beginning of the industrial revolution and an increase of about 300 percent between 1940 and the present. The presence of lead in such a remote place is a clear indication of the global nature of the problem.

Acute lead poisoning results in brain damage, mental deficiency, and serious behavior problems. Kidney damage is seen both in acute and chronic cases of lead poisoning. Acute poisoning is observed in some people who work in industries which expose them to high concentrations of lead. Even more tragic is the acute lead poisoning of children who live in urban slum areas. These children often eat chips of paint that contain lead as a pigment. Although titanium dioxide has replaced lead pigments in most white paints today, older houses may still have the older toxic paints, perhaps a layer or two below the current paint. Surveys in major United States cities show that lead poisoning of children is a major problem—one that can be prevented with sufficient knowledge and care. We know that many children who live in cities have levels of lead in their blood that are one-half those found in children who have definite symptoms of lead poisoning. For an excellent discussion of this and other aspects of lead poisoning, see the article by Chisholm in the list of readings at the end of this chapter.

We do not know the effects on the general population of exposure to the lead from automobile exhausts. As with ionizing radiation, none of our senses warn us that we are being poisoned. However, chemical analysis shows us that people accumulate more lead in their bones as they grow older, and thus lead can be considered a cumulative poison. Since the symptoms of chronic lead poisoning are similar to those caused by many diseases and other poisons, it is difficult to assess the long-term effect of low levels of exposure. Tests for the presence of lead in body tissues will have to be widely made and the results correlated with the cause of death in a large number of people before we can give definite answers. In contrast to the situation with the ancient Romans, in our civilization it is the poor who live in old houses near a great deal of traffic that are most likely to be poisoned.

Even in the absence of proof of the influences of low levels of lead, it seems logical to decrease the amount in our environment as much as possible. The widespread use of lead in plumbing, insecticides, and in paints is probably a thing of the past. It is clear that we must become more aware of the content of pottery glazes, and perhaps widespread testing will have to be instituted. The most important current effort to reduce lead in the environment is the introduction of low-lead and lead-free gasolines. These products are a response both to concern for the amount of lead in the atmosphere and to the fact that catalytic after-burners used to decrease other types of pollution from automobile exhausts cannot function efficiently if lead is present.

The story of the use of lead illustrates how chemical knowledge of things not perceived directly by our senses can lead to improvement in health and decrease in deaths by poison. It also illustrates how facts known as long ago as the age of the ancient Greeks may be rediscovered centuries later—often only after the senseless death of many people, especially children. Unfortunately, we lack knowledge and concern about many other poisons we are exposed to every day.

Mercury

The Mad Hatter in "Alice in Wonderland" is a classic example of a person poisoned by exposure to mercury. Felt hats were treated with compounds of mercury, and the expression "mad as a hatter" indicates a widespread knowledge of the symptoms but indicates no awareness that such behavior was due to mercury.

In 1969, the children of a New Mexico family fell ill with a variety of symptoms. Several weeks elapsed before urinalysis was carried out which showed that they were suffering from mercury poisoning. As the medical detective work later revealed, the father of the family fed his hogs some seed grain that had been treated with mercury to retard growth of fungi. Some of the hogs died of a condition described as "blind staggers," but this fact did not prevent the butchering of hogs from this group and the use of the pork for food. The symptoms observed in the afflicted children included dizziness, loss of muscular control, loss of vision, deafness, and long periods of unconsciousness or coma. A year later one child was confined to his wheelchair unable to walk more than a step or two and unable to control the bobbing of his head; another could walk a bit; and a third had just emerged from eight months of coma, blind and unable to talk. A fourth child who was born soon after the other children were affected is blind, and hope for him is slight. The effects on the children were greater than on the parents.

In addition to using mercury in thermometers, where it is safely encased in glass, we have used a variety of mercury compounds over the years. The compound $HgCl_2$ was once used as a germicide. Various compounds of mercury are used to prevent growth of fungi on seed grains. Many chemical processes, particularly the production of chlorine and sodium hydroxide, use mercury.

From 1953 until 1960 near Minimata Bay in Japan, more than 111 persons died or were seriously handicapped as a result of long-term consumption of mercury in fish and shellfish which were taken from industrially polluted water. Five more died and 25 were injured at Nigata in Japan in 1968. In Iraq, 35 people died and 321 were injured in 1961 from mercury poisoning, and there are unconfirmed reports of another poisoning incident involving hundreds of people in Iraq at the time this book is being written.

It has been known for some time that organic compounds of mercury, such as those containing the methylmercury ion, are water soluble, and that dimethyl mercury is fat soluble. It is also known that both of these compounds are extremely toxic. It has been assumed, however, that mercury itself and inorganic compounds of mercury are insoluble and heavy enough to settle to the bottom of bodies of water and thus remove themselves from circulation. In addition, we reclaim much of the mercury in industrial use because it is a valuable and rare metal. However, recent findings have led to reconsideration of the safety of our discard procedures. There is a bacterium that converts apparently insoluble inorganic mercury compounds to the toxic, soluble organic compounds. When we add to this knowledge the realization that much of the mercury used to treat seeds leaches out of the soil into streams and rivers, we have cause for concern.

These concerns led to the widespread analysis of various kinds of fish for mercury. Much of the American public was shocked in 1971 when analysis of deep-sea fish showed extensive amounts of mercury. Where are these large deep-sea animals getting the poison? Apparently the explanation involves food chains. Soluble mercury is present in water either as runoff from fields or from conversion of insoluble forms. The soluble mercury is ingested by plants or small animals, and these are eaten by larger animals. Each step can produce a greater concentration of mercury in the tissues, and it seems that the largest fish have the greatest amounts of mercury. Of course, a smaller fish or even a shellfish such as an oyster or clam could have high concentrations if it grew in waters with high mercury content.

Examination of preserved fish which lived fifty to a hundred years ago show that, even then, they contained a fairly high concentration of mercury. Even fossil fish contain small amounts of this element.

With mercury as with most cumulative poisons, we do now know whether there is a level low enough to be safe or what limits should be set on foods for human consumption. We do know that the substance is poisonous and that we must be cautious in its use and consumption. Recent analyses have shown that most foods, other than fish and sea foods, have very low levels of mercury.

Although some incidents have been reported, there are at present no evidences of widespread poisoning by the metals arsenic, beryllium, cadmium, chromium, and nickel. These elements are widely used for a variety of purposes in technologically advanced societies. We know they are toxic, but at present we have little information about levels of toxicity, amount and duration of retention in the human body, and effects of long-term exposure. Most chemical industries know of these toxicities and take care to protect their workers, but some firms are not sufficiently concerned about release of small amounts into the water or atmosphere. Less well-informed people use these elements in ways that could result in acute poisoning. Unfortunately, we usually do not become sufficiently concerned about such problems until they cause the death of several people.

Fluorine

Many ores, particularly those of aluminum, are fluorides. Chemical plants that process these ores often release fluorine in a volatile compound or small particles. Cattle, sheep, and other grazing animals have often been shown to die from fluoride toxicity. With continuous care, it is possible to remove this toxic element from both gaseous and liquid wastes of industry.

45-7 RESOURCES IN SHORT SUPPLY

So far, we have been concerned with the effects on the environment of *additions* that man and civilization have made. The opposite is also true. Continued farming has *depleted* many soils; our demands for energy for transportation and construction have made us use coal and petroleum at rates far exceeding their formation. The continuation of life to any degree approaching the style to which many of us have become accustomed or to which so many aspire demands that we give serious consideration to these problems.

Modern industrialized civilization, with its emphasis on buildings, automobiles, automatic dishwashers, television sets, and the myriad other things we use and feel we need places a great demand on metals. Those most critical in our society are iron, aluminum, and copper. Other metals such as mercury, zinc, lead, molybdenum, tin, tungsten, cobalt, nickel, and vanadium are needed in smaller amounts, and when the supply of these is small or located at relatively few places on the earth, supplies of them may also be critical.

At one time there were huge deposits of iron ore in the Mesabi range in Minnesota. Most of the high-grade ore there has now been used. Once copper ores containing nearly pure copper could be picked up at the surface of the earth. One hundred years ago in the United States we were mining ores in the 5% to 6% percent category. Now we have used these high-grade ores and are mining and refining ores that contain 0.4% to 0.8% of the metal. Mines once abandoned are being reexplored with the hope that they can now be worked profitably. Although there were once good deposits of high-grade aluminum ores in the United States, we are importing much of this ore today.

Almost all the industrialized nations are now importing the ores of iron, aluminum, copper, and the other metals needed to maintain a high production of consumer goods. In the past, mineral supplies have led to armed invasions and conquest of mineral-rich nations. We cannot disregard the relationship of mineral resources, including petroleum, to the wars of the twentieth century.

Even if political and transportation problems could be solved, the basic problem remains. We are using mineral resources at a rapid rate. Except for deposits in some of the nonindustrialized countries, the high-percentage ores have been used. Although technology can develop ways of using minerals having lower content of the desired metal, this is done only by using an increased amount of energy to separate the metal from the source in which it is found. One aspect of this problem is that the ores we are now using come from greater depths and we must remove more dirt and rock to get to the deposits. We also remove more waste material or slag as we refine lower and lower grade ores. In order to do all this, the price for the metal must rise to compensate for the greater cost of producing it. In terms of the available reserves of energy, particularly the fossil fuels, it is questionable that we should continue to use more of these to produce metals from low grade ores. Many critical metals might be extracted from sea water, but they exist there in extremely low levels and a great deal of energy is required to concentrate and refine them.

Combined with this shortage of many metals, we see the huge lots filled with discarded automobiles, our streams are littered with soft drink cans, and much of our metal waste is buried in disposal areas. While there are some losses that cannot be easily prevented such as loss of iron due to rusting, we can re-use much of our metals by the process we have come to call recycling. There is, of course, no shortage of iron atoms. It is just that most of them are widely dispersed in the earth's crust. We have concentrated some of them and used them to build skyscrapers and automobiles. If we are to build more skyscrapers and automobiles, we must increase the amount of iron that is recycled. The same is true of all the other metals we need. Except for corrosion, our use of them does not exhaust them in the same way our use of coal does; it merely requires an awareness and the development of facilities to use more effectively those atoms we have on board our spaceship earth.

Green plants need not only carbon dioxide and water to provide us with food, but they also require mineral elements from the soil. Continued farming with the removal of plants has seriously depleted the soils of Italy and the Near East. While the absence of sufficient rainfall is also serious, many soils are in urgent need of mineral elements, particularly nitrogen, phosphorus, and potassium. The Haber process for the production of ammonia from the nitrogen of the air and hydrogen provides a solution to the problem of lack of nitrogen in soils, provided that sufficient money and technological knowledge are available where the needs are great. Unfortunately, much of the nitrogen applied to soil is leached out by water; and it is lost to the soil while being of questionable use in streams. Supplies of potassium are relatively abundant. The major shortage is that of phosphate. Our sources of mineral phosphate are limited, and knowledge and technology are necessary if we are to use the supplies we have. Even in this period of phosphate shortage, much is being wasted. Since many phosphates are relatively soluble, they are easily leached out of fields by natural rainfall and by irrigation. Many soap powders and detergents have phosphates added to them. After use these phosphates go to streams, where they serve as excellent fertilizer for the growth of algae. This increase in growth (eutrophication) is found in many lakes and streams in populated areas. The two disadvantages of eutrophication are that the algae are not used as food, which is wasteful, and that fish and other animals die because the oxygen in the water has been used up by decaying algae.

The two problems, one the lack of sufficient nutrients in many soils combined with the eutrophication produced when these same nutrients, the nitrates and phosphates, flow into streams, seem to suggest possible solutions. Since there is no net deficiency of elements, merely a poor distribution of them, the solution must involve some form of recycling—taking from the areas of oversupply and returning to those of deficiency. While the theory is simple, practical recovery, particularly if the needed elements are widely dispersed in the oceans, is difficult.

45–8 ENERGY SOURCES

Modern civilization has been made possible because of our exploitation of fossil fuels. So long as our sources of power were animals who ate plants or the actual plants themselves as wood or charcoal, the amount of energy each individual could appropriate to his use was small. While some use could be made of water power, the industrial revolution came only when coal was available. The addition of petroleum products greatly increased the amount of available power and brought us from the age of the steam locomotive to the automobile and diesel tractor. Such use of energy poses several problems. As we have seen, much of the pollution in our atmosphere today is due to the burning of these fuels. Other problems dealing with the use of fossil fuels are associated with the effects on global concentrations of oxygen and carbon dioxide and the subsequent effects on the temperature of the earth. As indicated previously, these seem not to be crucial at the present, but they cannot be neglected.

The major problem posed involves our depletion of these sources of energy. The amount of coal and petroleum is limited. Other sources of energy, primarily that from nuclear reactions and from solar energy, must be explored. Energy from nuclear fission is now producing somewhat less than 1% of the electricity in the U.S., but rapid expansion is planned so this source will probably provide sufficient energy before we exhaust the fossil fuels; however, the possibilities of pollution of the environment by radioactive substances is great. The possibility of control of nuclear fusion offers great hope for the future. On a long term basis, we must begin to look to solar energy—more efficient ways of trapping it and ways of storing it for use on cloudy days or at a distance from the site of production.

45-9 SYNTHETIC ORGANIC COMPOUNDS IN THE ENVIRONMENT

The growth of knowledge about organic chemistry has enabled scientists to synthesize many chemical compounds not normally present on the earth. These compounds are used as plastics, drugs, dyes, and for thousands of other purposes. Two uses of these compounds, as detergents and pesticides, have led to specific ecological problems. Detergents and pesticides have been used in great quantities and the inability of natural processes to decompose, or degrade, them has resulted in their accumulation in our environment. The accumulation of plastics, which are also non-biodegradable, is a nuisance and presents a disposal or recycling problem, but some detergents and many of the insecticides have presented health problems both to humans and to the other species with whom we share our spaceship earth.

Detergents

The basic definition of a detergent is that it is any organic compound which has a cleansing action in water. However, in popular usage, we generally refer to synthetic chemical compounds such as alkylbenzenesulfonates as detergents and exclude the soaps which also belong in this category.

$$CH_3-(CH_2)_{11}-\bigcirc-SO_3^-\quad Na^+$$

sodium n-dodecylbenzenesulfonate

The widespread use of certain kinds of synthetic detergents led to the appearance of foam on many rivers and lakes in the United States in the middle of the twentieth century. This problem is one of the few ecological problems that have been largely solved by the use of molecules that are biodegradable. This change has been made by substituting straight chain alkyl groups for the highly branched ones used in the early synthetic detergents. The straight chain molecules are much more easily degraded since microorganisms contain enzymes that catalyze the degradation of these molecules, which resemble the natural fats. (See Section 27-7 for further details.)

DDT and Other Insecticides

The human species is in constant competition with other species for food and space. From the human standpoint, a pest is any competitor, such as rats, the fungi, insects, and weeds. Chemical agents used to destroy these competitors are called pesticides. If the enemy is a rat of other rodent, we call the poison used a rodenticide, whereas we use fungicides, herbicides, and insecticides to kill fungi, plants, and insects. Since we share many of our metabolic reactions with other living things, the poisons we use on them may be poisonous to humans also. The ideal pesticide is toxic to only one kind of life, but few even approach this ideal; most pesticides have a wide range of toxicity.

The estimated annual monetary loss of food crops in the United States due to rodents is one to two billion dollars, and insects and weeds each account for four to eleven billion dollar losses. Another estimate is that half of the food crops in India, a country where food is desperately needed, is lost to insects. In addition to destroying food, insects carry many diseases. Lice carry typhus and mosquitoes carry malaria, yellow fever, encephalitis, and other diseases. Well over 99% of the insect species are not harmful to humans, and many, such as bees, are necessary for our survival; however, there are perhaps three million species of insects, and we are truly struggling for existence with the harmful ones.

Several natural substances are poisonous to insects. The pyrethrins from chrysanthemums, nicotine, rotenone, and other plant products are still used to some extent as insecticides. We have also used inorganic chemicals such as sodium arsenite, arsenic trioxide, and lead arsenate to destroy insects; however, these substances are so toxic that their use is minimal. The most widely used insecticides are organic compounds produced by synthesis.

The compound d**ichlorod**iphenyl**t**richloroethane which we commonly call DDT was first reported in the doctoral dissertation of a German chemist in 1874. In 1939, the Swiss chemist Paul Mueller discovered that it was an effective insecticide.

$$\text{DDT structure: } Cl\text{-}C_6H_4\text{-}CH(CCl_3)\text{-}C_6H_4\text{-}Cl$$

DDT

The widespread use of DDT was stimulated by World War II. It was DDT that controlled an epidemic of typhus in Italy in 1943–44 by killing the lice that carried the disease. In fact, World War II was the first war in which disease did not kill more people than were killed in battle. This was due in part to DDT, but penicillin, the sulfa drugs, and better knowledge and medical care were also responsible. In 1953, it was estimated that on a worldwide basis DDT had saved five million lives and prevented 100 million illnesses. In Ceylon and Madagascar, death rates were cut in half after the introduction of DDT.

DDT is among the least toxic of the insecticides we use. Lice are controlled by dusting people's bodies with DDT powder, and according to the U.S. Public Health Service there are no well-described cases of fatal, uncomplicated DDT poisoning of humans resulting from proper use. The same is not true of other species. Although there is still some controversy as to the exact relationship of DDT to their deaths, several birds such as the peregrine falcon, the brown pelican, and the bald eagle are rapidly decreasing in number. These birds and their eggs contain relatively large amounts of DDT. In contrast, a controlled experiment in which pheasants, quail, and chickens were fed DDT indicated that they were resistant to its effects.

A major problem with DDT is that it persists in the environment. Indeed, part of its value lies in the fact that it is a persistent pesticide—it is not readily destroyed. Added to this is the fact that it is fat soluble and tends to be stored in the bodies of animals that eat it. The amount of DDT increases in each member of a food chain. Water with as little as 0.000001 parts per million of DDT may contain plants with 0.01 ppm. Fish that eat these plants contain 2 ppm, and birds of prey that eat fish contain as much as 10 ppm. While there is no evidence that these predators are developing any immunity, another problem in continued use of DDT is that many of the kinds of insects that were formerly sensitive to DDT are no longer affected by it.

DDT has been in widespread use only since the middle 1940's. We know that many organic substances take many years to produce effects; perhaps illnesses and deaths ascribed to other causes have been influenced by DDT or other organic chemicals. Although there is little evidence for ill effects of DDT on humans, it is a possibility that cannot be ignored.

Concern about the use of DDT and the availability of other insecticides has led the Environmental Protection Agency of the U.S. to prohibit most uses of DDT after January

1973. In general, other insecticides are both more toxic and more expensive than DDT. Poorer countries and those less technically developed continue to use large amounts of DDT.

Many of the alternative insecticides contain chlorine and thus are known as the organochlorine insecticides. Among these are dieldrin, aldrin, and methoxychlor. Like DDT, they are persistent insecticides that accumulate in the environment. Methoxychlor is a recently developed insecticide that is less toxic to experimental animals than DDT and may be somewhat more biodegradable under normal use.

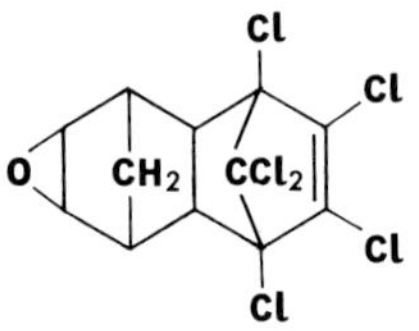

dieldrin 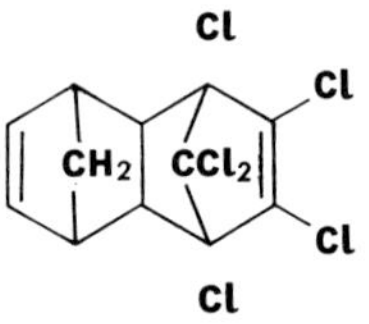   aldrin 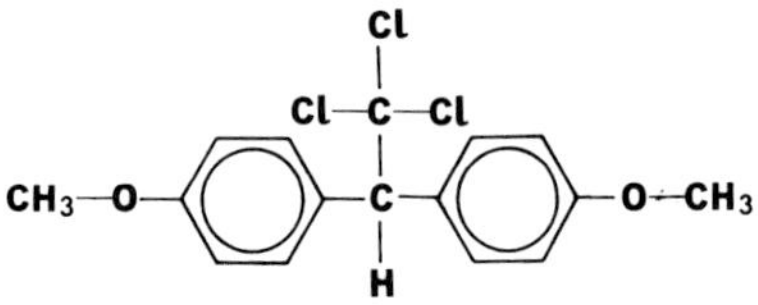   methoxychlor

Since these insecticides are new, fewer insects are immune to these organochlorine insecticides, and their levels in the environment are not so high as DDT. See Table 45–1 for the relative toxicities of these compounds.

TABLE 45–1 Relative toxicity of insecticides

Insecticide	Lethal Dose
DDT	113
aldrin	39
dieldrin	46
methoxychlor	6000
parathion	4
malathion	1000

The toxicity is expressed as the dose (mg per kg of body weight) that is lethal to 50% of the animals fed the insecticide. Low values indicate very toxic substances, high values, less toxic ones. Values are selected from data compiled by the U.S. Department of Health, Education, and Welfare. Other sources give slightly different values due to biological variability of the test animals and different experimental conditions.

Another class of compounds that are effective as insecticides are the organophosphorus compounds. Malathion and parathion are examples of this class.

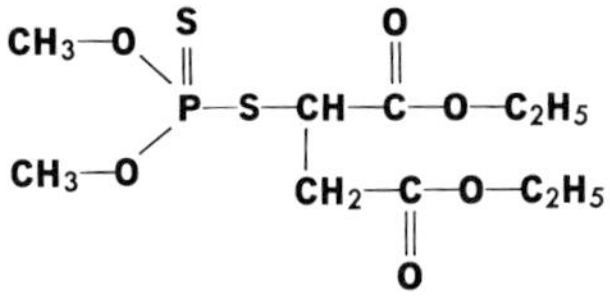

malathion 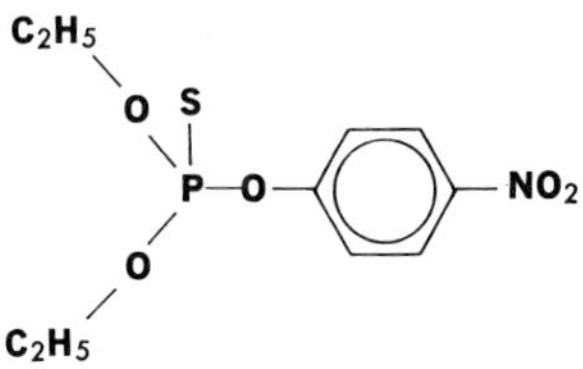   parathion

These substances, which are related to the nerve gases, work by inhibiting the enzyme cholinesterase. Since most animals use this enzyme in nerve conduction, these insecticides, especially parathion, are very toxic to humans. The organophosphates have the advantage of being relatively easily destroyed in the environment and apparently we need to be less concerned about their long-range effects. However, short-term exposure can kill. There were 103 deaths reported in India in 1958 due to parathion and 88 persons died from this poison in Colombia in 1967.

Other insecticides are continually being developed. Our experiences with DDT have shown us that we must be much more aware of the consequences of introducing any new substance into our environment, particularly if the new substance is introduced in large quantities. Our battle with the insects is not over, and methods other than the use of insecticides are being used. So far, biological control methods have been effective in a few cases. Crop rotation and control of winter breeding grounds of insects by selective burning are also effective.

Herbicides and the PCBs

Herbicides are chemical substances used to control undesirable plants. The two most common of these are 2,4-dichlorophenoxyacetic acid (2,4-D) and 2,4,5-trichlorophenoxyacetic acid (2,4,5-T).

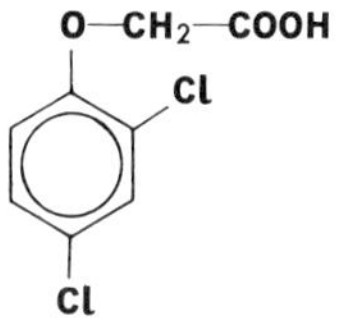

O—CH$_2$—COOH   O—CH$_2$—COOH

2,4-D 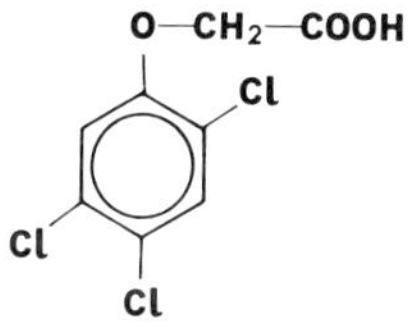   2,4,5-T

Most herbicides in use are relatively selective, some controlling only the broad-leaf plants (e.g., dandelions) and leaving grasses whereas others have opposite effects. Some work only on germinating seeds and do not affect established plants. We are just becoming aware of effects of these substances on animals and possibly on humans. At the time of writing, there is a question whether some of these effects, especially the production of abnormal or deformed offspring, are due to the herbicides themselves or to impurities usually contained in them.

Another class of compounds that is currently the subject of concern is the polychlorinated biphenyls (PCBs). These compounds are derivatives containing several chlorine atoms substituted on the biphenyl (or diphenyl) molecule.

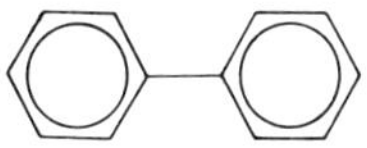

biphenyl

The PCBs are used as plasticizers in many plastic products, as a heat transfer agent, as lubricants, and as "carriers" in some insecticides. They have been shown to be highly toxic to shrimp and other marine animals. The PCBs are much like DDT in solubility and persistence; in fact, the normally used analytical methods make it difficult to distinguish between PCBs and DDT. More precise tests, however, show that PCBs are found in

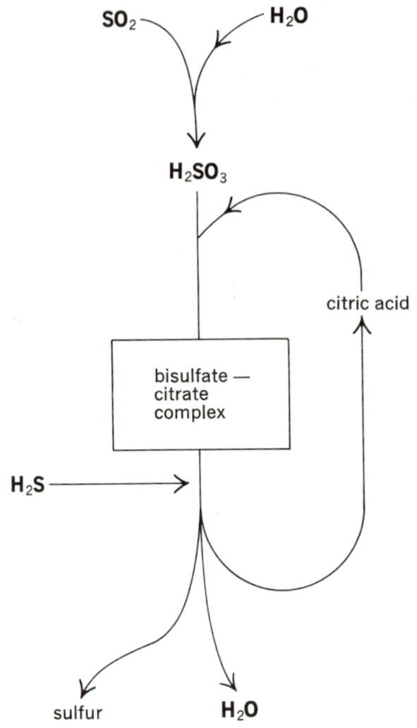

Fig. 45-6 The production of sulfur from sulfur dioxide. The diagram illustrates a process that is used to remove SO_2 gas from the flues of factories. SO_2 is first converted to the bisulfite ion by the addition of water. This ion then reacts with citric acid to give a bisulfite-citrate complex. When H_2S gas is added, sulfur is precipitated, and the citric acid is recycled to react with more bisulfite.

birds and humans. It is possible that some of the effects on birds that have been attributed to DDT are actually caused by PCBs.

45–10 SOLUTIONS TO ENVIRONMENTAL PROBLEMS

This chapter has been filled with a discussion of environmental problems. These problems are serious and require solutions, and one of the first steps in solution demands that we be aware of the nature of the problems.

However, if we stop with mere understanding of the problems, we can run into difficulties. First, we may be so impressed by the serious nature of the problems that we look for someone to blame, or we allow our frustration and anger to lead us into hopelessness and withdrawal. The second difficulty is that we may forget that modern technology has tremendous advantages as well as disadvantages. We can all complain about the sulfur dioxide and soot in the air, but we must realize that without steel mills and electrical power plants we would be forced to live in the age of lanterns, candles, and wooden wagons. Although some people are willing to try this, we must realize that such a return to the primitive would also mean an increase in famine, disease, death at early ages, and a condition of life long ago described as nasty, brutish, and short.

Many texts stop with a discussion of problems, but we feel that we must discuss some of the solutions that are being offered. Here we often leave the realm of facts and are forced to engage in speculation. Whereas many of the solutions proposed are technically simple, others require research and development that cannot be completed tomorrow or even next year. A major problem is convincing the majority of the people of the seriousness of the problems. A second problem is inducing people, in various ways, to take positive action.

There are many who blame our environmental problems on technology and insist that the only solution involves the destruction of technology. However, it has not been technology, as such, that lies at the root of our problems, but our inability to see the effects of technology. The answer is not a decrease in technology and knowledge, but the opposite. We need greater knowledge about our technology and increased human concern about the way it is used.

Recycling. The solution to many problems of shortages involves re-use of materials, or recycling. We must conserve our precious supplies of iron, copper, aluminum, and other metals by collecting and re-using discarded articles. This process has the added advantage of preventing many types of pollution.

The removal of sulfur-containing compounds from petroleum and the collection of sulfur dioxide from industrial discharges is already being done on a large scale (see Fig. 45–6). More than half of the elemental sulfur produced in the free world in 1970 was recovered from industrial processes. Only ten years previously the amount of recovered sulfur was less than 30% of total production. Another valuable product derived from the removal of sulfur dioxide from industrial wastes is sulfuric acid. This important chemical reagent is obtained by absorbing the sulfur dioxide in water in a process known as "scrubbing, followed by the oxidation of sulfurous acid to sulfuric acid.

Many gases can be prevented from polluting the air by the use of a "scrubber." Solids can be removed by building higher smokestacks that trap much of the ash, or by installing electrostatic precipitators that remove small charged particles. In both of these instances, the technology is already developed; it only remains to install the equipment— and pay for it. Even though we can sell some of the products derived from processes that clean up industrial discharges, the costs are high. (See Fig. 45–7.) This chapter is full of illustrations of the consequences we will suffer if we are not willing to pay the monetary costs.

Controlling Air Pollution. Even if industrial gas discharges are controlled, the air will still contain pollution from another major source—the internal combustion engine. The automobile is a major source of air pollution, and eliminating this pollution is very difficult to accomplish. We can develop more efficient internal combustion engines, ones that release fewer unburned hydrocarbons and less carbon monoxide, but these more efficient engines usually give off greater amounts of oxides of nitrogen. Although the amount of carbon monoxide in the atmosphere in Los Angeles county has decreased drastically in the past few years, the amount of oxides of nitrogen continues to increase. Nitrogen is present in the air that is the source of oxygen for combustion of the fuel. It is a basic chemical fact that oxygen and nitrogen will react under the conditions existing in an internal combustion engine. There are many experiments now being done in efforts to develop other forms of energy for the private motor car or for buses. Steam engines, gas turbines, and electric cars are promising solutions, but none is yet capable of giving the flexibility and freedom provided by the internal combustion engine.

An important suggestion in this area is to decrease the number of private motor vehicles and increase the use of public transportation. Although buses pollute the air, the pollution per person is less. Development and greater use of electric trains also shows promise. (See Fig. 45–8.)

Fig. 45-7 Two photographs of the Owens-Corning Fiberglas plant at Santa Clara, California, show the effects of pollution control. The photograph on the top was taken before addition of controls in February 1969 and that on the bottom was taken shortly thereafter. The remaining plumes of material shown on the bottom are only steam, which is not considered a pollutant. More recent changes have eliminated even these. The company has spent more than $3,000,000 for the equipment which reduced emissions and estimates that about 14% of their capital investment and 5% of operating expenses are used to control air pollution. These pictures illustrate what is being done by many industries to control air pollution. [Photos courtesy of Owens-Corning Fiberglas.]

Fig. 45-8 A great deal of the air pollution in urban areas is due to products given off by the internal combustion engines in automobiles and other motor vehicles. Modifications of engines and addition of control devices promise some decrease in pollution. Another way of decreasing pollution is by using electrically powered rapid transit vehicles. The photograph on the top shows a typical scene on a California freeway. On the bottom is a scene in Oakland, California, showing the track and cars of the Bay Area Rapid Transit System, commonly known as BART. This system began operation in 1972. Such systems can contribute greatly to the reduction of air pollution if enough people use them instead of automobiles. [Photos courtesy of BART.]

Solutions Involving Population and Rate of Consumption. Many of our ecological problems derive from the fact that there are more people on earth now than there ever were before and that many, if not all, of them are consuming energy and raw materials at a higher rate than ever before—and throwing more used things away. In the United States, the population growth is nearly 2% per year, but there is an increase of about 5% per year in the use of resources. The percentage increase in the gross national product shows an amazing correlation with the increase in the amount of waste we dispose of every year. Only 5% of the earth's population lives in the United States, but they consume 40% of the energy used in the world each year. The rate of power consumption in this country is increasing 8% every year.

It is obvious that one of the solutions to our ecological problems is to decrease the rate of population growth and to consider whether lower levels of population are not more consistent with a better environment and a fuller life. Another solution is for each person to consume less and recycle more of the resources of spaceship earth.

Population control is unpopular, and no one can deny that past efforts have sometimes been brutal and often discriminatory toward one race of other group. The fact we must face, however, is that there will be population control, if not by group or individual control, then by natural forces of pestilence, famine, war, and death. Hopefully, control will be individual, the result of educated people making wise choices. The situation is similar concerning the control of consumption of the world's material resources and energy. We cannot continue indefinitely to increase population and consumption at the rate that has prevailed in the industrialized countries during the past century. Although there may be controversy over how much longer we can continue to grow, the simple fact is that we cannot continue indefinitely. Whether the growth, consumption, and pollution will stop because of reasoned, orderly processes or by those harsher processes we think of as "natural" is the major problem facing each of us today. Education, concern, and the ability to act are our only hope.

The current concern for the environment has led to a new consciousness on the part of millions of people. It has also led to a great deal of research and better measurement of many substances in our environment. In a few cases, it has led to the beginning of a reversal of pollution. Attempting to discuss environmental chemistry in one chapter is extremely difficult. Whole books are needed to treat the subject adequately. We have added an extensive list of references for those who want more information. The fast pace of research, however, will soon make some of these obsolete, possibly even before publication of this book. We recommend that the interested reader become acquainted with current publications in the environmental area. An outstanding publication at a level consistent with the material in this text is the magazine *Environment*, which is listed in the references. It is generally available.

IMPORTANT TERMS

Carboxyhemoglobin Fossil fuels
Eutrophication   Smog
Feedback   Temperature inversion

SUGGESTED READING

Brown, J., J. Bonner, and J. Weir, *The Next Hundred Years*, The Viking Press, New York, 1963. This pioneering book discusses basic problems. Although some of the facts and figures are out of date, it remains a valuable book.

Cailliet, G. M., P. Y. Setzer, and M. S. Love, *Everyman's Guide to Ecological Living*, The Macmillan Co., New York, 1971. A practical guide to living on spaceship earth.

Chisholm, J. J., Jr., "Lead Poisoning," *Scientific American*, Feb. 1971 (Offprint #1211). A good, modern discussion of this killer and maimer of children—and of adults.

Clark, J. R., "Thermal Pollution and Aquatic Life," *Scientific American*, March 1969 (Offprint #1135). This article discusses the effects of heating the water of rivers and lakes by various industrial processes. These effects will become of greater importance if we use nuclear power for generation of electricity.

Edwards, C. A., "Soil Pollutants and Soil Animals," *Scientific American*, April 1969 (Offprint #1138). The effects of DDT and other pesticides on various soil organisms are discussed and a warning given about bad effects noted.

Goldsmith, J. R., and S. A. Landaw, "Carbon Monoxide and Human Health," *Science* **162**, 1352 (Dec. 20, 1968). This article is a bit technical, but contains a great deal of information.

Goldwater, L. J., "Mercury in the Environment," *Scientific American*, May 1971. An up-to-date discussion of this potentially deadly element and its significance in our lives.

Hutchinson, E. G., "The Biosphere," *Scientific American*, Sept. 1970 (Offprints #1188–1198). This is only the lead article in an issue packed with ecologically important articles such as those on the cycles of carbon, oxygen, energy, water, nitrogen, and minerals. The whole September 1970 issue is highly recommended.

Lippincott, W. T., editor, *The Journal of Chemical Education*, **49** (1), January 1972. A major part of this issue is devoted to environmental chemistry. Methods for measuring mercury and oxides of nitrogen are discussed. Highly recommended.

Merriman, D., "The Calefaction of a River," *Scientific American*, May 1970 (Offprint #1177). The author discusses the effects of discharging warm water into a river. Since he found no drastic effects, he prefers the term *calefaction* to the more widely used *heat pollution*.

Peakall, David B., "Pesticides and the Reproduction of Birds," *Scientific American*, April 1970 (Offprint #1174). The author tells about the effects of DDT and other pesticides on hawks and pelicans.

Starr, C., "Energy and Power," *Scientific American*, Sept. 1971. This entire issue of *Scientific American* is devoted to energy and power. It is "required reading" for those who want to be informed on this aspect of ecology.

Stoker, H. S., and S. L. Seager, *Environmental Chemistry: Air and Water Pollution*, Scott, Foresman and Co., Glenview, Ill., and London, 1972. This is an excellent paperback on the topics listed in the title. It gives the facts necessary for conclusions and generalizations in this area.

Swatek, P., *The User's Guide to the Protection of the Environment*, Friends of the Earth/Ballantine, 1970. The subtitle for this book is *The Indispensable Guide to Making Every Purchase Count*. A highly recommended paperback.

Tanner, J. T., et al., "Mercury Content of Common Foods Determined by Neutron Activation Analysis," *Science* **177**, 1102 (Sept. 22, 1972). A recent, optimistic report, except for sea foods.

Wilson, C. L., and others, *Man's Impact on the Global Environment*, Massachusetts Institute of Technology Press, Cambridge, Mass., 1970. A paperback filled with facts and authoritative discussions of ecological problems.

Anyone who is interested in keeping up with the rapidly advancing field of ecological chemistry will need to turn to the current literature. Several excellent magazines are available that contain authoritative and readable articles. *Environment* is devoted to publishing "information about the effects of technology on the environment and about the peaceful and military uses of nuclear energy." There are also excellent articles in *Science*, published by the American Association for the Advancement of Science, and, of course, in *Scientific American*.

GLOSSARY

Acetal: A compound produced by reaction of two molecules of an alcohol with an aldehyde; it has the structure

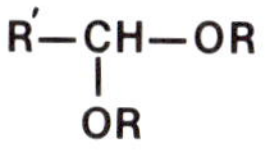

$$R'-\underset{\underset{OR}{|}}{CH}-OR$$

Acid: A proton donor. Common examples are HCl, H_2SO_4, and $HC_2H_3O_2$ (acetic acid).
Acid chloride: An organic compound having the functional group:

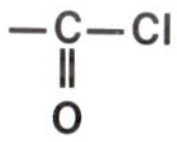

$$-\underset{\underset{O}{\|}}{C}-Cl$$

Addition reaction: One in which new atoms bond with available electrons in a compound, without displacing other atoms.
Adrenalin (epinephrine): A hormone of the adrenal gland that has a generally stimulating effect on the body.
Aerobic: Literally, in the presence of air. Aerobic reactions require oxygen in order to proceed.
Albumin: A protein that is water soluble, coagulated by heat, and is not precipitated in a solution that is 50% saturated with ammonium sulfate.
Alcohol: An organic compound having an — OH functional group on an alkyl skeleton, ROH.
Aldehyde: A compound in which one hydrogen and an organic group are bonded to a carbonyl group:

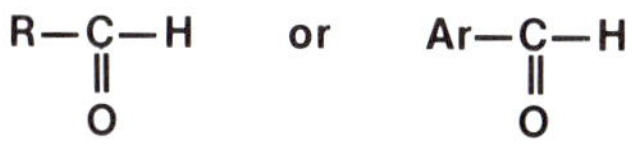

$$R-\underset{\underset{O}{\|}}{C}-H \quad \text{or} \quad Ar-\underset{\underset{O}{\|}}{C}-H$$

Compare with the definition of *ketone.*

Aldose: A carbohydrate that contains an aldehyde group. Glucose, galactose, and ribose are simple aldoses.
Alimentary canal: The food tube that carries the food through the body and serves for both digestion and elimination.
Alkane: The systematic name for saturated hydrocarbons (Section 22-3).
Alkanoic acid: The systematic name for RCOOH, an alkyl carboxylic acid.
Alkene: The systematic name for a hydrocarbon having a double bond, such as butene, $CH_3CH_2CH=CH_2$.
Alkyl group: A unit of organic structure equivalent to an alkane minus one hydrogen atom (Section 22-3). For example, CH_3CH_2- is an ethyl group.
Alkyne: Systematic name for a hydrocarbon having a triple bond, such as propyne, $CH_3C\equiv CH$.
Amide: A carboxyl derivative of the type

$$R-\underset{\underset{O}{\|}}{C}-NH_2 \quad \text{or} \quad R-\underset{\underset{O}{\|}}{C}-NHR$$

Amine: A derivative of ammonia in which one or more hydrogens have been replaced by organic groups; for example, $ArNH_2$ or R_2NH. Such compounds are weak bases.
Amino acid: An organic compound containing an amino ($-NH_2$) and an acid ($-COOH$) group. The majority of the naturally occurring amino acids have the carboxyl group and the amino group attached to the same carbon, and thus are α-amino acids.
Aminopeptidase: A peptidase that catalyzes the hydrolysis of the peptide bond nearest the end of a peptide chain that contains the free amino group.
Amphoteric: A substance which reacts as either an acid or a base. Amino acids and hydroxides of aluminum, tin, and lead, for example, are amphoteric.
Amylase: An enzyme that catalyzes the hydrolysis of starch.
Anabolic: A word used to describe reactions in living systems that combine simple molecules to make more complex molecules. Such reactions require energy in order to proceed.
Anabolism: Those reactions in a plant or animal that result in the synthesis of larger molecules from smaller ones. (See *Catabolism*.)
Anaerobic: A word used to describe reactions that occur in the absence of air. Actually, anaerobic reactions are those that do not require oxygen to proceed. Some anerobic reactions are inhibited by the presence of oxygen.
Anhydrous: Without water; e.g., hydrates from which the water of crystallization has been removed.
Anode: The electrode at which electrons leave the solution; the electrode to which negative ions migrate.
Anoxia: The condition in the body resulting from a deficiency of oxygen.
Antibiotic: A substance produced by a microorganism that inhibits the growth of another organism.
Antibody: A macromolecule produced by an animal in response to an antigen. The antibody reacts to overcome the effects of a specific antigen.
Antigen: A substance that initiates the production of antibodies.
Antimetabolite: A molecule that is similar to a normal metabolite but sufficiently different to interfere with reactions of the normal metabolite.
Aromatic compound: An organic substance whose carbon skeleton contains a benzene ring or closely related type of ring.
Assimilation: The process of distributing and using digested food.

Atherosclerosis: A condition in which fatty deposits are formed in the inner lining of blood vessels.
Atmosphere (the unit of pressure): The average atmospheric pressure at sea level, equaling 760 mm or 29.92 in. of mercury.
Atom: The smallest existing part of an element.
Atomic mass unit (amu): A unit of mass equaling 1/12 the mass of a carbon-12 atom; approximately the mass of a proton or neutron.
Atomic number: The number of protons in the nucleus of an atom; therefore, it also equals the number of electrons in the shells of a complete atom.
Atomic weight: The average weights of the atoms of an element based on carbon-12 being equal to twelve.
ATP: Adenosine triphosphate. A compound whose hydrolysis provide energy to drive energy-requiring reactions. ATP represents an energy-storing compound. (See Fig. 36-12 for the formula for ATP.)

Bacteriophage: A type of virus which infects bacteria. Many bacteriophages that attack the bacterial *E. Coli* have been studied. They contain DNA and protein.
Basal metabolic rate: The rate at which reactions are proceeding in an animal at rest. The BMR is determined by measuring oxygen consumed per unit of time.
Base: A proton acceptor. Common examples are OH⁻ and NH_3.
Bile: The fluid secreted from the liver which aids in digestion of lipids. It contains emulsifying agents but no enzymes.
Biochemistry: The science of the chemical composition structure, and reactions of plants and animals.
Boiling point: The temperature at which the vapor pressure of a liquid equals or just exceeds the pressure of the atmosphere above it.
Buffered solution: A solution whose pH changes only slightly with the addition of moderate quantities of either strong acids or strong bases. Stabilization of the pH results from the presence of a weak acid and the salt of the weak acid in the solution.

Calorie: The amount of heat required to raise the temperature of one milliliter of water by one degree centigrade. The "calorie" used to describe the energy content of food is actually a kilocalorie, which is 1000 calories.
Carbohydrate: Carbohydrates are polyhydroxy aldehydes or ketones or condensation products thereof (Sections 29-11 and 33-1).
Carbon black: A powdery form of carbon produced by incomplete burning of natural gas (primarily methane, CH_4), used as a pigment and in the manufacture of automobile tires.
Carboxyhemoglobin: The compound formed when carbon monoxide reacts with hemoglobin. This compound does not combine with oxygen.
Carboxylate salt: The salt related to a carboxylic acid; for example, RCOO⁻ Na⁺.
Carboxylic acid: An organic compound having a —COOH functional group. Such a compound is a weak acid compared to hydrochloric or sulfuric acid.
Carboxypeptidase: A peptidase that catalzes the hydrolysis of the peptide bond nearest the end of a peptide chain that contains the free carboxyl group.
Carcinogen: A substance that causes cancer.
Cardiovascular disease: Any disease of the heart or blood vessels.
Carnivore: An animal whose major source of food is the flesh of other animals.
Catabolic: A word used to describe reactions in living systems that degrade or oxidize large molecules to smaller or more highly oxidized molecules. Such reactions result in the release of energy.

Catabolism: Those reactions in a plant or animal that result in the degradation or oxidation of molecules.
Catalyst: A substance which alters the rate of a chemical reaction without being consumed in the reaction.
Cathode: The electrode at which electrons enter a solution; the electrode to which positive ions migrate.
Cellulose: A polysaccharide found in the structural parts of plants. It is composed of glucose units linked β 1-4.
Chemical change: Changes in a substance which involve composition and often energy content.
Chemical energy: Energy released as a result of chemical change.
Chemical properties: Those properties which describe the composition and reactivity of substances, including energy changes and changes in composition.
Chemistry: The science concerned with the composition of the substances of which the universe is composed, the properties of these substances, and the changes they undergo. It is also concerned with the energy relationships involved in the changes.
Chemotherapeutic agent: A chemical compound that is used to treat disease.
Chlorophyll: The green pigment found in plants. It is necessary for photosynthesis.
Chromatography: A process for separation of substances in a mixture by adsorption on paper or other materials. See Fig. 32-1 for illustrations of methods of chromatography.
Chromosome: Long, threadlike bodies that can be seen at certain stages in the life of a cell. The chromosomes are made of DNA and protein and are the site of hereditary information in the cell.
Chymotrypsin: A proteolytic (protein-digesting) enzyme secreted by the pancreas.
Cirrhosis: The formation of abnormal amounts of connective tissue accompanied by the disappearance of the normal tissue of the liver or other organs.
Citric acid cycle (Krebs cycle): A cyclic series of reactions by which acetic acid units are oxidized to carbon dioxide and water with the release of energy.
Coenzyme: A nonprotein substance that is required for activity of an enzyme.
Coke: The solid residue, primarily carbon, produced by heating coal intensely in the absence of oxygen.
Colloids: Particles of such small size that they may stay suspended in gases or in liquids for an indefinite period of time.
Combustion: An exothermic reaction which proceeds so rapidly that flames or light are produced.
Concentrated solution: A solution containing a relatively high proportion of solute.
Configuration: The arrangement in space of atoms within a molecule having a given structure.
Covalent bond: A pair of electrons shared between two atoms.
Cracking reaction: Decomposition of alkanes by heat energy, converting them to unsaturated hydrocarbons (Section 23-2); also called pyrolysis.
Cyanohydrin: A compound having the functional group

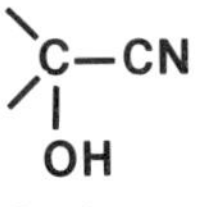

$$\begin{array}{c} \diagdown \\ \diagup \\ \end{array} \!\! C - CN \\ | \\ OH$$

Cyclo-, cyclic compound: A ring compound; for example, cyclopentane (Section 22-3).
Cystinuria: An abnormality in which large amounts of cystine are excreted in the urine. This condition may lead to formation of stones of cystine in the urinary bladder.
Cytochromes: A group of colored substances found in many cells. The cytochromes function in electron transport. See Fig. 37-9 for the structure of a cytochrome c.

Dark reaction: A series of the reactions of photosynthesis which can proceed without light.
Deamination: A process by which an amino group is removed from an organic compound. Deamination of an amino acid usually gives a keto acid.
Decarboxylation: The process of eliminating carbon dioxide from an organic molecule. Decarboxylation of an amino acid gives an amine.
Deficiency disease: A disease whose symptoms can be cured by administration of a specific element or compound. The most well-known deficiency diseases result from vitamin deficiencies.
Dehydrogenation: Reaction in which hydrogen atoms are removed from a molecule.
Denaturation: A change in the properties of a protein. Denaturation is believed to involve only the secondary linkages of proteins, not hydrolysis of peptide bonds. Denaturation may be reversible, in which case the starting properties are restored, or irreversible so that the protein is permanently altered.
Deoxyribonuclease (DNA-ase): An enzyme that catalyzes the hydrolysis of DNA.
Deoxyribonucleic acid (DNA): A nucleic acid that yields the pentose deoxyribose on hydrolysis. DNA is found primarily, but no exclusively, in the nuclei of cells and is a polymer of deoxynucleotides.
Depot lipids: The lipids that an animal accumulates or stores. The amount of these lipids may vary greatly as opposed to the tissue lipids which always make up a certain percentage of a cell or an animal.
Detergent: An organic compound which has cleansing action in water. It has a large nonpolar hydrocarbon group and at least one ionic or highly polar group.
Diabetes mellitus: A disease characterized by a high level of glucose in the blood and excretion of glucose in the urine. It is caused by a lack of insulin.
Dialysis: The separation of dissolved crystalline material from colloidal materials by use of a semipermeable membrane.
Diene: An unsaturated organic compound having two double bonds somewhere in the molecule; for example, butadiene, $CH_2=CH-CH=CH_2$.
Diffusion: The migration of molecules among other molecules. Diffusion takes place rapidly within gases and slowly within liquids.
Digestion: The process of converting ingested food to simpler forms that can be absorbed.
Dilute solution: A solution containing a relatively small proportion of solute.
Dipeptidase: An enzyme that catalyzes the hydrolysis of dipeptides.
Dipolar ion: A molecule having both a positive charge and a negative charge; also sometimes called an "inner salt." The most common example is an amino acid.

$$R-\underset{\underset{^+NH_3}{|}}{CH}-COO^-$$

Disaccharidase: An enzyme that catalyzes the hydrolysis of a disaccharide to give two monosaccharide molecules. Important disaccharidases are sucrase, maltase, and lactase.
Disaccharide: A carbohydrate composed of two glycose molecules (i.e., two monosaccharides) linked by a glycoside bond.
Disease: Any departure from health.

Edema: A condition in which there is an excess of water in many tissues of the body.
Electrolyte: A substance (acid, base, or salt) which conducts an electric current in a water solution.

Electron: A negatively charged subatomic particle with a mass 1/1837 that of a hydrogen atom.
Electronegativity: The tendency of an atom to attract electrons to itself. The electronegativity scale, devised by Dr. Linus Pauling, arranges the elements in a descending order of electronegativity.
Electrophoresis: A process in which components of solutions may be separated by being allowed to migrate in an electrical field. The process is used in the purification of proteins and other charged molecules.
Electrovalent bond (ionic bond): The bonding occurring in compounds formed by the transfer of electrons from one atom to another.
Empirical formula: The simplest formula representing the ratio of atoms in a molecule.
Emulsion: A finely divided suspension of two immiscible liquids.
Enantiomers: Compounds whose configurations are nonidentical mirror images. They are optically active.
Endergonic: A word used to describe an energy-requiring or energy-consuming reaction. Anabolic reactions are endergonic.
Endocrine gland: A gland which secretes directly into the blood stream. The secretions of the endocrine glands are called hormones.
Endopeptidase: A peptidase that catalyzes the hydrolysis of peptide bonds in the interior of a protein molecule.
Endothermic reaction: A chemical reaction in which heat is absorbed. (See also *Endergonic*.)
Energy: The ability or capacity to do work.
Energy of activation: The amount of energy required to start a reaction.
Enzyme: A protein that catalyzes a biochemical reaction. Some enzymes also contain factors in addition to the protein.
Epithelial: A word used to describe the cells and tissues on the surface and lining of the body. The skin and the lining of the digestive tract and of the lungs are made up of epithelial cells.
Equilibrium: A state of balance or equality between opposing forces; e.g., when opposing chemical reactions are proceeding at the same rate.
Essential amino acid: An amino acid that cannot be synthesized by an animal at a rate equal to his need for that amino acid.
Essential nutrient: A specific compound required by an animal which the animal cannot synthesize at a rate equal to his needs.
Ester: An organic compound having a functional group of the type

$$-\underset{\underset{O}{\|}}{C}-O-R \quad \text{or} \quad -\underset{\underset{O}{\|}}{C}-O-Ar$$

It is derived from a carboxylic group linked to an alcohol or phenol.
Esterification: The process of forming an ester group by direct reaction of a carboxylic acid with an alcohol.
Ether: An organic oxygen compound of the type ROR or ROAr.
Eutrophication: Increase in growth of certain kinds of plants in response to increased amounts of nutrients. Generally used to describe increased growth of algae when amounts of nitrates, phosphates, and organic substances are increased due to pollution.
Exergonic: A word used to describe a reaction that converts potential or stored energy to heat or other forms of energy. Catabolic reactions are exergonic.

Exopeptidase: A peptidase that catalyzes the hydrolysis of peptide bonds adjacent to the end of a peptide chain.
Exothermic reaction: A chemical reaction in which heat is released. (See also *Exergonic*.)

FAD: Flavin adenine dinucleotide. A substance that serves as an electron acceptor and donor in biological oxidations and electron transport. See Fig. 37-8 for the structure of this compound.
Fat: A triglyceride that is solid at room temperature. (See *Triglyceride* and Sections 27-1, 35-3.)
Fatty acid spiral: The sequence of reactions by which a fatty acid is degraded into acetate units.
Feedback: A process in which a product of a reaction influences a step in that reaction. We speak of positive feedback when the product accelerates the reaction and negative feedback when the reaction is slowed by a product.
Fermentation: The anaerobic metabolism of carbohydrates. Fermentations are catalyzed by the enzymes of bacterial and yeasts, and may result in the formation of lactic acid or of carbon dioxide and ethanol.
Fixation: The chemical conversion of a gaseous substance to a solid or liquid. Nitrogen is fixed by being incorporated into nitrates, mitrites, or ammonia. Although ammonia is a gas, it is extremely soluble in water, and is generally considered to be a fixed form of nitrogen.
Fog: A suspension or dispersion of tiny droplets of a liquid throughout a gas.
Formula weight: The sum of the atomic weights of the atoms in a molecule.
Fossil fuels: Substances that can be extracted from the earth and used for fuel. The most important fossil fuels are coal, petroleum, and natural gas.
Free energy: The energy available for doing useful work.
Freon: Any one of a number of compounds composed of carbon (one or two atoms), fluorine, and usually chlorine. They are used as refrigerants and as special aerosol propellants.
Functional group: A group of atoms (a unit of structure) within a molecule which reacts or functions. For example, the $-NH_2$ group in ethylamine, $CH_3CH_2NH_2$.

Galactosemia: The presence of galactose in appreciable amounts in the blood. A hereditary abnormality in which an individual cannot metabolize galactose.
Gamete: A cell capable of participating in fertilization and formation of a new individual. Ova and sperm are gametes.
Gamma globulin: A variety of globulin found in blood plasma. This fraction contains many antibodies against diseases.
Gel: A dispersion of a liquid within a solid.
Gene: The factor that influences the inheritance of a characteristic.
Genetic code: The sequence of organic bases in a nucleotide that specifies an amino acid sequence in a protein molecule.
Globulin: A protein that can be coagulated by heat, is soluble in a dilute salt solution but precipitates in solutions that are 50% saturated with ammonium sulfate. There are several different globulins in the blood, one of which, gamma globulin, contains antibodies to diseases.
Glucagon: A hormone of the pancreas that increases the level of glucose in the blood. See also *Insulin*.
Glucosuria: The presence of glucose in the urine.
Glyceride: An ester of glycerol.

Glycogen: A polysaccharide found in the muscles and liver of animals. It is made of glucose units.
Glycolysis: The process of degradation of glycogen and other carbohydrates. If the process is anaerobic the end product is lactic acid. In the presence of oxygen, glycolysis yields pyruvic acid.
Glycose: A polyhydroxy aldehyde or polyhydroxy ketone. The simplest type of carbohydrate; often called a monosaccharide.
Glycoside: A compound formed by reaction of an alcoholic —OH group with the carbonyl group of a glycose. It is a particular example of an acetal.
Gram-atomic weight: The atomic weight of an element expressed in grams; the weight in grams of Avogadro's number (6.023×10^{23}) of the atoms of the element.

Hard water: Water containing minerals (usually calcium and magnesium ions) which interfere with the detergent action of soaps.
Heat of fusion: The heat energy (calories) required to convert one gram of a solid to a liquid without a change in temperature of the substance being melted.
Heat of vaporization: The heat energy (calories) required to convert one gram of a liquid to vapor without a change in temperature of the substance which is being vaporized.
Hemiacetal: A compound resulting from addition of one molecule of an alcohol to an aldehyde, yielding the functional group

$$R'-\underset{\underset{OH}{|}}{CH}-OR$$

Hemoglobin: The red-colored protein which reversibly binds oxygen in the red blood cells.
Heterocyclic compound: A substance having a ring structure composed of carbon atoms and at least one different atom; for example, pyridine,

Heterozygote: An individual having unlike genes for any given characteristic.
Hexose: A monosaccharide containing six carbon atoms. Glucose, fructose, and galactose are important hexoses.
Homozygote: An individual having identical genes for a given characteristic.
Hormone: The product of an endocrine or ductless gland. Hormones are carried throughout the body by the blood stream.
Hydrocarbon: A compound composed only of hydrogen and carbon.
Hydrogen bond: The special attraction that exists between molecules containing hydrogen bonded to oxygen, nitrogen, or fluorine. The attraction is due to the polar nature of such bonds. Water is an outstanding example.
Hydrogenation: Addition of hydrogen to an unsaturated molecule.
Hydrolysis: The splitting (lysis) of a compound by a reaction with water. Examples are the reaction of salts with water to produce solutions which are not neutral, and the reaction of an ester with water (Section 26-6).
Hydroxyl group: The —OH functional group, covalently bonded, as in R—O—H. (*Not* the same as a hydrox*ide* ion, OH⁻.)
Hyperglycemia: A concentration of glucose in the blood that is higher than normal.
Hypertonic solution: A solution with a salt concentration greater than that of blood.

Hypervitaminosis: A disease due to excessive intake of a vitamin.
Hypoglycemia: A concentration of glucose in the blood that is lower than normal.

Imine: A compound having the functional group

$$\text{>C=NH} \quad \text{or} \quad \text{>C=NR}$$

Immiscible: Two liquids which will not dissolve in each other are said to be immiscible; e.g., water and oil.
Inborn error of metabolism: An inherited abnormality which results in failure to produce a normal protein. Many inborn errors of metabolism involve failure of reactions due to lack of the necessary enzyme.
Infection: The invasion of the body of a plant or animal by a microorganism.
Ingestion: The act of taking in. In simple animals it involves surrounding a food particle. In more complex animals we use the term eating to describe ingestion.
Inhibitor: A substance that decreases the rate of an enzymatic reaction.
Inorganic substances: All the elements and their compounds, except the compounds of carbon. (See *Organic substances*.)
Inner salt: See *Dipolar ion*.
Insulin: A hormone from the pancreas that promotes the utilization of glucose. Deficiencies of insulin result in hyperglycemia and the disease *diabetes mellitus*.
In vitro: Literally means "in glass." The term is used to describe reactions occurring in a reaction vessel as distinguished from those that occur *in vivo*, in the living animal.
In vivo: Literally means "in the living." The term is used to describe reactions that occur in a living animal.
Ionic bond: See *Electrovalent bond*.
Ions: Electrically charged atoms or groups of atoms. They may be positively or negatively charged, depending on whether the atoms from which they were formed lost or gained electrons.
Isolation artifact: A chemical compound or structure that does not occur as such in a living organism but is produced during the process of extraction and purification.
Isomers: Different compounds having the same molecular formula, For example, methyl ether, a gas, and ethyl alcohol, a liquid, each have formula C_2H_6O.
Isotonic solution: A solution with a salt concentration equal to that of blood.
Isotopes: Atoms of the same element containing different numbers of neutrons and therefore having different nuclear masses.

Kelvin scale: A temperature scale named after Lord Kelvin. It is also called the absolute temperature scale.
Ketal: A compound produced by reaction of two molecules of an alcohol with a ketone; it has the structure

$$\begin{array}{c} R'_2C-OR \\ | \\ OR \end{array}$$

Ketone: A compound in which two organic groups are bonded to a carbonyl group:

$$R-\underset{\parallel}{C}-R', \quad Ar-\underset{\parallel}{C}-R, \quad \text{etc.}$$

Compare with *Aldehyde*.

Ketose: A carbohydrate that contains a ketone group. The best known ketose is fructose.
Kilo-: A prefix meaning 1000; for example, a kilogram is 1000 grams.
Kinetic energy: The energy resulting from motion.
Kinetic theory of gases: A series of statements describing the characteristic behavior of gases, based on the theory that molecules of gases are tiny, elastic bodies in constant motion.
Krebs cycle: See *Citric acid cycle*.

Lanthanides: The fourteen elements (formerly called the rare earth elements) which follow lanthanum; elements having atomic numbers 57 to 71. These elements have properties resembling those of lanthanum.
Lecithinase: An enzyme that catalyzes the hydrolysis of a lecithin.
Light reaction: Those reactions of photosynthesis that stop when no light is present.
Lipase: An enzyme that catalyzes the hydrolysis of a lipid. The most important lipases are those that catalyze the hydrolysis of glycerides to give glycerol and fatty acids.
Lipid: A constituent of a plant or animal that is soluble in nonpolar solvents. A variety of solvents, particularly diethyl ether, ethanol, and chloroform, are used to dissolve lipids.
Liter: The unit of volume employed in the metric system. It is the volume of one kilogram of water at $4°C$.
Lymph: A fluid similar to blood except that it contains no red blood cells. It is derived from the tissues of the body and conveyed to the bloodstream by the lymphatic vessels. It is important in absorption of lipids and in responses of animals to infections.
Lysosome: A small body or organelle found in cells. Lysosomes contain hydrolytic enzymes.

Mass: The amount of matter in an object. Mass remains constant regardless of the location of the object or of the gravitational attraction.
Mass number: The sum of the numbers of protons and neutrons in the nucleus of an atom.
Mass-energy relationship: Theory conceived by Einstein that matter can be converted into energy and vice versa (This necessitated a slight revision in older versions of the law which stated that neither matter nor energy could be created or destroyed.)
Matter: That which has mass, inertia, and occupies space.
Meiosis: The process of cell division in which daughter cells contain half as many chromosomes as the parent cell. Meiosis is important in the formation of gametes (ova and sperm).
Messenger (template) RNA: A relatively high molecular weight RNA that carries information from DNA of the chromosomes in the nucleus of the cell to the ribosomes where protein is synthesized.
Metabolic (degenerative) disease: A noninfectious disease which results from the change in some aspect of metabolism. The causes and cures for most of these diseases are not well understood.
Metabolic water: Water that is produced as the result of reactions within an animal, specifically from the oxidation of hydrogen occurring in foods.
Metabolism: A word used to describe the total of all reactions that occur in plants and animals.
Metabolite: Any chemical substance that participates either as a reactant or product in a reaction in a plant or an animal.
Meter: The basic unit of length in the metric system, equivalent to approximately 39.37 in.
Micron: One thousandth of a millimeter; a unit of linear measure.

Microsomes: Small bodies separated from broken cells by centrifugation. Microsomes are probably ribosomes plus some of the materials that tie ribosomes together.
Milli-: One-thousandth part, i.e., a millimeter is one-thousandth of a meter.
Milli-equivalent: One-thousandth of an equivalent; the amount of acid or base in one milliliter of a 1 normal solution of an acid or base.
Mineral element: In biochemistry, any chemical element other than C,H,O,N, and S. That is, the elements not commonly found in carbohydrates, proteins, and lipids are called the mineral elements.
Miscible: A term used to describe liquids that will dissolve partially or completely in each other; e.g., water and alcohol.
Mitochondrion: A subcellular particle or organelle, Mitochondria are the site of much of the energy production of cells. (See Section 32-4.)
Mitosis: The process of cell division in which each daughter cell has the same number of chromosomes as the parent cell.
Mixed glyceride: A ester of glycerol in which the three fatty acids are not the same.
Molar solution: A solution containing one mole of solute in one liter of solution.
Molarity: The strength of a solution expressed in moles per liter; usually abbreviated M.
Mole: The weight in grams of Avogadro's number (6.023×10^{23}) of particles (ions, atoms, or molecules) of a substance.
Molecular formula: A formula which states the kind and number of all atoms in a molecule.
Molecular weight: The sum of the atomic weights of the atoms in a molecule; when expressed in grams it is the equivalent of a mole.
Monosaccharide: A polyhydroxy aldehyde or polyhydroxy ketone. The simplest type of carbohydrate. The more correct systematic name for such a compound is *glycose*.
Mutation: An abrupt change in an inheritable characteristic.

NAD: Nicotinamide adenine dinucleotide. A substance that serves as an electron acceptor and donor in biological oxidations. (See Fig. 36-13 for the formula of this compound.)
Neutron: A neutral subatomic particle with a mass approximately that of a hydrogen atom. Neutrons account for one-half or more of the mass of all atoms except hydrogen.
Nitration: Substitution of a nitro group, $-NO_2$, onto a molecule (Section 24-6).
Nitrogen fixation: The combining of elementary nitrogen with other elements to form usable compounds.
Nuclear fission: The transmutation in which an atom is converted into smaller atoms. Such reactions are accompanied by an enormous liberation of energy.
Nuclear fusion: The transmutation occurring naturally in the sun, or synthetically in a hydrogen bomb, in which small atoms combine to form larger atoms. Such reactions liberate greater amounts of energy than do comparable fission reactions.
Nuclear reactor: The term commonly applied to the device which is used to control the rate of nuclear reactions and to produce useful nuclear power.
Nucleic acid: A high-molecular-weight substance from natural sources which, on hydrolysis, yields organic bases, pentoses, and phosphate. Although originally found in the nuclei of cells, it is present throughout the cell.
Nucleoprotein: A combination of a nucleic acid and a protein. The proteins in such combinations contain many basic amino acids. The bonds between the acidic nucleic acids and basic proteins are saltlike and relatively weak.
Nucleosidase: An enzyme that catalyzes the hydrolysis of nucleosides to yield an organic base and a pentose.
Nucleoside: A compound which, on hydrolysis, yields one molecule of an organic base and a pentose.

Nucleotidase: An enzyme that catalyzes the hydrolysis of nucleotides to give nucleosides and phosphate.
Nucleotide: A compound which, on hydrolysis, yields one molecule of an organic base, a pentose, and a phosphate group.
Nutrient: A chemical substance that provides either energy or a necessary component of an animal.
Nutrition: The science that is concerned with foods and their effect on organisms, particularly on humans.

Oil: In biochemistry, a triglyceride that is a liquid at room temperature.
Olefin: The old common name for an alkene.
Oligosaccharide: A molecule containing from two to ten monosaccharide units. The most important oligosaccharides are the disaccharides.
Optical acitivity: The ability to rotate the plane of vibration of polarized light.
Organic substances: The compounds of carbon. They may be produced synthetically or by living organisms.
Osmosis: The selective passage of liquids through a semipermeable membrane in a direction which tends to make concentrations of all substances on one side of the membrane equal to those on the other side.
Oxidation state: A term used to indicate the valence of an atom in a compound. For example, the oxidation state of chromium in Cr_2O_3 is $(3+)$, and of sulfur in SO_3 is $(6+)$.
Ozone: An allotropic form of oxygen with a molecular formula of O_3.

Pentose: A monosaccharide containing five carbon atoms. Ribose, arabinose, and xylose are pentoses.
Pepsin: A gastric enzyme that catalyzes the hydrolysis of peptide bonds of proteins. It is an endopeptidase.
Peptidase: Any enzyme that catalyzes the hydrolysis of the peptide linkage between amino acids.
Peptide: A compound formed by a combination of the amino group of one amino acid with the acid group of another. Peptide is also used to described the linkage between the two amino acids.
Peptide bond: A group containing the carbonyl group from one amino acid and the N—H from another, for example:

$$\begin{matrix} & O & H \\ & \| & | \\ - & C - & N - \end{matrix}$$

This linkage is found between the α-amino acid units in a protein. It is just a particular example of an amide bond.
Periodic law: A law which states that when the elements are arranged in the order of their increasing atomic numbers, they exhibit a periodic recurrence of properties.
Periodic table: An arrangement of the chemical elements in table form to show the periodic recurrence of properties. See also *Periodic law*.
pH: A number denoting the hydrogen ion concentration in a solution.
Phagocyte: A cell of the body that can engulf foreign material.
Phenol: An organic compound having an —OH functional group on an aryl skeleton, ArOH. (The term is also used for the simplest compound of this class, C_6H_5OH. See Section 25-6.)

Phenyl group: The structural unit C_6H_5- or

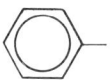

Phosphogluconate shunt: A series of reactions that provides for release of energy from glucose in fewer reactions than in glycolysis. This series of reactions may also result in the synthesis of ribose.

Phospholipid: A lipid that contains a phosphate group. The most abundant phospholipids are derivatives of phosphatidic acid in which the phosphate is esterified with one of the —OH groups of glycerol.

Phosphorylase: An enzyme that catalyzes the addition of a phosphate group to a molecule. The phosphate group comes from ATP. It is better to call such enzymes phosphotransferases.

Phosphorylation: The addition of a phosphate group to a molecule. In many phosphorylation reactions phosphoric acid functions as water does in hydrolysis; such a reaction is called a phosphorolysis.

Photolysis (of water): The lysis of water to give oxygen and hydrogen under the influence of light. This process is the source of the free oxygen formed in photosynthesis.

Photosynthesis: The process in which the energy of light is stored in chemical compounds of a plant.

Physical change: Changes in physical state or of shape with no change in composition. Change of ice to water and water to steam are physical changes.

Physical properties: Those properties of a substance (solubility, boiling point, freezing point, etc.) that can be measured without changing the composition of a substance.

Polar covalent bond: A covalent bond is polar when it is formed between two atoms having a considerable difference in electronegativity. In such a bond, the pair of shared electrons is shifted toward the more negative atom.

Polarized light: Light that has only one plane of vibration. (See Section 31-5.)

Polymer: A very large molecule built up when many ordinary size molecules are joined by covalent bonds. (See Section 30-2.)

Polypeptide: A substance containing many amino acids linked by peptide bonds. Polypeptides contain fewer amino acid units than do the proteins.

Polyprotic acid: A term indicating the compound has more than one potential acidic hydrogen; for example, H_3PO_4.

Polysaccharide: A polymer of monosaccharide (i.e., glycose) units. Prominent members of this group are the starches, cellulose, and glycogen.

Potential energy: The energy a body possesses by virtue of its position. Also the stored energy present in fuels and foods.

Pre-pepsin (Pepsinogen): The inactive precursor of the enzyme pepsin.

Primate: A classification of animals that includes man, apes, and monkeys.

Primer DNA: A molecule of DNA that must be present in a system capable of synthesizing DNA before the synthesis can start. The kind of DNA synthesized resembles the primer DNA.

Protein: A high-molecular-weight substance which, on hydrolysis, yields amino acids. A polymer of amino acids.

Proteolytic: A word used to describe those enzymes that catalyze the breakdown of proteins.

Proton: A hydrogen ion. The nucleus of a hydrogen atom. A positively charged subatomic particle with a mass approximating that of a neutron.

Protozoan: A member of a class of single-celled animals.
Purine bases: Important components of nucleic acids related to purine,

Pyrimidine bases: Important component of nucleic acids related to the organic base pyrimidine,

Quaternary ammonium salt: An ammonium salt having four organic groups (the same or different); for example, $R_4N^+Cl^-$.

Radioactive isotope: An isotope of an element so unstable that it decomposes to give α, β, γ or other radiations.
Radioactivity: The name suggested by Madame Curie for the disintegration of atomic nuclei in which α, β, or γ radiation occurs.
Reductive amination: The reaction of ammonia with a carbonyl group in the presence of a reducing agent (often hydrogen). The resultant product is an amine. Both chemical and biochemical examples are very important.
Ribonuclease (RNA-ase): An enzyme that catalyzes the hydrolysis of RNA.
Ribonucleic acid (RNA): A polymer composed of nucleotides which contain ribose. Three types of RNA are important in protein synthesis: messenger RNA, ribosomal RNA, and soluble (transfer) RNA. Many viruses contain RNA and protein.
Ribosomal RNA: The RNA found in ribosomes.
Ribosome: A small spherical body or organelle found in cells. The ribosomes are an important site of protein synthesis in a cell. (See Section 32-4.)

Saliva: The fluid secreted in the mouth by the salivary glands. It contains complex polysaccharides and the enzyme amylase.
Salt: A compound composed of the positive ion from a base and the negative ion from an acid; i.e., a metal ion and a nonmetal ion, such as KBr.
Salting out: The precipitation of a protein by addition of a salt, usually as a salt solution.
Saponification: The reaction of an ester with a metallic hydroxide (or certain other bases) to produce a carboxylate salt and an alcohol. The name originates from the fact that when a fat undergoes such a reaction, the salt formed is soap. (See Section 26-6 and 27-5.)
Saturated hydrocarbon: A carbon compound in which all carbon bonds, other than those required to hold the skeleton together, are occupied by hydrogen atoms.
Saturated solution: A solution in which the undissolved solute and the dissolved solute are in equilibrium. No greater amount of solute can be dissolved at the existing conditions.
Smog: Originally used to describe a mixture of smoke and fog. Now generally used to describe visible air pollution.
Smoke: A suspension of solid particles in a gas.
Soap: An alkali metal carboxylate salt having a total of twelve to about eighteen carbon atoms; for example, $C_{17}H_{35}COONa$.
Soluble (transfer) RNA: A relatively low-molecular-weight RNA that attaches to an amino acid during the process of protein synthesis. Transfer RNA is believed to match with a

specific part of messenger RNA known as the code. Following this, the amino acid is added to the growing peptide chain of the protein being synthesized.
Solute: A dissolved substance. In solutions composed of a solid and a liquid, the solid is the solute. In solutions composed of two liquids, the one present in the lesser proportion is considered to be the solute.
Solution: A homogeneous mixture of molecules, atoms, or ions of two or more substances.
Solvent: The dissolving substance. In a solution composed of a solid and a liquid, the liquid is considered to be the solvent. In a solution composed of two liquids, the one present in the greater proportion is considered to be the solvent. Water is the most common solvent since it is so abundant and so many substances are soluble in it.
Specific dynamic action: The increased heat production which accompanies digestion. It is especially noticeable during digestion of proteins.
Standard conditions (gases): Often abbreviated as S.T.P. meaning Standard Temperature and Pressure. Since the volume occupied by a sample of a gas is dependent upon its temperature and the pressure exerted upon it, some standards for measurement are necessary. Zero degrees centigrade and a pressure of 760 mm of mercury have therefore been adopted as standard conditions.
Starch: The polysaccharide found in seeds, roots, and tubers of plants. It is composed of glucose units linked α1-4 with some α1-6 branches.
Steroid: A compound containing a complex hydrocarbon nucleus. See Section 35-2 for a diagram of this nucleus. Among the important compounds having a steroid skeleton are bile acids, sex hormones, and adrenal hormones.
Structural formula: Way of writing the structure of a molecule to show which atoms are bonded to which other atoms and the types of bonds used.
Subcutaneous: Below the skin.
Substitution reaction: A reaction in which certain convalently bonded atoms of a compound are displaced by different atoms.
substrate: The substance on which an enzyme acts.
Sugar: A sweet-tasting monosaccharide or disaccharide.
Sulfonation: Substitution of a sulfonic acid group, $-SO_2OH$, onto a molecule (Section 24-5).

Teflon: A plastic material composed of carbon and fluorine. In terms of chemical structure it is a polymer.
Temperature inversion: Ordinarily the temperature of the air decreases with altitude. When there is a warm layer on top of a cooler layer, we call this a temperature inversion.
Template RNA: See *Messenger RNA*.
Ternary acid: An acid composed of three different elements, the third element usually being oxygen; for example, H_2SO_4.
Terpene: An oily substance having a distinct odor. A terpene has the formula $C_{10}H_{16}$ and is derived from plants. It is composed of isoprene units.
Tetrahedra: Four-sided. A term used to describe the regular arrangement of four covalent bonds at a carbon atom (Section 21-6).
Therapy: The treatment of disease.
Tissue lipids: Lipids that play an important role in the functioning of a cell and are found in all cells. The amount of tissue lipids in an animal does not vary appreciably as do the depot lipids.
Titration: The process of reacting a solution of unknown strength with one of known strength. This procedure is commonly used to determine concentrations of solutions of acids and bases.

Transamination: A reaction in which an amino group is transferred from an amino acid to a keto acid. This results in the synthesis of a new amino acid and the conversion of the original amino acid to a keto acid.
Transfer RNA: See *Soluble RNA.*
Triglyceride: An ester in which all three oxygens of glycerol have been esterified by carboxylic acids. Sometimes called more briefly a glyceride.
Trypsin: A proteolytic enzyme secreted by the pancreas.

Unsaturated: A term used to describe an organic compound having double or triple bonds between some atoms. The most common examples are $C=C$, $C\equiv C$, $C=O$, $C=N$, and $C\equiv N$. These unsaturated units of structure can be saturated by addition of hydrogen.
Unsaturated hydrocarbon: A carbon compound having one or more double or triple bonds, so that not all bonds are occupied by hydrogens. (Compare with *Saturated hydrocarbon*.)

Valence: A term used to indicate the combining ability of an element. Valence may be measured in terms of the number of electrons lost or gained by an atom or group of atoms, or by the number of pairs of electrons shared.
Virus: One type of agent that causes diseases in both plants and animals. Viruses contain large amounts of nucleic acids and protein. Viruses are generally smaller than bacteria and were originally distinguished from them because viruses could pass through filters.
Vitamin: An organic compound required in small amounts in the diet of an animal. A vitamin must be defined in terms of a given type of animal. Ascorbic acid is needed by all animals, but is synthesized by most animals and thus is not a vitamin for them. Ascorbic acid is a vitamin (vitamin C) for humans, other primates, and the guinea pig, since these animals do not synthesize it.

Wax: An ester of a fatty acid and a long-chain, monohydric alcohol. (See Section 26-9E.)
Weight: Commonly used to indicate an amount or mass of material. Actually, weight is dependent upon the gravitational attraction between bodies, and therefore is not constant, as is mass.

Zeolite: A silicate mineral used in water softeners.
Zymogen: The inactive precursor of an enzyme. Pre-pepsin is the zymogen of pepsin.

ANSWERS TO SELECTED WORK EXERCISES

Chapter 21:

3. a) four   c) seven

Chapter 22:

1. a) 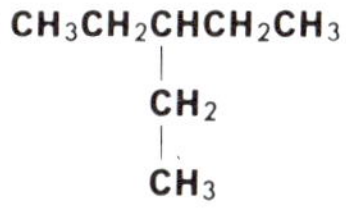 $CH_3CH_2CHCH_2CH_3$
       $\quad\quad\quad\;\;|$
       $\quad\quad\quad\;CH_2$
       $\quad\quad\quad\;\;|$
       $\quad\quad\quad\;CH_3$

c) 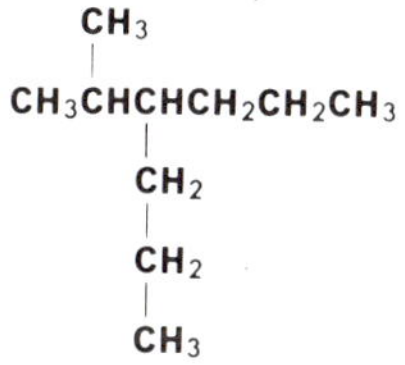 $CH_3CHCHCH_2CH_2CH_3$
   $\quad\quad|\quad\;|$
   $\quad CH_3\;\;CH_2$
   $\quad\quad\quad\;\;|$
   $\quad\quad\quad\;CH_2$
   $\quad\quad\quad\;\;|$
   $\quad\quad\quad\;CH_3$

e) 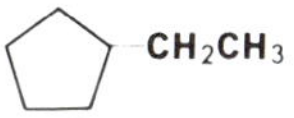 cyclopentane—CH_2CH_3

g) 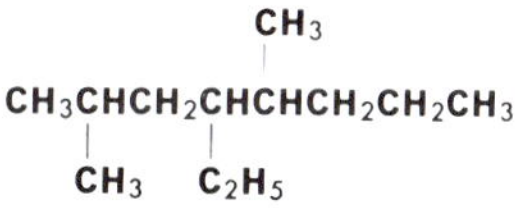 $CH_3CHCH_2CHCHCH_2CH_2CH_3$
   $\quad\quad|\quad\quad\quad|\quad|$
   $\quad CH_3\quad\;CH_3\;C_2H_5$

i) 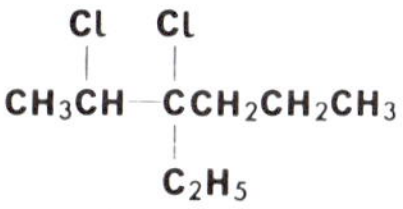 $\;\;\;\;\;\;\;\;Cl\;\;\;\;Cl$
   $\quad\quad\;\;|\quad\;\;|$
   $CH_3CH—CCH_2CH_2CH_3$
   $\quad\quad\quad\;|$
   $\quad\quad\;\;C_2H_5$

2. and 3.

a) $CH_3CH_2CH_2CH_2CH_2CH_3$ hexane

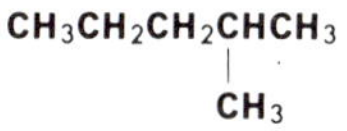

$CH_3CH_2CH_2CHCH_3$ 2-methylpentane
$\quad\quad\quad\quad\;|$
$\quad\quad\quad\;CH_3$

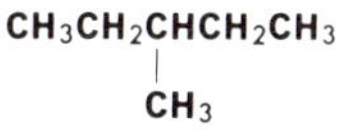

$CH_3CH_2CHCH_2CH_3$ 3-methylpentane
$\quad\quad\;\;|$
$\quad\quad CH_3$

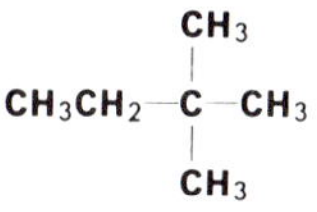

$\quad\quad\quad CH_3$
$\quad\quad\quad\;|$
$CH_3CH_2—C—CH_3$ 2,2-dimethylbutane 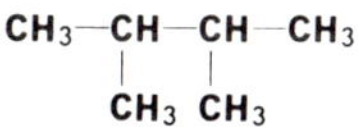
$\quad\quad\quad\;|$
$\quad\quad\quad CH_3$

$CH_3—CH—CH—CH_3$ 2,3-dimethylbutane
$\quad\quad\;|\quad\;\;|$
$\quad\;CH_3\;CH_3$

c) 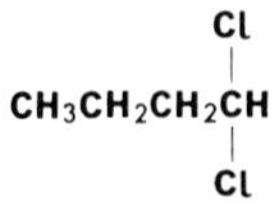 CH$_3$CH$_2$CH$_2$CH(Cl)(Cl) — 1,1-dichlorobutane ✓

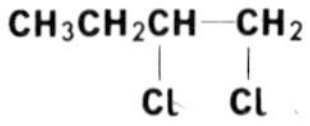

CH$_3$CH$_2$CH(Cl)—CH$_2$Cl — 1,2-dichlorobutane ✓

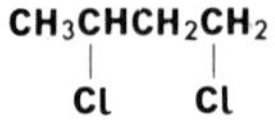

CH$_3$CH(Cl)CH$_2$CH$_2$Cl — 1,3-dichlorobutane

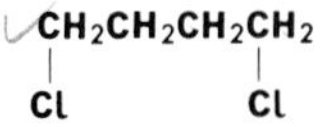

ClCH$_2$CH$_2$CH$_2$CH$_2$Cl — 1,4-dichlorobutane ✓

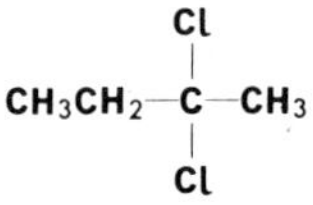

CH$_3$CH$_2$—C(Cl)(Cl)—CH$_3$ — 2,2-dichlorobutane ✓

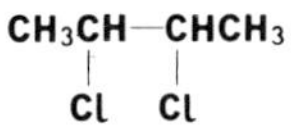

CH$_3$CH(Cl)—CH(Cl)CH$_3$ — 2,3-dichlorobutane ✓

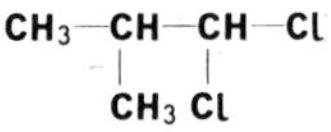

CH$_3$—CH(CH$_3$)—CH(Cl)—Cl — 1,1-dichloro-2-methylpropane

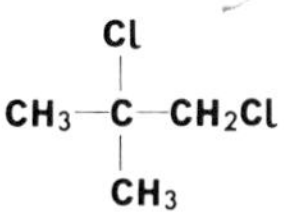

CH$_3$—C(Cl)(CH$_3$)—CH$_2$Cl — 1,2-dichloro-2-methylpropane

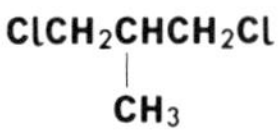

ClCH$_2$CH(CH$_3$)CH$_2$Cl — 1,3-dichloro-2-methylpropane

4. a) 2-methylpentane c) 2-chloro-4-ethylhexane e) cyclopentane
   g) methylcyclopentane i) 6-ethyl-2,3,5-trimethyloctane
5. a) chloroform: trichloromethane c) isobutane; methylpropane
   e) methyl bromide; bromomethane
6. a) CH$_3$CH$_2$CH$_3$ + 5O$_2$ → 3CO$_2$ + 4H$_2$O
   c) CH$_4$ + Cl$_2$ → CH$_3$Cl + HCl
   CH$_3$Cl + Cl$_2$ → CH$_2$Cl$_2$ + HCl
   CH$_2$Cl$_2$ + Cl$_2$ → CHCl$_3$ + HCl
   CHCl$_3$ + Cl$_2$ → CCl$_4$ + HCl

e) 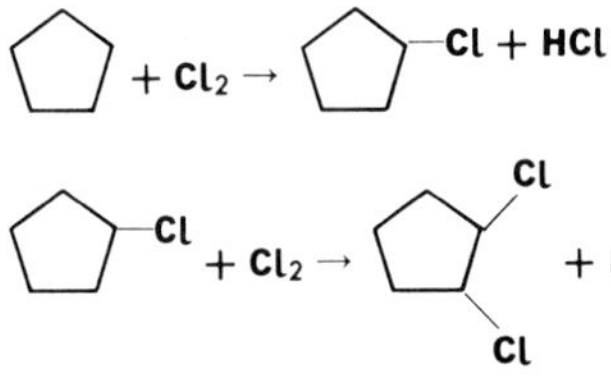 (cyclopentane) + Cl$_2$ → (cyclopentane)—Cl + HCl

(cyclopentane)—Cl + Cl$_2$ → (cyclopentane with two Cl) + HCl

and so forth

g) no reaction

i) CH$_2$Cl$_2$ + Br$_2$ → CHCl$_2$Br + HBr
CHCl$_2$Br + Br$_2$ → CCl$_2$Br$_2$ + HBr

Answers to Selected Work Exercises 737

9. a) carboxylic acid, ester, alkane, aromatic
   c) amine, alkane
   e) amine, alcohol, alkane, aromatic

Chapter 23:

1. c) 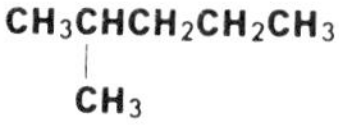 CH$_3$CHCH$_2$CH$_2$CH$_3$ C$_6$H$_{14}$
         |
         CH$_3$

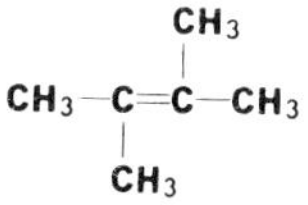

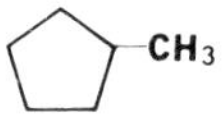

   CH$_3$
   |
   CH$_3$—C=C—CH$_3$ C$_6$H$_{12}$
   |
   CH$_3$

   ⬠—CH$_3$ C$_6$H$_{12}$

3. a) CH$_3$CH$_2$CH=CH$_2$

   CH$_3$ CH$_3$
      \\   /
       C=C
      /   \\
      H   H

   CH$_3$ H
      \\   /
       C=C
      /   \\
      H   CH$_3$

   CH$_3$—C=CH$_2$
          |
          CH$_3$

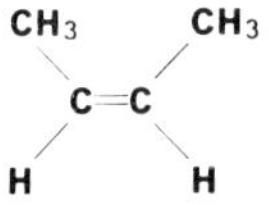

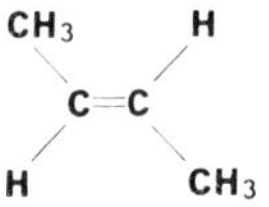

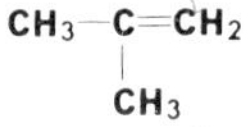

4. a) 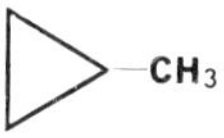 ☐   △—CH$_3$

5. a) six
6. a) 1-butene; cis-2-butene; trans-2-butene; methylpropene

7. a)  CH$_3$CHCH=CH$_2$
          |
          Cl

   c)  ⬡ (cyclohexene)

   e) 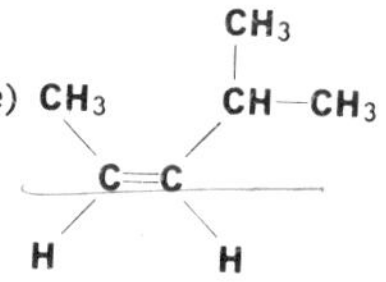 CH$_3$ CH$_3$
         \\ |
          C=C   CH—CH$_3$
         /
        H   H

10. a) CH$_3$CH$_2$CH=CH$_2$ + H$_2$ $\xrightarrow{Pt}$ CH$_3$CH$_2$CH$_2$CH$_3$
    c) 2CH$_3$CH$_2$CH=CHCH$_3$ + 15O$_2$ → 10CO$_2$ + 10H$_2$O

    e) 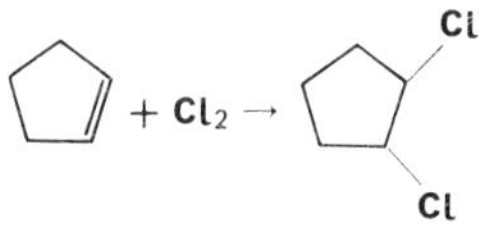 ⬠ + Cl$_2$ → ⬠ with Cl, Cl

    g)  CH$_3$CH=CHCH$_3$ + HOSO$_2$OH → CH$_3$CH$_2$CHCH$_3$
                                              |
                                              OSO$_2$OH

    i) 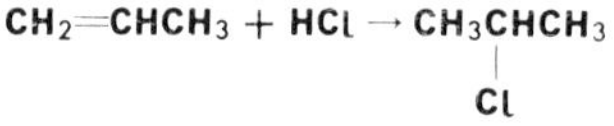 CH$_2$=CHCH$_3$ + HCl → CH$_3$CHCH$_3$
                                |
                                Cl

    k) 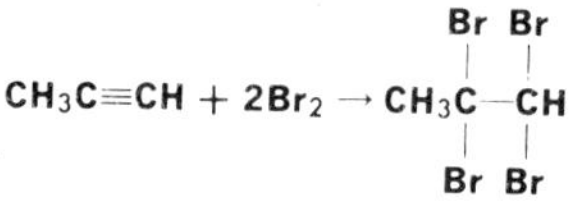 CH$_3$C≡CH + 2Br$_2$ → CH$_3$C—CH
                              | |
                              Br Br
                              (Br Br above and below)

    m) CH$_3$C≡CH + 4O$_2$ → 3CO$_2$ + 2H$_2$O

11. a) yes   c) no   e) yes   g) yes
13. a) petroleum
    b) CH$_3$CH$_3$ + Cl$_2$ $\xrightarrow{UV}$ CH$_3$CH$_2$Cl + HCl
    c) CH$_3$CH$_3$ $\xrightarrow{500°C}$ CH$_2$=CH$_2$ + H$_2$
       CH$_2$=CH$_2$ + HCl → CH$_3$CH$_2$Cl
    d) Method (c) gives only the desired product. Method (b) gives many other compounds, besides the desired product: dichloroethanes, trichloroethanes, etc. This result means that only a small amount of chloroethane is obtained, and it must be separated from all the other substances.

Chapter 24:

1. a) C$_6$H$_5$—CH$_3$ c) Cl—C$_6$H$_4$—Cl   e) Br,NO$_2$-disubstituted benzene   g) C$_6$H$_5$—SO$_3$Na

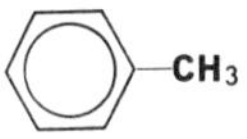

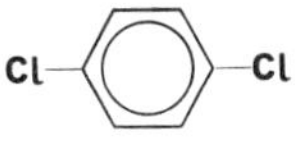

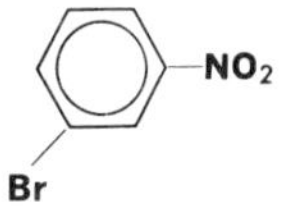

2. a) C$_6$H$_5$—SO$_3$H   benzenesulfonic acid   c) C$_6$H$_5$—NO$_2$ nitrobenzene

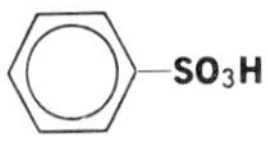

   e) no reaction

3. a) C$_6$H$_5$—COOH   c) benzene-1,2,4-tricarboxylic acid (COOH, COOH, COOH)

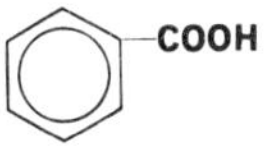

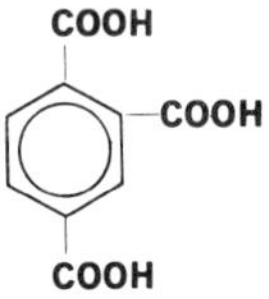

Chapter 25:

1. and 3.

   CH$_3$CH$_2$CH$_2$CH$_2$OH   primary

   CH$_3$CH$_2$CHCH$_3$ secondary
       |
      OH

   CH$_3$CHCH$_2$OH   primary
     |
    CH$_3$

      CH$_3$
       |
   CH$_3$—C—CH$_3$ tertiary
       |
      OH

5. a) CH$_3$CHCH$_3$ c) CH$_3$CHCH$_2$OH   e) CH$_3$OK   g) 3-nitrophenol (OH, NO$_2$ on benzene)
      | |
     OH   CH$_3$

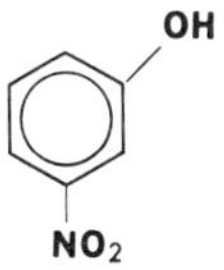

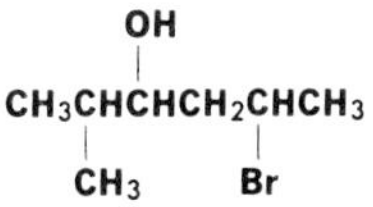

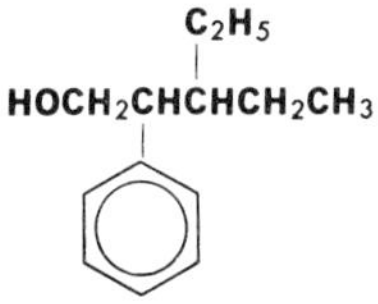

      OH   C$_2$H$_5$ OH
      | | |
   i) CH$_3$CHCHCH$_2$CHCH$_3$ k) HOCH$_2$CHCHCH$_2$CH$_3$ m) cyclohexanol
         | | |
       CH$_3$ Br   C$_6$H$_5$

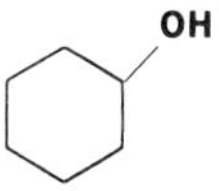

7. a) CH$_2$=CHCH$_2$CH$_3$ + HOSO$_2$OH → CH$_3$CHCH$_2$CH$_3$
                                                   |
                                                   OSO$_2$OH

   c) CH$_3$CH$_2$CHCH$_2$CH$_3$ $\xrightarrow{\text{Cu, 250°C}}$ CH$_3$CH$_2$CCH$_2$CH$_3$ + H$_2$
               | ‖
             OH   O

   e) CH$_3$CH$_2$CHCH$_3$ + [O] → CH$_3$CH$_2$CCH$_3$ + H$_2$O
             | ‖
         OH   O

9. a) amine, ether, alkane, aromatic

11. 2-methyl-2-butanol

14. a) aldehyde   b) ketone   c) no reaction

Chapter 26:

1. a) CH_3CH_2COOH c) [benzene ring]—COOH   e) [benzene ring]—COOK

    g) $CH_3CH{-}O{-}\underset{\underset{O}{\|}}{C}{-}CH_3$ i) $C_{27}H_{55}{-}\underset{\underset{O}{\|}}{C}{-}O{-}C_{30}H_{61}$
         |
         CH_3

2. a) $H{-}\underset{\underset{O}{\|}}{C}{-}O{-}H + Na^+OH^- \rightarrow H{-}\underset{\underset{O}{\|}}{C}{-}O^-Na^+ + HOH$

    c) [benzene]—C(=O)—O—H + HOC_2H_5 $\xrightarrow[\text{heat}]{H_2SO_4}$ [benzene]—C(=O)—O—C_2H_5 + HOH

    e) $CH_3CH_2CH_2\underset{\underset{O}{\|}}{C}{-}O{-}C_2H_5 + HOH \xrightarrow[\text{heat}]{H_2SO_4} CH_3CH_2CH_2\underset{\underset{O}{\|}}{C}{-}O{-}H + HOC_2H_5$

    g) [benzene]—C(=O)—O—CH_3 + NaOH $\rightarrow$ [benzene]—C(=O)—O⁻Na⁺ + $HOCH_3$

3. a) $CH_3(CH_2)_4CH_2OH + 2[O] \rightarrow CH_3(CH_2)_4\underset{\underset{O}{\|}}{C}{-}OH$

5. a) ethyl butanoate
    c) ammonium benzoate
7. a) oxidation ($KMnO_4$ or $Na_2Cr_2O_7$)
8. a) ester, phenol, alkane, aromatic

Chapter 27:

1. a) More than one structure is possible. This is a mixed glyceride.
    c) Two or more of the fatty acids chains would contain double bonds.
    e) See Section 27-7.
2. a) saponification   c) hydrolysis

Chapter 28:

1. a) diethylamine   c) isobutylamine; 1-amino-2-methylpropane
2. a) $CH_3{-}\underset{\underset{CH_3}{|}}{N}{-}CH_3$ c) $CH_3CH_2NHCH_2CH_2CH_3$

    e) $CH_3CHCH\underset{\underset{|}{NH_2}}{C}HCH_3$ with CH_3 and CH_3 branches   g) $(CH_3CH_2)_3\overset{+}{N}H$ Cl^-

4. a) $CH_3Br + NH_3 \rightarrow CH_3NH_2 + HBr$
    c) $CH_3CH_2{-}\underset{\underset{O}{\|}}{C}{-}OH + NH_3 \xrightarrow{\Delta} CH_3CH_2{-}\underset{\underset{O}{\|}}{C}{-}NH_2 + HOH$

    e) CH_3—[benzene]—$NO_2 + 6[H] \xrightarrow[HCl]{Sn} CH_3$—[benzene]—$NH_2 + 2H_2O$

    g) [benzene]—C(=O)—OH + $H_2NC_2H_5 \xrightarrow{heat}$ [benzene]—C(=O)—NHC_2H_5 + HOH

5. R—C—N—CH$_2$CH$_3$
    ‖ |
    O CH$_2$CH$_3$

6. a) 2-aminopropanoic acid
   c) 2-amino-3-phenylpropanoic acid
8. a) amine, ester, substituted ammonium salt, alkane, aromatic

Chapter 29:

1. a) propanal   c) ethyl phenyl ketone

2. a) CH$_3$CH$_2$CH$_2$CHCH$_3$ $\xrightarrow[250°C]{Cu}$ CH$_3$CH$_2$CH$_2$CCH$_3$ + H$_2$ c) CH$_3$CH$_2$CH$_2$OH $\xrightarrow[250°C]{Cu}$ CH$_3$CH$_2$—C—H
         | ‖   ‖
         OH   O   O

3. a) CH$_3$CH$_2$CHCH$_2$CH$_3$ c)   CH$_3$
         | |
         OH   CH
                                                        / \
                                                     H$_2$C   CH$_2$
                                                      | |
                                                     H$_2$C   CH$_2$OH
                                                      |
                                                     CH$_2$
                                                      |
                                                     CH
                                                    /   \
                                                  CH$_3$ CH$_3$

5. Benzoic acid is formed by exposure to air because the aldehyde is so easily oxidized.
7. For identification; see Eq. (29–23)
8. a) Cu, 250°
   b) First, add HCN; second, heat in water with HCl.
11. See Section 29–11A.
13. a) The glucoside, even when dissolved in water, has no aldehyde group available to react.
    b) Acid solution hydrolyzes the glucoside bond, releasing free glucose which then reacts with Tollen's reagent.
15. aldehyde, ether, phenol, alkane, aromatic

Chapter 30:

1. vinyl chloride   CH=CH$_2$ tetrafluoroethylene   CF$_2$=CF$_2$
                       |
                       Cl

   styrene   CH=CH$_2$ acrylonitrile   CH=CH$_2$
                       | |
                       C$_6$H$_5$ CN

   propylene   CH=CH$_2$ CH$_3$
                       | |
                       CH$_3$ methyl methacrylate   C=CH$_2$
                                                                    |
   ethylene   CH$_2$=CH$_2$ COOCH$_3$

2. See Eq. 30–1.

3.   CH$_3$ H   CH$_3$ H
           \   /   \   /
            C=C   C=C
           /   \   /   \
     ····CH$_2$ CH$_2$—CH$_2$ CH$_2$—CH$_2$ CH$_2$····
                         \   /
                          C=C
                         /   \
                      CH$_3$ H

Answers to Selected Work Exercises   741

Chapter 31:

1. a) active   c) inactive   e) active
3. a) *cis-trans* isomers   c) optical isomers   e) not isomers   g) optical isomers

Chapter 32:

none

Chapter 33:

2. a)

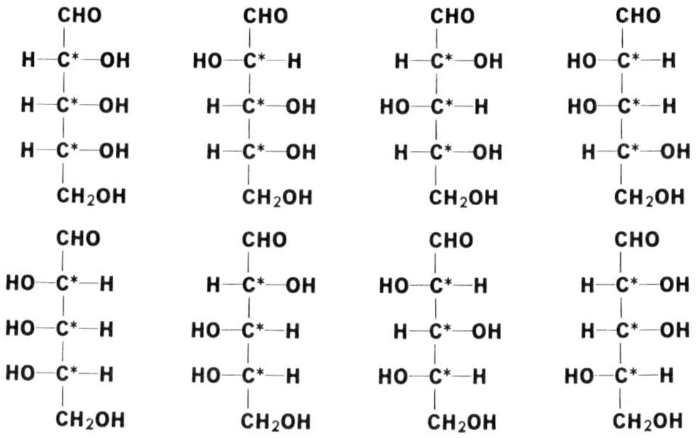

```
    CHO              CHO              CHO              CHO
H—C*—OH          HO—C*—H          H—C*—OH          HO—C*—H
H—C*—OH          H—C*—OH          HO—C*—H          HO—C*—H
H—C*—OH          H—C*—OH          H—C*—OH          H—C*—OH
   CH₂OH            CH₂OH            CH₂OH            CH₂OH

    CHO              CHO              CHO              CHO
HO—C*—H          H—C*—OH          HO—C*—H          H—C*—OH
HO—C*—H          HO—C*—H          H—C*—OH          H—C*—OH
HO—C*—H          HO—C*—H          HO—C*—H          HO—C*—H
   CH₂OH            CH₂OH            CH₂OH            CH₂OH
```

b)

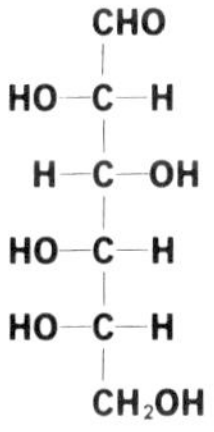

```
    CHO
HO—C—H
H—C—OH           or     [pyranose ring with CH₂OH, OH groups, showing OH (α) and OH (β)]
HO—C—H
HO—C—H
   CH₂OH
```

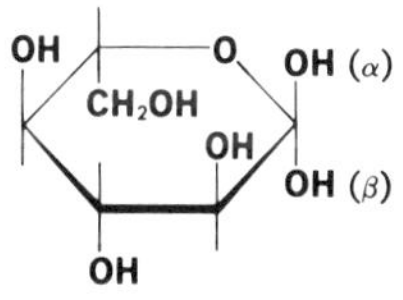

d) [two linked pyranose rings with CH₂OH and OH groups, connected by O]

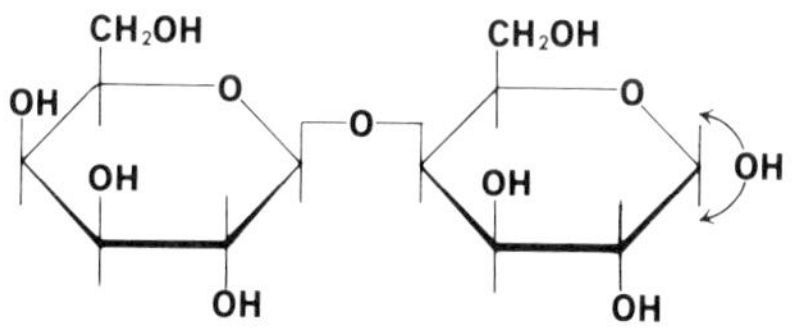

e) [pyranose ring with CH₂OH, OH, NH₂ linked via O to a ring with COOH, OH]

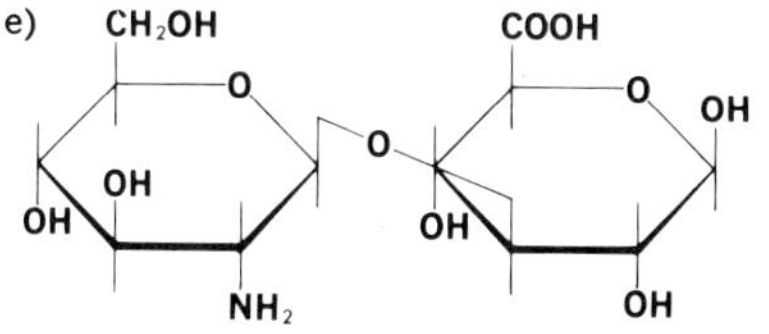

4. a) [pyranose ring with CH₂OH, OH] —O— [furanose ring with CH₂OH, OH, CH₂OH] + HOH ⟶ [pyranose with CH₂OH, OH] + [furanose with CH₂OH, OH, CH₂OH]

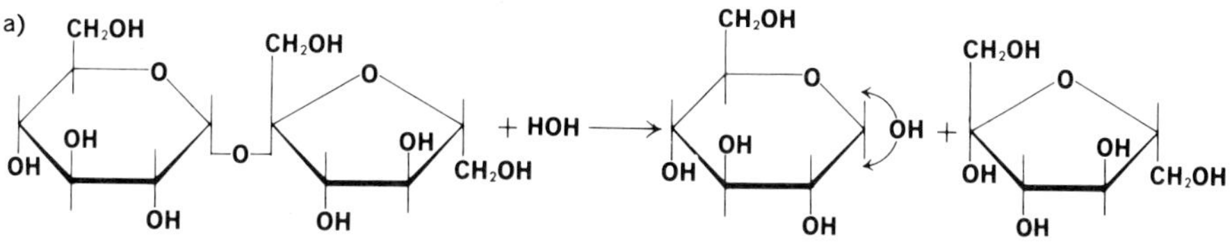

b)

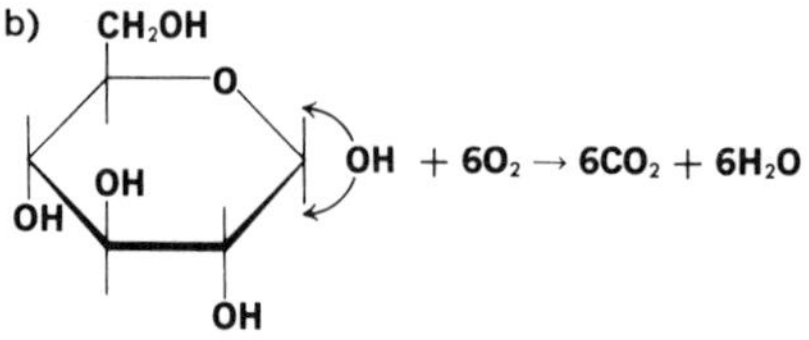

$$\text{glucose ring (C}_6\text{H}_{12}\text{O}_6\text{)} + 6O_2 \rightarrow 6CO_2 + 6H_2O$$

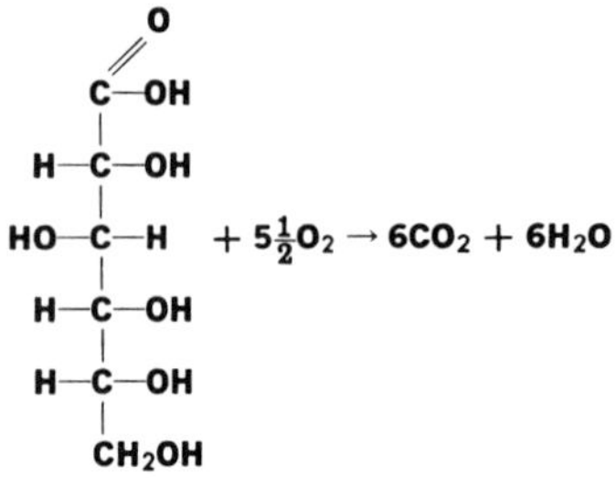

gluconic acid:

```
     O
     ‖
     C—OH
     |
  H—C—OH
     |
 HO—C—H    + 5½O₂ → 6CO₂ + 6H₂O
     |
  H—C—OH
     |
  H—C—OH
     |
     CH₂OH
```

```
     O
     ‖
     C—OH
     |
  H—C—OH
     |
 HO—C—H    + 4½O₂ → 6CO₂ + 5H₂O
     |
  H—C—OH
     |
  H—C—OH
     |
     C—OH
     ‖
     O
```

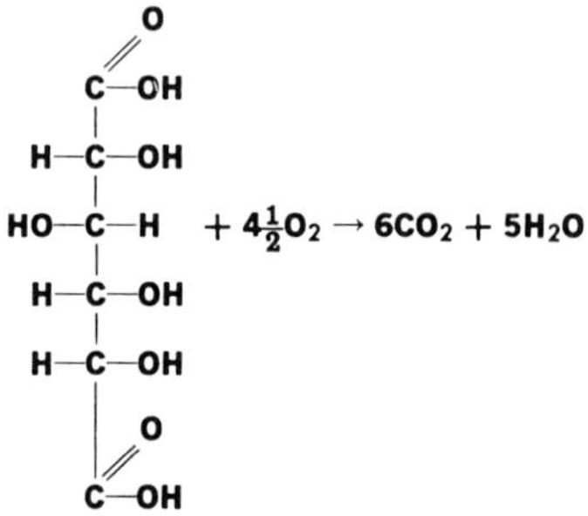

Chapter 34:

1. aspartic acid, glutamic acid
2. lysine, arginine, (histidine), (ornithine)
3. alcohol—serine, threonine
   phenol—tryosine
   sulfhydryl—cysteine
   guanido—arginine
4. eggs, milk, cheese, meats, leather
5. cystine:

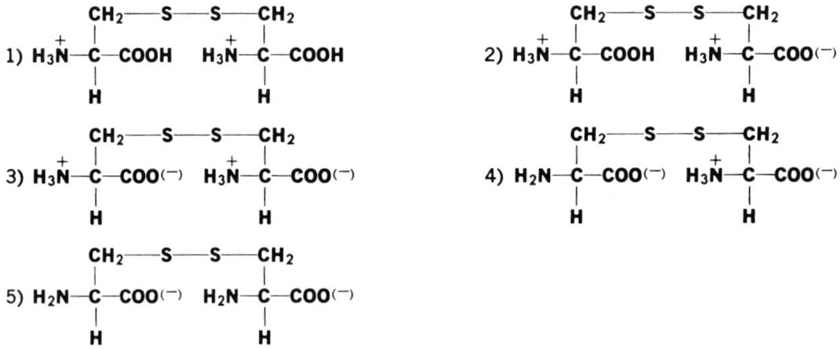

1) $H_3\overset{+}{N}$—CH(CH$_2$—S—S—CH$_2$—CH($\overset{+}{N}H_3$)—COOH)—COOH

2) $H_3\overset{+}{N}$—CH(CH$_2$—S—S—CH$_2$—CH($\overset{+}{N}H_3$)—COO$^{(-)}$)—COOH

3) $H_3\overset{+}{N}$—CH(CH$_2$—S—S—CH$_2$—CH($\overset{+}{N}H_3$)—COO$^{(-)}$)—COO$^{(-)}$

4) H_2N—CH(CH$_2$—S—S—CH$_2$—CH($\overset{+}{N}H_3$)—COO$^{(-)}$)—COO$^{(-)}$

5) H_2N—CH(CH$_2$—S—S—CH$_2$—CH(NH$_2$)—COO$^{(-)}$)—COO$^{(-)}$

glutamic acid:

1) $H_3\overset{+}{N}$—CH(COOH)—(CH$_2$)$_2$—COOH

2) $H_3\overset{+}{N}$—CH(COOH)—(CH$_2$)$_2$—COO$^{(-)}$

3) $H_3\overset{+}{N}$—CH(COO$^{(-)}$)—(CH$_2$)$_2$—COOH

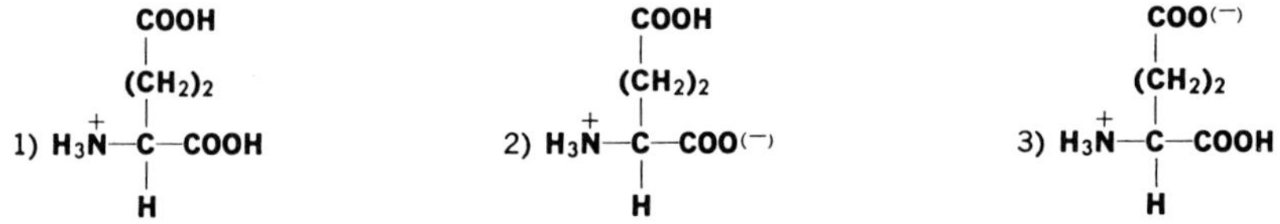

Answers to Selected Work Exercises 743

4)
$$\text{H}_3\overset{+}{\text{N}}-\underset{\underset{\text{H}}{|}}{\overset{\overset{\text{COO}^{(-)}}{|}}{\underset{|}{\text{C}}}}-\text{COO}^{(-)}$$
$(\text{CH}_2)_2$

5)
$$\text{H}_2\text{N}-\underset{\underset{\text{H}}{|}}{\overset{\overset{\text{COO}^{(-)}}{|}}{\underset{|}{\text{C}}}}-\text{COO}^{(-)}$$
$(\text{CH}_2)_2$

lysine:

1) $\text{H}_2\text{C}-\overset{+}{\text{NH}}_3$ / $(\text{CH}_2)_3$ / $\text{H}_3\overset{+}{\text{N}}-\text{C}(\text{H})-\text{COOH}$

2) $\text{H}_2\text{C}-\overset{+}{\text{NH}}_3$ / $(\text{CH}_2)_3$ / $\text{H}_3\overset{+}{\text{N}}-\text{C}(\text{H})-\text{COO}^{(-)}$

3) $\text{H}_2\text{C}-\text{NH}_2$ / $(\text{CH}_2)_3$ / $\text{H}_3\overset{+}{\text{N}}-\text{C}(\text{H})-\text{COO}^{(-)}$

4) $\text{H}_2\text{C}-\overset{+}{\text{NH}}_3$ / $(\text{CH}_2)_3$ / $\text{H}_2\text{N}-\text{C}(\text{H})-\text{COO}^{(-)}$

5) $\text{H}_2\text{C}-\text{NH}_2$ / $(\text{CH}_2)_3$ / $\text{H}_2\text{N}-\text{C}(\text{H})-\text{COO}^{(-)}$

It should be realized that all the formulas given above represent possible structures. In any given solution two or more of the forms will probably be in equilibrium with each other.

6. a)
$$\text{H}_3\overset{+}{\text{N}}-\underset{\text{H}}{\overset{\text{H}}{\text{C}}}-\overset{\text{O}}{\overset{||}{\text{C}}}-\text{O}^{(-)} + \text{H}_3\overset{+}{\text{N}}-\underset{\text{H}}{\overset{\text{CH}_2\text{OH}}{\text{C}}}-\overset{\text{O}}{\overset{||}{\text{C}}}-\text{O}^{(-)} \rightarrow \text{H}_3\overset{+}{\text{N}}-\underset{\text{H}}{\overset{\text{H}}{\text{C}}}-\overset{\text{O}}{\overset{||}{\text{C}}}-\underset{\text{H}}{\overset{\text{H}}{\text{N}}}-\underset{\text{H}}{\overset{\text{CH}_2\text{OH}}{\text{C}}}-\overset{\text{O}}{\overset{||}{\text{C}}}-\text{O}^{(-)} + \text{H}_2\text{O}$$
glycine   serine   glycylserine

Serylglycine is also formed.

b) $\text{H}_3\overset{+}{\text{N}}-\underset{\text{H}}{\overset{\text{CH}_3}{\text{C}}}-\overset{\text{O}}{\overset{||}{\text{C}}}-\text{O}^{(-)} +$ $\underset{\underset{\text{O}=\text{C}-\text{COO}^{(-)}}{\text{CH}_2}}{\overset{\text{COO}^{(-)}}{\text{CH}_2}}$ $\rightarrow \text{O}=\overset{\text{CH}_3}{\text{C}}-\text{COO}^{(-)} +$ $\underset{\underset{\text{H}}{\text{H}_3\overset{+}{\text{N}}-\text{C}-\text{COO}^{(-)}}}{\overset{\text{COO}^-}{\underset{\text{CH}_2}{\text{CH}_2}}}$

alanine / ketoglutaric acid / pyruvic acid / glutamic acid

c) $\underset{\text{N}\diagup\text{NH}}{\boxed{}}-\text{CH}_2-\underset{\overset{+}{\text{NH}_3}}{\text{CH}}-\text{COO}^{(-)} \rightarrow \text{CO}_2 + \underset{\text{N}\diagup\text{NH}}{\boxed{}}-\text{CH}_2-\text{CH}_2-\text{NH}_2$

histidine / histamine

$$\underset{\text{ornithine}}{\overset{\text{CH}_2-\text{NH}_2}{\underset{\text{H}_3\overset{+}{\text{N}}-\text{CH}-\text{COO}^{(-)}}{(\text{CH}_2)_2}}} \rightarrow \text{CO}_2 + \text{H}_2\text{N}-(\text{CH}_2)_4-\text{NH}_2$$
1,4-diaminobutane or putrescine

744 Answers to Selected Work Exercises

d) Lysylphenylalanylasparagine structure:

$$H_3\overset{+}{N}-\underset{H}{\overset{(CH_2)_4-NH_3^+}{C}}-\underset{H}{\overset{O}{\overset{\|}{C}}}-N-\underset{H}{\overset{CH_2-C_6H_5}{C}}-\underset{H}{\overset{O}{\overset{\|}{C}}}-N-\underset{H}{\overset{CH_2-C(=O)NH_2}{C}}-C-O^{(-)} + 3H_2O$$

lysylphenylalanylasparagine

$\rightarrow H_3\overset{+}{N}-\underset{H}{\overset{(CH_2)_4-NH_3^+}{C}}-COO^{(-)} + H_3\overset{+}{N}-\underset{H}{\overset{CH_2-C_6H_5}{C}}-COO^{(-)} + H_3\overset{+}{N}-\underset{}{\overset{CH_2-COO^{(-)}}{C}}-COO^{(-)} + NH_4^+$

lysine   phenylalanine   aspartic acid   ammonium ion

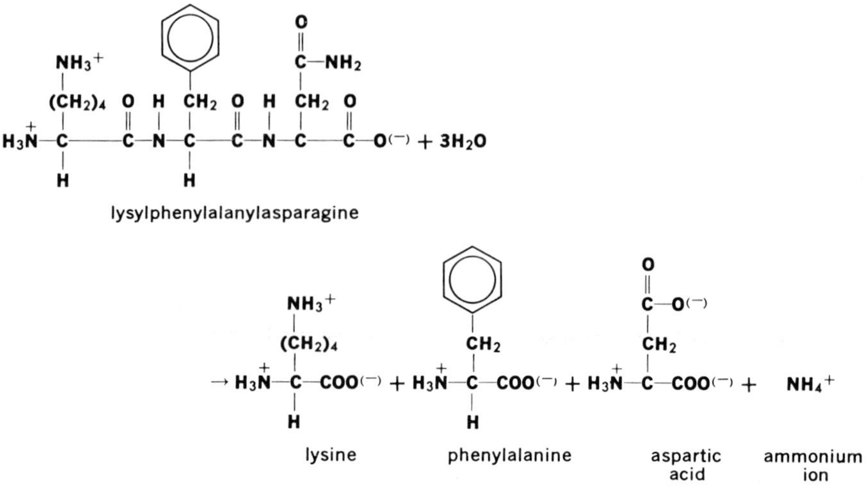

e) Aspartic acid → Oxaloacetic acid:

$$H_3\overset{+}{N}-\underset{H}{\overset{CH_2-COO^{(-)}}{C}}-COO^{(-)} + [O] \rightarrow NH_3 + O=\underset{}{\overset{CH_2-COO^{(-)}}{C}}-COO^{(-)} + H^+$$

aspartic acid   oxaloacetic acid

$$H_3\overset{+}{N}-\underset{H}{\overset{CH_3}{C}}-COO^{(-)} + [O] \rightarrow NH_3 + O=\overset{CH_3}{C}-COO^{(-)} + H^+$$

alanine   pyruvic acid

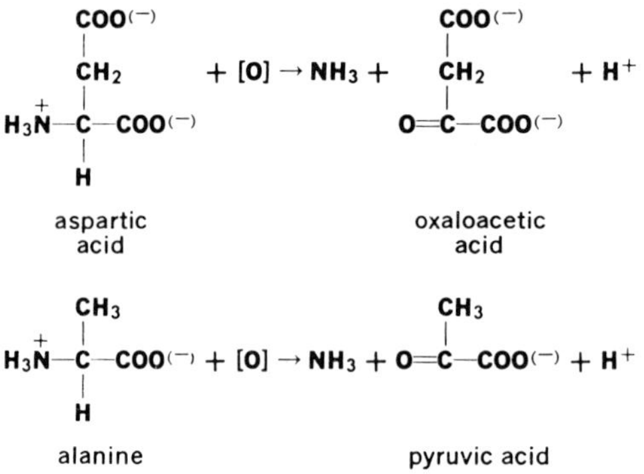

7. Tripeptide/polypeptide structure showing one disulfide bond and several hydrogen bonds between:

$H_3N-\underset{H}{\overset{H}{C}}-\underset{OH}{\overset{}{C}}-N-\underset{CH_2}{\overset{H}{C}}-C-N-\underset{CH_3}{\overset{H}{C}}-C-O^{(-)}$
              with side chains including SH (disulfide), CH₂, CH₃, and a second chain:
$CH_3-\underset{OH}{\overset{}{C}}-H$, $H_3N-\underset{H}{\overset{}{C}}-C-N-\underset{SH}{\overset{}{C}}-C-N-\underset{CH_2}{\overset{}{C}}-C-C-O^{(-)}$

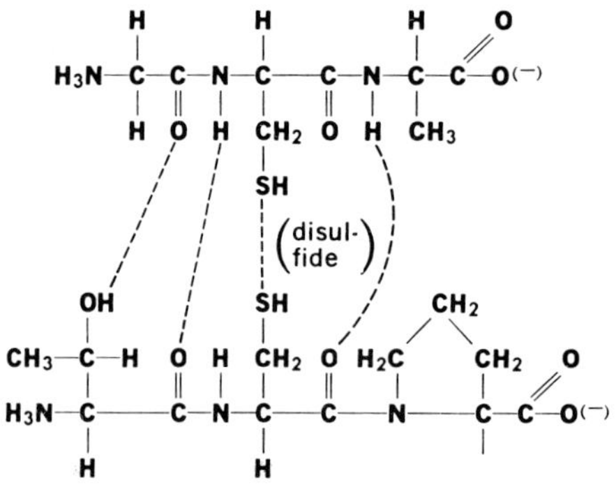

One disulfide and several hydrogen bonds are shown. Other hydrogen bonds are possible. Their formation would depend on the nearness of N—H and C—O groups to each other.

Answers to Selected Work Exercises   745

Chapter 35:

1. a) glycerol, fatty acids
   b) phosphate, glycerol, choline, ethanolamine, serine, fatty acids, possibly sphingol
   c) a fatty acid and a long chain alcohol
   d) the steroid nucleus, perhaps fatty acids
   e) terpenes are not decomposed by hydrolysis.
3. a) $C_3H_5(OH)_3 + C_{13}H_{27}COOH + C_{17}H_{33}COOH + C_{11}H_{23}COOH$ (products only are listed)
   b) $C_{15}H_{29}Br_2COOH$
   c) $C_3H_5(OH)_3 + RCOONa + R'COONa + R''COONa$

Chapter 36:

1. a) [structure: cytidine — ribose with CH₂OH, OH, H substituents linked to cytosine base with NH₂ group]

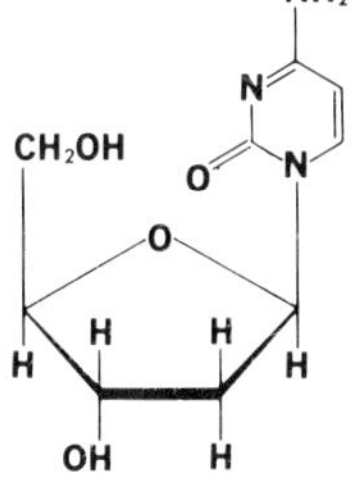

b) [structure: uridine 5'-monophosphate — HO-P(=O)(OH)-O-CH₂ linked to ribose with two OH groups, attached to uracil base]

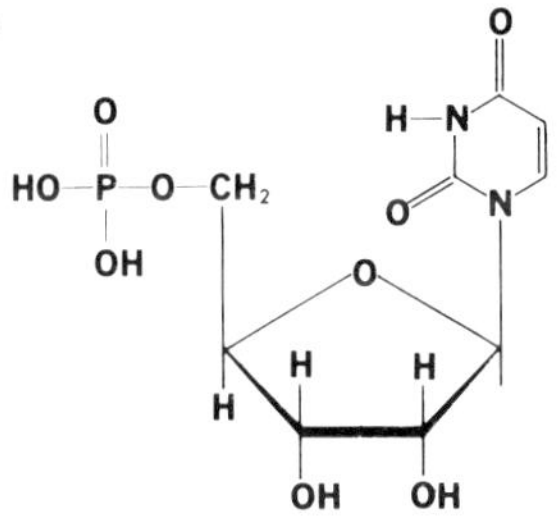

c) [structure: thymidine 5'-monophosphate — HO-P(=O)(OH)-O-CH₂ linked to deoxyribose with OH, attached to thymine base with CH₃ group]

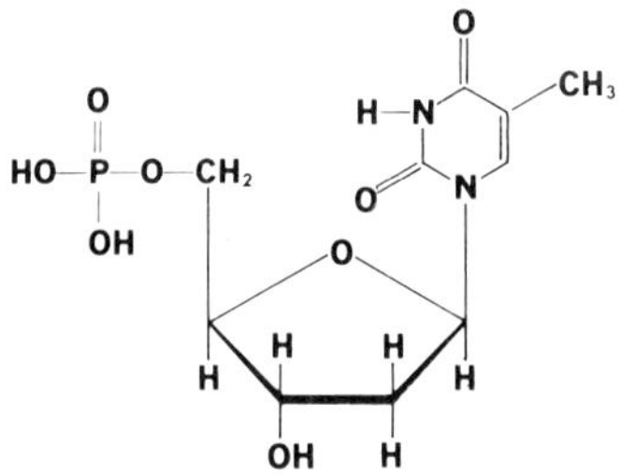

d) [structure: guanosine deoxyribonucleoside — HOCH₂ linked to deoxyribose with OH, attached to guanine base with H₂N group]

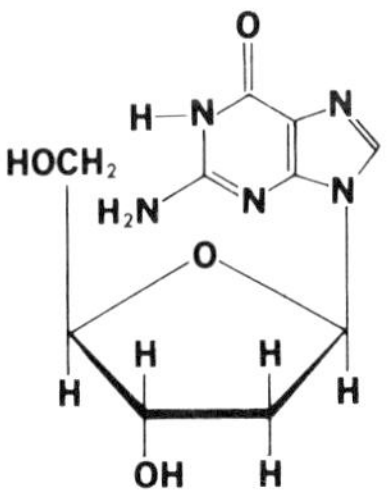

e) [structure: ADP — HO-P(=O)(OH)-O-P(=O)(OH)-O-CH₂ linked to ribose with two OH groups, attached to adenine base with NH₂ group]

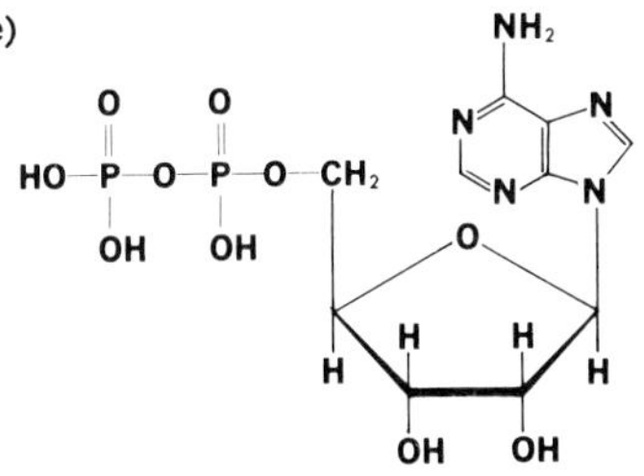

746 Answers to Selected Work Exercises

3.

[Structural diagram showing cyclic arrangement of four nucleotides with hydrogen bonding between cytosine–guanine (three bonds) and adenine–thymine-like pairs (two bonds), with ribose sugars and phosphate groups.]

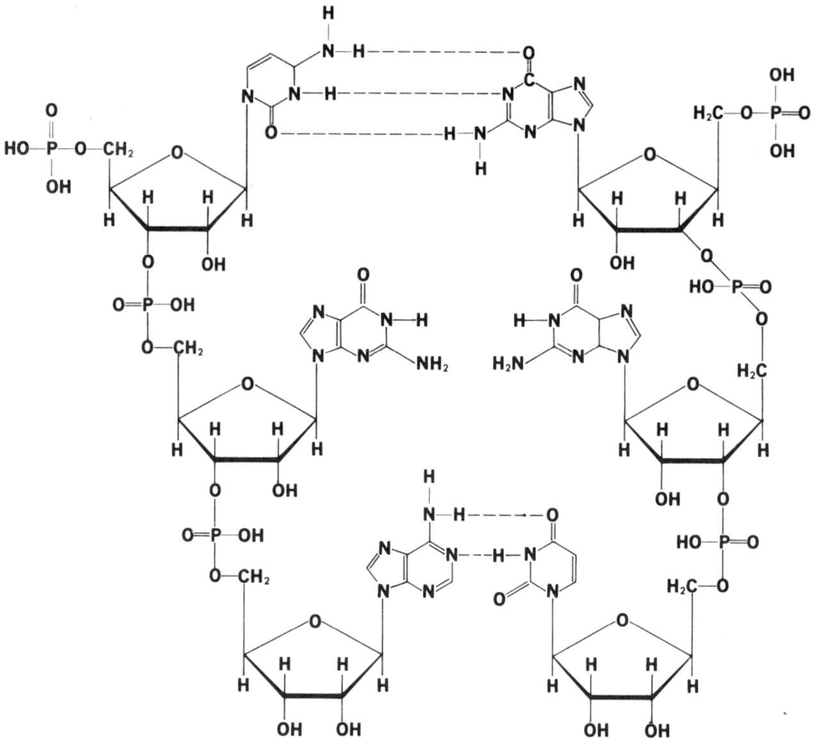

There are three possible bonds between cytosine and guanine, and two between adenine and thymine. There are probably no bonds between the two guanine groups. They are too large, the wrong shape, and do not contain matching groups.

4. a) RNA should be left intact and the DNA converted to its component nucleotides.
   b) no reaction

Chapter 37:

1. [Structural diagram showing a disaccharide + HOH → two monosaccharides]

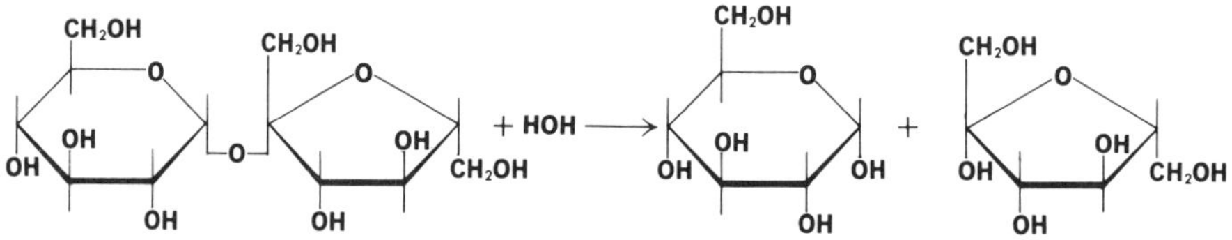

b) 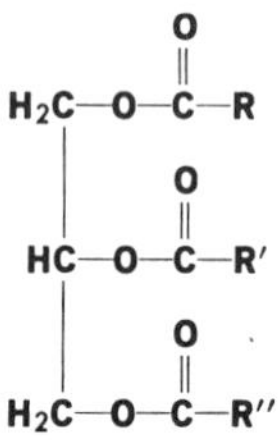
$$H_2C-O-\overset{O}{\underset{\|}{C}}-R$$
$$HC-O-\overset{O}{\underset{\|}{C}}-R' + 3HOH \rightarrow C_3H_5(OH)_3 + RCOOH + R'COOH + R''COOH$$
$$H_2C-O-\overset{O}{\underset{\|}{C}}-R''$$

a triglyceride   glycerol

Answers to Selected Work Exercises   747

c) $\underset{\text{ethyl acetate}}{CH_3\overset{\overset{O}{\|}}{C}-O-C_2H_5}$ + HOH → $\underset{\underset{\text{acid}}{\text{acetic}}}{CH_3COOH}$ + $\underset{\text{ethanol}}{C_2H_5OH}$

3. a) a substituted amide   b) an ester   c) a semiacetal—a special kind of ether

Chapter 38:

1. In metabolism the oxidation of glucose is catalyzed by enzymes and follows a definite pathway. Some of the energy of glucose is converted to the energy in ATP. When burned in a flame the oxidation of glucose is random. The energy of oxidation of glucose is released as heat. Both processes, however, give the same end products, CO_2 and H_2O, and both yield the same amount of energy.
3. Glucose can be metabolized by being converted to glycogen or by being oxidized. Insulin promotes the conversion of glucose to glycogen when the blood glucose is high. Other hormones, particularly glucagon and adrenalin, promote the formation of glucose when blood glucose levels are low.
4. Glycogen represents the most likely source. In severe carbohydrate deprivation, glucose can be formed from amino acids and the glycerol of fatty acids.
6. Other conditions can cause glucose in the urine, in addition to *diabetes mellitus*. A low kidney threshold for glucose and emotional stress may produce glucosuria.
9. Only 12 of the possible 38 molecules of ATP are produced in a poisoned animal.
11. CO_2, pyruvic acid, lactic acid, glucose, glycogen, or any intermediate in the glycolytic scheme. The pyruvate and lactate would be labeled in the carboxyl carbon, the glucose and glycogen in carbons 3 and 4.

Chapter 39:

6. a) aminopeptidase   b) carboxypeptidase   c) an endopeptidase   d) a dipeptidase

Chapter 40:

3. It is phosphorylated with ATP to yield glycerol phosphate which can then be reduced to glyceraldehyde phosphate. The latter is an intermediate in glycolysis and can yield glucose, glycogen, or be oxidized to pyruvic acid and eventually to CO_2 and H_2O in the citric acid cycle.
4. Although the series of reactions is the same, degradation of a fatty acid results in a loss of two carbons each time the series of reactions is repeated. Thus each reactant is less by two carbons than the one preceding it, and the series is not truly a cycle.
5. Those animals that are the most mobile contain the highest amount of calories per gram. Mobile fish, for example, store fat, whereas clams and oysters store glycogen.
6. 79 moles of ATP/mole of capric acid. Efficiency is 948/1500 or 63%.
7. **alanine → pyruvate → acetyl CoA → acetoacetate → β hydroxybutyric acid**
9. **glucose or starch → pyruvate → acetyl CoA → fatty acids → triglycerides**

Chapter 41:

2. Many plant proteins are deficient in some amino acids.
3. Minimal caloric requirements are found by multiplying weight in kilograms by number of hours in a day (24). A person should eat one gram of protein per kilogram of body weight per day. If foods are 70% water, the formula becomes: weight in kilograms × (100/30) = grams of protein food per day.
5. They supply energy and serve as a source of essential fatty acids. They may help prevent atherosclerosis.

9. a) nicotinamide   b) riboflavin   c) pantothenic acid
14. Nitrogen is found in thiamin, riboflavin, nicotinamide, pyridoxal, pantothenic acid, biotin, folic acid, and vitamin B_{12}. Sulfur is found in thiamin and biotin. Cobalt is found in vitamin B_{12}.

Chapter 42:

1. The theory states that a gene governs the synthesis of one protein and the effects of the gene are due to the action of the protein. The theory is based on evidence that hereditary defects are known to be caused by protein abnormalities.
3. The person would suffer from hypothyroidism. Such deficiencies could result in death of a fetus and a miscarriage.
4.

Hemoglobin Type	Normal Codes	Altered Codes
Hgb S	GAA, GAG	GUU, GUC, GUA, GUG
Hgb C	GAA, GAG	AAA, AAG
Hgb G San José	GAA, GAG	GGU, GGC, GGA, GGG
Hgb E	GAA, GAG	AAA, AAG
Hgb M Saskatoon	CAU, CAC	UAU, UAC
Hgb M Milwaukee	GUU, GUC, GUA, GUG	GAA, GAG
Hgb D Punjab	GAA, GAG	CAA, CAG
Hgb Zurich	CAU, CAC	CGU, CGC, CGA, CGG, AGA, AGG
Hgb O Arabia	GAA, GAG	AAA, AAG

Chapter 43:

1. magnesium
2. There would be more O_2 and plant material, and less CO_2 and water on the earth. Less CO_2 would result in less absorption of solar energy, and the temperature would probably decrease. Plants oxidize materials as well as photosynthesize, so a new equilibrium would probably be established.

Chapter 44:

1. Antibiotics and antibodies both serve to protect against infections.
   Antibiotics are produced by simple plants principally by microorganisms and are simpler in structure than antibodies. Apparently antibiotics are produced without the need for prior exposure to the infectious agent.
   Antibodies are complex molecules (proteins with other additions possible) and are produced by animals. Antibodies are produced in response to an antigen.

INDEX

ABS, 456
Acetal: hydrolysis of, 427
  structure of, 427
Acetaldehyde, 418, 420
  in biosynthesis of ethanol, 589
  from oxidation of ethanol, 568
Acetamide, 402
Acetanilide, 403
Acetic acid, 378
  glacial, definition, 379
  origin of name, 369
  structure, 369
  uses, 379
  in vinegar, 369, 379
Acetoacetic acid, 596
Acetone: excretion by diabetics, 596
  structure, 418
  uses, 420
Acetyl chloride, 307, 402
Acetyl CoA, formula, 556
Acetyl salicylic acid, 381
Acetylene (ethyne), 334-335
  combustion, 334
  industrial production, 334-335
  structure, 334
  use in synthesis, 334-335
Acid chlorides, 377, 402-403
cis-Aconitic acid, 592
Acrilan; see Polyacrylonitrile
Acrylonitrile, 335

ACTH (adrenocorticotropic hormone), 683
  structure, 526
Active site of enzymes, 528
Addition reactions of alkenes, 328
Adenine, formula, 546
Adenosine, structure, 443
Adenosine triphosphate, 443
  see also ATP
Adipic acid, 457
ADP (adenosine diphosphate), 502
Adrenalin; see Epinephrine
  and blood glucose, 580
Aerobic glycolysis, energetics, 587
Aerobic metabolism, 582
Aging, 687-688
Alanine, formula, 406, 513
Albinism, 662
Albumin, definition, 511
Albumins: importance, 529
  precipitation of, 529
Alcohol: absolute, 356
  denatured, 356
  structure assignment, 280
  see also Ethanol
Alcoholism and thiamine, 641
Alcohols: boiling points, 357-358
  classifiaction as primary, secondary, tertiary, 355

749

dehydration, 362
dehydrogenation, 360-361
nomenclature, 354-355
reaction with carboxylic acids, 373-374
solubility, 358-359
as solvents, 359
structure, 353
for the synthesis of carbonyl compounds, 417-418
Aldehydes: oxidation of, 421-422
structure, 417
synthesis from alcohols, 360
Tollen's test for, 422
see also Carbonyl compounds
Aldose: definition, 497
examples of, 431
Aldosterone, 540
Aldrin, 710
Alkaloids, 411-412
Alkanes: combustion, 307
cracking of, 324
definition, 300
flammability, 307
names of, 300
nature of reactions, 311
nonpolar structure, 307
reaction with bromine, 310
reaction with chlorine, 308
reactivity, 307
solubility, 304
as solvents, 396
substitution reactions, 309
Alkenes: addition of halogen, 329
addition of hydrogen chloride, 330
addition of sulfuric acid, 331
bond angles in, 323
conversion to polymers, 449, 450
definition, 323
hydrogenation, 328
industrial source, 324
nomenclature, 324-326
oxidation by permanganate, 332
reactivity, 328
Alkyl group, definition, 302
Alkyl halides, 362, 400
Alkyl hydrogen sulfate, 332
Alkylbenzenesulfonates, 394

Alkynes, 333,334
Alpha linkage between monosaccharides, 501
Aluminum, importance of, 705
nutrional significance, 637
Amides: hydroylsis of, 400-401
N-substituted, 403
structure, 402-403
synthesis of, 402-403
Amination, reductive, 520
Amines: alkyl, preparation, 400-401
aryl, preparation, 402
basicity, 399
nomenclature, 398
odor and solubility, 399
reactions with acids, 399
structure, 398
Amino acids, 404-405, 513-522
charged forms, 409, 518
chemical properties, 408-410, 519
essential, 520, 633
examples of natural, 405
excretion of, 611
names, 512
physical properties, 518
sources, 517
summary of metabolism of, 605-611
synthesis from primitive gases, 518
Amino group, definition, 398
Aminoethanol, 535
Amino-peptidases, 604
Aminopterin, 677
Ammonia: conversion to urea, 412
excretion of, 612
reaction with alkyl halides, 400
reaction with HCl, 399
Ammonium cyanate, 276, 277
Ammonium ions, excretion of, 612
Ammonium salts: quaternary, 401, 411
substituted, 398
Amobarbital, 413
Amphetamine (Benzedrine), formula, 682
Amphetamines, 681-682
in amino acids, 409
Amyl acetate; see Isopentyl acetate
Amylase, action action on starch (diagram), 578

Amylase, salivary, 560
Amylopectin, 507
Amylose starch, 507
Anabolism (anabolic reactions), 490, 562
Anaerobic glycolysis, energetics, 587
Anemia, iron deficiency, 637
Anesthetics, 365
Aniline: manufacture of, 413
   origin of name, 413
   reaction with alkyl halide, 401
   relation to dyes, 413
   structure, 398
Animals, nutritional requirements, 631
Antibiotics, 677–678
   formulas, 678
   synthesis, 676
Antibodies, 675
   in mother's milk, 603
Antidepressants, 680–681
Antimetabolite, 676
Arabinose, formulas, 426, 497, 498
Arginine: formula, 517
   hydrolysis to give urea, 521
   precursor of urea, 613
Aromatic acids, preparation, 346, 347
Aromatic compounds: definition, 338
   fused ring, 341
   isolation from coal tar, 348, 350
   nomenclature, 340–342
   origin of name, 338
   synthesis from petroleum, 351
Aromatic hydrocarbons, in coal tar, 350
Arthritis, 674, 686–687
Aryl group, definition, 342
Asbestos, as air pollutant, 700
Ascorbic acid (vitamin C), 644
   formula, 644
Asparagine, formula, 516
Aspartic acid, formula, 516
Asphalt, 296
Aspirin, 381
Asymmetric carbon, 475
Atherosclerosis, 626
Atmosphere, composition of, 692
   pollution of, 692
ATP (adenosine triphosphate), 502, 565
   energy of hydrolysis of, 565
   formula, 554
   functions, 555
   synthesis of, 567
   synthesis from primitive gases and phosphate, 518
ATP storage (illustration), 564
ATP synthesis in fatty acid oxidation, 623
Autolysis, 493

B

Bacteria: in gastrointestinal tract and vitamin synthesis, 640
   in intestine, 561
Bacteriophages, 549, 555
Bakelite, 461
Barbital, 413
Barbiturate drugs, 412–413
Bases, amines as, 399
   ammonia as, 399
Beeswax, 539
Benedict test for sugar, 422, 500
Benzaldehyde in food, 400
Benzamide, 403
Benzedrine; see Amphetamines
Benzene: in coal tar, 348–349
   formulas for, 340
   industrial synthesis, 351
   origin of name, 339
   reaction: with halogen, 343–344
      with nitric acid, 345–346
      with sulfuric acid, 344–345
   reactivity, 339, 343
   structure, 338–340
Benzenehexachloride, 314
Benzenesulfonic acid: chemical properties, 345
   formula, 344
   preparation from benzene, 344
Benzoic acid: in food, 408
   industrial synthesis, 378
   origin of name, 369
   preparation from toluene, 346
   structure, 369
   uses, 379
Benzoyl chloride, 403
Beriberi, 640, 679

Berzelius, J.J., 275, 511
Beta linkage between mono-
    saccharides, 501
BHC, 314
BHT (butylated hydroxytoluene), 364
Bile, 561
    function in lipid digestion, 620
Bile acids, 541
Biochemistry, definition, 487, 493
Biological half-life: lipids, 625
    proteins, 605
Biotin, 643
    formula, 643
Biuret test, 528
Blood clotting and vitamin K, 647
Bond angles of carbon, 282–284
Bond strain in ring compounds, 292
Boron, and plant growth, 637
Branched chain, 289
Bromine: biochemical importance, 637
    test for unsaturation, 329
Bromobenzene, 340
Bromotoluene, 341
Buchner, Eduard and Hans, 484, 487
Buffers: citric acid-citrate, 381
1,3-Butadiene, 452, 455
Butane, cracking of, 312
Butanone: structure, 418
    uses, 420
Butter: composition, 386
    processing of, 388
Butylbenzene, 340
Butyric acid, 369

C

Cadaverine, 399, 410
Caffeine, 411
Calcite crystals, 467
Calcium: in nutrition, 635
Calcium carbide, 334
Calcium carbonate, crystalline, 467
Calorie, requirement to maintain life, 631–632
Calories: in nutrition, 631
    relation of activity to, 631–632
Calvin, Dr. Melvin, 669
Camphor, 541

Cancer, 685–686
    from aromatic hydrocarbons, 342
    from cigarette tar, 342
Cancer and viruses, 555
Carbamide, 412
Carbamyl phosphate, 613
Carbohydrate metabolism: diseases:
    related to, 598
    in plants, 577
    summary, 594
Carbohydrates: caloric content, 632
    caloric value of, 627
    common names, 498
    definition, 432, 496
    digestion, summary of, 577–579
    in foods (table), 489
    nutritional importance, 633
Carbolic acid, 363, 364
Carbon compounds: chains in, 279, 289
    rings in, 291
Carbon cycle, 695
Carbon dioxide, in the atmosphere, 693
    conversion to urea, 412
    in the oceans, 693
Carbon monoxide, in air, 697, 698
    toxicity of, 698
Carbon tetrachloride, 309
Carbonium ion, 331
Carbonyl compounds: hydrogenation
    of, 420–421
    in nature, 419–420
    nomenclature, 417–418
    reductive amination of, 401
    synthesis, 417
Carbonyl group, structure, 417
Carboxyhemoglobin, 698
    and cigarettes, 698
Carboxyl group, structure, 368
Carboxylate salts: names, 371
    reaction with mineral acids, 372
    as soaps, 389, 391
    solubility, 372, 391
    structure, 371
Carboxylic acids: acidity, 370
    conversion to acid chlorides, 377
    formulas, 368
    ionization, 370
    names, 369

in nature, 368-369
odors, 374
preparation by oxidation, 346, 378
preparation from cyanides, 378
reaction with alcohols, 373-374
reaction with carbonates, 371
relation to carbonyl compounds, 417
salt formation, 369-371
solubility, 372
structure, 368-369
Carboxy-peptidases, 604
Carcinogen, 685
Carcinoma, 685
Cardiovascular disease: see Disease, cardiovascular
β-Carotene, formula, 536
Carotenes in photosynthesis, 542
Cartilage, 448
  mucopolysaccharides in, 509
Catabolism (catabolic reactions), 490, 562
Catalytic reforming, 350
Cell: diagram, 492
  photomicrograph, 491
Cellobiose, formula, 501
Cellulose: caloric content, 632
  formula, 441, 506
  indigestibility, 578
  a polymer, 448
Centrifugation: of cell fragments, 490, 492
  in purification of proteins, 429
Cephalins, 538
Chemotherapeutic agents, 676-677
Children, susceptibility to pollutants, 698, 700, 703, 704
Chloramphenicol, formula, 678
Chlorine: in nutrition, 636
Chlorophyll, 666-667
  structure, 667
Chlorpromazine, 680, 682
  formula, 681
Chlorpromide (diabinese), 597
Cholesterol, 539
  absorption of, 621
  formula, 536
  possible stereoisomers of, 476
  synthesis of, 626

Cholesterol and cardiovascular disease, 626
Cholesterol esterase, 561, 621
Cholic acid, 541
Choline, formula, 535
Chlorobutanes, 310
Chloroethane, 314
Chloroform, 309
  as an anesthetic, 365
Chloromethane, 309
Chloroprene, 335, 454
Chromatography, 485, 486
Chromosomes, 654, 655
  human, 654
Chymotrypsin, 561
Cinnamaldehyde, 419
Cinnamon, flavor, 419
Cis-trans isomers of alkenes, 326-327
Citral, 419
Citric acid, 381
  occurrence, 381
  in preservation of blood, 381
  structure, 381
  uses, 381
Citric acid cycle, 582, 591-594
  (diagram), 591
  energetics of, 593, 594
Citronellol, 328
Citrulline, 613
Coal, conservation of, 462
Coal tar: composition of, 348-349
  drugs from, 381
  manufacture (photograph), 348
  origin, 348
Cobalamin (vitamin B_{12}), 644
  structure, 645
Cobalt: biochemical significance, 637
Cocaine, 375, 411
Coconut fat, composition, 386
Codeine, 412
Coenzyme, function of, illustration, 569
Coenzyme A (CoA), 643
  in fatty acid oxidation, 622
  formula, 556
Coenzymes, 568, 569
Coke, production of (photograph), 348
  source, 348
  in the synthesis of acetylene, 334

Collagen, relationship to aging, 687
Colloidal solutions, soap, 392
Combustion, of alkanes, 307-308
  of alkenes, 333
Composition of biochemical substances (table), 489
Condensation reaction, 457
Condensed structural formula, 288
Configuration, definition, 465
Configurations, nonsymmetric, 469-470
Coniine, 411
Conservation, 462
Contraceptives, oral, 540, 679
Copolymer, 455-456
Copper, importance of, 705, 706
  in biochemistry, 637
Copper ion in Fehling's and Benedict's reagents, 422
Cortisol, 540
Cortisone, 676
Cotton, 448
Cottonseed oil: composition, 386
  in manufacture of oleomargarine, 388
Covalent bonds, basis of organic structure, 277
  in organic compounds, 448
  tetrahedral, 283
Cracking reaction, 324
Creosote, 364
Cronar, 459
Cross linkage of proteins, 523
Crystallization of sugars, 500
Cyanide, organic, see Nitrile
Cyanohydrin, 422
  conversion to hydroxy acid, 424
  conversion to hydroxy aldehyde, 424
  structure of, 423
Cycles, 695
  carbon, 695
  oxygen, 695
Cycloalkanes, 302
Cyclohexane in the synthesis of benzene, 351
Cyclohexanone, 420
Cyclopropane as an anesthetic, 365
Cysteine: formula, 515
  oxidation, 521
Cystinuria, 611

Cytochrome c, structure, 566
Cytochromes in electron transport, 565
Cytosine, formula, 546

D

D-isomers of sugars, 497, 498
Dacron, 459
Dark reaction (photosynthesis), 667, 668-670
DDT, 708, 709
Deamination: of amino acids, 610
  oxidative, 520
Decarboxylation and thiamine, 641
Decarboxylation of amino acids, 520, 611
Deficiency disease, 638
Degree of polymerization, 449
Dehydrogenation of alcohols, 360, 418
Denaturation: of DNA, 547
  of proteins, 528
Deoxyribonuclease (DNA-ase), 553
Deoxyribonucleic acid; see DNA
Deoxyribose, formula, 503
De-repressors, 656
Detergents: biodegradable, 394, 707, 706, 708
  definition, 394
  in hard water, 395
  non-ionic, 395
  required structure, 392, 393
  synthetic, 393-395
Dexedrine, 681: see also Amphetamines
Dextrorotatory sugars, 497-498
Dextrose; see Glucose
Diabetes mellitus, 500, 595-597, 683
1,6-diaminohexane, 457
Diastereomers, definition, 476
Dibromobenzene, 341
Dichloricide, 314
p-Dichlorobenzene, 314
Dichloroethane, 287, 314
Dicholoromethane, 282-286
Dicoumarol: anticoagulant action, 647
  formula, 647
Dieldrin, 710
Diene, 452
Diesel fuel, 308
Diethylamine, 400

Digestion, 558-562
  of carbohydrates, 577-579
  summary, 579
Digestive system, 559, 560
  human (diagram), 559
Dihydroxyacetone, formula, 597
Dihydroxyacetone phosphate, 584
2,4-Dinitrophenylhydrazine, 430
Dipeptidases, 604
Dipeptides, 519
1,3-Diphosphoglyceric acid, 585
Dipolar ions, 409, 518
Disaccharidases, 561, 577
Disaccharides, 432, 438, 503
  chemical properties, 504
  sources, 503, 504
Diseases: cardiovascular, 626-627, 639
  endocrine malfunction, 674, 683-685
  general, 674
  infectious, 674, 675
  mental, 674
  metabolic or degenerative, 674
  noninfectious, 674
  nutritional, 674
Disulfide bonds in proteins, 523
DNA (deoxyribonucleic acid), 443, 544
  deoxyribose in, 502
  function, 443-444, 544
  as hereditary substance, 444, 656
  melting or denaturation, 547
  in protein synthesis, 443, 526, 605-606
  relation to aging, 687
  relationship in cancer, 686
  in sperm cells, 656
  structure, 545, 548, 549
  synthesis (replication), 444, 552-553
Drying oils for paints, 396
Dyes, synthetic, 414
Dynamic equilibrium: in biology, 625
  illustration of, 574
Dynel, 456

E

Egg cell: drawing of, 550
  human, 657
Elastomers, 448, 455
Electron transport, 565-568
  summary of (diagram), 567
Electrophoresis, 485
  illustration of, 518
  proteins, 527
  in purification of proteins, 529
Electrophoretic diagrams of serum, 527
Electrovalent compounds, structure in, 277-278
Elements: found in human body, 488
  toxicity, 702, 705
Enantiomers: behavior of, 473, 474
  definition, 470
  formulas for, 471-472
Endocrine glands, 572, 683-685
Endopeptidases, 604
Energetics, of metabolism, 562-565
Energy, balance in nature, 691
  from fission, 707
  from foods, 632
  from fusion, 707
  in fossil fuels, 693
  solar, 707
Energy of activation, 562
  illustrations of, 563, 570
Environment, global, 692
Enzymes, 568-572
  classification, 570, 571
  definition, 490, 568
  discovery, 487
  effect on optical activity, 474-479
  mode of action, 569
    illustration, 569, 571
  naming of, 560, 570
  proteolytic, classification of, 603
Epinephrine (adrenaline): biosynthesis, 660
  synthesis, 526
EPR, 456
Equanil, 646; see also Meprobamate
Ergot, 411
Erlich, Paul, 676
Erythrose, formula, 425
Esophagus, 560
Essential amino acids, 633
Essential fatty acids, 625, 634
Essential foods, 632
Esterification, 373-374
Esters: formation from phenols, 378

hydrolysis, 376
in nature, 374-375
nomenclature, 374
odors, 374
preparation, 373, 377-378
saponification, 376-377
structure, 373
use in perfumes and flavors, 374-375
Estradiol, 540
Estrone, 419
Ethanol: from fermentation, 353, 356
oxidation of, 568
production, 356, 357
properties, 356
reaction with HBr, 280
synthesis by yeast, 590
uses, 357
Ethene: from alkane cracking, 312
as an anesthetic, 356
use in synthesis of ethanol, 357
Ether (ethyl ether): as an anesthetic, 365
flammability, 360
petroleum, 306
physical properties, 360
as a solvent, 360
Ethers: chemical properties, 365
preparation from alcohols, 362
structure, 353
Ethyl acetate, 373, 374
Ethyl alcohol; see Ethanol
Ethyl chloride, 314
Ethyl ether; see Ether
Ethylamine, 400
Ethylbenzene, 340, 350
Ethylene glycol: structure, 355
in synthetic fibers, 459
systematic name, 355
uses, 357
Ethyne; see Acetylene
Eutrophication, 701, 707
Excretion of nitrogenous waste products, 611-616
Exopeptidases, 604

F

FAD: reduction and oxidation of, 565–566
structure of oxidized and reduced forms, 566
Fats, 537
composition, 386
definition, 384, 387
hydrolysis, 384
rancidity, 387
relation to body temperature, 387
solubility in ether, 360
structure, 384
Fatty acids: biochemical synthesis, 624–626 (diagram), 624
essential, 625, 634
names, formulas, and sources, 535
natural, 386
oxidation, 622-624 (diagram), 623
Feathers, 448
Feedback: in ecological processes, 694, 695
in hormone action, 573
Fehling's test, 500
for sugar, 422
Fermentation, 484, 487
alcoholic, 590
Fibrinogen, 530
Fingernails, 448
Fischer, Emil, 499
Fixation: of CO_2, 668
of nitrogen, 671-672
Fleming, Dr. Alexander, 677
Fluorides, toxicity, 705
biochemical significance, 637
Fluothane, 314
Fog, 186
Folic acid, 644
formula, 644
and sulfonamides, 676, 677
Food chain, DDT in, 709
Foods, essential, 632
Formaldehyde: in polymers, 461
structure, 418
uses, 420, 422
Formica, 461

Formula: condensed structural, 288
  molecular, 278
  structural, 278
  valence bond, 288
Free energy, 562
Freon, 313
Fructose: formula, 431, 498, 499
  metabolism of, 586
  ring structure, 436
  sources, 500
Fructose-1,6-diphosphate formula, 502
Fructose-6-phosphate, 584
Fuels, fossil, 693, 707
Fumaric acid, 593
  in urea cycle, 614
Functional groups(s); characteristic behavior of, 328
  definition, 315

G

Galactose; formula, 431, 498, 499
  metabolism of, 586
Galactose 1-phosphate, conversion to glucose 1-phosphate, 586
Galactosemia, 597
Gallstones, 539
Gamete, 652
Gamma globulin, 675
  importance, 530
Gammexane, 314
Gas, liquefied petroleum (LPG), 306
  natural, 296
  for heat, 308
  occurrence, 308
  in petroleum, 306
Gasoline, 306, 324
Gastrointestinal tract, 560
Gene, 651
Genetic code (table), 609
Geneva rules, 299
Globulin(s): definition, 511
  importance, 530
  precipitation of, 528
Glucaric (saccharic) acid, 501
Glucogenic substances, 629
Gluconic acid, formula, 501
Glucosamine, 503
  in chitin, 508
Glucose: in blood control of, 580
  blood levels and effects (diagram), 580
  concentration in blood, 579
  energy: from biochemical oxidation, 593–594
  released in oxidation of, 582
  formula, 419, 431, 497, 498, 499
  isomers, 435, 497, 499
  optical isomers, 497
  oxidation, 562
    energetics of, 564
  in plating of mirrors, 422
  ring structure of, 433–434
  sources, 500
  summary of metabolism, 582
  in Tollen's, Fehling's and Benedict's tests, 422
  trapping of energy from, 587
Glucose 1-phosphate, 581
  formula, 502
Glucose 6-phosphate, 584
  formation of, 502
  formula, 419, 502
Glucose 6-phosphate dehydrogenase, disorders involving, 598
Glucose tolerance curves, 595, 596
Glucoside, 438, 442
Glucuronic acid, formula, 501
Glutamic acid: formula, 406, 516
  from reductive amination, 429
Glutamine: in ammonia excretion, 612
  formula, 516
Glyceraldehyde: formula, 425, 497
  optical isomers, 426, 497
Glyceraldehyde 3-phosphate, 584
  in carbohydrate metabolism, 425
  in photosynthesis, 425
Glyceride, mixed, 384
  structure, 384
Glycerin; *see* Glycerol
Glycerol: from fats, 384
  in polymers, 459–461
  production, 357
  properties, 357
  solubility in ether, 360

structure, 355
systematic name, 355
uses, 357
Glyceryl trinitrate (nitroglycerin),
    production and use, 363
Glycine, formula, 513
Glycocholic acid, 541
Glycogen, 508, 577
    synthesis and breakdown, 581-582
Glycogen storage diseases, 598
Glycol, definition, 355
Glycolysis, 582, 583-587
    (diagram), 583
    energetics, 587
    significance of, 587
Glycoproteins, 530
Glycoses: conversion to glycosides, 437
    definition of, 431
    ring structure of, 432
Glycosides: definition, 437-438
    hydrolysis of, 444
    natural, 442
    structure of, 437
Glyptal, 461
Goldberger, Dr. Joseph, 612
Group, alkyl, 300
Guanine, formula, 546
Gypsum, 127, 287

H

Hair, 448
Half-life: metabolic, 574
    of radioactive materials, 257
Halides: alkyl, 312
    organic: biological effects, 315
        examples of, 313
        as solvents, 315
Hallucinogenic drugs, formulas, 682
Hallucinogenic substances, 682
Halothane, 314
Hard water, effect on soap and detergents, 395
Haworth structures, 499
Heat attacks and transaminases, 610
Helix in protein structure, 524
Hemiacetal, structure of, 426

Hemoglobin, 523
    structure, 526
    model of, 525
Hemoglobins, abnormal, 664
Hemp, 448
Heparin, 509
Heptane, 308, 351
Herbicides, 708, 711
Heredity, biochemical control of, 664
Heroin, 412
Hetero atoms, 347
Heterocycles in alkaloids, 411
Heterocyclic compounds, definition, 347
Heteropolysaccharides, 509
Heterozygote (heterozygous), definition, 651
Heterozygotes, detection of, 661
Hexachlorocyclohexane, 314
Hexane, for the synthesis of benzene, 351
1,6-hexanedioic acid, 457
Hexosans, 506
Hexoses, 498
Hexylresorcinol, 364
Hippuric acid, 408
Histamine, 410, 520
Histidine: formula, 406, 516
    decarboxylation of, 410
Homogenate, cellular, 490
Homologous series, 298
Homozygote (homozygous), definition, 651
Hormones, 572-573, 683
    protein, 573
    relation of RNA to action of, 573
    sex, 419
    steroid, 573
Horns, 448
Hyaluronic acid, 509
Hydrocarbons: in air pollution, 697
    definition, 298
    saturated: definition, 298
        in petroleum, 298
        rings in, 291-293, 300
    unsaturated, 323
        bromine test for, 329
Hydrochloric acid, in stomach, 560

Hydrogen bonds: effect on properties
of alcohols, 357, 358
between nucleic acids, 547
in proteins, 523, 526
Hydrogen chloride, addition to an
alkene, 330
Hydrogenation: of alkenes, 328
of carbonyl compounds, 420
of oils, 388
Hydrogen cyanide, addition to carbonyls, 422
Hydrolases (table), 571
Hydrolysis: of a fat, 384
β-Hydroxybutyric acid, 596
Hydroxylamine, 430
Hydroxyl group, definition, 360
Hydroxyproline, formula, 515
Hyperglycemia, 500, 579-580
Hypervitaminosis, 638
Hypoglycemia, 500, 579

I

in vitro, definition, 484
in vivo, definition, 484
Inborn errors of metabolism, 657-663
Indican, 442
Indigo, 413, 414, 442
Infection, 675
Inflammation, 676
Inflammation-promoting hormones, 683
Inflammation-suppressing hormones, 683
Inhibitors of enzymes, 572
Inorganic compounds, origin of name, 275
Insect repellent, 416
Insecticides, 708
Insulin, 683
and blood glucose, 580
importance, 530
mode of action, 596
structure, 526, 530
International Union of Chemistry (ICU), 299
International Union of Pure and Applied Chemistry (IUPAC), 299

Inversion layers, 696, 697
Inversion of sucrose; *see*
Sucrose, hydrolysis
Invert sugar, 504
Invertase; *see* Sucrase
Iodine: biochemical significance, 637
Iodoform, 313
Iproniazid, 680, 681
formula, 681
Iron, importance, 705, 706
in nutrition, 636
for reduction of nitro compounds, 402
Isobutyl alcohol, 354
Isocarboxazid, 680
formula, 681
Isocitric acid, 592
Isoelectric point, 518
of proteins, 527
Isolation artifacts, 484
Isoleucine, formula, 513
Isomer(s): *cis-trans*, of ring compounds, 464
definition, 277
relation to rotational forms, 290
types of, 477
Isomerases (table), 571
Isopentyl acetate, 374
Isopentyl butanoate, 374
Isoprene: in copolymers, 456
formula, 536
manufacture of, 455
poymerization of, 435
relation to rubber, 459
structure, 459
Isopropyl alcohol; *see* 2-Propanol

J

Jelly, petroleum, 306

K

Kerosene, 306
for heat, 308
Ketal, 428
Ketogenic substance, 629
α-Ketoglutaric acid, 429, 593
Ketones: oxidation of, 421

from oxidation of an alkene, 332
structure, 417
synthesis from alcohols, 360-361
see also Carbonyl compounds
Ketose, definition, 431, 497
Kornberg, Arthur, 553
Krebs cycle, 582; see also Citric acid
Kwashiorkor, 679

L

L-isomers of sugars, 497
Lactic acid: formulas, 471
in muscles, 587
occurrence, 380
optical isomers, 465-466, 471, 474
origin of name, 369
racemic, 472
structure, 369, 465
synthetic, 472
uses, 380
Lactose: formula, 439, 503
sources, 503
Lard, composition, 386
Lauric acid, 385
Lead, in pottery, 703
toxicity of, 702, 703
le Bel, Joseph, 469
Lecithin, 538
formula, 538
Lecithinases, 621
Legumes, in nitrogen fixation, 671
Leucine, formula, 405, 513
Leukemia, 677, 685
Levorotatory sugars, 497
Levulose; see Fructose
Ligases, 571
Light reaction (photosynthesis), 667-668
Lignin, 448
Lime, in the synthesis of acetylene, 334
Limonene, 541
Lindane, 314
Linolenic acid, formula, 385
Linseed oil, composition, 386
Lipases, 537, 620
Lipid, 534
caloric values of, 627

Lipid metabolism, summary, 627-628
Lipids: absorption, 621
classification, 534
depot (storage), 625
digestion, 620-621
in foods, 489
nutritional importance, 633-634
plants, 666
tissue, 625
Lipofucsin, 687
London pollution disaster, 696
Los Angeles air pollution, 700
LSD (lysergic acid diethylamide), 412, 682
Lucite, 420, 453
Lyases, 571
Lymph, absorption of lipids by, 621
Lymphatic system (diagram), 622
Lysergic acid, 411
Lysine: decarboxylation of, 410
formula, 517
Lysosomes, 492, 493

M

Magnesium, in nutrition, 636
Malathion, 710
Maleic acid, hydrogenation, 329
Malic acid, 593
Malnutrition, 679
Maltose: formula, 501, 503
sources, 504
Manganese, in nutrition, 637
Mannose, formula, 498
Marijuana, 682
Markovnikov rule, 330, 331, 333
Mauve, 414
Meat tenderizers, 605
Meiosis, 654
(diagram), 655
Melanin, 660
Melmac, 461
Mendel, Gregor, 651
experiments of (diagram), 652
Menthol, 421, 541
Menthone, 421
Meprobamate, 680
formula, 681

Mercury, in fish, 705
　methyl compounds, 704
　toxicity, 702, 704
Mescaline, 682
　formula, 682
Messenger RNA, 549
　in protein synthesis, 606
　stability, 609
　synthesis (diagram), 607
Mestranol, 540
Metabolic half-life, 574
　of lipids, 625
Metabolism: definition, 490
　energetics, 564
　general summary, 573-574
　interrelationships of carbohydrates, proteins, and lipids in, 629
　in plants, 666
　summary of, 629
　　(diagram), 628
Metabolite, definition of, 562
Methane: combustion, 307
　reaction with chlorine, 308
　in the synthesis of acetylene, 334
Methanol, 356
Methionine, formula, 515
Methoxychlor, 710
Methyl alcohol, 356
Methyl benzoate, 374
Methyl chloride, 309
Methyl ether, structure assignment, 284
Methyl ethyl ketone, *see* Butanone
Methyl salicylate, 381
Methylaniline, 398
Microsomes, 492, 493
Milk, lactose content, 503
Miller, Stanley, 517
Miltown, 680; *see also* Meprobamate
Mineral elements, 488
　in enzyme action, 635
　in nutrition, 634-635
　and osomotic pressure, 635
　requirements and functions, 635
　see also individual elements
　toxicity, 635

Mineral spirits, 396
Mint, compounds in, 421
Mirror images, 469, 470
Mirrors, plating of, 422
Mitochondria, 490, 491, 492, 493
　electron transport in, 565
Mitosis, 653
Models: ball-and-peg, 285-287
　relation to formulas, 283, 290
　relation to structures, 287
　space-filling, 285-286
Molecular formula, definition 280
Molybdenum: biochemical importance, 637
Monoamine oxidase (MAO), 681
Monomer, definition, 449
Monosaccharides, 431, 497-502
　chemical properties, 436, 500
　oxidation, 500
　physical properties, 500
　sources, 500
　see also Glycoses
Morphine, 412
Mucopolysaccharides, relationship to aging, 687
Mulder, G.M., 511
Muscular dystrophy and vitamin E, 646
Mutations, 656
Mylar, 459
Myoglobin: structure, 526
　model of, 525
Myristic acid, formula, 385

N

NAD (nicotinamide adenine dinucleotide): formula, 555
　in glycolysis, 585, 586
　reduction and oxidation of, 565-567
　structure of oxidized and reduced forms, 566
NADP (nicotinamide adenine dinucleotide phosphate), 565
　in fatty acid synthesis, 624
　in lipid synthesis, 588

in phosphogluconic acid shunt, 588
in photosynthesis, 667
Naphtha, 306
Naphthalene, 341
2-Naphthylamine, 402
Natta, Giulio, 459
Neoprene, 458–459, 462
Neurosis (neurotic), 680
Niacin, structure, 411
Nickel, catalyst for hydrogenation, 328
Nicotinamide (niacin), 642
  formula, 642
Nicotine, 411, 709
Night blindness, 639
Nitration of benzene, 345
Nitrile (organic cyanide), conversion to an acid, 378
  hydrolysis of, 378
  synthesis of, 379
Nitro group, formula, 345
Nitrobenzene, 345, 402
Nitrogen heterocycles: in coal tar, 349
  oxides and ozone, 699
  oxides in air, 697, 715
  oxides, toxicity, 699
  in soils, 707
Nitrogenous waste products, 611–616
  in chicken embryo, 614
Nitroglycerin; see Glyceryl trinitrate
2-Nitronaphthalene, 402
Nitrous oxide as an anesthetic, 365
Nobel, Alfred, 363
Noradrenalin; see Norepinephrine
Norepinephrine, 682
  formula, 682
  synthesis, 521
Norethindrone, 540
Nucleic acids, 489
  chemical properties, 553
  general, 544, 549, 551
  physical properties, 553
  as polymers, 448
  synthesis, 551–553

Nucleoproteins, 551
Nucleosidases, 561
Nucleoside, definition, 545
Nucleotidases, 561
Nucleotide(s): definition, 545
  names, 548
  nonpolymerized, 555
Nucleus of cell, 492, 493
Nylon, 457–458, 462

O

Oil: crude (petroleum), 296, 304
  fuel, 306
  lubricating, 306
  mineral, 306
Oil of wintergreen, 381
Oils (glyceride), 537
  composition, 386
  definition, 387
  drying, for paints, 396
  "hardening," 388
  rancidity, 387
Oleic acid, formula, 385
Oleomargarine, manufacture from plant oils, 388
Oligosaccharides, 503–505
Olive oil, composition, 386
One-gene–one-enzyme theory, 663
Optical activity: cause of, 469
  definition, 468
  examples of, 469
Optical isomers: number possible, 476–477
  origin of term, 466
Oral contraceptives, 540
Organelles: definition, 490
  subcellular, 490, 492
Organic acids; see Carboxylic acids
Organic compounds: chains in, 279, 289
  combustibility, 308
  definition, 277
  elements in, 279
  models, 283–284, 286, 288, 291, 292, 293
  origin of name, 275
  rings in, 291
  structural formulas for, 284–

285, 288, 289-292
  structure in, 278
  systematic names, 299-301
Orlon; see Polyacrylonitrile
Ornithine in urea cycle, 613
Oxalic acid, 380
Oxaloacetic acid, 593
Oxalosuccinic acid, 592
Oxidoreductases, 571
Oximes, 430
Oxyacetylene torch, 334
Oxygen in the atmosphere, 692, 693
  cycle, 695
Oxygen debt, 594
Oxytocin, 530
Ova (egg cells), biochemical composition, 656
Ozone: in air, 696, 697, 700
  oxides of nitrogen, 699
  toxicity of, 696, 700

P

PABA (para-aminobenzoic acid), 676
Paint, composition and drying of, 396
Palm oil, composition, 386
Palmitic acid formula, 385
PAN (peroxyacetyl nitrate) in air, 700
Pantothenic acid, 643
Papain, 605
Paracelsus, 483
Parasitism, 675
Parathion, 710
Paris green, 379
Particulates in air, 700
Pasteur, Louis, 469, 483
PCB (polychlorinated biphenyls), 711
Peanut oil, composition, 386, 679
Pellagra, 642
Penicillin, 677
  formula, 678
Penicillin sensitivity, 678
n-Pentane, 289
Pentosans, 506
Pentoses, 490
Pepsin, 560
  action of, 604
Peptidases, 561, 603

Peptide bonds, 519
Peptides, 519
Peroxides, catalysts for polymerization, 449-450
Pesticides, 708
Petroleum: common products from, 306
  conservation of, 462
  distillation of, 304
  occurrence, 296
  origin of name, 296
  refinery, 305
  use in ancient cultures, 296
Phagocytes, 675
Phenobarbital, 413
Phenol(s): acidic properties, 364
  in coal tar, 349
  as a disinfectant, 364
  in polymers, 461
  properties and uses, 364-365
  solubility, 359
  structure, 353, 363
Phenyl acetate, structure, 375
Phenyl group, 342
Phenylalanine: formula, 406, 514
  metabolism, 661
  (diagram), 661
Phenylhydrazine, 430
Phenylketonuria, 526, 662
  treatment, 662
Phenylpyruvic acid, 660
Phosphate esters of sugars, 502
Phosphates in detergents, 394
Phosphatidic acid, 538
Phosphoenolpyruvic acid, 585
Phosphogluconate shunt (diagram) 588
6-Phosphogluconic acid, 588
Phosphogluconic acid shunt, 588
2-Phosphoglyceric acid, 585
3-Phosphoglyceric acid, 585
Phospholipids, 538-539
Phosphorus, human requirements, 634, 635
Photolysis of water, 668
Photosynthesis, 588, 667-671
  dark reaction (diagram), 669
  formation of pyruvic oxaloacetic

malic, and aspartic acids, (diagram), 670
  relation to glycolysis, 671
  relation to respiration, 671
  as a source of oxygen, 693
Phthalic acid: in polymers, 459–461
  preparation by oxidation, 346
Picric acid, 364
Pinene, 541
Plants, nutrional requirements, 631
Plasmalogens, 538
Plastics, 448
Platinum as a catalyst, 351
Pleated sheet, protein structure, 524
Plexiglas, 420, 453, 462
Poison ivy, 364
Poison oak, 364
Poisons, cumulative, 702
Polarimeter, 467
Polarized light: nature of, 467
  rotation of, 465, 468
Pollution: air, 696
  as cause of death, 696
  effect on plants, 701
  thermal, 702
Polyacrylonitrile, use of, 453
Polyamide, definition, 457
Polybutadiene, 455
Polyethylene, a thermoplastic, 459
  use of, 451, 453
Poly(ethylene terephthalate), 439
Polymers: addition, defined, 449
  condensation, 457–459
  cross-linked, 459–461
  definition, 448
  formation of, 449–450
  natural, 448, 456
  size of, 448, 449
  synthetic, 448
  thermoplastic, 459
  thermoset, 459–461
Poly(methyl methacrylate), 453
Polypropylene, 451, 452, 453
Polysaccharides, 432, 440, 505–508
  digestion of, 560
  molecular weight, 505

Polystyrene, 452, 453
Poly(vinyl chloride), 449, 453
Population control, 715–716
Population growth, 715–716
Potassium: in nutrition, 636
  in soils, 707
Potassium permanganate, test for unsaturation, 332
Precipitators, electrostatic, 714
Pre-pepsin, 604
  activation of, 604
Primitive gases, 517
Prochlorperazine, formula, 681
Progesterone, 540
Proline, formula, 515
2-Propanol (isopropyl alcohol):
  production, 357
  properties, 357
  structure, 354
  synthesis from propene, 332, 357
  uses, 357
Propene: reaction with hydrogen chloride, 330
  reaction with sulfuric acid, 332
  structure, 323
  for synthesis of isopropyl alcohol, 357
Propionaldehyde, 418
Propionic acid, 369
Protein metabolism, summary, 617
Protein synthesis, 606–609
Protein(s): biological half-life, 605
  caloric content, 632
  caloric value, 627
  chemical properties, 528
  classification, 511
  cross linkages, 523, 526
  definition, 511
  denaturation, 528
  digestion of, 407, 560, 603–605
  in foods, 489
  hydrolysis of, 407
  nutritional importance, 633
  peptide (amide) bonds in, 407
  physical properties, 527
  plant, nutritional value, 633

as polymers, 448
purification of, 529
requirements for, 633
sources, 522
specific dynamic action, 632
structure, 404–408
synthesis of, 526, 605–609
Proteolytic enzymes, 603
in medical practice, 605
Prothrombin, 530
Psilocybin, formula, 682
Psychedelic drugs, 682
Psychic energizers, 680, 681
formulas, 681
Psychosis (psychotic), 680
Psycotogenic substances; see Hallucinogenic substances
Purine, 545
metabolism (diagram), 615
structure, 347
Putrescine, 399
PVC, 453
Pyrethrins, 709
Pyridine: as a base, 400
formulas, 347
Pyridinium chloride, 400
Pyridoxal (vitamin B_6), 642–643
formula, 642
Pyrimidine, 545
structure, 347
Pyruvic acid: conversion to lactic acid, 472, 474
metabolism of, 589–591

Q

Quinine, an alkaloid, 412
synthesis of, 413
Quinoline, formula, 347

R

Racemate: definition, 472
examples of, 473, 476
separation of, 473
Radiation of sun, trapping of, 631
Radicals: in alkane reactions, 311
in combustion, 311
in cracking reactions, 312
methyl, structure, 312

in polymerization, 449
propyl, 312
Radioactive isotopes, 486
in biochemistry, 486
Radioisotopes, biochemical use of, 600
Rancidity, 537
in fats and oils, 387
Rayon, 457
Reactions, biochemical, 490
Recycling, 706, 713
Red blood cells, normal, 658
Reducing sugars, 500
Reductive amination, 428
biochemical, 429
of carbonyls, 401, 428
Renal threshold, 580
Repressors, 656
Reserpine, 680, 682
formula, 681
Respiration, relation to photosynthesis, 671
Rhodopsin, 639
Rhubarb, oxalic acid in, 380
Riboflavin (vitamin B_2), 641–642
formula, 641
Riboflavin and FAD, 641
Ribonuclease (RNA-ase), 553
Ribose, formula, 426, 498, 499
Ribosomal RNA, 549
Ribosomes, 492, 493
Ribulose diphosphate, 668, 669
Rickets, 645
Ring compounds: bond angles in, 291–292
formulas of, 291–292
models of, 291
names, 300
puckered, 292
RNA (ribonucleic acid), 544
functions, 544
and heredity, 656
in protein synthesis, 526, 605
relation to aging, 687
Rotation: on carbon bonds, 287–288
at a double bond, 326, 327
Rotational forms: of dichloroethane, 288

of pentane, 290
Rubber, 536
　butyl, 456
　natural, structure, 459
　　use, 448
　synthetic, 448, 452, 459–460
Ruminants, 579

S

Safflower oil, composition, 386
Salicylic acid: origin of name, 369
　structure, 369, 381
　uses, 381
Saliva, 560
　function, 560, 561
Salivary amylase, 560, 577
Salting out of proteins, 528
Salvarsan (Dr. Erlich's compound 606), 676
Saponification: definition, 377
　of fats, 388–391
Saran, 456
Sarcoma, 687
Sardine oil, composition, 386
Saturated hydrocarbons, names, 298–301
SBR, 456
Schizophrenia, 680
Scrubbing, 714
Serendipity, 487
Serine, formula, 406, 514
Serotonin, 682
　formula, 682
Serum, electrophoretic patterns, 527
Sickle cell anemia, 532, 658–660
　sickled cells, 658
Sickle cell hemoglobin, 659
Sickle cell trait, 659
　and malaria, 660
Side chain: on an aromatic ring, 346
　oxidation of, 346
Silk, 448, 458
Silver, in Tollen's reagent, 422
Smaze, 697
Smog, 697
Smoke, 185

Soap: cleansing action, 393
　in hard water, 395
　history of use, 391
　invert, 395
　manufacture, 389–391
　relation of structure to solubility, 391–392
Soap micelle, 392, 393
Sodium: in nutrition, 636
　reaction with alcohols, 281
Sodium benzoate: as a preservative, 379
　structure, 372
Sodium stearate, structure and use as a soap, 389
Soil, depletion of, 707
Soluble RNA, 549
　in protein synthesis, 608
　structure, 607
Soybean oil: composition, 386
　in manufacture of oleomargarine, 388
Specific dynamic action, 632
Specific rotation, 468
Sperm, composition, 551
　human, 656, 657
Sperm cells: biochemical composition, 656
　drawing of, 550
Spinach, oxalic acid in, 380
Stanley, Wendell, 555
Starch: formula, 440, 505
　digestion of, 446
　hydrolysis of, 445
　a polymer, 448
Starch grains, 506
Starches, 496
Starvation, 679
Stearic acid, formula, 385
Stereoisomers: definition, 465
　types of, 477–478
Sterility and vitamin E, 646
Steroid nucleus, 536
Steroids, 539–541
Sterols, 539
Straight chain, 290
Stratosphere, aircraft in, 694
Streptomycin, formula, 678

Structural formula, definition, 278
Structure, molecular: definition, 465
  organic, assignment of, 279
  of organic compounds, 277-279
  relation to organic properties, 287
  valence bond, 288
Strychnine, 412
Styrofoam, 453
Substitution reaction(s): of alkanes, 309
  of benzene, 343
Succinic acid, 593
  occurrence, 380
  synthesis from maleic acid, 329
  use, 380
Sucrase, 505
Sucrose: formula, 439, 503
  hydrolysis, 503, 504
  sources, 503
Sugars, 496
  solubility in ether, 360
Sulfa drugs, 676, 677
Sulfanilamide, 676, 677
Sulfonamides, 676
Sulfonation of benzene, 344
Sulfur dioxide in air, 696
Sulfuric acid: addition to an alkene, 331
  in air, 696
  structure, 331
Sulfurous acid in air, 696
Sunshine vitamin, 646
"Superball," 455
Survival value of lipid storage, 627
Sweetness of sugars, 504
Systematic names of organic compounds, 299

T

Tallness in pea plants, 652, 653
Tallow: beef, composition, 386
  mutton, composition, 386
Tar (asphalt), 296
Tars, cancer and, 697
Tartaric acid, 380

Technology, social and economic effects of, 414, 461-462
Teflon, use of, 452, 453
Temperature, of earth, 691, 694
Terephthalic acid, 459
Termites, digestion of cellulose by, 579
Terpenes, 541-542
Terylene, 458
Testosterone, 419, 540
Tetrachloroethene, 314
Tetrachloromethane, 309, 313
Tetracycline, formula, 678
Tetrahedral bonding, 283
Tetrahedral carbon, relation to, optical activity, 468-471
Thiamine (vitamin B_1), 640-641
  formula, 640
  structure, 411
Thiazole, structure, 347
Thinners for paint, 396
Threonine, formula, 514
Threose, formula, 425
Thymine, formula, 546
Thymol, 353-354
Thyroxine, 683
  biosynthesis, 660, 661
Tin, for reduction of nitro compounds, 402
TNT (trintrotoluene), synthesis and structure, 346
Tobacco mosaic virus, 550, 551
α-Tocopherol, 646
Tolbutamide (Orinase), 597
Tollen's reagent, 422
Tollen's test, 501
Toluene: industrial synthesis, 351
  origin of name, 338
  structure, 340
Toxicity of mineral elements, 635
Tranquilizers, formulas, 681
Tranquilizing drugs, 680
Transaminases in diagnosis of disease, 610
Transamination, 520
  experiments indicating, 609
  and pyridoxal, 643

Transfer RNA, 549
  structure, 607
Transferases, 571
Trichloroethene, uses, 314
Trichloromethane: formula, 309
  uses, 313
Triethylamine, 401
Triglycerides, 537
  caloric content, 632
  chemical properties, 537
  digestion of, 560
  importance, 537
  physical properties, 537
  separation from natural
    products, 537
  summary of metabolism, 622
Trilene, 314
Triose, 497
Tripeptides, 519
Trypsin, 561
Tryptophane, formula, 514
Tung oil, composition, 386
Turpentine, in paint, 396
Tyrosine, formula, 514

U

UDP-glucose, in glycogen
  synthesis, 581
Unbranched chain, 289
Unsaturated hydrocarbon,
  definition, 323
Uracil, formula, 546
Urea: excretion of, 612–614
  as fertilizer, 412
  formation, 521
  isomer of ammonium cyanate,
    277
  laboratory synthesis, 296
  manufacture of, 412
  normal blood level, 612
  in polymers, 461
  in shark blood, 613
  structure, 412
  synthesis, 412
Urea cycle, 612–614
  (diagram), 612
Urey, Harold, 517
Uric acid: excretion of, 614–616
  by humans, 615
  formation from purines, 615
Uricase and gout, 616
Uridine triphosphate (UTP) in
  glycogen synthesis, 581
Urinalysis, 562
Urine, 561
Urushiol, 364

V

Valence bond structure, 288
Valine, formula, 406, 513
Vanillin, 353, 354, 400, 442
van't Hoff, Jacobus, 469–471
Vasopressin, 530
Vinyl acetate, 335
Vinyl chloride: in copolymers,
    456
  structure, 449
  synthesis from acetylene, 334
Vinyl ether, 365
Vinyon, 456
Viruses: action of, 552
  composition, 551
  mechanism of action, 609
  relationship to cancer, 555, 686
  structure, 550, 551
Visual cycle, 639
Vital force, 276–277, 483
Vitamin: deficiencies; see
    specific vitamin
  definition, 638
  function; see specific vitamin
  requirements; see specific
    vitamin
Vitamin A, 638–639
  formula, 542
  hypervitaminosis, 639
Vitamin B_1 (thiamine), 640–641
Vitamin B_2 (riboflavin), 641–642
Vitamin B_6 (pyridoxal), 642–643
Vitamin B_{12} (cobalamin), 644
  structure, 645
Vitamin C (ascorbic acid),
    644–645
  structure, 375
Vitamin D, 541, 645–646
  formula, 646

Vitamin E (tocopherols), 646–647
  as inhibitor of rancidity in fats, 387
Vitamin K, 647
Vitamins, classification, 638
Vitamins B, general functions, 640
Volatility, relation to molecular weight, 304
Von Gierke's disease, 598

W

Waksman, Selman, 677
Warfarin: anticoagulant action, 647
  formula, 647
Water, pollution, 701
  structure of, 278

Wax, paraffin, 306
Waxes, 382, 539
Whale oil, 539
Wohler, Friedrich, 276
Wood, 448
Wool, 448

X

Xanthoproteic reaction, 528
Xerophthalmia, 639
X-ray diffraction, 283
Xylene, 350
Xylose, formula, 498

Z

Ziegler, Karl, 454
Zinc: biochemical importance, 637
Zymogens, 604

LIST OF THE ATOMIC WEIGHTS OF THE ELEMENTS

Element	Symbol	Atomic Number	Atomic Weight		Element	Symbol	Atomic Number	Atomic Weight
Actinium	Ac	89	(227)		Mercury	Hg	80	200.59
Aluminum	Al	13	26.9815		Molybdenum	Mo	42	95.94
Americium	Am	95	(243)		Neodymium	Nd	60	144.24
Antimony	Sb	51	121.75		Neon	Ne	10	20.189
Argon	Ar	18	39.948		Neptunium	Np	93	237.0482
Arsenic	As	33	74.92		Nickel	Ni	28	58.71
Astatine	At	85	(210)		Niobium	Nb	41	92.91
Barium	Ba	56	137.34		Nitrogen	N	7	14.007
Berkelium	Bk	97	(247)		Nobelium	No	102	(254)
Beryllium	Be	4	9.012		Osmium	Os	76	190.2
Bismuth	Bi	83	208.9804		Oxygen	O	8	15.9994
Boron	B	5	10.81		Palladium	Pd	46	106.4
Bromine	Br	35	79.909		Phosphorus	P	15	30.9738
Cadmium	Cd	48	112.40		Platinum	Pt	78	195.09
Calcium	Ca	20	40.08		Plutonium	Pu	94	(242)
Californium	Cf	98	(249)		Polonium	Po	84	(210)
Carbon	C	6	12.011		Potassium	K	19	39.09
Cerium	Ce	58	140.12		Praseodymium	Pr	59	140.91
Cesium	Cs	55	132.9054		Promethium	Pm	61	(147)
Chlorine	Cl	17	35.453		Protactinium	Pa	91	231.0359
Chromium	Cr	24	52.00		Radium	Ra	88	226.0254
Cobalt	Co	27	58.93		Radon	Rn	86	(222)
Copper	Cu	29	63.54		Rhenium	Re	75	186.23
Curium	Cm	96	(247)		Rhodium	Rh	45	102.91
Dysprosium	Dy	66	162.50		Rubidium	Rb	37	85.47
Einsteinium	Es	99	(254)		Ruthenium	Ru	44	101.1